Intellectual Property Issues
in Nanotechnology

Intellectual Property Issues in Nanotechnology

Edited by
Chetan Keswani

CRC Press
Taylor & Francis Group
Boca Raton London New York

CRC Press is an imprint of the
Taylor & Francis Group, an **informa** business

First edition published 2020
by CRC Press
6000 Broken Sound Parkway NW, Suite 300, Boca Raton, FL 33487-2742

and by CRC Press
2 Park Square, Milton Park, Abingdon, Oxon, OX14 4RN

Library of Congress Cataloging-in-Publication Data

Names: Keswani, Chetan, editor.
Title: Intellectual property issues in nanotechnology / edited by Chetan Keswani.
Description: First edition. | Boca Raton : CRC Press, 2020. | Includes bibliographical references and index.
Identifiers: LCCN 2020012018 (print) | LCCN 2020012019 (ebook) | ISBN 9780367482305 (hardback) | ISBN 9781003052104 (ebook)
Subjects: LCSH: Patent laws and legislation. | Nanobiotechnology--Law and legislation. | Nanotechnology--Research--Law and legislation. | Nanostructured materials industry--Law and legislation. | Nanobiotechnology--Law and legislation.
Classification: LCC K1517 .I58 2020 (print) | LCC K1517 (ebook) | DDC 346.048/6--dc23
LC record available at https://lccn.loc.gov/2020012018
LC ebook record available at https://lccn.loc.gov/2020012019

ISBN: 978-0-367-48230-5 (hbk)
ISBN: 978-1-003-05210-4 (ebk)

Typeset in Times
by Deanta Global Publishing Services, Chennai, India

Contents

SECTION 1   Introduction

SECTION 2   Food and Agricultural Nanotechnology

SECTION 3   *Pharmaceutical Nanotechnology*

SECTION 4   *Industrial Nanotechnology*

Foreword

The extremely fast technological progress in the field of nanotechnology has brought about newle gislativeandjudicialattemptstorestructuretheglobalregulatoryregimeespeciallywith regard to intellectual property rights, trying to balance the interests of both rights-holders and consumers. The success of the nanobiotechnology industry was not achieved spontaneously. On the contrary, it was the result of several influencing and favorable government policies toward the sector. Debates and concerns about the need to regulate the disruptive potential of biological applications were apparent almost from the moment when nanotech-based technologies became available. The emergence of all these concerns has also shed light on the problem of how to regulate the use, access, distribution, and appropriation of essential public knowledge assets in the lifesciences.

This book focuses on the integrated approach for sustained innovation in various areas of nanotechnology. The outlook of this book is based to a great extent on industrial and socio-legal implications of intellectual property rights (IPR) in nanotechnological advances. The book takes a comprehensive look not only on the implications of IPR in omics-based research but on what are the ethical and intellectual standards and how these can be developed for sustained innovation. I congratulate Dr.Keswani for synchronizing with global authorities on the subject to underline the upcoming challenges and presenting the most viable options for translating commercially viable ideas into easily affordable products and technologies.

Prof. Renata de Lima
University of Sorocaba
Bioactivity Assessment and Toxicology of Nanomaterials
Lab Rodovia Raposo Tavares
Sorocaba, SP Brazil

Preface

The current era of incredible innovations and the zeal to chase the heights of development has made nanotechnology one of the most powerful tools to accomplish the task of incremental prosperity for human welfare and sustainable development. The development of nanotechnology-based industries in any given country is shaped by the characteristics of the technology, particularly its close relation to scientific knowledge and by country-specific factors, the level and nature of the scientific knowledge base, the institutional set-up, and the role assumed by the government, which influence the country's ability to exploit the new opportunities and appropriate the respective results.

This volume focuses on the integrated approach for sustained innovation in various areas of nanotechnology. The outlook of this book is based to a great extent on the industrial and socio-legal implications of IPR in nanotechnology-based advances. The book takes a comprehensive look not only on the implications of IPR in omics-based research but also on the ethical and intellectual standards and how these can be developed for sustained innovation. This book attempts to collate and organize information on current attitudes and policies in several emerging areas of nanotechnology.

Adopting a unique approach, this book integrates science and business for an inside view on the industry. Peering behind the scenes, it provides a thorough analysis of the foundations of the present-day industry for students and professionals alike.

Chetan Keswani

Editor

Chetan Keswani is a Post-Doctoral Fellow in the Department of Biochemistry, Institute of Science, Banaras Hindu University, Varanasi, India. He has a keen interest in regulatory and commercialization issues of agriculturally important microorganisms. He is an elected Fellow of the Linnaean Society of London, U.K. He received Best Ph.D. Thesis Award from the Uttar Pradesh Academy of Agricultural Sciences, India in 2015. He is an editorial board member of several reputed agricultural microbiology journals.

List of Contributors

Rakesh A. Afre
Ajeenkya D.Y. Patil University
Pune, India

Anees Ahmad
Department of Chemistry
Aligarh Muslim University
Aligarh, India

Adelaide Maria de Souza Antunes
School of Chemistry
Federal University of Rio de Janeiro
Rio de Janeiro, Brazil

and

Academy of Intellectual Property
National Institute of Industrial Property – INPI
Rio de Janeiro, Brazil

Arash Ketabforoush Badri
Industrial Nanotechnology Research Center
Islamic Azad University
Tabriz, Iran

Parsa Ketabforoush Badri
Industrial Nanotechnology Research Center
Islamic Azad University
Tabriz, Iran

Eraldo Antonio Bonfatti Júnior
Department of Forest Engineering and
    Technology (DETF)
Federal University of Paraná (UFPR)
Curitiba, Brazil

Cristine Marie B. Brown
National Institute of Molecular Biology and
    Biotechnology (BIOTECH)
University of the Philippines
Los Baños, Philippines

Marilyn B. Brown
National Institute of Molecular Biology and
    Biotechnology (BIOTECH)
University of the Philippines
Los Baños, Philippines

Aline Caldonazo
Postgraduate Program in Pharmaceutical
    Sciences (PPGCF)
Federal University of Paraná (UFPR)
Curitiba, Brazil

Kumari Divyanshu
Department of Botany
Banaras Hindu University
Varanasi, India

Saeideh Ebrahimiasl
Department of Chemistry
Islamic Azad University
Ahar, Iran

and

Industrial Nanotechnology Research Center
Islamic Azad University
Tabriz, Iran

Fernández-Luqueño Fabián
Sustainability of Natural Resources and Energy
    Program
Cinvestav-Saltillo, Mexico

Dominick Fazarro
Department of Technology
The University of Texas at Tyler
Tyler, Texas

Susan K. Finston
President
Finston Consulting, LLC
Jerusalem, Israel

Medina-Pérez Gabriela
Transdisciplinary Doctoral Program in
    Scientific and Technological Development
    for the Society
Cinvestav-Zacatenco
Mexico City, Mexico

Aniket Gade
Department of Biotechnology
Sant Gadge Baba Amravati University
Amravati, India

Priyanka Gautam
Department of Neurology
Banaras Hindu University
Varanasi, India

Amirah Mohd. Gazzali
School of Pharmaceutical Sciences
Universiti Sains Malaysia
Pulau Pinang, Malaysia

G. Ambarasan Govindasamy
Ann Joo Integrated Steel Sdn.
Pulau Pinang, Malaysia

Kelvii Wei GUO
Department of Mechanical and Biomedical
    Engineering
City University of Hong Kong
Kowloon, Hong Kong

Ankush Gupta
Department of Biochemistry
Banaras Hindu University
Varanasi, India

Craig Hanks
Philosophy Department
Texas State University
San Marcos, Texas

Ahmed S. Hashem
Agricultural Research Center (ARC)
Plant Protection Research Institute
Sakha, Egypt

Pérez-Hernández Hermes
El Colegio de la Frontera Sur
Agroecología
Unidad Campeche
Campeche, Mexico

Lav Kumar Jaiswal
Department of Biochemistry
Banaras Hindu University
Varanasi, India

Sathish Kumar Kamaraj
Department of Engineering
Technology Institute of El Llano
    Aguascalientes (ITEL) / National
Technology Institute of México (TecNM)
Aguascalientes, México

Chetan Keswani
Department of Biochemistry
Banaras Hindu University
Varanasi, India

Yashfeen Khan
Department of Chemistry
Aligarh Muslim University
Aligarh, India

Elaine Cristina Lengowski
Faculty of Forestry Engineering
Federal University of Mato Grosso (UFMT)
Cuiabá, Brazil

Marcos Emiliano Lima Alves Hir
Centre for Technological Innovation
Farmanguinhos, Oswaldo Cruz Foundation/
    FIOCRUZ
Rio de Janeiro, Brazil

and

Federal University of Rio de Janeiro State
Rio de Janeiro, Brazil

Jorge Lima de Magalhães
Centre for Technological Innovation
Farmanguinhos, Oswaldo Cruz Foundation/
    FIOCRUZ
Rio de Janeiro, Brazil

and

Global Health and Tropical Medicine (GHMT)
NOVA University of Lisbon
Lisbon, Portugal

Maria Simone de Menezes Alencar
Graduate Program in Nursing and Biosciences
Federal University of Rio de Janeiro State
    – UNIRIO
Rio de Janeiro, Brazil

Tatiana Minkina
Academy of Biology and Biotechnology
Southern Federal University
Rostov-on-Don, Russia

Lakshmipathy Muthukrishnan
Saveetha Dental College and Hospitals
Saveetha Institute of Medical and Technical
    Sciences
Chennai, India

Rabiatul Basria S.M.N. Mydin
Oncological and Radiological Sciences Cluster
Universiti Sains Malaysia
Kepala Batas, Malaysia

Tanmayee Nayak
Department of Biochemistry
Banaras Hindu University
Varanasi, India

Robert A. Nepomuceno
National Institute of Molecular Biology and
    Biotechnology (BIOTECH)
University of the Philippines
Los Baños, Philippines

Deb Newberry
Newberry Technology Associates
Burnsville, Minnesota

Siti Hawa Ngalim
Regenerative Medicine Cluster
Universiti Sains Malaysia
Kepala Batas, Malaysia

Evrim Özkale
Manisa Celal Bayar University
Manisa, Turkey

Abhishek Pathak
Department of Neurology
Institute of Medical Sciences
Banaras Hindu University
Varanasi, India

Luc Quoniam
Department of Competitive Intelligence
Université Du Sud-Toulon Var
La Garde, France

and

Law School – FADIR
Federal University of Mato Grosso do Sul
Vila Olinda, Brazil

Mahendra Rai
Department of Biotechnology
Sant Gadge Baba Amravati University
Amravati, India

Vishnu Rajput
Academy of Biology and Biotechnology
Southern Federal University
Rostov-on-Don, Russia

Al-Kazafy Hassan Sabry
Pests and Plant Protection Department
National Research Centre
Cairo, Egypt

Kestur Gundappa Satyanarayana
Poornaprajna Institute of Scientific
    Research (PPISR)
Bangalore, India

Suzanne de Oliveira Rodrigues Schumacher
School of Chemistry
Federal University of Rio de Janeiro
Rio de Janeiro, Brazil

Sudhir Shende
Department of Biotechnology
Sant Gadge Baba Amravati University
Amravati, India

Tahsin Shoala
College of Biotechnology
Misr University for Science and Technology
Giza, Egypt

Juhi Singh
Department of Gastroenterology
Banaras Hindu University
Varanasi, India

Rakesh Kumar Singh
Department of Biochemistry
Banaras Hindu University
Varanasi, India

Sandeep Kumar Singh
Indian Scientific Education and Technology
    Foundation

and

Centre of Biomedical Research SGPGI Campus
Lucknow, India

Shishu Pal Singh
Department of Agriculture
Uttar Pradesh, India

Theivasanthi Thirugnanasambandan
International Research Centre
Kalasalingam Academy of Research and
    Education
Krishnankoil, India

Yashoda Nandan Tripathi
Department of Botany
Banaras Hindu University
Varanasi, India

Walt Trybula
Ingram School of Engineering
Texas State University
Austin, Texas

Ram Sanmukh Upadhyay
Department of Botany
Banaras Hindu University
Varanasi, India

Valle-García Jessica Denisse
Laboratory of Soil Ecology
ABACUS
Cinvestav, Mexico

Vera-Reyes Ileana
Department of Plastics in Agriculture
Centro de Investigación en Química Aplicada
Saltillo, Mexico

Walia Zahra
Department of Biochemistry
Banaras Hindu University
Varanasi, India

Section 1

Introduction

1 Intellectual Property Issues in Nanotechnology

Rabiatul Basria S.M.N. Mydin, Siti Hawa Ngalim, Amirah Mohd. Gazzali, and G. Ambarasan Govindasamy

CONTENTS

1.1 INTRODUCTION

Nature has shown that properties at nanoscale are appealing for advances in photonics, bioenergetics, catalysis, anti-dirt material, and anti-microbial function, among others. Unlike other types of intellectual property (IP), patents supersede trade secrets in terms of the rights of a company over other competing companies, to exclusive utilization, production, exportation, and financial benefit from the patented item for a maximum of 20 years. Patents can be very expensive; it takes several years for full documentation and is only applicable in the countries where the patent is filed and documented (Zhang et al. 2013). It gives power and authority to the investors to stop others from copying, manufacturing, selling, or importing the invention without their approval. The patent system also allows the investor to maintain a monopoly on the innovation for their personal profit.

Patents for any fundamental discoveries and innovations at these minuscule sizes are future for at least USD 4.4 trillion revenue (Flynn et al. 2014). Globally, the World Intellectual Property Organization (WIPO) reported that patent filing on micro- and nanotechnology was low (259 files) prior to 1992 and there was a steep increase from 2000 (1,332 applications filed) until recently (19,154 applications filed in 2017). Meanwhile, patents granted were only 57–70% of those patents filed between the year 2000 until 2017 (Genieser and Gollin, 2007). The statistics suggest the Patent and Trade Office (PTO), irrespective of the country, does not only analyze hard evidence or 'enablement', but the institution also considers subjective perspectives like 'unstated limitation' or 'inherent anticipation' and 'obviousness' before a nanotechnology patent can be granted or rejected (Koppikar et al. 2004; Barpujari, 2010). More issues on nanotechnology are uncovered as there is

3

a growing trend of nanotechnology patents being granted. Importantly, naturally occurring phenomena and abstract ideas such as mathematical equations are not patentable.

1.2 STANDARD DEFINITION OF NANOTECHNOLOGY

Advances in nanotechnology have improved the turnover of product from raw materials, functionality with superior performance, and a growing knowledge of quantum effects within the range of 1–100 nm. Lack of a standard definition for nanoscale products poses a challenge in obtaining IP for nanotechnology. The presence of different terms such as nanoagglomerates has further complicated this issue. In addition, the overlapping range of the products or particles has also proved to be a challenge. The patentee would have to mention a specific size range that is covered by the invention, outside which it is not covered by the patent. This was the case between BASF and Orica Australia (T-0547/99). A prior patent granted to BASF mentioned the nanoparticles of 111 nm, with certain technical properties. Orica Australia filed a patent application for a nanoparticle formulation for a similar application but with smaller particle sizes (less than 100 nm). Orica also demonstrated that these particles had improved technical properties when compared with BASF, hence this was not considered as a patent breach (Foster, 2015).

1.3 PHYSICAL AND CHEMICAL PROPERTIES OF THE TECHNOLOGY

Generally, an IP application, like a patent on a nanotechnology issue, often falls under any of the four different categories: a constitutive of a substance, an engineered item, a process, and machinery. Any naturally occurring entity is not patentable even though its size fits into nano-scale conformation. For example, a 60-unit carbon with a 'soccer ball' structure called Buckminsterfullerene (the first fullerene to be discovered) is 0.7 nm in diameter and can be found in carbon black. However, 'unnatural' fullerene that can be synthesized using special apparatus – though the size and shape of the synthetic fullerene may not be exactly the same as the original C60 fullerene (Leigh and Julie, 1991) – is patentable (William, 1991). Likewise, a product from modifications made to the original C60 is also patentable, such as endohedral fullerene whereby an atom is added inside the C60 scaffold (Yakobson et al. 2007). Requirement for non-obviousness for a patent to be granted may be limited to the technology or understanding at the time. As an example, the decision to grant a patent on tires containing fullerene would have been different in later years (Genieser and Gollin, 2007; Lukich et al. 1998). Similarly, broad claims were given leeway, such as focusing on generic fullerene in some claims (Lei et al. 2004). Additionally, a patent application will be rejected if the claim is obvious; an example is a process claiming to increase the yield of endohedral by using soot as the source of fullerene (Chai et al. 1991; Nabou and Hisanori, 1992). Obviousness is subjective, as seen in retrospect, with a patent of a known water-soluble molecule like hyaluronic acid and its binding to fullerene (2012).

Naturally occurring genetic and epigenetic codes are also non-patentable despite laborious efforts to isolate and analyze them. For instance, a well-known cell adhesion tripeptide, arginine-glycine-aspartic acid (RGD), appeared to be too short for makeover use in nanomedicine fit for patenting. Challenges in patenting short peptides can be overcome (Erica, 2015) by claiming the changes of functional groups at the peptide N- or C- terminus when in vitro (Erkki and Michael, 1984), the conjugation of the peptide with other motifs (Joerg et al. 2002; Lahann et al. 2003), and the conformational changes attributable to additional amino acids or the retaining matrix (Michael and Erkki, 1988). Though patents for nanoscale devices may be novel and non-obvious (Anderson, 2017), details in the application may be less convincing. For example, despite employment of nanometer-scale particles bio-functionalized with RGD, patent claims made for nanobioelectronics suggest claims were made on natural phenomena and obviousness, such as electrical flow phenomenon and expected cellular responses (Fernando et al. 2007; James, 2000), although a similar concept at macro-scale had its patent granted (Martin et al. 2011).

An obvious issue in nanotechnology is when scaling down machinery and article size to nanometer-scale alone is not a convincing factor in granting a patent over existing patents of macro sized articles or machinery. Alternatively, claims could be filed on changes of nature due to the fact that chemical and physical properties at nanometer scale are different than for bulk material. The process, equipment, materials, and reagents used in scaling down are dissimilar to processes in production of macro size articles and machinery; hence, this gives a chance for nanotechnology innovations to be patentable. Another challenge in patenting an article or machinery at nanometer-scale would be the feasibility of reverse engineering. As an example, a cantilever tip for scanning probe microscopy can be made top-down via a photolithographic process (Albrecht et al. 1993) or bottom-up via epitaxy (Cohen et al. 2009). The same is true for quantum dots (2014).

1.4 MULTI-DISCIPLINARY NATURE

The multidisciplinary nature of nanotechnology causes difficulties in appointing expert patent examiners with suitable academic background and experience. This, in turn, risks the patent processing in which the patent requested may be mistakenly rejected due to inaccurate or immature evaluation, overlook any prior application, and may also produce patents that are not legally robust. Conversely, a patent may also be granted with an overly broad aspect, giving excessive control over areas that are not meant to be covered (Bastani et al. 2005; Brown, 2002).

1.5 ABILITY OF NANO-PRODUCTS TO SOLVE SPECIFIC PROBLEMS

Among the main criteria in patent filing is the ability of the patented product to solve specific problems. In certain products such as nanobiotechnology, the unpredictability of the products is considered significant, which may at times render them inactive. Under this condition, any patent granted will be invalid because it did not serve its purpose of application.

1.6 DIFFICULTIES TO PROTECT NANOTECHNOLOGY AS TRADE SECRET

The material and shape of nano-based products could be easily reverse engineered using different methods, which means that they could not be protected under trade secret rules. As a trade secret is usually applied to processes, designs, patterns, and commercial methods, there is no one exact method or material to produce certain nano-based products. This becomes a hurdle for IP protection.

1.7 ENFORCEMENT AND POLICING

The enforcement and policing of patents related to nanotechnology is also a significant challenge for the patentee and the IP authorities. There is no guarantee that the patentee has the monopoly and exclusive right to their invention, especially if the patent is available in open media such as online databases. There are always chances that any competitor could adopt and produce the patented product, and subsequently commercialize the products without anyone knowing. Enforcing is very difficult because to determine the infringement would necessitate the utilization of expensive instruments such as electron microscopes and particle size analyzers, and this, in practice, is impossible to do for each company suspected of IP infringement (Singh et al. 2016a; Singh et al. 2019).

1.8 LEVEL OF TRANSPARENCY IN IP FILING

The concept of nanotechnology is not accepted by all sectors of society. There are people who are scared of the ideas due to the uncertainties involved, such as safety and other ethical issues.

Hence, there are suggestions of increasing transparency in nanotechnology applications and this may include the secret in the innovations, which are usually protected under IP (Mata et al. 2015).

1.9   IP REGULATORY BODIES AND LAW PROVISIONS

The regulatory office for IP on nanotechnology is not available in all countries. Specifically, these offices are available in the United States (U. S. Patent and Trademark Office), Japan (Japan Patent Office), and Europe (European Patent Office), which are not easily accessible to all parties, especially to innovators from developing countries (Anuar et al. 2013). The innovators struggle to draft and submit their applications according to the requirements of these regulatory offices which are difficult to reach and may have different requirements from the IP office in their own country (Singh et al. 2016b).

In the absence of specific law provisions, it is difficult to determine the novelty criteria, the related invention, the eligibility of the subject matter, and the ability of the invention to solve specific problems. It is hence very important to work on producing the required law to aid and ease IP filing.

1.10   PATENT VS. PUBLICATION

Current industries are not really interested in spending money on providing laboratories and equipment for the purpose of research analysis on new nanotechnology inventions. This is because of the high cost of the equipment and laboratory setup, the need for trained employees to run the equipment, and because the costly equipment is utilized less on a daily basis. Industry and government/ private institution collaboration is crucial in order to cut down all unnecessary cost, for example by working together with universities by hiring graduate students as researchers and lecturers to conduct their new research projects. In this case students and lecturers are brainstorming to create an outstanding invention using university laboratories and equipment according to the needs of industry, but the innovative ideas of graduate students and lecturers is ignored and industry uses the patent system to protect those ideas and restrict them from publication (Cooper and Rines, 2004). Accordingly, industry gets full credit for the new invention and the idea is sold for the benefit of industrial business interest and growth, while the students and lecturers cannot promote their invention and advancement of the useful technology due to the patent. Overall it can be concluded, that the patent system implemented by industry does not benefit the students at all.

1.11   PATENT BY THIRD PARTY

Sometimes an ordinary man also can invent something new for the sake of humanity and those new nanobased products can be more effective with low cost, but once the invention is secured with a patent by a third party the product value becomes higher and only rich people can afford to buy and benefit from it. The new product does not reach the maximum number of rural villagers at minimum cost if the product formulation is filed in the patent system and only the investor has the opportunity to earn more money. The product should not be a money-making machine for personal interests focused on making society sustainable and free of problems. It is an investor's duty to take a 'reciprocal philanthropy' approach and give something back to society. The patent system is a modern tool which prevents all of society from benefiting.

1.12   CONCLUSION

The aforementioned are simplified examples of many more patenting issues in nanotechnology. It all comes down to ways to satisfy the PTO, in ways of writing and presenting claims, limitations, non-obviousness, novelty, innovation, utility, synthetic, and ideas that are non-abstract. In short, the challenges described could reduce commitment from different parties, the patentee, the IP office, and the public alike, in the filing and protection of IP for nanobased works. Understanding the challenges is an important step towards addressing them and it is the duty of all parties to improve the process and make IP filing for nanotechnology more appealing and favorable to the inventors. Patenting innovations of nanotechnology – though the finer details regarding controllable phenomena and risks at the scale of molecules and atoms are less known – still promises trillion dollar business opportunities.

ACKNOWLEDGMENT

The authors would like to acknowledge the Ministry of Education (MOE) Malaysia for funding this work under Prototype Research Grant Scheme (PRGS), project code 203.CIPPT.6740054.

REFERENCES

A water soluble fulleren and its application patent CN102898542B. 2012.

Albrecht TR, Akamine S, Carver TE, Zdeblick MJ, inventors; Leland Stanford Junior University, assignee. Method of forming microfabricated cantilever stylus with integrated pyramidal tip. United States patent US 5,221,415. 1993 June 22.

American Chemical Society National Historic Chemical Landmarks. Discovery of fullerenes. http://www.acs.org/content/acs/en/education/whatischemistry/landmarks/fullerenes.html. Assessed 16 August 2019.

Anderson C. Small can be inventive: The patentability of nanoscale reproductions of macroscale machines. *Wm. & Mary Bus. L. Rev.* 2017;9:285.

Anuar HS, Nawi MN, Osman WN, Mahidin NO. Patent protections, challenges and applications of nanotechnology. *J. Tech. Oper. Manage.* 2013;8:17–29.

Barpujari I. The patent regime and nanotechnology: Issues and challenges. *J. Intellectual Property Rights* 2010;15: 206–213.

Bastani B, Fernandez D. Intellectual property rights in nanotechnology. *Inform. Technol. J.* 2005;4:69–74.

Brown D. US patent examiners may not know enough about nanotech. *Small Times.* 2002 February;4.

Chai Y, Guo T, Jin C, Haufler RE, Chibante LF, Fure J, Wang L, Alford JM, Smalley RE. Fullerenes with metals inside. *J. Phys. Chem.* 1991 October;95(20):7564–7568.

Cohen GM, Hamann HF, inventors; International Business Machines Corp, assignee. Monolithic high aspect ratio nano-size scanning probe microscope (SPM) tip formed by nanowire growth. United States patent US 7,572,300. 2009 August 11.

Cooper B, Rines R. Intellectual property issues associated with the emerging field of nanotechnology. 2004. https://ocw.mit.edu/courses/electrical-engineering-and-computer-science/6-901-inventions-and-patents-fall-2005/projects/24501_nanotech.pdf. Accessed on October 25, 2019.

Erica P. Patents for proteins & peptides: Threading the needle between patent eligibility and written description. 2015 November 20. https://www.linkedin.com/pulse/patents-proteins-peptides-threading-needle-between-patent-pascal. Assessed on 15 August 2019.

Erkki R., Michael P, inventor; Sanford-Burnham Prebys Medical Discovery Institute, assignee. Cell adhesion peptide patent CA1277265C. 1984.

Fernando P., Brian PT., Guihua Y, Charles ML, inventor; Harvard College, assignee. Nanobioelectronics patent US20090299213A1. 2007.

Fluorescent carbon quantum dot as well as preparation method and application thereof patent CN104357049A 2014.

Flynn H, Hwang D, Holman M. Nanotechnology update: Corporations up their spending as revenues for nano-enabled products increase. *Lux Research.* 2014 September 9.

Foster S. 2015. *Patent Protection for Inventions in the Field of Nanotechnology.* Swindle & Pearson Ltd. http://www.patents.co.uk/News/Article.aspx?ArticleID=142d6338-7804-4a18-9fc3-b47c6161c2d8. Assessed 15 August 2019.

Genieser L, Gollin M. Intellectual property issues in nanotechnology. *J. Commer. Biotechnol.* 2007 May 1;13(3):195–198.

James JH. An apparatus for the analysis of the electrophysiology of neuronal cells and its use in high-throughput functional genomics patent WO2000071742A2. 2000.

Joerg L., Samir SM., Robert SL, inventor; Massachusetts Institute of Technology, assignee. Switchable surfaces patent US7020355B2. 2002.

Koppikar V, Maebius SB, Rutt JS. Current trends in nanotech patents: A view from inside the patent office. *Nanotech. L. & Bus.* 2004;1:24.

Lahann J, Mitragotri S, Tran TN, Kaido H, Sundaram J, Choi IS, Hoffer S, Somorjai GA, Langer R. A reversibly switching surface. *Science.* 2003 January 17;299(5605):371–374.

Lei HY, Chou CK, Luh TY; National Health Research Institutes. Fullerene pharmaceutical compositions for preventing or treating disorders. United States patent US 6,777,445. 2004 August 17.

Leigh D. and Julie J. Technology: Buckyballs 'covered by old patent', 1991 July 6.https://www.newscientist.com/article/mg13117764-600-technology-buckyballs-covered-by-old-patent/. Assessed on 15 August 2019.

Lukich LT, Duncan TE, Lansinger CM, inventors; Goodyear Tire, Rubber Co, assignee. Use of fullerene carbon in curable rubber compounds. United States patent US 5,750,615. 1998 May 13.

Martin DC, Richardson-Burns S, Kim D, Hendricks JL, Povlich L, Abidian MR, Meier M, inventors; University of Michigan, assignee. Biologically integrated electrode devices. United States patent US 8,005,526. 2011 August 23.

Mata MT, Martins AA, Costa AV, Sikdar KS. Nanotechnology and sustainability-current status and future challenges. In: *Life Cycle Analysis of Nanoparticles. Risk, Assessment, and Sustainability.* DEStech Publications, 2015 March 30, pp. 271–306.

Michael DP, Erkki IR, inventor; Sanford-Burnham Prebys Medical Discovery Institute, assignee. Conformationally stabilized cell adhesion peptides patent EP0394326A4. 1988.

Nabou S., Hisanori S., inventor; Manufacture of metal stored fullerene and the like patent JPH05282938A. 1992.

Singh HB, Jha A, Keswani C. Biotechnology in agriculture, medicine and industry: An overview. In: *Intellectual Property Issues in Biotechnology*, Eds. H. B. Singh, C. Keswani, A. Jha. CABI, Wallingford, UK, 2016a, pp. 1–4.

Singh HB, Jha A, Keswani C (Eds.) *Intellectual Property Issues in Biotechnology.* CABI, Wallingford, UK, 2016b, 304 p. ISBN-13: 978–1780646534

Singh HB, Keswani C, Singh SP (Eds.) *Intellectual Property Issues in Microbiology.* Springer-Nature, Singapore, 2019, 425 p. ISBN- 978-981-13-7465-4.

William MB., inventor; Tetley Manufacturing Pty Ltd Ij and La, assignee. Metallic vapour patent AU589578B2. 1991.

Yakobson BI, Avramov PV, Mickelson ET, Hauge RH, Boul PJ, Huffman CB, Smalley RE, Margrave JL, inventors; William Marsh Rice University, assignee. High-yield method of endohedrally encapsulating species inside fluorinated fullerene nanocages. United States patent US 7,252,812. 2007 August 7.

Zhang Y, Sulfab M, Fernandez D. Intellectual property protection strategies for nanotechnology. *Nanotechnol. Rev.* 2013 December 1;2(6):725–742.

Section 2

Food and Agricultural Nanotechnology

2 Edible Crop Production by Nanotechnology

Is It a Sustainable Technology for Healthy Soil?

Pérez-Hernández Hermes, Medina-Pérez Gabriela,
Valle-García Jessica Denisse, Vera-Reyes Ileana, and
Fernández-Luqueño Fabián

CONTENTS

2.1 INTRODUCTION

Currently, thousands of engineered nanomaterials (ENMs) or engineered nanodevices (ENDs) with specific properties never seen before are being spread worldwide to increase the production of healthy, affordable, and innocuous food to shape sustainable development (Fernández-Luqueño et al. 2018). However, daily new evidence regarding the toxicity of ENMs is being published in the most reputable journals, i.e., it looks like the cutting-edge knowledge regarding nanoscience and nanotechnology jeopardizes human and environmental health or hampers the pursuit of sustainable development goals.

It has to be remembered that nanoscience and nanotechnology have significantly increased the performance of hitherto unknown materials, which have allowed synthesis of ENMs with applications in the health, energy, environment, and agriculture sectors, among others, i.e., nanoscience and nanotechnology have changed lives with enormous benefits for a better and more comfortable existence (Medina-Pérez et al. 2019).

However, no one can deny that the nanotechnological advances in the environmental and agricultural sectors are linked to concerns based on toxicological studies with soil microorganisms, animals, or crops in different experimental conditions at laboratory, greenhouse, or field scale. Fortunately, several advantages regarding the characteristics of the crops, pest or disease resistance, or better yield, among others, have also been reported in studies where ENMs or ENDs have been used (León-Silva et al. 2016; 2018).

The objective of this chapter is to discuss updated evidence published by worldwide scientists and by our research team to balance the advantages and disadvantages regarding the use of ENMs or ENDs in the agricultural sector to increase the quality, innocuity, and affordability of crops without jeopardizing soil quality, or human or environmental health.

2.2   NATURAL AND ENGINEERED NANOMATERIALS IN THE SOIL

According to the European Commission (2014), nanomaterials (NMs) are defined as natural, incidental, or manufactured materials that contain particles in a non-bound state or as an aggregate whereby 50% or more of the particles have a size distribution in the range of 1–100 nm. The origin, size, and shape of NMs define their properties; hence, NMs are different from existing microscale materials. Nanoparticles (NPs) can be formed naturally through processes that occur in the atmosphere, hydrosphere, lithosphere, and even the biosphere (Sharma et al. 2015). There are numerous examples of NPs that emanate from natural sources such as volcanoes, minerals, springs, and living organisms (Griffin et al. 2017). Natural nanomaterials (NNMs) can be found anywhere in nature, and researchers have been able to locate, isolate, characterize, and classify a wide range of NNMs (Figure 2.1).

However, the aggregation of NNMs with natural organic matter (NOM) is a complex material to investigate (Loosli et al. 2019; von der Heyden et al. 2019). Some examples are silver (Ag), gold (Au), iron (Fe), metal oxides that include Fe_2O_3, MnO_2, and SiO_2 and metal sulfides such as Fe_2S_2, Zn_3S_3, and Cu_3S_3, which are among the most abundant in nature (Findlay et al. 2019; Mansor et al. 2019). NOM, under certain conditions, such as high temperatures, exposure to high solar radiation, and changes in soil pH, will promote the formation of inorganic NPs (Chakre and Sharon 2019; Yin et al. 2012; Pédrot et al. 2011).

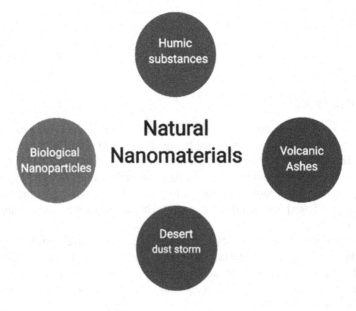

FIGURE 2.1   Natural sources of nanoparticles.

Inorganic NPs can be produced through biogeochemical processes, such as the NNMs that are present in the clouds of volcanic ash, which contain a wide variety of scattered NPs and microparticles that are primarily of silicate and iron origin (Griffin et al. 2017). The formation of NNMs occurs through the combination of several mechanical processes, such as weathering. It is noteworthy that NNMs are formed predominantly at phase boundaries, for example, solid-gas wind erosion, liquid-gas evaporation of seawater, or solid-liquid rock wear, among others (Sharma et al. 2015).

There is an additional type of NNMs that are synthesized by microorganisms. These microbes are composed of molecules of nanometric size, such as DNA, RNA, or proteins. There are reports in the literature that describe the various biological processes that are associated with the production of NPs, which include the redox conditions that some microorganisms use to store and transport electrons through the transitory production of inorganic NNMs, or the output of nanominerals based on Fe and silicon (Si), calcium carbonate, and calcium phosphate (Wu et al. 2019; Shrama et al. 2015; Faramarzi and Sadighi 2013). Therefore, various bacteria, fungi, and plants have been used to reduce metals like silver, gold, palladium, selenium, etc. (Ali et al. 2019; Rajeshkumar et al. 2018). It is noteworthy that in the soil the NNMs form colloids that consist of silicate clay minerals, Aluminum (Al) or Fe oxides/hydroxides, and organic colloids, which are made up of humic organic matter and biopolymers such as carbohydrates (Shrivastava et al. 2019).

Engineered nanoparticles (ENPs) are nanoparticles that are human-made for industrial and medical purposes, and currently they are also used in agriculture. It is noteworthy that naturally generated NPs are not comparable to ENPs, because NNMs are generally coated on the surface with polymers and surfactants that are found in the environment where they were produced (Chakre and Sharon 2019). Eventually, ENPs will enter the environment through water, soil, and air as part of human-based activities. However, ENPs are also used in the treatment of polluted water and soil by deliberately injecting ENPs into the soil or aquatic systems (Yang et al. 2019; Dong et al. 2014). NNMs are of great importance in the biogeochemical cycles, weathering processes, bioavailability, transport, and ecotoxicity of metals. In addition, these factors have contributed to the evolution and adaptation of higher organisms (Steinberg et al. 2006), hence, there is some concern because ENPs are made of specific structures that microorganisms may not recognize, and as a consequence, may induce toxic effects that are not observed in the micron-sized counterparts (Khan et al. 2017). Among the ENPs commonly used are the metals and oxides of Ag, Fe_2O_3, Al_2O_3, ZnO, and TiO_2. Interactions of the soil microbiota with ENPs play a crucial role in the transport and destination of the ENPs; even though living organisms require these metals as micronutrients, higher doses of some of these metals can be cytotoxic (Dhand et al. 2015). ENPs are most likely to undergo significant modifications when exposed to soils, because of aggregation caused by the NOM interaction. This interaction changes the reaction speed, adsorption capacity, redox state, shape, and modifies the reactivity of the NPs, which consequently will attenuate or augment their fate in the environment and their bioactivity with plants (Liu et al. 2019; Luo et al. 2018).

ENPs can affect soil characteristics such as pH and organic matter, influence the production of exudates from plant roots and affect the growth of microbial communities in the rhizosphere. The results of a study undertaken by Kibbey and Strevett (2019) showed that ENPs (depending on their nature) could decrease or increase the bacteria in the rhizosphere (Table 2.1). Consequently, ENPs can cause changes in the lengths of the root and stem of the plant. Kibbey and Strevett (2019) hypothesized that ENPs prevent bacteria from adhering so that they can carry out synergistic interactions between the bacteria and the root to assimilate nitrogen and carbon, and other essential nutrients for plants. Other studies have revealed that the concentration and exposure of ENPs may have different effects on the soil microbial community (Moll et al. 2017; Shcherbakova et al. 2017; Shah et al. 2014). Wang et al. (2017) investigated the effect of different concentrations of silver nanoparticles (AgNPs) and observed that although the content of *Acidobacteria*, *Actinobacteria*, *Cyanobacteria*, and *Nitrospirae* decreased when increasing the doses of AgNPs, concomitantly, several other phyla, which included *Proteobacteria* and *Planctomycetes* were increased in number. Exudates from soil microbiota can induce the dissolution of ENPs and promote the absorption of

TABLE 2.1

Some Effects of Engineering Nanoparticles on Soil Microorganisms

Engineering nanoparticle	Concentration in soil	Effects	References
Ag	0.01, 0.10 and 1.00 mg kg^{-1} dry weight	Exposure time had significant effects on the relative variation of the microbial community.	Grün et al. (2019)
CuO, ZnO	10 mg kg^{-1}	ENPs did not show an effect on the growth of microorganisms and their activity.	Joško et al. (2019)
ZnO	0, 10, 20, 30, 40, 60, 90, 120, 150, 180, 240 Zn mg L^{-1}	Zn bioaccumulation was found for the earthworm *Eisenia Next time andrei*.	He et al. (2019)
Zerovalent iron (ZVI), ferrous sulfide (FeS) ferriferous oxide (Fe$_3$O$_4$)	3% (w/w)	FeS-NPs have significantly changed microbial community richness and diversity, followed by ZVI and Fe$_3$O$_4$.	Peng et al. (2019)
Zero-valent iron particles (ZVI)	0, 1, 5, 10 and 20 mg g^{-1} DW soil	The toxic effects of ZVI on soil microbial communities are soil dependent. In sandy-loam soil, bacterial biomass and diversity were negatively affected.	Gómez-Sagasti et al. (2019)
ZnO, TiO$_2$, CeO$_2$, Fe$_3$O$_4$	0.5, 1.0, and 2.0 mg g^{-1}	ZnO-NP in saline-alkali soil showed a higher effect on variance in their bacterial community composition, eg., Bacilli, Alphaproteobacteria, and Gammaproteobacteria class.	You et al. (2018)
CeO$_2$	1 mg L^{-1}	Decreased the denitrifying bacteria, such as *Flexibacter* and *Acinetobacter*.	Wang et al. (2018)
Single-walled carbon nanotubes (SWCNTs) and multiwalled carbon nanotubes (MWCNTs)	0.05%, 0.1%, or 0.5% (w/w)	CNT caused fluctuations in microbial community function, differential tolerance size, and dose-dependence among bacterial genera.	Wu et al. (2019)
MWCNTs, reduced graphene oxide (rGO) and ammonia-functionalized graphene oxide (aGO)	One ng, 1 μg or 1 mg kg^{-1} dry soil	The composition of bacterial communities was significantly influenced over time.	Forstner et al. (2019)
Fullerene (C$_{60}$), reduced graphene oxide (rGO) and MWCNTs	50 and 500 mg kg^{-1}	rGO had greater changes in the bacterial composition, especially at the taxonomic level of the genus, particularly at the higher concentration.	Hao et al. (2018)

metal ions by plants (Zhao et al. 2016). In addition to the transformation in the rhizosphere, it has been found that the change in ENPs occurs within or on the surface of plant tissues and has different levels of accumulation that depend on the plant. ENPs such as ZnO, CuO, NiO, CeO_2, Yb_2O_3, and La_2O_3 are capable of transformation that results in changes in speciation and accumulation in the tissues of plants (Lv et al. 2015). Zinc oxide nanoparticles (ZnO-NPs) were able to transform zinc phosphate when they were supplementing nutrition in wheat that was grown in the substrate. This is due to the uptake of zinc in the rhizosphere in its ionic form, and its subsequent translocation to the tissues of the plant. It has also been demonstrated for zinc citrate in the same way and zinc nitrate in soybean and cowpea (for example, Hernández-Viezcas et al. 2013, López-Moreno et al. 2010).

In soils, ENPs interact with the rhizosphere processes that influence the plants, which affect those processes and concerns factors that include the complexity and properties of the soil, the biological species found, and the intrinsic properties of the ENPs, including their nature, dose, and exposure time.

2.3 NANOSCIENCE AND NANOTECHNOLOGY IN THE AGRICULTURAL SECTOR

Nanotechnology has importance in a wide range of key and revolutionary applications, including medicine, the bioremediation of polluted soils, and agriculture wastewater treatment, among others (Dasgupta et al. 2016). Generally, agriculture utilizes chemicals like pesticides and fertilizers in extensive amounts that contribute to the acceleration of soil degradation. In the same way, nanotechnology can devise a way to offset those chemicals and improve the utilization of new products to achieve efficiency in plants and crops (Kumar et al. 2018). In the last 20 years a lot of research work has specifically focused on the applications of nanotechnology to solve issues and to improve agricultural research (Mishra et al. 2014a; 2016; 2017). The above could be achieved through the release of nanofertilizers to enhance plant growth, and increase food production, taking into account food preservation, the detection of bacteria and contaminants, and the reduction of nutrient loss, to shape sustainable agriculture. Other studies could also show the importance of these cutting-edge knowledge areas (nanoscience and nanotechnology) that suggest the solution to the problem of deficiency and availability of micro- and macronutrients to produce healthy, affordable, and innocuous food (Paine and Paine 2012).

The researchers have to consider the different aspects in the use of nanotechnology in agriculture that allow the elimination of the use of chemicals as fertilizers, meet the demand for the production of food and crops due to the increase in the global population, and care for the environment. The only purpose is to deliver the correct quantity of nutrients and pesticides that promote productivity (Singh et al. 2017; Yata et al. 2018) and favor agricultural practices without causing damage.

2.3.1 Applications of Nanotechnology in Agriculture

The issue of integrating nanotechnology in agriculture provides a mode of sensing, detection, and bioremediation (Pandey 2018) through:

- Controlling the release of water and nutrients,
- Sustained delivery of herbicide, pesticides, and insecticide, and
- Monitoring environmental conditions.

Nowadays a wide range of applications of nanotechnology in agriculture exists: nanoremediation, nanofertilizers, nanocomposites, and nanosensors, among others. (Pandey 2018; Paine and Paine 2012) The above has become a reality because science is looking to stop the use of inorganic and organic compounds such as chlorinated phenols or halogenated hydrocarbons, among others, which could contribute to soil degradation and environmental contamination (Figure 2.2).

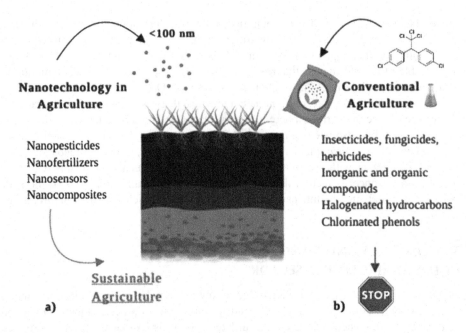

FIGURE 2.2 Modern and traditional agricultural supplies. a) Different applications of nanotechnology in agriculture. b) Chemical products used in conventional agriculture.

2.3.2 NANOPESTICIDES

Until 2015, there were no registered nanopesticides on the market due to the risks they represented (Kah 2015); nevertheless, some metal NPs can be used as pesticides because of such characteristics as large surface area, stability, and biodegradability (Khot et al. 2012). For example, application of nanosilver is utilized to fight against pathogens (Fraceto et al. 2018; Mishra et al. 2014b; Yasur and Rani 2013). Larue et al. (2014) reported the possible safe use of nanosilver for disease control in lettuce according to the neutral effects found on the growth of the plant (100 µg/g). It is well known that silver is antimicrobial, and it has positive effects on *Bacopa monnieri* plants, thus it may be considered as an excellent candidate as a nanopesticide compared to bulk silver compounds (Achari and Kowshik 2018).

2.3.3 NANOFERTILIZERS

To facilitate plant growth and crop productivity it is useful to deliver nanoparticles in a safe dosage. In the same way, it is important to implement new strategies to amend macronutrient and micronutrient deficiencies in agricultural soils (Achari and Kowshik 2018; Pandey 2018). NPs of essential and non-essential elements can affect the growth of different plants. Even when some NPs are not essential, elements have a positive effect on plants directly related to growth, physiological processes, and disease control (Table 2.2). These studies suggest that some elements such as Ag, Au, or titanium (Ti) present a hormesis effect (Achari and Kowshik 2018; Arora et al. 2012; Fageria et al. 2009; Khot et al. 2012; Rico et al. 2011).

Some reports present the effectiveness of soil application of nanofertilizers of Zn, Fe, N, P, and K because these are more effective than conventional fertilizers due to the efficient plant uptake of nanoparticles as a function of their small size (Gunaratne et al. 2016).

2.3.4 NANOSENSORS AND NANOCOMPOSITES

A nanosensor is a powerful tool employed in the detection of pollutants in contaminated environments based on nanomaterials that provide several advantages. The innovation of these

TABLE 2.2

Overview of the Effects of Nanoparticles with Essential and Nonessential Elements for Terrestrial Plants

Nanoparticles	Effects on edible or non-edible plants	References
	With essential elements	
Cu	At concentrations of 100, 200, 400 and 600 mg/L copper nanoparticles present significant inhibition of seed germination and root elongation.	Moon et al. (2014)
Zn	Zinc nanoparticles promote germination rate at 10 mg/L for corn on hydroponic culture.	Zhang et al. (2015)
Fe	Fe_2O_3 nanoparticles reduced chlorophyll content and growth in *Arabidopsis thaliana* at 4 ppm. The study suggests that nanoparticles were not taken up and therefore not bioavailable.	Marusenko et al. (2013)
Co	Co_3O_4 nanoparticles at 5 g/L improve the root elongation of radish seedlings.	Wu et al. (2012)
	With non-essential elements	
Ti	An increase in shoot length, leaf area, and root dry weight on *Vicia faba* after seven days of exposure time (0.01%).	Abdel Latef et al. (2018)
Ag	A positive response in the growth of *Brassica juncea* and *V. unguiculate* at 50–75 ppm.	Pallavi et al. (2016)
Al	At 50 ppm of Al_2O_3 nanoparticles, soybean presents enhanced growth under flooding stress.	Ma et al. (2010)
Ce	At 2000 mg/L, seed germination of corn, tomato, and cucumber reduced, and root elongation of corn and cucumber increased when roots were exposed over eight days.	López-Moreno et al. (2010)
Cd	Photosynthesis was the main process affected by the addition of Cd in *Helianthus annuus* L. at 50 mg.	Lopes Júnior et al. (2015)

nanomaterials is presented according to their highly sensitive response, selectivity, and portability. Great advances have been made in the detection of heavy metal ions in water, and they are better than conventional instruments. These nanosensors have been reported for applications in providing crop production that promises to change the agricultural sector to enhance food production because they allow monitoring in real time (Kim et al. 2018; León-Silva et al. 2018; Ullah et al. 2018).

Nanosensors can also monitor temperature, traceability, and humidity, among others. Therefore, nanoparticles in nanosensors might help farmers to maintain precise control. On the other hand, with the approach of polymers and biopolymers and their special properties such as mechanical strength, thermal stability, and resistance, nanocomposites play a role in the conventional technology to take advantage of polymer properties. For example, a nanocomposite of biopolymers has a promising future in the food industry, but it also enables the mediated delivery of nutrients as part of a smart system to reduce nutrient loss and increase uptake in the plant (Kumar et al. 2017; Yata et al. 2018).

2.3.5 IMPACT ON HEALTH, SOCIETY, AND ECONOMY

Agro-nanotechnology, with the implementation of nanofertilizers, nanosensors, and nanopesticides, has changed and revolutionized research worldwide (Yata et al. 2018). In the future, nanotechnology can provide solutions to address crop production, disease prevention, food quality, and sustainable agriculture in a green environment, such as the case of India which has proposed creating an Indian Institute of Nanotechnology in Agriculture (Pandey 2018).

However, all the effects caused by application of NPs are unique, and there is not a universal property in the response of plants. So, there is an urgent need to study the different effects of these nanomaterials according to plant species, soil properties, environmental characteristics, physicochemical factors, characterization of nanoparticle properties, analysis of their interactions, assessment of human exposure, testing of different sizes and concentrations of nanoparticles, and investigations of the effects on human health (Achari and Kowschik 2018; Nelson 2008). The above, through nanotoxicological studies, seeks to establish standard procedures for the safe use of nanoparticles in agriculture to preserve human and environmental health.

The main cause for the lack of application of nanoagriculture is because of the need for more studies, the unsuitable regulatory strategies, laws, normativity, and civic/social opinions. For these reasons, the acceptance of use nanoparticles depends mainly on safety concerns and the social implications. Future research has to focus on carrying out studies under controlled conditions but also long-term studies to evaluate a more realistic approach, the implications of nanotechnology on the growth of plants, and toxicological studies to prevent the risk to food quality (Du et al. 2017). Some researchers suggest that agro-nanotechnology is the best option in agricultural innovation because it allows avoiding the use of chemicals, implements sustainable practices, takes care of the environment, and prevents future shortages due to food demand and global population growth. However, it has to be taken into account that nanoscience and nanotechnology in agriculture have not been studied enough, so if this cutting-edge knowledge spreads its technological developments, humanity could soon be facing environmental problems.

2.4 NANOTECHNOLOGY TO PRODUCE EDIBLE CROPS

Agricultural research has documented the potential benefit in applications of nanotechnology in fertilization (Hossain et al. 2008; Liu and Lal 2014), crop protection (Liu et al. 2006; Sahayaraj et al. 2016), germination, growth promotion in plants, sensors of pollution, and other substances, recovery, and treatment of soil and water (Haghighi et al. 2015; Ibrahim et al. 2016; Stoimenov et al. 2002).

There are four categories of engineered nanomaterials (ENMs) according to Lin and Xing, (2007): a) carbon-based materials, usually including fullerene, single-walled carbon nanotubes (SWCNT) and multiwalled carbon nanotubes (MWCNT); b) metal-based materials such as quantum dots, nanogold, nanozinc, nanoaluminum, and nanoscale metal oxides like TiO_2, ZnO, and Al_2O_3; c) dendrimers which are nanosized polymers built from branched units, capable of being tailored to perform specific chemical functions; and d) composites which combine nanoparticles with other nanoparticles or with larger bulk-type materials. All ENMs indicated above, present different morphologies such as spheres, tubes, rods, or prisms (Ju-Nam and Lead 2008). Since plants are the first link in food chains, they strongly interact with the environment, and that is the principal concern, because of the ability of ENMs to move through cell membranes and walls (Nowack and Bucheli 2007).

The current uptake and accumulation of different ENMs by plants is the motivation for some scientific studies. Zhu et al. (2008) did the first study about iron oxide (Fe_3O_4) nanoparticles, they reported that the ENPs were taken up by pumpkin (*Cucurbita maxima*) roots and translocated in several plant tissues. When they measured the amount of NPs at the end of the experiment, they found that 45.5% of applied NPs were accumulated in roots, and approximately 0.6% of the NPs were found in leaves. However, this phenomenon is not common to every plant species. The uptake of NPs by vegetal cells could happen in several ways, through aquaporins, binding to carrier proteins, endocytosis, ion channels, creating new channels by carbon nanotubes (CNTs), or binding to organic chemicals. The NPs can form complexes associated with membrane transporters or root exudates, and then NPs are transported into the internal plant structure (Watanabe et al. 2008). The uptake depends on the method of NP application (root entry or foliar entry), then the exposed tissue of the plant describes different transport and defense mechanisms, and the chemical and physiological responses are specific in each site of contact.

The diameter of the cell structure pores or transport channels determines the ability for NPs to enter; in roots, for example, there is a size exclusion barrier: a) cell walls for the apoplastic transport pathway (5–20 nm); b) the Casparian strip transport (<1 nm); or c) symplastic transport (3–5 nm) (Wang et al. 2016). However, there are several reports where NP dimensions were bigger than 20 nm uptake and translocation was prevented. There are possible explanations for the entry of larger NPs than the exclusion barriers: the formation of new large pores in the cell wall; the rupture of membranes; and some interactions between cations, proteins, and viruses, among others, can cause changes in cellular structure (Wang et al. 2016). Proseus and Boyer (2005) explained the internalization of copper nanoparticles (CuNPs) of different sizes into and across algal cell walls, and they found that AuNPs (10 nm or above) were unable to enter through algal cell walls even under pressured conditions. Another study by Birbaum et al. (2010), demonstrated that the uptake did not depend on closed or open stomata, or under dark and light exposure conditions, and an interesting finding, there was no translocation into newly grown leaves of cultivated maize plants after the foliar particle exposure. These results may indicate that the natural entry barriers of some plants could be more resistant against nanoparticle translocation than mammalian barriers. Lee et al. (2008) reported the accumulation of CuNPs in bean and wheat plant tissues.

In *Medicago sativa* cells, bioaccumulation of quantum dots (CdSe/ZnS) occurred specifically in the cytoplasm and the nucleus (Santos et al. 2010). Leaves that were exposed to the ENPs first accumulated the nanosized materials in the stomata, instead of the vascular bundle, and finally, they were translocated to different parts via the phloem. When the NPs are absorbed by the root they are transferred and accumulated in the mature leaves because they are closer to the root, so it is the mature leaves that are usually more exposed than the young ones, therefore the time of exposure is higher; a few other papers document the translocation of NPs to grains, fruits, and flowers. Lin et al. (2009) showed that C70 fullerene is capable of accumulating in *Oryza sativa* seeds. The translocation and bioaccumulation also seem to be species-specific in the case of nano cerium oxide (nano-CeO$_2$); Schwabe et al. (2015), found that there was a greater accumulation of Ce-ion in *Helianthus annuus*, compared with the insignificant amount accumulated by *Cucurbita maxima* and *Triticum aestivum*.

The effect of NPs on plants can be both positive and negative. There are important findings of the impact of ENPs, and it depends on the concentration, composition, size, physical, and chemical characteristics, and the plant species (Ma et al. 2010). The effect of NPs on germination depends on concentrations of NPs but it also varies from plant to plant. Lee et al. (2008) found that SiO$_2$-NPs and Al$_2$O$_2$-NPs do not affect germination and growth of *Arabidopsis thaliana,* while ZnO-NPs hinder their germination. Nano-SiO$_2$ at low concentration improved seed germination of tomato (Siddiqui et al., 2015). When exogenous nano-SiO$_2$ was applied, it enhanced seed germination of soybean, raising the activity of nitrate reductase enzyme, absorption potential, and antioxidant system activity (Lu et al. 2002). These results were the same when the authors added nano-titanium dioxide (nano-TiO$_2$).

Suriyaprabha et al. (2012) found in maize seeds that nano-SiO$_2$ increased seed germination by increasing nutrient availability, and by changing pH and conductivity to the growing medium. In tomato, nano-SiO$_2$ enhanced seed germination and stimulated the antioxidant system under NaCl stress (Haghighi et al. 2012). De la Rosa et al. (2013) administered several concentrations of ZnO-NP on alfalfa, tomato, and cucumber, but the seed germination was enhanced only in cucumber. Other NPs, such as silica, palladium, gold, and copper, were studied by Shah and Belozerova (2009). They found that all of them had a significant influence on lettuce seeds. Platinum NPs exhibited a high germination index and had no negative effects on tomato and radish (Shiny et al. 2013). Several reports exist where AuNPs improved the seed germination in cucumber, lettuce, and *Brassica juncea* (Arora et al., 2012; Barrena et al., 2009). When CNTs were applied, there were some positive effects; MWCNTs at the concentration of 10–40 mg/L enhanced the seed germination and growth of tomato plants. The MWCNTs penetrated the thick seed coat of tomato (Khodakovskaya et al. 2009), and increased water uptake by seeds.

Lin and Xing (2007) experimented with five types of ENMs: MWCNTs, AlNPs, Al_2O_3-NPs, ZnNPs, and ZnO-NPs on six plant species: lettuce, ryegrass, radish, oilseed rape (*Brassica napus*), maize (*Zea mays*), and cucumber. They reported significant inhibition of germination of ryegrass by ZnNPs, and maize by ZnO-NPs or Al_2O_3-NPs, but there was no inhibition for MWCNTs. Under abiotic stress, nano-Si increased seed germination. Bao-Shan et al. (2004) applied exogenous nano-SiO_2 on changbai larch (*Larix olgensis*) seedlings, and they found that nano-SiO_2 enhanced seedling growth by improving root length, mean height, root collar diameter, and the number of lateral roots, and nano-SiO_2 almost induced the synthesis of chlorophyll. Nano-TiO_2 did not affect germination and root elongation of oilseed rape, wheat, and arabidopsis (Larue et al. 2011) cucumber, lettuce, and radish (Wu et al. 2012). Contradictory results were found by Asli and Neumann (2009). They found that TiO_2-NPs inhibited leaf growth and transpiration via damaging root water transport. It has also shown that MWCNTs enhanced root growth and peroxidase and dehydrogenase enzyme activity (Smirnova et al. 2012; Tripathi and Sarkar 2015). Aluminum oxide NPs were probed in five crop species, cabbage (*Brassica oleracea*), maize, cucumber, soybean, and carrot (*Daucus carota*). The NPs had effects on root elongation in hydroponic culture conditions (Yang and Watts 2005). Wang et al. (2014) experimented with rice plants supplying quantum dots (QDs), and with silica coated with QDs, these promoted rice root growth. Nanosized TiO_2 sprayed onto the leaves had a positive effect on the growth of spinach (Yang et al. 2006; Zheng et al. 2005). Nano-SiO_2 NPs slightly improved leaf biomass, proline content, and chlorophyll, improving plants' tolerance to abiotic stress (Haghighi et al. 2012; Kalteh et al. 2018; Li et al. 2012; Ramesh et al. 2014).

Tomato plants (*Lycopersicon esculentum*) were exposed to carbon nanotube and showed aquaporins regulation as a response to the stress caused by MWCNTs (Khodakovskaya et al. 2009). Nano-SiO_2 enhanced plant growth because it increased the gas exchange and the photosynthetic rate, transpiration rate, stomatal conductance, photosystem II potential activity, effective photochemical efficiency, actual photochemical efficiency, electron transport rate, and photochemical quench (Ramesh et al. 2014; YinFeng et al. 2011). In another study, exposure to SiO_2-NPs did not cause effects in zucchini (Stampoulis et al. 2009). CuNPs did not affect the plant system interactions in mung bean (Pradhan et al. 2015) but CuO and NiO-NPs showed harmful impacts on the growth of lettuce, radish, and cucumber, showing an increased lipid peroxidation, oxidized glutathione, reactive oxygen species (ROS), peroxidase and catalase activities, and decreased chlorophyll content. Negative effects of nano-CuO were assessed in soybean (*Glycine max*), and chickpea (*Cicer arietinum*) (Adhikari et al. 2012), and in rice (Shaw and Hossain 2013).

The effect on ryegrass and Indian mustard (*Brassica juncea*) indicated that plant growth was inhibited by ZnO NP (Lin and Xing 2008). Du et al. (2011) reported that TiO_2 and ZnO-NPs negatively affected rice and wheat growth. Raliya and Tarafdar (2013), reported that ZnO-NPs induced a significant improvement in *Cyamopsis tetragonoloba* plant biomass, shoot and root growth, root area, chlorophyll and protein synthesis, rhizospheric microbial population, acid phosphatase, alkaline phosphatase, and phytase activity in cluster bean rhizosphere. AgNPs have been frequently reported as detrimental to plant growth, but several studies have demonstrated the stimulatory effects of them in plant growth (Salama 2012; Syu et al. 2014; Yin et al. 2012). MWCNTs have enhanced the growth of tobacco (*Nicotiana tabacum*) cell culture (Khodakovskaya et al. 2012). Dimkpa et al. (2012) found that Cu and Zn levels in the shoot were similar, whether NPs or bulk materials were used. They also stated that the oxidative stress in the NP-treated plants increased lipid peroxidation and oxidized glutathione in roots and decreased chlorophyll content, and higher peroxidase and catalase activities were present in roots. Also, some authors reported the inhibitory effect of MWCNTs on plant growth (Ikhtiari et al. 2013; Tiwari et al. 2014). Giraldo et al. (2014), reported that SWCNTs augmented photosynthetic activity three times higher than controls by enhancing maximum electron transport rates, SWCNTs also enabled plants to sense nitric oxide.

2.5 DOES NANOTECHNOLOGY SHAPE SUSTAINABLE DEVELOPMENT?

The world population increases the demand for basic products, finished products, and services, as well as facing the challenge of reducing and minimizing the impact of human activities on the global environment and climate (Diallo et al. 2013). Nevertheless, several scientists claim that nanotechnology holds huge potential in several fields and is envisaged as a technology to lead the way toward sustainable environmentally friendly development in the coming years (Kumar et al. 2018). Currently, the use of this advanced technology is discussed as a form of renewable energy and environment (Nejad and Asadpour 2019), water treatment (Bishoge et al. 2018; Araújo et al. 2015), salinization process (Ray et al. 2018), remediation of soil by heavy metals (Cecchin et al. 2017; Michálková et al. 2016), extraction of minerals (Chugh et al. 2018), pharmaceuticals (Samad et al. 2009; Robles-García et al. 2016), automobiles (Presting and König 2003), healthcare (Sahoo et al. 2007), agriculture, and food systems (Duhan et al. 2017; Scott et al. 2018; Prasad et al. 2017).

In the area of remediation of soil and water, the application of nanotechnology, as an innovative method for contaminated site remediation, has received greater attention recently (Ibrahim et al. 2016; Feizi et al. 2018; Zhang et al. 2018). Nanomaterials present enhanced reactivity, and thus are more effective when compared to their bulkier counterparts due to their higher surface-to-volume ratio. Nanomaterials also offer the potential to leverage unique surface chemistry as compared to traditional approaches, such that they can be functionalized or grafted with functional groups that can target specific molecules of interest (pollutants) for efficient remediation. Furthermore, the intentional tuning of the physical properties of the nanomaterials (such as size, morphology, porosity, and chemical composition) can confer additional advantageous characteristics that directly affect the performance of the material for contaminant remediation (Guerra et al. 2018). For example, Su et al. (2016) showed that a biochar-supported nanoscale zero-valent iron (nZVI@BC) material with nZVI to BC mass ratio of 1:1 exhibited better stability and mobility than of bare-nZVI, for the immobilization efficiency of Cr (VI) and Cr_{total}. Rathor et al. (2017) showed that with nZVI (10–100 nm), the nickel contamination was removed as much as 85% after the treatment of nZVI (100 mg 10 g^{-1} of soil). Likewise, Tafazoli et al. (2017), reported that nZVI NPs reduced the bioavailability of lead (Pb) and cadmium (Cd). On the other hand, Arenas-Lago et al. (2016), found that $Ca_3(PO_4)_2$ nanoparticles (30 nm) decrease the contents of Cu, Pb, and Zn in polluted soil.

In water remediation, Chen et al. (2019) showed that a cross-linked poly acrylic acid (PAA) modified poly(ether sulfone) nanofibrous membrane (NFM), can separate emulsified oil/water emulsion and a wide range of oil/water mixtures. The research by Posati et al. (2019) demonstrated that polysulfone granules (PS) prepared by using industrial scraps of polysulfone hollow fiber membranes (PS-HPG) could be efficiently coated with polydopamine (PD) NPs to provide an eco-sustainable and low-cost material to be used as adsorbents for drinking water treatment. In order to evaluate the multi-functionality engineering on nanocomposites by combining one dimensional (1D) ZnO nanorod (NR) and two dimensional (2D) reduced graphene oxide (rGO) for efficient water remediation, the results of Ranjith et al. (2017) revealed that the nanocomposite rGO and ZnO NR show a 4-fold enhancement in the visible photocatalytic activity and effective removal of Cu (II) and Co (II) ions from aqueous solution respectively. Hu et al. (2018), showed that the carbon nanofibers (CNFs)@I-doped $Bi_2O_2CO_3$–MoS_2 membranes nanocomposite exhibited excellent photodegradation for eliminating the refractory pollutant Rhodamine B (RhB) from wastewater.

Nevertheless, nanotechnological techniques are currently being developed which use nano zero-valent iron (nZVI), carbon nanotubes, sponges, aerogels and nanocomposites, metal, and non-metal nanostructured oxides, nitrides, salts, and zeolites. Some of these nanomaterials and methods are being applied on a small scale due to insufficient knowledge of their toxicity, lack of more detailed investigations, or higher costs (Kharisov et al. 2014), as well as the ecological risks associated with their use. In agriculture, although there are enough practices to produce food, they cannot continue to do so in a sustainable manner without technological intervention (Willet et al. 2019). On the other hand, it has been argued and demonstrated that the agricultural practices that have emerged since the green revolution, and continue to be applied to this day, continue to be inefficient and unsustainable and have led to unacceptable environmental degradation (Willet et al. 2019). Currently there is deterioration in the quality of resources (water, soil, and air, among others), excessive and inefficient use of pesticides and herbicides, the appearance of diseases and pests, and the need to resort to a healthy diet so that several authors agree that nanotechnology is an opportunity to mitigate environmental impact, save costs and increase crop yields (Jägermeyr et al. 2015; Siddiqui et al. 2015; Fraceto et al. 2016; León-Silva et al. 2018). These authors stated that nanotechnologies could also improve the efficiency of cropping systems concerning all inputs through more efficient delivery mechanisms, improving the efficiency of nutrient utilization, and better management of disease and crop loss (Alexander et al. 2017). In particular, development and application of new fertilizers using nanotechnology is one of the potentially effective options to achieve sustainable agriculture (Cieschi et al. 2019). NPs are found within CNTs, Cu, Ag, Mn, Mo, Zn, Fe, Si, and Ti, and nanoformulations of conventional agricultural inputs like phosphorus, urea, sulfur, validamycin, tebuconazole, and azadirachtin have been converted into nanopesticide and nanofertilizer form (Liu and Lal 2015). Even the term nanoagriculture is utilized to refer to nanoencapsulated conventional fertilizers, pesticides, and herbicides in the slow and sustained release of nutrients and agrochemicals, resulting in a precise dosage to the plants (Duhan et al. 2017).

In the last three years a variety of information related to the effect of nanomaterials on crop production continues to be published. For example, Anderson et al. (2017) revealed that CuO and ZnO nanoparticles at doses of ≥ 10 mg metal kg^{-1} change the production of key metabolites involved in plant protection in a root-associated microbe, *Pseudomonas chlororaphis* O6. It was found that altered synthesis occurs in the microbe for phenazines, which function in plant resistance to pathogens, the pyoverdine-like siderophore that enhances Fe bioavailability in the rhizosphere and indole-3-acetic acid affecting plant growth. Rizwan et al. (2019), showed that with FeNP (50–100 nm, 15–20 mg L^{-1}) positively affected the photosynthesis, reduced the electrolyte leakage and superoxide dismutase and peroxidase activities in leaves of Cd-stressed wheat. The concentrations of Cd in roots, shoots, and grains were also significantly decreased with NP application. In other studies, Priyanka and Venkatachalam (2016), reported that ZnO-NP promoted the growth rate, biomass, photosynthetic pigment levels, protein content, the activity of antioxidant enzymes, and up-regulated the expression level of superoxide dismutase (SOD) and peroxidase (POX) in cotton plants. In another study, the improvement of the physicochemical properties of plants alleviated by saline stress conditions utilizing nanomaterials as biostimulants was evaluated, as reported by Juárez-Maldonado et al. (2019). They also found that the application of CuNPs increased the yield and fruit quality of the tomato crop (Juárez-Maldonado et al. 2016). Juárez-Maldonado et al. (2016) reported that the application of CuNPs and potassium silicate was effective in reducing the severity of *Clavibacter michiganensis*. Cumplido-Nájera et al. (2019) stated that the loss of yield due to the *C. michiganensis* was reduced in 16.1% in tomato crops when NPs were added.

Rodrigues et al. (2017) proposed five promising opportunities for nanotechnology to improve the sustainability of agri-food systems: (1) sensors for monitoring chemicals, for assessing physical, chemical, or biological properties and processes, and for detecting pathogens or toxins; (2) technologies for controlling pathogens to increase food safety and minimize food waste; (3) membrane and sorbent technologies for distributed water treatment and resource recovery; (4) materials for timed and targeted delivery of agrochemicals; and (5) materials for monitoring and improving animal health. Table 2.3 shows the effectiveness and application of nanomaterials in soil and water remediation, in crop production, and food systems.

Kah et al. (2018) argued that some nanoagrochemicals could exhibit significantly superior properties relative to non-nano analogs under laboratory conditions. In addition, differences in toxicological effectiveness or efficacy gains observed in the laboratory were not always confirmed for different target organisms/plants and are not guaranteed to take place under field

TABLE 2.3

Effect of Nanomaterials in Water and Soil Remediation, Crop Production, and Food Systems

Nanomaterial type	Effects on soil and water	References
C_{60}, MWCNT and fullerene soot	Decreased the population of phenanthrene degrading bacteria and reduced phenanthrene catabolic activity.	Oyelami and Semple (2015)
TiO_2/CuO nanoneedle arrays	These nanomaterials are potentially useful in practical oil/water separation.	Yuan et al. (2017)
Iron phosphate NPs stabilized by a sodium carboxymethyl cellulose composite	The immobilization efficiency of Cd was 81.3% after 28 days of soil remediation.	Qiao et al. (2017)
SiO_2 and zeolite, and two bacteria; *Bacillus safensis* FO-036b (T) and *Pseudomonas fluorescens* p.f.169 in the rhizosphere of sunflower	Pb and Zn were immobilized in the soil.	Mousavi et al. (2018)
Interactions of polyvinylpyrrolidone (PVPs) coated magnetite NPs and oil-degrading bacteria for enhanced oil removal	The combination of NP and oil-degrading bacterial strains worked effectively to remove 100% of oil within 48 h or less from water.	Alabresm et al. (2018)
Modified carbon black nanoparticles (MNCBs)	Petroleum degradation increased by 50% in petroleum-Cd co-contaminated soil and 65% in petroleum-Ni co-contaminated soil in plant-microbe combined remediation.	Chen et al. (2019)
Nanomaterial type	**Effects on crop production and food systems**	**References**
Chitosan coated solid lipid NPs (NC)-based nanodelivery systems	To enhance the stability and oral bioavailability of curcumin and help incorporate it in functional foods and nutraceuticals. NC has been shown to increase the bioavailability of curcumin above that of curcumin suspensions after oral administration.	Ramalingam et al. (2016)
Synthesized $Cu_3(PO_4)_2 \cdot 3H_2O$ nanosheets and commercial CuO-NPs	Reduced disease presence by an average of 31%, resulting in greater plant biomass of tomato.	Ma et al. (2019)
γ-Fe_2O_3 and Fe_3O_4 NPs	Promote plant growth, increase chlorophyll content at a certain stage of exposure. Besides, fruit weight of muskmelon significantly increased by 9.1%, 9.4%, and 11.5%.	Wang et al. (2019)
ZnO-NPs	In coffee, positively affected the fresh weight and dry weight (FW and DW) of roots and leaves, increasing the FW by 37% (root) and 95% (leaves) when compared to control. The DW increase was 28%, 85%, and 20% in roots, stems, and leaves, respectively.	Rossi et al. (2018)

conditions. Accordingly, other authors showed negative effects on non-target plants and organisms (Cao et al. 2016; Moll et al. 2016; Zaytseva and Neumann 2016; Zuverza-Mena et al. 2016; Brami et al. 2017; McKee et al. 2017; Romero-Freire et al. 2017; Rajput et al. 2017; Medina-Pérez et al. 2018). Some authors ask, do the nanotechnology supplies outweigh the traditional ones in the production of crops? The answer is that there has not been enough evidence published regarding the benefits of agro-nanotechnologies without disadvantages in long-term field experiments, although several scientists stated that these new developments are eco-friendly materials for more acceptable nanoagriculture.

Moreover, plants may absorb the required materials and leave the rest in the soil (Elemike et al. 2019); however, León-Silva et al. (2016), Terekhova et al. (2017), and Prasad et al. (2017) affirmed that the pollution of soils with nanomaterials presents a serious risk of entering the human organism and the tissues of land plants and animals. NP entry into any biocenosis component can also lead to their introduction into the system and transfer them through the food chain (Terekhova et al. 2017). Researchers emphasize the search for potential methods of assessing the effect of synthetic nanotechnology products on natural complexes, and the functioning of the main links of the trophic chain and separate organisms. The complexity of the soil organo-mineral composition and the unpredictable dynamics of soil properties in time and space create problems in the structural and functional analysis of the biotic complex of soil under the impact of conventional pollutants, whose chemical transformations are well understood (Terekhova et al. 2017).

Nevertheless, currently the integration of green chemistry principles into nanotechnology is becoming relevant in nanoscience research. In this sense, biological methods exist to synthesize metal and metal oxide nanoparticles of specific shape and size since they enhance the properties of nanoparticles used in green chemistry. Particularly, plant-mediated methods which are devoid of the use of toxic chemicals, which has adverse effects on the environment (Anderson et al. 2017), contrary to conventional NP production which improves impacts on human health and the environment, whether by the physical or chemical assembly. However, mainly investigations have been realized in laboratory conditions.

2.6 WHAT DOES HEALTHY SOIL AND THE THREAT POSED BY NANOTECHNOLOGY MEAN?

The advance of nanotechnology has progressed by leaps and bounds. Nanotechnological innovations have been used to cure diseases in biomedical instrumentation, in electronics, cosmetics, the textile industry, agricultural technology, and environmental protection, among others (Verma et al. 2018; Mishra et al. 2017). Nanoscale materials have been applied broadly to the remediation of contaminated water and soil due to their unique physicochemical properties, such as large surface areas, highly active reaction sites, strong adsorption, and unique catalytic and chelating abilities (Cheng et al. 2019). Nanotechnology research in the agricultural sector has also become necessary, even a key factor for sustainable development. This technology has proved to be as good in resource management, in the field of agriculture, and drug delivery mechanisms in plants, and helps to maintain the fertility of the soil (Prasad et al. 2017). Figure 2.3 shows the progress of publications related to nanotechnology and sustainable agriculture. Specifically, when searching the database in the field of 'web science' topics, it is found that there is an increase in the number of publications (the words nanotechnology, agriculture, and sustainability were used), however, the reports date from 2015, i.e. the advanced technology proposal is recent and in most of the investigations a section is used to discuss the potential risks. Despite the potential benefits and widespread uses of nanotechnology, several contradictory effects have been reported in the environment. Scientists have also raised concerns regarding the unregulated use of engineered nano-sized materials or devices.

Soil health is a measure of the state of natural capital that reflects the capacity of the soil, relative to its potential, to respond to agricultural management by maintaining agricultural production (human nutrition) (Barrios et al. 2015). The soil is the basis of multiple ecosystem services, climate regulation, and the nutrient cycle, among others (Pachapur et al. 2016). Therefore, sustainable agriculture is inherently dependent on soil health. Many ecosystem processes have the soil as their regulatory center, and soil biota plays a key role in a wide range of ecosystem services that underpin the sustainability of agro-ecosystems (Barrios et al. 2015). Due to its importance, there is a growing concern about the presence of ENP in the soil, as there is little evidence of the effect on the soil regarding the impacts of ENPs on the physical, chemical (Pachapur et al. 2016), mechanical (Bayat et al. 2018), and biologic properties of soil (Pachapur et al. 2016).

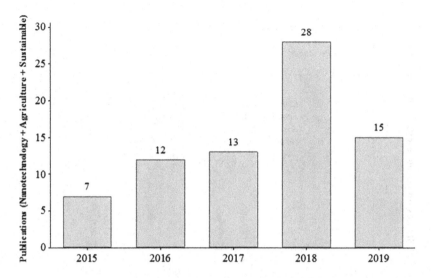

FIGURE 2.3 Publications that address and integrate the topics of nanotechnology, agriculture, and sustainability (data obtained from the Web Science, 2015–2019 [the search was carried out in June 2019]).

ENPs can enter soil through various pathways, such as agricultural amendments of sewage sludge, atmospheric deposition, landfills, or accidental spills during industrial production (Simonin and Richaume 2015). Therefore, soils are considered as a major sink for ENPs because of their inevitable release to the subsurface environment during production, transportation, use, and disposal processes (He et al. 2019; Parada et al. 2018). Hence, a detailed understanding of the processes governing the transport and retention of AgNPs in soil (Rahmatpour et al. 2018) as well as in other NPs such as Ti, Al, Au, Cu, Ce, Zn, Fe, Mn, and CNTs (Duhan et al. 2017) is required. Its knowledge could accurately assess the fate and distribution of NPs to design effective strategies for reducing their toxicity in the environment (Rahmatpour et al. 2018).

The recent development of a nanoencapsulated pesticide formulation has slow releasing properties with enhanced solubility, specificity, permeability, and stability (Bhattacharyya et al. 2016). For example, CuNPs have been proposed as an antimicrobial agent in agriculture. However, little is known about the combined effects of CuNPs and pesticides in soil and the changes in microbial community profiles (Parada et al. 2018). In the last decade, nanofertilizers have been freely available on the market, but the agricultural fertilizers in particular have not been promoted by the major chemical companies (Prasad et al. 2017). Studies of the uptake, biological fate, and toxicity of some metal oxide NPs, such as Al_2O_3, TiO_2, CeO_2, FeO, and ZnO, have been carried out intensively in the present decade for agricultural production (Zhang et al. 2016). Many advantages of the physical-chemical properties of nanoscale materials are also exploitable in the field of biosensor development in human activities such as health care, agriculture, genome analysis, food and drink, the processing industries, environmental monitoring, defense, and security, principally for pathogen detection and diagnosis (Fogel and Limson et al. 2016). In agriculture, other nanotechnological advantages have been presented and disseminated in terms of weed or disease control (Chhipa 2017; Duhan et al. 2017). Many of these and other advantages, and risk perspectives are discussed in Iavicoli et al. (2017) and Rastogi et al. (2017). Nevertheless, based solely on the review articles by Tourinho et al. (2012), Mckee and Felser (2016), Liné et al. (2017), and Hou et al. (2018), Figure 2.4 was drawn up showing the negative effects on soil organisms.

Currently, the only way to obtain information on existing levels of NPs in the environment is to model predicted environmental concentrations. In this sense, authors such as Keller et al. (2013) and Sun et al. (2014) suggest that exposure modeling strongly indicates that soil could be a major sink of NP compared to air and water ecosystems. The experiments of Yang et al. (2016) and (2017)

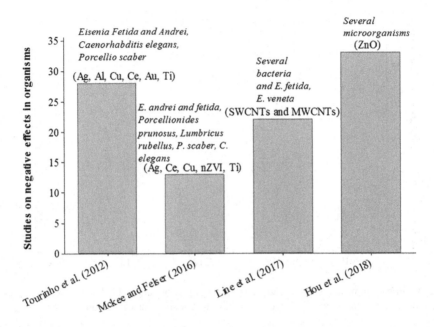

FIGURE 2.4   Studies on toxic effects of different NPs on soil macro-invertebrates and microorganisms.

confirmed the effectiveness of *C. elegans* for estimating soil risk metrics which could help to develop methods of management for mitigating the toxicity induced by NPs on terrestrial ecosystems. According to Sanivada et al. (2017), the examination of current models of the soil community having contact with nanotechnology agricultural interventions affirms the profoundly integrative prototype of associations inside each of these functions and prompts the conclusion that measurement of individual groups of organisms, processes, or soil properties is not sufficient to specify the state of soil health.

Conversely, Kah et al. (2018) found that from around 78 recently published papers, some nanoformulations can alter the properties of pesticides and fertilizers; however, not all cause such conditions. Most of the studies evaluated were also performed in the laboratory, and little is known about the effectiveness of various nanomaterials under field conditions. In recent research based in a meta-analysis study, Wang et al. (2018) concluded that the risk of Ag ENPs to terrestrial plants and fauna (including humans through trophic transfer) is small. However, the usage of Ag ENP products may potentially pose a risk to soil microbial communities and in fresh water, and more research is needed in these areas. Likewise, for more than a decade Chang et al. (2013) investigated the toxicity of TiO_2-NPs and the potentially harmful impact on human health using meta-analysis of in vitro and short-term animal studies. The results showed that of all the experimental studies, more than 50% showed positive statistical significance except the apoptosis group, and the cytotoxicity was dose-dependent but was not clear if it was size-dependent, and TiO_2-NPs were retained in several organs. Therefore, the authors concluded that TiO_2 could induce cell damage related to exposure size and dose. However, studies will be needed to demonstrate that nanoparticles have toxic effects on the human body, especially in epidemiological studies.

Despite considerable advances in identifying possible applications of nanotechnology in agriculture, many issues remain to be resolved shortly before this technology may make significant contributions to the area of agriculture. Some of the main aspects that require further attention are: (i) development of specific hybrid carriers for delivering active agents including nutrients, pesticides, and fertilizers in order to maximize their efficiency following the principles of green chemistry and environmental sustainability (De Oliveira et al. 2014; Fraceto et al. 2016); (ii) design of processes that are easy to upscale at the industrial level; (iii) comparison of effects of nanoformulations/nanosystems with existing commercial products, in order to demonstrate real practical advantages; (iv) acquisition of knowledge and developments of methods for risk and life-cycle assessment of nanomaterials, nanopesticides, and nanofertilizers, as well as assessment of the impacts (e.g., phytotoxic effects) on non-target organisms, e.g. other plants, soil microbiota, and bees; and (v) advances in the regulations on the use of nanomaterials (Amenta et al. 2015; Fraceto et al. 2016).

2.7 AGRO-NANOBIOTECHNOLOGY FOR SUSTAINABLE DEVELOPMENT

The principal aim of agriculture is to increase food production, and this applies to several strategies that affect the growth and, thus, the physiology of plants. Agro-nanotechnology, which promises better resource management through novel tools and technology platforms within the limited resources of land and water (Mishra et al. 2014c), has several challenges. Some of them are focused on: a) pesticides and fertilizers formulated for crop improvement; b) applications of nanosensors and nanobiosensors to detect pathogens or hazard residues; c) nanocarriers to improve genetic manipulation of plants and benefic microorganisms; d) plant disease diagnostic; e) animal health and production; and f) post-harvest handling (Medina-Perez et al. 2019).

New technologies for the intelligent release of active compounds and prolonging the effect of fertilizers and agrochemicals have been developed by bionanotechnology, an example is the nano-encapsulation of nutrients in polysaccharides, minerals, or liposomes, nanocapsules (urea, phosphorus, sulfur, among others) and other pesticides (validamycin, tebuconazole, and azadiractin) managing to optimize the costs of traditional doses, for crop protection (Chhipa 2017). Some products declared as nanofertilizers are already marketed worldwide. Some examples are Nano-GroTM (plant growth regulator and immunity enhancer), Nano-Ag AnswerR (microorganism, mineral electrolyte, and seaweed), TAG NANO (NPK, PhoS, Zn, Ca, etc.)., nano protein-lactogluconate and micronutrients, probiotics, vitamins, seaweed extracts, and humic acid, among others, which can be bought (Prasad et al. 2017). Currently, very promising nanosensors are being developed for use in the field without the need for a complex laboratory, highly qualified personnel, or expensive and difficult to manage equipment. Nanosensors have sensitive and efficient detection tools for a large variety of agrochemicals, phytopathogens, soil moisture, and soil pH, among others, to further increase productivity (Saxena et al. 2017). For the detection of genetically modified crops, a method using a new electrochemical unlabeled carbon nanosphere immunosensor with ultrasensitive quantification of Cry1Ab cultures detecting up to 3.0 pg mL^{-1} was developed. Its use can also be extended to other proteins that detect toxins in food (Gao et al. 2017).

An immunofluorescent biosensor was also developed for genetically modified organisms using novel nano-composite and iron-polymer-graphene nanocomposites for the sensitive detection of 5-enolpyruvylshikimate-3-phosphate synthase of the CP4 strain of *Agrobacterium tumefaciens* (CP4-EPSPS), a biomarker of genetically modified cultures with a limit of detection of 0.34 ng mL^{-1} (Yin et al. 2017). On the other hand, for the cucumber green mottle mosaic virus, a simple and sensitive label-free colorimetric detection method was developed with gold nanoparticles modified as colorimetric probes detecting up to 30 pg μL^{-1} (Wang et al. 2017). The use of carbon nanotubes has also become one of the most popular in the development of a sensor to monitor environmental samples such as methyl parathion, paraoxon, and their metabolites (Baker et al. 2017). Recently, gold nanoparticles have been used as labels where they amplify the analytical signal, and the sensitivity of the immunological assay is greatly improved, where *Pantoea stewartii* subsp. *stewartii* can be detected. Pathogen biosensors use receptors such as antibodies, DNA probes, phages, peptides, and proteins. DNA biosensors in hybridization including electroluminescence, fluorescent, colorimetric, and voltammetric without labels, the disadvantage they have is the treatment of the sample by DNA extraction. Therefore, more portable nanosensors are expected to be used directly in the field and detected in real time to have an efficient diagnosis and achieve better production (Khater et al. 2017).

While the optimization of crop production is necessary, the uses of nanomaterials have raised concerns due to the possible dangerous effects on the environment and human health. The plants interact closely with the environment, playing an important link in the trophic chains. The greatest risk of exposing the plants to the ENPs is the possibility that they are absorbed by air or root entries and subsequently transferred and stored in the tissues of the plants, causing biomagnifications (Holbrook et al. 2008). To assure the health of consumers of plant products that could have been exposed to ENPs, accidentally or intentionally, several studies have been conducted on the behavior and fate of nanoparticles within plants. Biotransformation studies have been of great importance in the development of sustainable alternatives in the use of ENPs in agriculture. In the biotransformation process in which pollutants are biochemically modified by reducing their possible toxicity, detoxification of ENPs, it has been shown that they can also be altered by biological systems or environmental media (Van Hoecke et al. 2011). Among the most studied reactions can be listed: redox reaction, sulfuration, phosphorylation, and macromolecular or molecular modification (Zang et al. 2012, Lowry et al. 2012). Biotransformation, which is defined as biochemical modification by living organisms, has been studied for a long time on the common pollutants. In the process of biotransformation, either enhanced toxicity or detoxification is possible. In terms of ENPs, they may also be modified by the ambient environmental media and biological systems, and their final fate and toxicity to organisms may be altered (Van Hoecke et al. 2011; Table 2.4). Typical transformations of nanomaterials studied most recently include redox reaction, sulfidation, phosphorylation, and macromolecular/molecular modification (Zhanget al. 2012).

A new approach of sustainable studies is the research in plant and environmental microbiomes to increase the crop productivity for food and nutrients in a safe way (Figure 2.5). It becomes a challenge to adapt agricultural biotechnologies to climate change and the restricted availability of nutrients; efforts are focused today on optimizing the natural resources present in the environment. Agriculture will increase exponentially in the future driven by sophisticated technologies (e.g., the next generation, sequencing, nanotechnology, synthetic biology) to analyze and manipulate the environment and microbiomes (Singh 2017). Research in the efficiency of agrochemical inputs, the improvement of smart crops through targeted delivery, and improving the relationship between soil and plants through microbiome enhancement (Lowry et al. 2019). For example,

TABLE 2.4
Examples of Reported Studies about Biotransformation of Nanoparticles in Edible Crops

Nanoparticle	Reaction	Effect	References
AgNP	Sulfidation	Decreased toxicity to *Escherichia coli* growth due to the lower solubility of silver sulfide.	Reinsch et al. (2012)
AgNP	Oxidation	Silver speciation in the roots of AgNP exposed *Lolium multiflorum* was oxidized as Ag(I).	Yin et al. (2011)
CuO-NP	Reduction	Reduced to Cu_2O and Cu_2S in maize plants.	Wang et al. (2012)
Rare earth oxide NP	Oxidation	Root elongations of cucumber seedlings were severely inhibited by both La_2O_3 and Yb_2O_3-NPs.	Ma et al. (2015; 2011)
ZnO-NP	Oxidation	Zn was found only in $Zn_2þ$ oxidation state and was not present as ZnO-NP in soybean.	López-Moreno et al. (2010)
$Ni(OH)_2$	Oxidation	Biotransformation of $Ni(OH)_2$ to $Ni_2þ$ in plant shoots and leaves while no biotransformation occurred in the roots of mesquite plants.	Parsons et al. (2010)
CuO-NP	Reduction	Transported from the roots to the leaves, and that Cu(II) combined with cysteine, citrate, and phosphate ligands and was even reduced to Cu(I), in rice (*Oryza sativa* L.).	Peng et al. (2015)

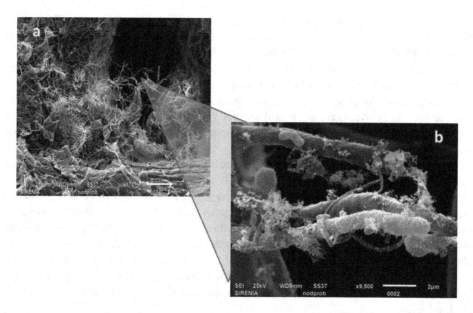

FIGURE 2.5 Microbiome and nanoparticles. SEM micrograph of the external surface of secondary roots of common bean plant (*Phaseolus vulgaris* L.) treated with TiO_2-NP (0.30 g kg^{-1}). a) 170×; b) 9500×.

plants secrete many organic compounds through their roots (polysaccharides, proteins, enzymes, phytohormones, and some secondary metabolites) with which they interact with microorganisms in the rhizosphere. The ENPs, according to their state of aggregation, impact on the destination, and the arrangement of the particles is probably significant but remains unknown. The rhizosphere microbiome will produce proteins and other biomolecules that can affect the fate of the ENPs. A study by Maurer-Jones et al. (2013) showed that amino acids, such as cysteine present in proteins and natural organic matter, increase the initial aggregation rates of ENPs but not the size of long-term aggregation.

Understanding the impacts of plant root exudation and the microbial activity of the rhizosphere in soil-based conditions and realistic exposure scenarios is essential for a meaningful assessment of the fate of the ENPs in the environment (Servin et al. 2016). Other advances are worth mentioning concerning crops that are irrigated with desalinated water where they act as a nanofilter and solar energy so that their crops grow in arid lands (Ghermandi et al. 2017). There are also studies with increasing tolerance to salinity where cerium oxide nanoparticles improve the photosynthesis of plants stressed with salt; it is thought that it modifies the anatomy of the root and thus improves the tolerance to salt (Rossi et al. 2017). On the other hand, there are studies with nanoclay, nanochitosan, and nanozeolite where such carbon-based formulations of nanocomposites make their way into the agricultural sector because there was a positive response from the health of the plants (Khati et al. 2017).

2.8 CONCLUSION

Nanoscience and nanotechnology have progressed by leaps and bounds in several cutting-edge knowledge areas. However, in the agricultural sector, there are still many studies to conduct. In particular, the long-term experiments in field conditions regarding agricultural nanotechnologies are scarce. However, hundreds of nanotechnological products are being spread on agricultural land worldwide.

The soil quality, also referred to as soil health, is a measure of the state of natural capital that reflects the capacity of the soil, relative to its potential, to respond to agricultural management by maintaining the agricultural production and sustain plants, animals, and humans. The soil is also the basis of multiple ecosystem services, climate regulation, and the nutrient cycle, among others. Therefore, sustainable agriculture is inherently dependent on soil health.

Many ecosystem processes have the soil as their regulatory center, and soil biota plays a key role in a wide range of ecosystem services that underpin the sustainability of agriculture. Due to the importance of healthy soil, there is a growing concern about the presence of engineered nanoparticles in the soil, as their effects on the soil are not yet well described. However, several studies have stated that manmade nanoparticles have an impact on the physical, chemical, mechanical, or biological soil properties.

There is a huge technological challenge regarding the quantification and identification of engineered nanoparticles in natural environments such as soils, air, or water. Similar challenges also have to be faced with the identification of these materials in cells or tissues from plants, microorganisms, or animals. Fortunately, scientists and technologists from several cutting-edge knowledge areas are working together to increase the yield of crops and improve the quality, affordability, and innocuity of food to underpin the sustainability of agroecosystems worldwide. The above takes into account the quality of the soil, the necessity of food, and the urgent duty of increasing human wellbeing to shape sustainable development, without hampering the pursuit of sustainable development goals, through the use of the safest and best available technologies.

ACKNOWLEDGMENTS

This research was funded by 'Ciencia Básica SEP-CONACyT' project 287225, by the COAH-2019-C13-C006_FONCYT-COECYT project, by the Sustainability of Natural Resources and Energy Programs (Cinvestav-Saltillo), and by Cinvestav Zacatenco.

REFERENCES

Abdel Latef AAH, Srivastava AK, El-sadek MSA et al. (2018) Titanium dioxide nanoparticles improve growth and enhance tolerance of broad bean plants under saline soil conditions. *Land Degrad Dev* 29(4):1065–1073.

Achari GA, Kowshik M (2018) Recent developments on nanotechnology in agriculture: Plant mineral nutrition, health, and interactions with soil microflora. *J Agric Food Chem* 66(33):8647–8661. doi: 10.1021/acs.jafc.8b00691

Adhikari T, Kundu S, Biswas AK et al. (2012) Effect of copper oxide nano particle on seed germination of selected crops. *J Agric Sci Technol* 2:815–823.

Alabresm A, Chen YP, Decho AW, Lead J (2018) A novel method for the synergistic remediation of oil-water mixtures using nanoparticles and oil-degrading bacteria. *Sci Total Environ* 630:1292–1297. doi: 10.1016/j.scitotenv.2018.02.277

Alexander P, Brown C, Arneth A et al. (2017) Losses, inefficiencies and waste in the global food system. *Agric Syst* 153:190–200. doi: 10.1016/j.agsy.2017.01.014

Ali J, Ali N, Wang L et al. (2019) Revisiting the mechanistic pathways for bacterial mediated synthesis of noble metal nanoparticles. *J Microbiol Methods* 159:18–25. doi: 10.1016/j.mimet.2019.02.010

Amenta V, Aschberger K, Arena M et al. (2015) Regulatory aspects of nanotechnology in the agri/feed/food sector in EU and non-EU countries. *Regul Toxicol Pharmacol* 73(1):463–476. doi: 10.1016/j.yrtph.2015.06.016

Anderson AJ, McLean JE, Jacobson AR, Britt DW (2017) CuO and ZnO nanoparticles modify interkingdom cell signaling processes relevant to crop production. *J Agri Food Chem* 66(26):6513–6524. doi: 10.1021/acs.jafc.7b01302

Araújo R, Castro ACM, Fiúza A (2015) The use of nanoparticles in soil and water remediation processes. *Mater Tod* 2(1):315–320. doi: 10.1016/j.matpr.2015.04.055

Arenas-Lago D, Rodríguez-Seijo A, Lago-Vila M (2016) Using $Ca_3(PO_4)_2$ nanoparticles to reduce metal mobility in shooting range soils. *Sci Total Environ* 571:1136–1146. doi: 10.1016/j.scitotenv.2016.07.108

Arora S, Sharma P, Kumar S et al. (2012) Gold-nanoparticle induced enhancement in growth and seed yield of *Brassica juncea*. *Plant Growth Regul* 66(3):303–310. doi: 10.1007/s10725-011-9649-z

Asli S, Neumann PM (2009) Colloidal suspensions of clay or titanium dioxide nanoparticles can inhibit leaf growth and transpiration via physical effects on root water transport. *Plant Cell Environ* 32(5):577–584. doi: 10.1111/j.1365-3040.2009.01952.x

Baker S, Volova T, Prudnikova SV et al. (2017) Nanoagroparticles emerging trends and future prospect in modern agriculture system. *Environ Toxicol Pharmacol* 53:10–17. doi: 10.1016/J.ETAP.2017.04.012

Bao-shan LIN, Chun-hui LI, Li-jun F et al. (2004) Effect of TMS (nanostructured silicon dioxide) on growth of Changbai larch seedlings. *J For Res* 15(2):138–140.

Barrena R, Casals E, Colón J et al. (2009) Evaluation of the ecotoxicity of model nanoparticles. *Chemosphere* 75(7):850–857. doi: 10.1016/J.CHEMOSPHERE.2009.01.078

Barrios E, Shepherd K, Sinclair F (2015) Soil health and agricultural sustainability: The role of soil biota. In: FAO ed. *Agroecology for Food Security and Nutrition: Proceedings of the FAO International Symposium*. FAO, Rome. pp. 104–122.

Bayat H, Kolahchi Z, Valaey S et al. (2018) Novel impacts of nanoparticles on soil properties: Tensile strength of aggregates and compression characteristics of soil. *Arch Agron Soil Sci* 64(6):776–789. doi: 10.1080/03650340.2017.1393527

Bhattacharyya A, Duraisamy P, Govindarajan M et al. (2016) Nano-biofungicides: Emerging trend in insect pest control. In: Prasad R ed. *Advances and Applications through Fungal Nanobiotechnology*. Springer International Publishing, Cham. pp. 307–319. doi: 10.1007/978-3-319-42990-8_15

Birbaum K, Brogioli R, Schellenberg M et al. (2010) No evidence for cerium dioxide nanoparticle transloca-tion in maize plants. *Environ Sci Technol* 44(22):8718–8723. doi: 10.1021/es101685f

Bishoge OK, Zhang L, Suntu SL et al. (2018) Remediation of water and wastewater by using engineered nanomaterials: A review. *J Environ Sci Health A* 53(6):537–554. doi: 10.1080/10934529.2018.1424991

Brami C, Glover AR, Butt KR, Lowe CN (2017) Effects of silver nanoparticles on survival, biomass change and avoidance behaviour of the endogeic earthworm *Allolobophora chlorotica*. *Ecotox Environ Safe* 141:64–69. doi: 10.1016/j.ecoenv.2017.03.015

Cao J, Fen Y, Lin X, Wang J (2016) Arbuscular mycorrhizal fungi alleviate the negative effects of iron oxide nanoparticles on bacterial community in rhizospheric soils. *Front Environ Sci* 4:10. doi: 10.3389/fenvs.2016.00010

Cecchin I, Reddy KR, Thom A et al. (2017) Nanobioremediation: Integration of nanoparticles and bioreme-diation for sustainable remediation of chlorinated organic contaminants in soils. *Int Biodeter Biodegr* 119:419–428. doi: 10.1016/j.ibiod.2016.09.027

Chakre M, Sharon M (2019) How old is nanotechnology? *Hist Nanotechnol*:1–19. doi: 10.1002/9781119460534.ch1

Chang X, Zhang Y, Tang M, Wang B (2013) Health effects of exposure to nano-TiO_2: A meta-analysis of experimental studies. *Nanoscale Res Lett* 8(1):51. doi: 10.1186/1556-276x-8-51

Chen S, Lv C, Hao K et al. (2019) Multifunctional negatively-charged poly (ether sulfone) nanofibrous mem-brane for water remediation. *J Colloid Interface Sci* 538:648–659. doi: 10.1016/j.jcis.2018.12.038

Cheng J, Sun Z, Yu Y et al. (2019) Effects of modified carbon black nanoparticles on plant-microbe reme-diation of petroleum and heavy metal co-contaminated soils. *Int J Phytoremediation*:1–9. doi: 10.1080/15226514.2018.1556581

Chhipa H (2017) Nanofertilizers and nanopesticides for agriculture. *Environ Chem Lett* 15(1):15–22. doi: 10.1007/s10311-016-0600-4

Chugh K, Kapur A, Jerath K (2018) Nanotechnology for sustainable raw mineral extraction and use. In: Sridharan K eds. *Emerging Trends of Nanotechnology in Environment and Sustainability*. Springer Briefs in Environmental Science, Springer, Cham, pp. 27–34. doi: 10.1007/978-3-319-71327-4_4

Cieschi MT, Polyakov AY, Lebedev VA et al. (2019) Eco-friendly iron-humic nanofertilizers synthesis for the prevention of iron chlorosis in soybean (*Glycine max*) grown in calcareous soil. *Front Plant Sci* 10:413. doi: 10.3389/fpls.2019.00413

Cumplido-Nájera CF, González-Morales S, Ortega-Ortíz H et al. (2019) The application of copper nanopar-ticles and potassium silicate stimulate the tolerance to *Clavibacter michiganensis* in tomato plants. *Sci Hort* 245:82–89. doi: 10.1016/j.scienta.2018.10.007

Dasgupta N, Ranjan S, Chakraborty AR et al. (2016) Nanoagriculture and water quality management. In: Ranjan S, Dasgupta N, Lichtfouse E eds. *Nanoscience in Food and Agriculture 1: Sustainable Agriculture Reviews*, vol. 20. Springer, Cham. doi: 10.1007/978-3-319-39303-2_1

de la Rosa G, López-Moreno ML, de Haro D et al. (2013) Effects of ZnO nanoparticles in alfalfa, tomato, and cucumber at the germination stage: Root development and X-ray absorption spectroscopy studies. *Pure Appl Chem* 85(12):2161.

De Oliveira JL, Campos EVR, Bakshi M et al. (2014) Application of nanotechnology for the encapsulation of botanical insecticides for sustainable agriculture: Prospects and promises. *Biotechnol Adv* 32(8):1550–1561. doi: 10.1016/j.biotechadv.2014. 10.010

Dhand C, Dwivedi N, Loh XJ et al. (2015) Methods and strategies for the synthesis of diverse nanoparticles and their applications: A comprehensive overview. *RSC Adv* 5(127):105003–105037. doi: 10.1039/c5ra19388e

Diallo MS, Fromer NA, Jhon MS (2013) Nanotechnology for sustainable development: Retrospective and outlook. *J Nanopart Res* 15(11):2044. doi: 10.1007/s11051-013-2044-0

Dimkpa CO, McLean JE, Latta DE et al. (2012) CuO and ZnO nanoparticles: Phytotoxicity, metal speciation, and induction of oxidative stress in sand-grown wheat. *J Nanopart Res* 14(9):1125. doi: 10.1007/s11051-012-1125-9

Dong F, Koodali RT, Wang H, Ho W (2014) Nanomaterials for environmental applications nanomaterials. *J Nanomater* 2014:1–4. doi: 10.1155/2014/276467

Du W, Sun Y, Ji R et al. (2011) TiO_2 and ZnO nanoparticles negatively affect wheat growth and soil enzyme activities in agricultural soil. *J Environ Monit* 13(4):822–828. doi: 10.1039/C0EM00611D

Du W, Tan W, Peralta-Videa J et al. (2017) Interaction of metal oxide nanoparticles with higher terrestrial plants: Physiological and biochemical aspects. *Plant Physiol Biochem* 110:210–225.

Duhan JS, Kumar R, Kumar N et al. (2017) Nanotechnology: The new perspective in precision agriculture. *Biotechnol Rep* 15:11–23. doi: 10.1016/j.btre.2017.03.002

Elemike E, Uzoh I, Onwudiwe D, Babalola O (2019) The role of nanotechnology in the fortification of plant nutrients and improvement of crop production. *Appl Sci* 9(3):499. doi: 10.3390/app9030499

European Commission (2014) European Commission, Joint Research Centre: Towards a review of the EC recommendation for a definition of the term "Nanomaterial" part 1 : Compilation of information concerning the experience with the definition. *JRC Scientific and Policy Report EUR 265.*

Fageria NK, Filho MB, Moreira A, Guimarães CM (2009) Foliar fertilization of crop plants. *J Plant Nutr* 32(6):1044–1064.

Faramarzi MA, Sadighi A (2013) Insights into biogenic and chemical production of inorganic nanomaterials and nanostructures. *Adv Colloid Interface Sci* 189–190:1–20. doi: 10.1016/j.cis.2012.12.001

Feizi M, Jalali M, Renella G (2018) Nanoparticles and modified clays influenced distribution of heavy metals fractions in a light-textured soil amended with sewage sludges. *J Hazard Mater* 343:208–219. doi: 10.1016/j.jhazmat.2017.09.027

Fernández-Luqueño F, Medina-Pérez G, López-Valdez F et al. (2018) Chapter 1. Use of agronanobiotechnology in the agro-food industry to preserve environmental health and improve the welfare of farmers. In: López-Valdez, Fernando, Fernández-Luqueño, Fabián eds. *Agricultural Nanobiotechnology, Modern Agriculture for a Sustainable Future.* Springer. pp. 3–16. ISBN 978-3-319-96718-9.

Findlay AJ, Estes ER, Gartman A et al. (2019) Iron and sulfide nanoparticle formation and transport in nascent hydrothermal vent plumes. *Nat Commun* 10(1):1–7. doi: 10.1038/s41467-019-09580-5

Fogel R, Limson J (2016) Developing biosensors in developing countries: South Africa as a case study. *Biosensors* 6(1):5. doi: 10.3390/bios6010005

Forstner C, Orton TG, Wang P et al. (2019) Science of the total environment effects of carbon nanotubes and derivatives of graphene oxide on soil bacterial diversity. *Sci Total Environ* 682:356–363. doi: 10.1016/j.scitotenv.2019.05.162

Fraceto LF, Grillo R, de Medeiros et al. (2016) Nanotechnology in agriculture: Which innovation potential does it have? *Front Environ Sci* 4. doi: 10.3389/fenvs.2016.00020

Fraceto LF, Maruyama CR, Guilger M, et al. (2018) *Trichoderma harzianum* based novel formulations: Potential Applications for Management of Next-Gen agricultural challenges. *J Chem Technol Biotechnol* 93(8):2056–2063. doi: 10.1002/jctb.5613

Gao H, Wen L, Wu Y et al. (2017) An ultrasensitive label-free electrochemiluminescent immunosensor for measuring Cry1Ab level and genetically modified crops content. *Biosens Bioelectron* 97:122–127. doi: 10.1016/J.BIOS.2017.04.033

Ghermandi A, Naoum S, Alawneh F et al. (2017) Solar-powered desalination of brackish water with nanofiltration membranes for intensive agricultural use in Jordan, the Palestinian Authority and Israel. *Desalin Water Treat* 76:332–338. doi: 10.5004/dwt.2017.20624

Giraldo JP, Landry MP, Faltermeier S et al. (2014) Plant nanobionics approach to augment photosynthesis and biochemical sensing. *Nat Mater* 13(4):400–408.

Gómez-Sagasti MT, Epelde L, Anza M et al. (2019) The impact of nanoscale zero-valent iron particles on soil microbial communities is soil dependent. *J Hazard Mater* 364:591–599. doi: 10.1016/j.jhazmat.2018.10.034

Griffin S, Masood M, Nasim M et al. (2017) Natural nanoparticles: A particular matter inspired by Nature. *Antioxidants* 7(1):3. doi: 10.3390/antiox7010003

Grün AL, Manz W, Kohl YL et al. (2019) Impact of silver nanoparticles (AgNP) on soil microbial community depending on functionalization, concentration, exposure time, and soil texture. *Environ Sci Eur.* doi: 10.1186/s12302-019-0196-y

Guerra F, Attia M, Whitehead D, Alexis F (2018) Nanotechnology for environmental remediation: Materials and applications. *Molecules* 23(7):1760. doi: 10.3390/molecules23071760

Gunaratne GP, Kottegoda N, Madusanka N et al. (2016) Two new plant nutrient nanocomposites based on urea coated hydroxyapatite: Efficacy and plant uptake. *Indian J Agric Sci* 86:494–499.

Haghighi M, Afifipour Z, Mozafarian M (2012) The effect of N-Si on tomato seed germination under salinity levels. *J Biol Environ Sci* 6:87–90.

Haghighi M, Afifipou Z, Mozafarian M (2015) The effect of N-Si on tomato seed germination under salinity levels. *J Biol Enviromental Sci* 6:87–90.

Hao Y, Ma C, Zhang Z et al. (2018) Carbon nanomaterials alter plant physiology and soil bacterial community composition in a rice-soil-bacterial ecosystem. *Environ Pollut* 232:123–136. doi: 10.1016/j.envpol.2017.09.024

He E, Qiu H, Huang X et al. (2019) Different dynamic accumulation and toxicity of ZnO nanoparticles and ionic Zn in the soil sentinel organism *Enchytraeus crypticus*. *Environ Pollut* 245:510–518. doi: 10.1016/j.envpol.2018.11.037

He J, Wang D, Zhou D (2019) Transport and retention of silver nanoparticles in soil: Effects of input concentration, particle size and surface coating. *Sci Total Environ* 648:102–108. doi: 10.1016/j.scitotenv.2018.08.136

Hernández-Viezcas JA, Castillo-Michel H, Andrews JC et al. (2013) In situ synchrotron X-ray fluorescence mapping and speciation of CeO$_2$ and ZnO nanoparticles in soil cultivated soybean (*Glycine max*). *ACS Nano* 7(2):1415–1423. doi: 10.1021/nn305196q

Holbrook RD, Murphy KE, Morrow JB, Cole KD (2008) Trophic transfer of nanoparticles in a simplified invertebrate food web. *Nat Nanotechnol* 3(6):352–355. doi: 10.1038/nnano.2008.110

Hossain K-Z, Monreal CM, Sayari A (2008) Adsorption of urease on PE-MCM-41 and its catalytic effect on hydrolysis of urea. *Colloids Surf B Biointerfaces* 62(1):42–50. doi: 10.1016/J.COLSURFB.2007.09.016

Hou J, Wu Y, Li X et al. (2018) Toxic effects of different types of zinc oxide nanoparticles on algae, plants, invertebrates, vertebrates and microorganisms. *Chemosphere* 193:852–860. doi: 10.1016/j.chemosphere.2017.11.077

Hu J, Chen D, Li N et al. (2018) Recyclable carbon nanofibers@Hierarchical I-Doped Bi$_2$O$_2$ CO$_3$–MoS$_2$ membranes for highly efficient water remediation under visible-light irradiation. *ACS Sustain Chem Eng* 6(2):2676–2683. doi: 10.1021/acssuschemeng.7b04270

Iavicoli I, Leso V, Beezhold DH, Shvedova AA (2017) Nanotechnology in agriculture: Opportunities, toxicological implications, and occupational risks. *Toxicol Appl Pharmacol* 329:96–111. doi: 10.1016/j.taap.2017.05.025

Ibrahim RK, Hayyan M, AlSaadi MA et al. (2016) Environmental application of nanotechnology: Air, soil, and water. *Environ Sci Pollut Res Int* 23(14):13754–13788. doi: 10.1007/s11356-016-6457-z

Ikhtiari IR, Begum P, Watari F et al. (2013) Toxic effect of multiwalled carbon nanotubes on lettuce (*Lactuca sativa*). *Nano Biomed* 5:18–24. doi: 10.11344/nano.5.18

Jägermeyr J, Gerten D, Heinke J et al. (2015) Water savings potentials of irrigation systems: Global simulation of processes and linkages. *Hydrol Earth Syst Sci* 19(7):3073–3091. doi: 10.5194/hess-19-3073-2015

Joško I, Oleszczuk P, Dobrzyńska J et al. (2019) Long-term effect of ZnO and CuO nanoparticles on soil microbial community in different types of soil. *Geoderma* 352:204–212. doi: 10.1016/j.geoderma.2019.06.010

Juárez-Maldonado A, Ortega-Ortíz H, Morales-Díaz A et al. (2019) Nanoparticles and nanomaterials as plant biostimulants. *Int J Mol Sci* 20(1):162. doi: 10.3390/ijms20010162

Juárez-Maldonado A, Ortega-Ortíz H, Pérez-Labrada F et al. (2016) Cu nanoparticles absorbed on chitosan hydrogels positively alter morphological, production, and quality characteristics of tomato. *J Appl Bot Food Qual* 89:183–189. doi: 10.5073/JABFQ.2016.089.023

Ju-Nam Y, Lead JR (2008) Manufactured nanoparticles: An overview of their chemistry, interactions and potential environmental implications. *Sci Total Environ* 400(1–3):396–414. doi: 10.1016/J.SCITOTENV.2008.06.042

Kah M (2015) Nanopesticides and nanofertilizers: Emerging contaminants or opportunities for risk mitigation? *Front Chem* 3:1–6. doi: 10.3389/fchem.2015.00064

Kah M, Kookana RS, Gogos A, Bucheli TD (2018) A critical evaluation of nanopesticides and nanofertilizers against their conventional analogues. *Nat Nanotechnol* 13(8):677–684. doi: 10.1038/s41565-018-0131-1

Kalteh M, Alipour ZT, Ashraf S et al. (2018) Effect of silica nanoparticles on basil (*Ocimum basilicum*) under salinity stress. *J Chem Heal Risks* 4. doi: 10.22034/jchr.2018.544075

Keller AA, McFerran S, Lazareva A, Suh S (2013) Global life cycle releases of engineered nanomaterials. *J Nanopart Res* 15(6):1692. doi: 10.1007/s11051-013-1692-4

Khan I, Saeed K, Khan I (2017) Nanoparticles: Properties, applications and toxicities. *Arab J Chem.* doi: 10.1016/j.arabjc.2017.05.011

Kharisov BI, Dias HVR, Kharissova OV (2014) Nanotechnology-based remediation of petroleum impurities from water. *J Petrol Sci Eng* 122:705–718. doi: 10.1016/j.petrol.2014.09.013

Khater M, Escosura-muñiz A De, Merkoçi A (2017) Biosensors and bioelectronics biosensors for plant pathogen detection. *Biosens Bioelectron* 93:72–86. doi: 10.1016/j.bios.2016.09.091

Khati P, Sharma A, Gangola S et al. (2017) Impact of agri-usable nanocompounds on soil microbial activity: An indicator of soil health. *CLEAN Soil Air Water* 45(5). doi: 10.1002/clen.201600458. http://www.ncbi.nlm.nih.gov/pubmed/1600458

Khodakovskaya M, Dervishi E, Mahmood M et al. (2009) Carbon nanotubes are able to penetrate plant seed coat and dramatically affect seed germination and plant growth. *ACS Nano* 3(10):3221–3227. doi: 10.1021/nn900887m

Khodakovskaya MV, de Silva K, Biris AS et al. (2012) Carbon nanotubes induce growth enhancement of tobacco cells. *ACS Nano* 6(3):2128–2135. doi: 10.1021/nn204643g

Khot LR, Sankaran S, Maja JM et al. (2012) Applications of nanomaterials in agricultural production and crop protection: A review. *Crop Prot* 35:64–70.

Kibbey TCG, Strevett KA (2019) The effect of nanoparticles on soil and rhizosphere bacteria and plant growth in lettuce seedlings. *Chemosphere* 221:703–713. doi: 10.1016/j.chemosphere.2019.01.091

Kim DY, Kadam A, Shinde S et al. (2018) Recent developments in nanotechnology transforming the agricultural sector: A transition replete with opportunities. *J Sci Food Agric* 98(3):849–864.

Kumar A, Gupta K, Dixit S et al. (2018) A review on positive and negative impacts of nanotechnology in agriculture. *Int J Environ Sci Technol* 16(4):2175–2184. doi: 10.1007/s13762-018-2119-7

Kumar N, Kaur P, Bhatia S (2017) Advances in bio-nanocomposite materials for food packaging: A review. *Nutrit Food Sci* 47(4):591–606.

Larue C, Castillo-Michel H, Sobanska S et al. (2014) Foliar exposure of the crop *Lactuca sativa* to silver nanoparticles: Evidence for internalization and changes in Ag speciation. *J Hazard Mater* 264:98–106.

Larue C, Khodja H, Herlin-Boime N et al. (2011) Investigation of titanium dioxide nanoparticles toxicity and uptake by plants. *J Phys Conf Ser* 304:012057. doi: 10.1088/1742-6596/304/1/012057

Lee W-M, An Y-J, Yoon H, Kweon HS (2008) Toxicity and bioavailability of copper nanoparticles to the terrestrial plants mung bean (*Phaseolus radiatus*) and wheat. *Environ Toxicol Chem* 27(9):1915–1921.

León-Silva S, Arrieta-Cortes R, Fernández-Luqueño F et al. (2018) Design and production of nanofertilizers. In: López-Valdez F, Fernández-Luqueño F eds. *Agricultural Nanobiotechnology*. Springer, Cham. pp. 7–31. doi: 10.1007/978-3-319-96719-2

León-Silva S, Fernández-Luqueño F, López-Valdez F (2016) Silver nanoparticles (AgNP) in the Environment: A review of potential risks on human and environmental health. *Water Air Soil Pollut* 227(9):306. doi: 10.1007/s11270-016-3022-9

Li B, Tao G, Xie Y et al. (2012) Physiological effects under the condition of spraying nano-SiO_2 onto the Indocalamus barbatus McClure leaves. *J Nanjing For Univ Nat Sci Ed* 36:161–164.

Lin D, Xing B (2007) Phytotoxicity of nanoparticles: Inhibition of seed germination and root growth. *Environ Pollut* 150(2):243–250. doi: 10.1016/J.ENVPOL.2007.01.016

Lin D, Xing B (2008) Root uptake and phytotoxicity of ZnO nanoparticles. *Environ Sci Technol* 42(15):5580–5585. doi: 10.1021/es800422x

Lin S, Reppert J, Hu Q et al. (2009) Uptake, translocation, and transmission of carbon nanomaterials in rice plants. *Small* 5(10):1128–1132. doi: 10.1002/smll.200801556

Liné C, Larue C, Flahaut E (2017) Carbon nanotubes: Impacts and behaviour in the terrestrial ecosystem - A review. *Carbon* 123:767–785. doi: 10.1016/j.carbon.2017.07.089

Liu J, Louie SM, Pham C et al. (2019) Aggregation of ferrihydrite nanoparticles: Effects of pH, electrolytes, and organics. *Environ Res* 172:552–560. doi: 10.1016/j.envres.2019.03.008

Liu R, Lal R (2014) Synthetic apatite nanoparticles as a phosphorus fertilizer for soybean (Glycine max). *Sci Rep* 4:5686.

Liu R, Lal R (2015) Potentials of engineered nanoparticles as fertilizers for increasing agronomic productions. *Sci Total Environ* 514:131–139. doi: 10.1016/j.scitotenv.2015.01.104

Liu X, Feng Z, Zhang S et al. (2006) Preparation and testing of cementing and coating nano-subnano-composites of slow or controlled release fertilizer. *Agric Sci China* 39:1598–1604. doi: 10.1016/S1671-2927(06)60113-2

Loosli F, Yi Z, Wang J, Baalousha M (2019) Dispersion of natural nanomaterials in surface waters for better characterization of their physicochemical properties by AF4-ICP-MS-TEM. *Sci Total Environ* 682:663–672. doi: 10.1016/j.scitotenv.2019.05.206

Lopes Júnior CA, de Sousa Barbosa H, Moretto Galazzi R et al. (2015) Evaluation of proteome alterations induced by cadmium stress in sunflower (*Helianthus annuus L.*) cultures. *Ecotoxicol Environ Saf* 119:170–177.

López-Moreno M, De la Rosa G, Hernández-Viezcas JA et al. (2010) X-ray absorption spectroscopy (XAS) corroboration of the uptake and storage or CeO_2 nanoparticles and assessment of their differential toxicity in four edible plant species. *J Agric Food Chem* 58(6):3689–3693.

López-Moreno ML, De La Rosa G, Hernández-Viezcas JA et al. (2010) Evidence of the differential biotransformation and genotoxicity of ZnO and CeO_2 nanoparticles on soybean (*Glycine max*) plants. *Environ Sci Technol* 44(19):7315–7320. doi: 10.1021/es903891g

Lowry GV, Avellan A, Gilbertson LM (2019) Opportunities and challenges for nanotechnology in the agritech revolution. *Nat Nanotechnol* 14(6):517–522. doi: 10.1038/s41565-019-0461-7

Lowry GV, Gregory KB, Apte SC, Lead JR (2012) Transformations of nanomaterials in the environment. *Environ Sci Technol* 46(13):6893–6899. doi: 10.1021/es300839e

Lu C, Zhang C, Wen J et al. (2002) Research of the effect of nanometer materials on germination and growth enhancement of *Glycine max* and its mechanism. *Soybean Sci* 21:168–171.

Luo M, Huang Y, Zhu M et al. (2018) Properties of different natural organic matter influence the adsorption and aggregation behavior of TiO_2 nanoparticles. *J Saudi Chem Soc* 22(2):146–154. doi: 10.1016/j.jscs.2016.01.007

Lv J, Christie P, Zhang S (2019) Uptake, translocation, and transformation of metal-based nanoparticles in plants: Recent advances and methodological challenges. *Environ Sci Nano* 6(1):41–59. doi: 10.1039/C8EN00645H

Ma C, Borgatta J, De La Torre Roche R et al. (2019) Time-dependent transcriptional response of tomato (*Solanum lycopersicum L.*) to Cu nanoparticle exposure upon infection with *Fusarium oxysporum* f. sp. *lycopersici*. *ACS Sustain Chem Eng*. doi: 10.1021/acssuschemeng.9b01433

Ma X, Geiser-Lee J, Deng Y, Kolmakov A (2010) Interactions between engineered nanoparticles (ENPs) and plants: Phytotoxicity, uptake and accumulation. *Sci Total Environ* 408(16):3053–3061. doi: 10.1016/J.SCITOTENV.2010.03.031

Ma Y, He X, Zhang P et al. (2011) Phytotoxicity and biotransformation of La_2O_3 nanoparticles in a terrestrial plant cucumber (*Cucumis sativus*). *Nanotoxicology* 5(4):743–753. doi: 10.3109/17435390.2010.545487

Ma Y, Zhang P, Zhang Z et al. (2015) Origin of the different phytotoxicity and biotransformation of cerium and lanthanum oxide nanoparticles in cucumber. *Nanotoxicology* 9(2):262–270. doi: 10.3109/17435390.2014.921344

Mansor M, Berti D, Hochella MF et al. (2019) Phase, morphology, elemental composition, and formation mechanisms of biogenic and abiogenic Fe-Cu-sulfide nanoparticles: A comparative study on their occurrences under anoxic conditions. *Am Mineral* 104(5):703–717. doi: 10.2138/am-2019-6848

Marusenko Y, Shipp J, Hamilton GA et al. (2013) Bioavailability of nanoparticulate hematite to *Arabidopsis thaliana*. *Environ Pollut* 174:150–156.

Maurer-Jones MA, Gunsolus IL, Murphy CJ, Haynes CL (2013) Toxicity of engineered nanoparticles in the environment. *Anal Chem* 85(6):3036–3049. doi: 10.1021/ac303636s

McKee MS, Engelke M, Zhang X et al. (2017) Collembola reproduction decreases with aging of silver nanoparticles in a sewage sludge-treated soil. *Front Environ Sci* 5:19. doi: 10.3389/fenvs.2017.00019

McKee MS, Filser J (2016) Impacts of metal-based engineered nanomaterials on soil communities. *Environ Sci Nano* 3(3):506–533. doi: 10.1039/c6en00007

Medina-Pérez G, Fernández-Luqueño F, Campos-Montiel RG et al. (2018) Effects of nanoparticles on plants, earthworms, and microorganisms. In: López-Valdez F, Fernández-Luqueño F eds. *Agricultural Nanobiotechnology*. Springer, Cham. pp.161–181. doi: 10.1007/978-3-319-96719-9

Medina-Pérez G, Fernández-Luqueño F, Campos-Montiel RG et al. (2019) Nanotechnology in crop protection: Status and future trends. *NANO-Biopesticides Today Futur Perspect*:17–45. doi: 10.1016/B978-0-12-815829-6.00002-4

Michálková Z, Komárek M, Veselská V, Číhalová S (2016) Selected Fe and Mn (nano) oxides as perspective amendments for the stabilization of As in contaminated soils, Veselskál V et al. *Environ Sci Pollut Res Int* 23(11):10841–10854. doi: 10.1007/s11356-016-6200-9

Mishra S, Keswani C, Abhilash PC et al. (2017) Integrated approach of Agri-nanotechnology: Challenges and future trends. *Front Plant Sci*. doi: 10.3389/fpls.2017.00471

Mishra S, Keswani C, Singh A et al. (2016) Microbial nanoformulation: Exploring potential for coherent nanofarming. In: Gupta VK Sharma GD, Tuohy MG, Gaur R eds. *The Handbook of Microbial Bioresources*. CABI-UK, Wallingford, UK. pp. 107–120.

Mishra S, Singh A, Keswani C, Singh HB (2014a) Nanotechnology: Exploring potential application in agriculture and its opportunities and Constraints. *Biotech Today* 4(1):9–14. doi: 10.5958/2322-0996.2014.00011.8

Mishra S, Singh BR, Singh A (2014b) Biofabricated silver nanoparticles act as a strong fungicide against *Bipolaris sorokiniana* causing spot blotch disease in wheat. *PLoS One* 9(5):e97881. doi: 10.1371/journal .pone.0097881

Mishra V, Mishra RK, Dikshit A et al. (2014c) Interactions of nanoparticles with plants: An emerging prospective in the agriculture industry: An emerging prospective in the agriculture industry. In: *Emerging Technologies and Management of Crop Stress Tolerance: Biological Techniques*, Ahmad P, Rasool S. (eds). Elsevier Inc. pp. 159–180.

Moll J, Klingenfuss F, Widmer F et al. (2017) Effects of titanium dioxide nanoparticles on soil microbial communities and wheat biomass. *Soil Biol Biochem* 111:85–93. doi: 10.1016/j.soilbio.2017.03.019

Moll J, Okupnik A, Gogos A et al. (2016) Effects of titanium dioxide nanoparticles on red clover and Its rhizobial symbiont. *PLoS One* 11(5). doi: 10.1371/journal.pone.0155111. http://www.ncbi.nlm.nih.gov/ pubmed/0155111

Moon YS, Park ES, Kim TO et al. (2014) SELDI-TOF MS-based discovery of a biomarker in *Cucumis sativus* seeds exposed to CuO nanoparticles. *Environ Toxicol Pharmacol* 38(3):922–931.

Mousavi SM, Motesharezadeh B, Hosseini HM et al. (2018) Root-induced changes of Zn and Pb dynamics in the rhizosphere of sunflower with different plant growth promoting treatments in a heavily contaminated soil. *Ecotox Environ Safe* 147:206–216. doi: 10.1016/j.ecoenv.2017.08.045

Nejad JDM, Asadpour F (2019) A strategy for sustainable development: Using nanotechnology for solar energy in buildings (case study Parand Town). *J Eng Technol Sci* 50(1):103–120. doi: 10.5614/j.eng .technol.sci.2019.51.1.7

Nelson LM (2008) Plant growth promoting rhizobacteria (PGPR): Prospects for new inoculants. Crop Manage 3(1). doi: 10.1094/cm-2004-0301-05-rv

Nowack B, Bucheli TD (2007) Occurrence, behavior and effects of nanoparticles in the environment. *Environ Pollut* 150(1):5–22. doi: 10.1016/J.ENVPOL.2007.06.006

Oyelami AO, Semple KT (2015) The impact of carbon nanomaterials on the development of phenanthrene catabolism in soil. *Environ Sci Process Impacts* 17(7):1302–1310. doi: 10.1039/C5EM00157A

Pachapur VL, Larios AD, Cledón M et al. (2016) Behavior and characterization of titanium dioxide and silver nanoparticles in soils. *Sci Total Environ* 563–564:933–943. doi: 10.1016/j.scitotenv.2015.11.090

Paine FA, Paine HY (2012) *A Handbook of Food Packaging*. Springer, Berlin Parisi.

Pallavi Mehta CM, Srivastava R, Srivastava R et al. (2016) Impact assessment of silver nanoparticles on plant growth and soil bacterial diversity. *Biotech* 6(2):254.

Pandey G (2018) Challenges and future prospects of agri-nanotechnology for sustainable agriculture in India. *Environ Technol Innov* 11:299–307. doi: 10.1016/j.eti.2018.06.012

Parada J, Rubilar O, Diez MC et al. (2018) Combined pollution of copper nanoparticles and atrazine in soil: Effects ondissipation of the pesticide and on microbiological community profiles. *J Hazard Mater* 361:228–236. doi: 10.1016/j.jhazmat.2018.08.042

Parsons JG, Lopez ML, Gonzalez CM et al. (2010) Toxicity and biotransformation of uncoated and coated nickel hydroxide nanoparticles on mesquite plants. *Environ Toxicol Chem* 29(5):1146–1154. doi: 10.1002/etc.146

Pédrot M, Boudec A Le, Davranche M et al. (2011) How does organic matter constrain the nature, size and availability of Fe nanoparticles for biological reduction? *J Colloid Interface Sci* 359(1):75–85. doi: 10.1016/j.jcis.2011.03.067

Peng C, Duan D, Xu C et al. (2015) Translocation and biotransformation of CuO nanoparticles in rice (*Oryza sativa* L.) plants. *Environ Pollut* 197:99–107. doi: 10.1016/J.ENVPOL.2014.12.008

Peng D, Wu B, Tan H et al. (2019) Effect of multiple iron-based nanoparticles on availability of lead and iron, and micro-ecology in lead contaminated soil. *Chemosphere* 228:44–53. doi: 10.1016/j. chemosphere.2019.04.106

Posati T, Noccetti M, Kovtun A et al. (2019) Polydopamine nanoparticle-coated polysulfone porous granules as adsorbents for water remediation. *ACS Omega* 4(3):4839–4847.

Pradhan S, Patra P, Mitra S et al. (2015) Copper nanoparticle (CuNP) nanochain arrays with a reduced toxicity response: A biophysical and biochemical outlook on *Vigna radiata*. *J Agric Food Chem* 63(10):2606–2617. doi: 10.1021/jf504614w

Prasad R, Bhattacharyya A, Nguyen QD (2017) Nanotechnology in sustainable agriculture: Recent developments, challenges, and perspectives. *Front Microbiol* 8:1014. doi: 10.3389/fmicb.2017.01014

Presting H, König U (2003) Future nanotechnology developments for automotive applications. *Mater Sci Eng C* 23(6–8):737–741. doi: 10.1016/j.msec.2003.09.120

Priyanka N, Venkatachalam P (2016) Biofabricated zinc oxide nanoparticles coated with phycomolecules as novel micronutrient catalysts for stimulating plant growth of cotton. *Adv Nat Sci Nanosci Nanotechnol* 7(4):045018. doi: 10.1088/2043-6262/7/4/045018

Proseus TE, Boyer JS (2005) Turgor pressure moves polysaccharides into growing cell walls of Chara corallina. *Ann Bot* 95(6):967–979. doi: 10.1093/aob/mci113

Qiao Y, Wu J, Xu Y et al. (2017) Remediation of cadmium in soil by biochar-supported iron phosphate nanoparticles. *Ecol Eng* 106:515–522. doi: 10.1016/j.ecoleng.2017.06.023

Rahmatpour S, Mosaddeghi MR, Shirvani M, Šimůnek J (2018) Transport of silver nanoparticles in intact columns of calcareous soils: The role of flow conditions and soil texture. *Geoderma* 322:89–100. doi: 10.1016/j.geoderma.2018.02.016

Rajeshkumar S, Veena P, Santhiyaa RV (2018) *Synthesis and Characterization of Selenium Nanoparticles Using Natural Resources and Its Applications*, pp. 63–79. doi: 10.1007/978-3-319-99570-0_4

Rajput VD, Minkina T, Sushkova S et al. (2017) Effect of nanoparticles on crops and soil microbial communities. *J Soils Sediments* 18(6):2179–2187. doi: 10.1007/s11368-017-1793-2

Raliya R, Tarafdar JC (2013) ZnO nanoparticle biosynthesis and Its effect on phosphorous-mobilizing enzyme secretion and gum contents in clusterbean (*Cyamopsis tetragonoloba* L.). *Agric Res* 2(1):48–57. doi: 10.1007/s40003-012-0049-z

Ramalingam P, Yoo SW, Ko YT (2016) Nanodelivery systems based on mucoadhesive polymer coated solid lipid nanoparticles to improve the oral intake of food curcumin. *Food Res Int* 84:113–119. doi: 10.1016/j.foodres.2016.03.031

Ramesh M, Palanisamy K, Kumar Sharma N (2014) Effects of bulk & nano-titanium dioxide and zinc oxide on physio-morphological changes in *Triticum aestivum* Linn. *J Glob Biosci ISSN* 3:2320–1355.

Ranjith KS, Manivel P, Rajendrakumar RT, Uyar T (2017) Multifunctional ZnO nanorod-reduced graphene oxide hybrids nanocomposites for effective water remediation: Effective sunlight driven degradation of organic dyes and rapid heavy metal adsorption. *Chem Eng J* 325:588–600. doi: 10.1016/j.cej.2017.05.105

Rastogi A, Zivcak M, Sytar O et al. (2017) Impact of metal and metal oxide nanoparticles on plant: A critical review. *Front Chem* 5:0078. doi: 10.3389/fchem.2017.00078

Rathor G, Chopra N, Adhikari T (2017) Remediation of nickel ion from soil and water using nano particles of zero-valent iron (nZVI). *Orient J Chem* 3(2):1025–1029. doi: 10.13005/ojc/330259

Ray SS, Chen SS, Sangeetha D et al. (2018) Sustainable desalination process and nanotechnology. In: *Nanotechnology, Food Security and Water Treatment*. Springer, Berlin. pp. 185–228. doi: 10.1007/978-3-319-70166-0_6

Reinsch BC, Levard C, Li Z et al. (2012) Sulfidation of silver nanoparticles decreases *Escherichia coli* Growth Inhibition. *Environ Sci Technol* 46(13):6992–7000. doi: 10.1021/es203732x

Rico CM, Majumdar S, Duarte-Gardea M et al. (2011) Interaction of nanoparticles with edible plants and their possible implications in the food chain. *J Agric Food Chem* 59(8):3485–3498.

Rizwan M, Ali S, Ali B et al. (2019) Zinc and iron oxide nanoparticles improved the plant growth and reduced the oxidative stress and cadmium concentration in wheat. *Chemosphere* 214:269–277. doi: 10.1016/j.chemosphere.2018.09.120

Robles-García MA, Rodríguez-Félix F, Márquez-Ríos E et al. (2016) Applications of nanotechnology in the agriculture, food, and pharmaceuticals. *J Nanosci Nanotechnol* 16(8):8188–8207. doi: 10.1166/jnn.2016.12925

Rodrigues SM, Demokritou P, Dokoozlian N et al. (2017) Nanotechnology for sustainable food production: Promising opportunities and scientific challenges. *Environ Sci Nano* 4(4):767–781. doi: 10.1039/c6en00573j

Romero-Freire A, Lofts S, Peinado FJM, van Gestel CA (2017) Effects of aging and soil properties on zinc oxide nanoparticle availability and its ecotoxicological effects to the earthworm *Eisenia andrei*. *Environ Toxicol Chem* 36(1):137–146. doi: 10.1002/etc.3512

Rossi L, Fedenia LN, Sharifan H et al. (2018) Effects of foliar application of zinc sulfate and zinc nanoparticles in coffee (*Coffea arabica* L.) plants. *Plant Physiol Biochem* 135:160–166. doi: 10.1016/j.plaphy.2018.12.005

Rossi L, Zhang W, Ma X (2017) Cerium oxide nanoparticles alter the salt stress tolerance of Brassica napus L. by modifying the formation of root apoplastic barriers. *Environ Pollut* 229:132–138. doi: 10.1016/J.ENVPOL.2017.05.083

Sahayaraj K, Madasamy M, Radhika SA (2016) Insecticidal activity of bio-silver and gold nanoparticles against Pericallia ricini Fab (Lepidaptera: Archidae). *J Biopestic* 9:63–72.

Sahoo SK, Parveen S, Panda JJ (2007) The present and future of nanotechnology in human health care. *Nanomed Nanotechnol* 3(1):20–31. doi: 10.1016/j.nano.2006.11.008

Salama HMH (2012) Effects of silver nanoparticles in some crop plants, Common bean (*Phaseolus vulgaris* L.) and corn. *Int Res J Biotechnol* 3:190–197.

Samad A, Alam M, Saxena K (2009) Dendrimers: A class of polymers in the nanotechnology for the delivery of active pharmaceuticals. *Curr Pharm Des* 15(25):2958–2969. doi: 10.2174/138161209789058200

Sanivada SK, Pandurangi VS, Challa MM (2017) Nanofertilizers for sustainable soil management. In: Ranjan S, Dasgupta N, Lichtfouse E eds. *Nanoscience in Food and Agriculture 5: Sustainable Agriculture Reviews*, Springer, Cham vol. 26, pp. 267–307.

Santos AR, Miguel AS, Tomaz L et al. (2010) The impact of CdSe/ZnS Quantum dots in cells of *Medicago sativa* in suspension culture. *J Nanobiotechnology* 8:24. doi: 10.1186/1477-3155-8-24

Saxena R, Kumar M, Singh Tomar R (2017) Nanobiotechnology: A new paradigm for crop production and sustainable agriculture. *Res J Pharm Biol Chem Sci* 8:823–832.

Schwabe F, Tanner S, Schulin R et al. (2015) Dissolved cerium contributes to uptake of ce in the three crop plants. *Metallonomics* 7(3):466–477. doi: 10.1039/c4mt00343h

Scott NR, Chen H, Cui H (2018) Nanotechnology applications and implications of agrochemicals toward sustainable agriculture and food systems. *J Agri Food Chem* 66(26):6451–6456. doi: 10.1021/acs.jafc.8b00964

Servin AD, White JC (2016) Nanotechnology in agriculture: Next steps for understanding engineered nanoparticle exposure and risk. *NanoImpact*. doi: 10.1016/j.impact.2015.12.002

Shah V, Belozerova I (2009) Influence of metal nanoparticles on the soil microbial community and germination of lettuce seeds. *Water Air Soil Pollut* 197(1–4):143–148. doi: 10.1007/s11270-008-9797-6

Shah V, Jones J, Dickman J, Greenman S (2014) Response of soil bacterial community to metal nanoparticles in biosolids. *J Hazard Mater* 274:399–403. doi: 10.1016/j.jhazmat.2014.04.003

Sharma VK, Filip J, Zboril R, Varma RS (2015) Natural inorganic nanoparticles-formation, fate, and toxicity in the environment. *Chem Soc Rev* 44(23):8410–8423. doi: 10.1039/c5cs00236b

Shaw AK, Hossain Z (2013) Impact of nano-CuO stress on rice (*Oryza sativa* L.) seedlings. *Chemosphere* 93(6):906–915. doi: 10.1016/J.CHEMOSPHERE.2013.05.044

Shcherbakova EN, Shcherbakov AV, Andronov EE et al. (2017) Combined pre-seed treatment with microbial inoculants and Mo nanoparticles changes composition of root exudates and rhizosphere microbiome structure of chickpea (*Cicer arietinum* L.) plants. Symbiosis:57–69. doi: 10.1007/s13199-016-0472-1

Shiny PJ, Mukerjee A, Chandrasekaran N (2013) Comparative assessment of the phytotoxicity of silver and platinum nanoparticles. In: *International Conference on Advanced Nanomaterials & Emerging Engineering Technologies*. pp. 391–393.

Shrivastava M, Srivastav A, Gandhi S et al. (2019) Monitoring of engineered nanoparticles in soil-plant system: A review. *Environ Nanotechnol Monit Manag* 11:100218. doi: 10.1016/j.enmm.2019.100218

Siddiqui MH, Al-Whaibi MH, Firoz M et al. (2015) Role of nanoparticles in plants. In: Siddiqui MH, Al-Whaibi MH, Firoz M et al. eds. *Nanotechnology and Plant Sciences: Nanoparticles and Their Impact on Plants*. Springer International Publishing, Switzerland. pp. 19–35. doi: 10.1007/978-3-319-14502-0_2

Simonin M, Richaume A (2015) Impact of engineered nanoparticles on the activity, abundance, and diversity of soil microbial communities: A review. *Environ Sci Pollut R* 22(18):13710–13723. doi: 10.1007/s11356-015-4171-x

Singh BK (2017) Creating new business, economic growth and regional prosperity through microbiome-based products in the agriculture industry. *Microb Biotechnol* 10(2):224–227. doi: 10.1111/1751-7915.12698

Smirnova E, Gusev A, Zaytseva O et al. (2012) Uptake and accumulation of multiwalled carbon nanotubes change the morphometric and biochemical characteristics of Onobrychis arenaria seedlings. *Front Chem Sci Eng* 6(2):132–138. doi: 10.1007/s11705-012-1290-5

Steinberg CEW, Kamara S, Prokhotskaya VY et al. (2006) Dissolved humic substances - Ecological driving forces from the individual to the ecosystem level? *Freshw Biol* 51(7):1189–1210. doi: 10.1111/j.1365-2427.2006.01571.x

Stoimenov PK, Klinger RL, Marchin GL, Klabunde KJ (2002) Metal oxide nanoparticles as bactericidal agents. *Langmuir* 18(17):6679–6686. doi: 10.1021/la0202374

Su H, Fang Z, Tsang PE et al. (2016) Stabilisation of nanoscale zero-valent iron with biochar for enhanced transport and in-situ remediation of hexavalent chromium in soil. *Environ Pollut* 214:94–100. doi: 10.1016/j.envpol.2016.03.072

Sun TY, Gottschalk F, Hungerbühler K, Nowack B (2014) Comprehensive probabilistic modelling of environmental emissions of engineered nanomaterials. *Environ Pollut* 185:69–76.

Suriyaprabha R, Karunakaran G, Yuvakkumar R et al. (2012) Silica Nanoparticles for Increased Silica Availability in Maize (*Zea mays*. L) seeds under hydroponic conditions. *Curr Nanosci* 8:902–908.

Syu Y, Hung J-H, Chen J-C, Chuang HW (2014) Impacts of size and shape of silver nanoparticles on Arabidopsis plant growth and gene expression. *Plant Physiol Biochem* 83:57–64. doi: 10.1016/J. PLAPHY.2014.07.010

Tafazoli M, Hojjati SM, Biparva P et al. (2017) Reduction of soil heavy metal bioavailability by nanoparticles and cellulosic wastes improved the biomass of tree seedlings. *J Plant Nutr Soil Sci* 180(6):683–693. doi: 10.1002/jpln.201700204

Terekhova V, Gladkova M, Milanovskiy E et al. (2017) Engineered nanomaterials' effects on soil properties: Problems and advances in investigation. In: Ghorbanpour M, Manika K, Varma A eds. *Nanoscience and Plant–Soil Systems: Soil Biology*, vol 48. Springer, Cham. doi: 10.1007/978-3-319-46835-8_4

Tiwari DK, Dasgupta-Schubert N, Villaseñor Cendejas LM et al. (2014) Interfacing carbon nanotubes (CNT) with plants: Enhancement of growth, water and ionic nutrient uptake in maize (Zea mays) and implications for nanoagriculture. *Appl Nanosci* 4(5):577–591. doi: 10.1007/s13204-013-0236-7

Tourinho PS, van Gestel CAM, Lofts S et al. (2012) Metal-based nanoparticles in soil: Fate, behavior, and effects on soil invertebrates. *Environ Toxicol Chem* 31(8):1679–1692. doi: 10.1002/etc.1880

Tripathi S, Sarkar S (2015) Influence of water soluble carbon dots on the growth of wheat plant. *Appl Nanosci* 5(5):609–616. doi: 10.1007/s13204-014-0355-9

Ullah N, Mansha M, Khan I, Qurashi A (2018) Nanomaterial-based optical chemical sensors for the detection of heavy metals in water: Recent advances and challenges. *TrAC Trends Anal Chem* 100:155–166.

Van Hoecke K, De Schamphelaere KAC, Van Der Meeren P et al. (2011) Aggregation and ecotoxicity of CeO_2 nanoparticles in synthetic and natural waters with variable pH, organic matter concentration and ionic strength. *Environ Pollut.* doi: 10.1016/j.envpol.2010.12.010

Verma SK, Das AK, Patel MK et al. (2018) Engineered nanomaterials for plant growth and development: A perspective analysis. *Sci Total Environ* 630:1413–1435. doi: 10.1016/j.scitotenv.2018.02.313

von der Heyden B, Roychoudhury A, Myneni S (2019) Iron-rich nanoparticles in natural aquatic environments. *Minerals* 9(5):287. doi: 10.3390/min9050287

Wang A, Zheng Y, Peng F (2014) Thickness-controllable silica coating of CdTe QDs by reverse microemulsion method for the application in the growth of rice. *J Spectrosc* 2014:1–5. doi: 10.1155/2014/169245 Research

Wang J, Shu K, Zhang L, Si Y (2017) Effects of silver nanoparticles on soil microbial communities and bacterial nitrification in suburban vegetable soils. *Pedosphere* 27(3):482–490. doi: 10.1016/S1002-0160(17) 60344-8

Wang L, Liu Z, Xia X et al. (2017) Colorimetric detection of Cucumber green mottle mosaic virus using unmodified gold nanoparticles as colorimetric probes. *J Virol Methods.* doi: 10.1016/j.jviromet.2017.01.010

Wang P, Lombi E, Menzies N et al. (2018) Engineered silver nanoparticles in terrestrial environments: A meta-analysis shows that the overall environmental risk is small. *Environ Sci Nano.* doi: 10.1039/c8en00486b

Wang P, Lombi E, Zhao F-J, Kopittke PM (2016) Nanotechnology: A New opportunity in plant sciences. *Trends Plant Sci* 21(8):699–712. doi: 10.1016/J.TPLANTS.2016.04.005

Wang X, Zhu M, Li N et al. (2018) Effects of CeO_2 nanoparticles on bacterial community and molecular ecological network in activated sludge system. *Environ Pollut* 238:516–523. doi: 10.1016/j.envpol.2018.03.034

Wang Y, Wang S, Xu M et al. (2019) The impacts of γ-Fe_2O_3 and Fe_3O_4 nanoparticles on the physiology and fruit quality of muskmelon (*Cucumis melo*) plants. *Environ Pollut* 249:1011–1018. doi: 10.1016/j. envpol.2019.03.119

Wang Z, Xie X, Zhao J et al. (2012) Xylem- and phloem-based transport of CuO nanoparticles in Maize (*Zea mays* L.). *Environ Sci Technol* 46(8):4434–4441. doi: 10.1021/es204212z

Watanabe T, Misawa S, Hiradate S, Osaki M (2008) Root mucilage enhances aluminum accumulation in Melastoma malabathricum, an aluminum accumulator. *Plant Signal Behav* 3(8):603–605. doi: 10.4161/ psb.3.8.6356

Willett W, Rockström J, Loken B et al. (2019) Food in the Anthropocene: The EAT–Lancet Commission on healthy diets from sustainable food systems.*Lancet* 393(10170):447–492. doi: 10.1016/ s0140-6736(18)31788-4

Wu S, Cajthaml T, Semerád J et al. (2019) Nano zero-valent iron aging interacts with the soil microbial community: A microcosm study. *Environ Sci Nano* 6(4):1189–1206. doi: 10.1039/c8en01328d

Wu SG, Huang L, Head J et al. (2012) Phytotoxicity of metal oxide nanoparticles is related to both dissolved metals ions and adsorption of particles on seed surfaces. *J Petrol Environ Biotechno* 3(4):2–6. doi: 10.4172/2157-7463.1000126

Yang F, Hong F, You W et al. (2006) Influence of nano-anatase TiO_2 on the nitrogen metabolism of growing spinach. *Biol Trace Elem Res* 110(2):179–190. doi: 10.1385/BTER:110:2:179

Yang L, Watts DJ (2005) Particle surface characteristics may play an important role in phytotoxicity of alumina nanoparticles. *Toxicol Lett* 158(2):122–132. doi: 10.1016/J.TOXLET.2005.03.003

Yang L, Yang L, Ding L et al. (2019) Principles for the application of nanomaterials in environmental pollution control and resource reutilization. *Micro Nanotechnologis*:1–23. doi: 10.1016/B978-0-12-814837-2.00001-9

Yang YF, Cheng YH, Liao CM (2016) In situ remediation-released zero-valent iron nanoparticles impair soil ecosystems health: A *C. elegans* biomarker-based risk assessment. *J Hazard Mater* 317:210–220. doi: 10.1016/j.jhazmat.2016.05.070

Yang YF, Cheng YH, Liao CM (2017) Nematode-based biomarkers as critical risk indicators on assessing the impact of silver nanoparticles on soil ecosystems. *Ecol Indic* 75:340–351. doi: 10.1016/j.ecolind.2016.12.051

Yasur J, Rani PU (2013) Environmental effects of nanosilver: Impact on castor seed germination, seedling growth, and plant physiology. *Environ Sci Pollut Res* 20(12):8636–8648.

Yata VK, Tiwari BC, Ahmad I (2018) Nanoscience in food and agriculture: Research, industries and patents. *Environ Chem Lett* 16(1):79–84. doi: 10.1007/s10311-017-0666-7

Yin K, Liu A, Shangguan L et al. (2017) Construction of iron-polymer-graphene nanocomposites with low nonspecific adsorption and strong quenching ability for competitive immunofluorescent detection of biomarkers in GM crops. *Biosens Bioelectron*. doi: 10.1016/j.bios.2016.11.070

Yin L, Cheng Y, Espinasse B et al. (2011) More than the Ions: The effects of silver nanoparticles on *Lolium multiflorum*. *Environ Sci Technol* 45(6):2360–2367. doi: 10.1021/es103995x

Yin L, Colman BP, McGill BM et al. (2012) Effects of silver nanoparticle exposure on germination and early growth of eleven wetland plants. *PLoS One* 7(10):1–7. doi: 10.1371/journal.pone.0047674

Yin Y, Liu J, Jiang G (2012) Sunlight-induced reduction of ionic Ag and Au to metallic nanoparticles by dissolved organic matter. *ACS Nano* 6(9):7910–7919. doi: 10.1021/nn302293r

YinFeng X, Li B, Zhang Q, et al. (2011) Effects of nano-TiO$_2$ on photosynthetic characteristics of Indocalamus barbatus. *J Northeast For Univ* 39:22–25.

You T, Liu D, Chen J et al. (2018) Effects of metal oxide nanoparticles on soil enzyme activities and bacterial communities in two different soil types. *J Soils Sediments* 18(1):211–221. doi: 10.1007/s11368-017-1716-2

Yuan S, Chen C, Raza A et al. (2017) Nanostructured TiO$_2$/CuO dual-coated copper meshes with superhydrophilic, underwater superoleophobic and self-cleaning properties for highly efficient oil/water separation. *Chem Eng J* 328:497–510. doi: 10.1016/j.cej.2017.07.075

Zaytseva O, Neumann G (2016) Carbon nanomaterials: Production, impact on plant development, agricultural and environmental applications. *Chem Biol Technol Agric* 3(1):17. doi: 10.1186/s40538-016-0070-8

Zhang P, Ma Y, Zhang Z et al. (2012) Biotransformation of ceria nanoparticles in cucumber plants. *ACS Nano* 6(11):9943–9950. doi: 10.1021/nn303543n

Zhang Q, Han L, Jing H et al. (2016) Facet control of gold nanorods. *ACS Nano* 10(2):2960–2974. doi: 10.1021/acsnano.6b00258

Zhang R, Zhang H, Tu C et al. (2015) Phytotoxicity of ZnO nanoparticles and the released Zn (II) ion to corn (*Zea mays* L.) and cucumber (*Cucumis sativus* L.) during germination. *Environ Sci Pollut Res* 22(14):11109–11117.

Zhang S, Gao H, Huang Y et al. (2018) Ultrathin g-C$_3$N$_4$ nanosheets coupled with amorphous Cu-doped FeOOH nanoclusters as 2D/0D heterogeneous catalysts for water remediation. *Environ Sci Nano* 5(5):1179–1190. doi: 10.1039/c8en00124c

Zhao L, Ortiz C, Adeleye AS et al. (2016) Metabolomics to detect response of lettuce (*Lactuca sativa*) to Cu(OH)$_2$ Nanopesticides: Oxidative stress response and detoxi fi cation mechanisms. *Environ Sci Technol* 50(17):9697–9707. doi: 10.1021/acs.est.6b02763

Zheng L, Hong F, Lu S, Liu C (2005) Effect of nano-TiO$_2$ on strength of naturally aged seeds and growth of spinach. *Biol Trace Elem Res* 104(1):83–91. doi: 10.1385/BTER:104:1:083

Zhu H, Han J, Xiao JQ, Jin Y (2008) Uptake, translocation, and accumulation of manufactured iron oxide nanoparticles by pumpkin plants. *J Environ Monit* 10(6):713–717. doi: 10.1039/b805998e

Zuverza-Mena N, Armendariz R, Peralta-Videa JR, Gardea-Torresdey JL (2016) Effects of silver nanoparticles on radish sprouts: Root growth reduction and modifications in the nutritional value. *Front Plant Sci* 7:90. doi: 10.3389/fpls.2016.00090

3 Optimistic Influences of Nanotechnology on Food Security and Agriculture

Tahsin Shoala and Ahmed S.Hashem

CONTENTS

3.1 INTRODUCTION

Nanotechnology has become one of the most important vital sciences of the current era. Applications of nanotechnology have grown rapidly and cover many fields and different aspects. It has started being integrated into agriculture and food manufacturing since 2003 (US DOA, 2003). Research studies on the application of nanotechnology in the agriculture and food industries has increased rapidly over the last decade. It covers virtually every aspect in water purification, food processing and packaging, animal fodder, and aquiculture (Dasgupta and Ranjan, 2018; Sozer and Kokini, 2009). The food and beverage division is a universal multi trillion dollar manufacturing industry (Cushen et al. 2012). The global economic power of nanotechnology is estimated to be at least USD 3 trillion by 2020, which may employee 6 million in nanotechnology manufacturing worldwide (Roco et al. 2011). This is very attractive and has motivated much food creativity involved in the growing and promoting of innovative nanomaterial-based products, and enhancing production efficacy, food features, taste, and safekeeping. Amazingly, there are several products that have already been advertised and used in commercial food production. These products are mainly designed out-of-food but integrated in the food industry, i.e. materials attached to food but not directly consumed by people. Novel products containing nanomaterial have not yet been directly integrated into human food. The essential purpose is that rules and legislation are restricted regarding nanofood, especially due to the complications of nanomaterials and case-by-case authorizing actions (Kavitha et al. 2018; Marrani, 2013).

3.2   APPLICATION OF NANOTECHNOLOGY IN THE AGRICULTURAL SECTOR (AT THE PRODUCTION SITE)

The important role of nanomaterials in agricultural science is demonstrated by their application to fertilizer compounds in supplying necessary nutrition to growing plants, including using nanoscale fertilizer inputs, nanomaterial additives, nanocoatings, and using nanomaterials as material carriers for fertilizers. Furthermore, this modern technology has made some progress in other agricultural categories, such as reproductive science and technology, conversion of agricultural and food wastes to energy, and other useful byproducts through enzymatic nanobioprocessing, disease prevention, and the treatment of plants using various nanoscale devices with novel properties. Recently, smart agricultural delivery systems have attracted the attention of many researchers. These delivery systems can specifically target a tissue, can be highly controlled, remotely regulated and preprogrammed, and have self-regulating and multifunctional characteristics that ensure they avoid biological barriers to successful targeting (Moraru et al. 2003; Li et al. 2005; Rai and Ingle, 2010). On top of all these applications, nanotechnological devices and tools, such as nanocapsules, nanoparticles, and even viral capsids, are examples of possible uses for the detection and treatment of diseases, enhanced absorption of nutrients by plants, delivery of active ingredients to specific sites, and water treatment processes. The use of target-specific nanoparticles can reduce damage to nontargeted plant tissues and volumes of chemicals released to the environment (Owolade et al. 2008; Prasad et al. 2012). Moreover, nanomaterials in the agriculture sector aim to reduce the application of plant protection products by minimizing fertilizer nutrient losses, and increasing yields to achieve optimized nutrient management schemes. Other applications of nanotechnology are in the field of agricultural machinery for machine body coatings, tools, and even glass for increasing stability against corrosion and ultraviolet radiation. In addition, nanotechnology can be used for the production of more robust mechanical components using nanocovers, and the use of intelligent machines and spraying equipment to chemically and mechanically identify weeds, allowing farmers to reduce the volumes of pesticides they use and to specifically target weeds. Furthermore, nanocoatings can be developed to reduce friction and develop agricultural machines that are resistant to corrosion and heat. Nanofuels can be used to decrease environmental pollution. Nanofertilizers and nanopoisons can be developed to directly target weeds, insects, and other agricultural pests (Table 3.1).

3.2.1   NANOPESTICIDES

Chemical pesticides, such as insecticides, fungicides, rodenticides, nematicides, acaricides, molluscicides, and herbicides (or weed killers) are mainly used in agriculture to improve crop yield and efficiency, and in general to protect plants from damaging influences such as weeds, plant diseases, or insects. As a consequence of their harmful effects on humans, animals, and the environment, they were restricted by international authorities, which led to the development and testing of new pest control technologies that are less harmful. Nanotechnology and nanomaterials have been demonstrated to have great potential in providing novel and improved solutions (Gopal et al. 2012; Chowdappa and Gowda, 2013; Kah et al. 2013; Sekhon, 2014; Agrawal and Rathore, 2014; Mishra et al. 2014a; Rai et al. 2015; Alghuthaymi et al. 2015; Unsworth et al. 2015; Fraceto et al. 2018). Nanotechnology is one of the major technologies in recent times. Nanomaterials change their physical and chemical properties. A variety of industrial, pharmaceutical, and medical products have been improved and innovated (Hussain et al. 2005; Burm et al. 2006; Frohlich, 2013; Dulis, 2015; Unsworth et al. 2015). The application of nanotechnology in the agriculture sector is relatively new compared to its use in the medical and pharmacy fields (drug delivery) (Garcia et al. 2010; Sekhon, 2014).In general, it can be stated that the ability of nanoparticles (NPs) 'to permeate anywhere' relates primarily to their particle size and shape (Buzea et al. 2007; Hagens et al. 2007; Keck and Müller, 2013). Three types of NPs were defined based on particle size: (1) coarse-mode particles >2.5 µm in diameter (largely mechanically generated particles); (2) accumulation-mode particles between 100 nm and 2.5 µm in diameter (from aggregation of ultrafine particles or vapors); (3) ultrafine

TABLE 3.1
Effect of Different Nanomaterials on Plants

Nanomaterial	Plant	Plant Parts	Effect	References
MWCNTs Multiwalled carbon nanotubes	Tomato	Seeds	Enhanced seed germination and vegetative biomass	Khodakovskaya et al. (2009)
MWCNTs Multiwalled carbon nanotubes	Tomato	Roots	Efficient water and nutrient uptake	Villagarcia et al. (2012)
MWCNTs Multiwalled carbon nanotubes	Rape, rye grass, corn	Root	Root elongation	Lin and Xing (2007)
SWCNTs	Tomato	Tissue	DNA-dye molecule delivery	Srinivasan and Saraswathi (2010)
SWCNTs	Cucumber, onion	Root	Root elongation	Canas et al. (2008)
Ag NPs	Cabbage, maize	Root	Root elongation	Pokhrel and Dubey (2013)
Ag NPs	*Bacopa monnieri*	Whole plant	Increased carbohydrate, protein, and decreased total phenol content	Krishnaraj et al. (2012)
Ag NPs	*Trigonella foenum*	Seeds	Enhanced growth, diosgenin content, and secondary metabolite production	Jasim et al. (2016)
Au NPs	*Arabidopsis thaliana*	Seeds	Enhanced seed germination	Kumar et al. (2013)
Au NPs	*Boswellia ovalifoliata*	Seeds	Enhanced seed germination	Savithramma et al. (2012)
Au NPs	*Brassica juncea*	Seeds	Increase in net productivity and yield	Arora et al. (2012)
Nanoanatase	*Petroselinum crispum*	Seeds	Enhanced shoot length, chlorophyll, and yield	Dehkourdi and Mosavi (2013)
Zn NPs	Pearl millet	Whole plant	Enhanced shoot length, chlorophyll, and yield	Tarafdar et al. (2014)
TiO$_2$ NPs	*Foeniculum vulgare* Mill.	Seeds	Enhanced seed germination	Feizi et al. 2013
SiO$_2$ NPs	*Zea mays* L.	Seeds	Enhanced seed germination	Suriyaprabha et al. (2012)
SiO$_2$ NPs	Changbai larch	Seeds	SiO$_2$ Enhanced seed germinationand chlorophyll biosynthesis	Bao-shan et al. (2004)
Al	*Raphanus sativus* and *Brassica napus*	Root	Improved root growth	Lin and Xing, 2007
CeO$_2$	*Cucumis sativus* and *Zea mays*	Root and stem	Increased root and stem growth	López-Moreno et al. 2010
CeO$_2$	*Coriandrum sativum*	Biomass	Increased shoot and root length, biomass, catalase activity in shoots and ascorbate peroxidase activity in roots	Morales et al. 2013
TiO$_2$	*Triticum aestivum* var. *Pishtaz*	Shoot and seedling	Increased shoot and seedling lengths	Feizi et al. 2012
Ag	*Trigonella foenum-graecum*	Growth	Enhanced plant growth and diosgenin synthesis	Jasim et al. 2017
ZnO	*Arachis hypogaea*	Growth	Improved growth and yield	Prasad et al. 2012
SiO$_2$	*Lycopersicum esculentum*	Seed	Improved seed germination	Siddiqui, and Al-Whaibi, 2014

particles, <100 nm in diameter (largely consisting of primary combustion products) (Sioutas et al. 2005). NPs are extremely small in particle size, giving them the advantage of penetrating any bio-membrane and the ability to cover the largest possible area, in addition to their ability to bind, absorb, and carry compounds (Nehoff et al. 2014). Therefore, nanosystems consist of two basic components: an active ingredient and a nanocarrier that stabilizes the active ingredient in the nanoform. Nanoformulations usually consist of several surfactants, polymers, or inorganic (e.g. metal) NPs in the nanometer size range and therefore cannot be considered as a single entity (Perlatti et al. 2013). Current NP platforms can be classified into three major categories including: (1) inorganic-based (solid) NPs (nonbiodegradable), for example, nanocrystals, shells, quantum dots based on gold, silver, copper, iron, various semiconductors, ceramic (silica-based NPs), and carbon nanotubes or fuller-enes; (2) organic-based NPs (frequently biodegradable), for example, liposomes, solid lipid NPs, polymeric NPs, micelles, and capsules; and (3) hybrid NPs (a combination of inorganic and organic components) (Jampílek and Kráľová, 2017). Insect pests are a major issue in the agriculture sector, as they destroy crops in field and infest products during storage (Ragaei and Sabry, 2014). The botanical insecticides include those based on active agents isolated from plant extracts, as well as essential oils derived from certain plants. The use of botanical insecticides associated with nanotechnology offers considerable potential for increasing agricultural productivity, while at the same time reducing impacts on the environment and human health (de Oliveira et al. 2014). Fungi cause many crop losses worldwide. Fungal diseases have a significant economic impact on plant yield and quality, thus managing such diseases is an essential component of production for most crops. A fungicide is a specific type of pesticide that controls fungal disease by specifically inhibiting or killing the fungus that causes the disease (Mishra et al. 2014b; Mishra et al. 2016). Not all diseases caused by fungi can be adequately controlled by fungicides only, such as *Fusarium* and *Verticillium*. Fungicides are used to control disease during the establishment and development of a crop, to increase the productivity of a crop, and to reduce blemishes and toxicity. Therefore, fungicides are extensively used in agriculture systems to control soilborne, seedborne, or airborne fungal pathogens (McGrath, 2004). Moreover, nematodes are common soilborne organisms. Plant-parasitic nematodes are known to cause significant damage to crops. For many years, nematodes were controlled using inexpensive chemical nematicides that effectively kill nematodes in the soil. Recently most nematicides have been banned or restricted by the United States Environmental Protection Agency as environmental toxins. Since nematicides are expensive to develop, new approaches have been tested, including application of NPs (Lambert and Bekal, 2002) (Tables 3.1–3.3).

3.2.2 Nanonutrients and Nanofertilizers

Nanoparticles may also be applied to plants using injection or spraying methods. One of the main drawbacks of the injection method is that it cannot be used for large-scale field application. Attempts have also been made to concentrate the particles at the desired area in plants by using magnets. This opens an area of smart delivery to target locations. An observation of bioferrofluids along with metallic NPs in the xylem vessels (Melendi et al. 2008) is one such attempt. Additionally, spraying seems to be a viable option because NPs have been reported to penetrate plant tissue and reach regions distant from the point of application even though movements over short distances are favored. It is important to note here that most of the studies in the literature provide limited information because they have been performed at the germination stage when the vascular system is not fully developed. Titanium dioxide (TiO_2) NPs of 30 mm size could not pass through the pore size (6.6 mm) of maize roots and hence did not register any entry (Asli and Neumann, 2009). In addition, the selective permeability determined by the pore diameter of the cell wall ranging from 5 to 20 nm, as well as interaction of biomolecules encountered in the pathway, are also important. However, once the NPs have entered the cell wall it also exercises a certain amount of specificity as well as regulating the entry on the basis of the size of their pores (Fleischer et al. 1999; Navarro et al. 2008; Mishra et al. 2017). After entering they may move apoplastically and symplastically. NPs come into

TABLE 3.2

List of Encapsulated Nano-Systems against Some Pests and Pathogens

Types of Nano- Formulation	Active Ingredient	Polymer / Ingredient	Activity	Target Organism (S)	References
Nano-gels	Carum copticum EO	Chitosan	Insecticidal	Sitophilus granaries and Tribolium confusum adults	Ziaee et al. (2014)
	L. sidoides EO	CS + cashew gum	Ticks and Mites		Abreu et al. (2012)
	Cu(II) ions	CS		Third-instar larvae of S. aegypti	Brunel et al. (2013)
Nano-capsulation	Garlic EO	PEG	Insecticidal	F. graminearum	Yang et al. (2009)
	Piperonyl butoxide and deltamethrin	High density and low density polyethylene	Insecticidal	Tribolium castaneum Mosquitoes	Frandsen et al. (2015)
	Cuminum cyminum EO	Poly(ureaformaldehyde)	Insecticidal	Tribolium castaneum	Negahban et al. (2012)
	β-Cyfluthrin	PEG	Insecticidal	Callosobruchus maculatus	Loha et al. (2012)
	Fipronil in oil	Silica shell	Insecticidal	Coptotermes acinaciformis	Wibowo et al. (2014)
	Methomyl	Azidobenzaldehyde and carboxymethyl chitosan	Insecticidal	Armyworm	Sun et al. (2014)
	Granulovirus of Cydia pomonella L.	Lignin	Insecticidal	Virus solar protection	Arthurs et al. (2006)
	Artemisia arborescens L. EO	lipid and Poloxamer 188 or as surfactants	Enhance pesticides	Reduced evaporation of the essential oil	Lai et al. (2006)
	Photodegradable nano IMI	CHI–ALG	Insecticidal	M. dermestoides	Guan et al. (2008)
	LB	Poly(MMA-co-St)	Nematodes		Yin et al. (2012)
	Atrazine, Ametryn, Simazine	PCL		Brasica sp.	Pereira et al. (2014) Grillo et al. (2012)
	Neem oil (Aza)	PCL	Insecticidal	P. xylostella	Forim et al. (2013)
	Acetamiprid	ALG–CHI	Insecticidal	Insects	Kumar et al. (2015)
	Onopordopicrin	PEG–PBA–PEG		Fungi and bacteria	Esmaeili and Saremnia (2012)
	IMI	PCA–PEG–PCA		G. pyloalis	Memarizadeh et al. (2014)
	Acephate	PEG	Insecticidal	Insects	Choudhury et al. (2012)
	Methomyl	Photocrosslinkable CMCS	Insecticidal	Armyworm larvae	Sun et al. (2014)

(Continued)

TABLE 3.2 (CONTINUED)

List of Encapsulated Nano-Systems against Some Pests and Pathogens

Types of Nano- Formulation	Active Ingredient	Polymer / Ingredient	Activity	Target Organism (S)	References
	IMI	ALG	Insecticidal	Leafhopper	Kumara et al. (2014)
	Deltamethrin	Beeswax solid lipid nanoparticle coated with chitosan	Enhance pesticides	Deltamethrin is protected against photodegradation	Nguyen et al. (2012)
	Carbendazim, and tebuconazole	Myritol	Fungicidal	Reduced toxicity and controlled release of fungicides	Campos et al (2015)
	Pelargonium graveolens EO	Stearic acid	Insecticidal	Phthorimaea operculella	Adel et al. (2015)
Mesoporous nanopartic	Imidacloprid	Mesoporous silica nanoparticles	Insecticidal	Termites	Popat et al. (2012)
	Large surface area to volume ratio, active surface sites, and open channel pores	Mesoporous alumina sphere nanoparticles	Fungicidal	Fusarium oxysporium	Shenashen et al. (2017)
Nano-emulsion	β-Cypermethrin	Poly(oxyethylene) lauryl ether, ethylene oxide and methyl decanoate, water	Insecticidal	Reduced pesticide application	Wang et al. (2007)
	Triazophos	Water, polyoxyethylene ether/ noctyl-2-pyrrolidone	Insecticidal	Improved stability of the pesticide	Song et al. (2009)
	Neem oil	Tween 20, water	larvicidal	Culex quinquefasciatus	Anjali et al. (2012)
	Glyphosate isopropylamine	Esterified vegetable oils, surfactants, and water	Insecticidal	Asystasia gangetica, Diodia ocimifolia, Paspalum conjugatum	Lim et al. (2013)
	Pimpinella anisum EO	Tween 80, water	Insecticidal	Tribolium castaneum	Hashem et al. 2018
	Eucalyptus oil	Tween 20, water	Larvicidal	Pectinophora gossypiella and Earias insulana	Moustafa et al. (2015)
Nano-spheres	Aza	PCL	Insecticidal		da Costa et al. (2014)
	Bifenthrin	Poly(styrene)-b-poly(ethylene oxide) (PSt-b-PEG)			Liu et al. (2008)
Micelles	Carbofuran	Poly[poly(oxyethylene 18000)- oxy-5-decanyloxyisophthaloyl]		Insects and Nematodes	Shakil et al. (2010)
	β-Cyfluthrin	Poly-(oxyethylene- 600/1000/1500/2000)-oxyisophthaloyl]	Insecticidal		Loha et al. (2011)
	Aza	Ricinoleic acid-grafted CMCS (R-CMCS)	Insecticidal		Feng and Peng (2012)
	Rotenone	N-(octadecanol-1-glycidyl ether)-O-sulfate chitosan (NOSCS)	Insecticidal		Lao et al. (2010)
	IMI	CS-co-PLA	Insecticidal		Li et al. (2011)
					(Continued)

TABLE 3.2 (CONTINUED)

List of Encapsulated Nano-Systems against Some Pests and Pathogens

Types of Nano- Formulation	Active Ingredient	Polymer / Ingredient	Activity	Target Organism (S)	References
	CPF	CS-co-PLA-DPPE	Insecticidal		Zhang et al. (2013)
Nanofibers	(Z)-9-dodecenyl acetate pheromone	Polyamide 6 Cellulose acetate	Insecticidal	European grape berry moth	Hellmann et al. (2011)
Polymeric chelate nanoparticles	Cu(II) ions	CS	Fungicidal	A. solani and F. oxysporum	Saharan et al. (2013, 2015)
Nanosuspension	Pyridalyl	Sodium alginate	Insecticidal	Helicoverpa armigera	Saini et al. (2014)
Nanomicelle	Carbofuran	PEG with dimethyl 5-hydroxyisophthalate	Enhance pesticides	Improve release of pesticides	Shakil et al. (2010)
	Imidacloprid	PEG and aliphatic diacids	Enhance pesticides	Enhance the quality pesticides	Adak et al. (2012)
	Thiamethoxam	PEG and aliphatic and aromatic diacids	Enhance pesticides	Protection, safety, and effectiveness of pesticides	Sarkar et al. (2012)
	Diethylphenylacetamide	PEG	Enhance pesticides	Increased stability and bioefficacy of poorly soluble	Balaji et al. (2015)
	Carbofuran	Poly[poly(oxyethylene 18000)-oxy-5-decanyloxyisophthaloyl]		Insects and Nematodes	Shakil et al. (2010)
	β-Cyfluthrin	Poly-(oxyethylene-600/1000/1500/2000)-oxyisophthaloyl]	Insecticidal		Loha et al. (2011)
	Aza	Ricinoleic acid-grafted CMCS (R-CMCS)	Insecticidal		Feng and Peng (2012)
	Rotenone	N-(octadecanol-1-glycidyl ether)-O-sulfate chitosan (NOSCS)	Insecticidal		Lao et al. (2010)
	IMI	CS-co-PLA	Insecticidal		Li et al. (2011)
	CPF	CS-co-PLA-DPPE	Insecticidal		Zhang et al. (2013)
Nanoliposomes	Etofenprox	Chitosan-coated lecithinforming liposome	Enhance pesticides	Enhance the quality pesticides	Hwang et al. (2011)

TABLE 3.3

List of Nanoparticles as Active Ingredient against Some Pests and Pathogens

Nanoparticles Name	Activity	Target Organism (S)	References
Silver	Antimicrobial	*Bipolaris sorokiniana* and *Magnaporthe grisea*	Jo et al. (2009)
Silver	Antimicrobial	Oak wilt pathogen *Raffaelea* sp.	Kim et al. (2009)
Silver	Antimicrobial	*Alternaria alternata, Sclerotinia sclerotiorum, Macrophominaphaseolina, Rhizoctonia solani, Botrytis cinerea,* and *Curvularia lunata*	Krishnaraj et al. (2012)
Silver	Antimicrobial	*A. alternata, Alternaria brassicicola, Alternaria solani, B.cinerea, Cladosporium cucumerinum, Corynespora cassiicola, Cylindrocarpon destructans, Didymella bryoniae, Fusarium oxysporum* f. sp. *cucumerinum, F. oxysporum* f. sp. *lycopersici, F. oxysporum, Fusarium solani, Glomerella alternata, Monosporascus cannonballus, Pythium aphanidermatum, Pythium spinosum, Stemphylium lycopersici*	Kim et al. (2012)
Silver	Antimicrobial	*Xanthomonas campestris* pv. *malvacearum*	Rajesh et al. (2012)
Silver	Antimicrobial	*Colletotrichum coccodes, Monilinia* sp., *Pyricularia* sp.	Lee et al. (2013)
Silver	Antimicrobial	*Citrobacter freundii, Erwinia cacticida*	Paulkumar et al. (2014)
Ag, Cu, Fe, Zn, Mn	Herbicides	*Allium cepa*	Konotop et al. (2014)
Cu	Herbicides	*Cucurbita pepo*	Hawthorne et al. (2012) Musante and White (2012)
CuO	Herbicides	*Raphanus sativus, Lolium perenne* and *Lolium rigidum*	Atha et al. (2012)
CuO and ZnO	Herbicides	*Fagopyrum esculentum*	Lee et al. (2013)
Cu	Herbicides	*Elodea densa*	Nekrasova et al. (2011)
CuO and ZnO	Herbicides	*Cucumis sativus*	Kim et al. (2012)

contact with the plant cell surface where they influence the physical and chemical nature of the plant organs. Additionally, they may also influence subcellular organization. Aggregation of zinc oxide (ZnO) NPs at the root surface has been observed in sequence through scanning electron microscopy (Lin and Xing, 2008). Initially, they aggregate in adjacent parenchyma cells reflecting cell-to-cell movements. In another study, through transmission electron microscopy, it was confirmed that ZnO NPs could pass through the plant cell wall and were present in the roots of *Schoenoplectus tabernaemontani*, but the potential for its translocation to the shoot was limited (Zhang et al. 2015). The NPs are internalized by wrapping a membrane around them to form a closed vesicular structure primarily because of a localized decrease in Gibb's free energy (Gao et al. 2005). This may be caused by interaction between ligands on NPs and receptors on the cell surface wherein shortage of either the receptor or the ligands does not result in optimum endocytosis (Yuan et al. 2010). The net accumulation of NPs in the cell will always be the sum total effect of endocytosis and exocytosis, the two processes that have been reported to be influenced by both size and shape (Chithrani and Chan, 2007). Furthermore, NPs that are positively charged are taken up at a faster rate than neutral or negatively charged ones because membranes are invariably slightly negatively charged (Thorek and Tsourkas, 2008). This is probably caused by strong electrostatic attraction between the biomembrane and positively charged NPs. Once in the cytosol, NPs may disrupt mitochondrial function resulting in the production of reactive oxygen species (ROS) and consequent activation of the cascade of oxidative stress-mediated reactions (Rani et al. 2008). NPs may also either remain in the cytosol or may be exocytosed. In the former case, upon mitosis they are also distributed within daughter cells (Rees et al. 2011). Exposure to TiO_2 NPs adversely influences interaction between pea and rhizobium, leading to poor nodule development and N_2 fixation that could be caused by alteration in the polysaccharide composition of the cell wall, leading to a change in host response. This reflects the possible negative potential of TiO_2 NPs with respect to rhizobium–legume interaction (Fan et al. 2014). In our laboratory, too, nitrogen fixation was differentially influenced depending on the concentration of ZnO NPs: 10 ppm was deleterious, while 1.5 ppm improved in clusterbean, cowpea, and green gram. In mothbean, however, either concentration was deleterious for nitrogen fixation. Additionally, an interactive effect of concentration and duration of exposure time was also observed in green gram (Kumar et al. 2015). In addition, Lee et al. (2013) also reported that the biomass of buckwheat seedlings was significantly reduced in response to ZnO NPs compared to ZnO macroparticles at concentrations between 10 and 2,000 mg/L. Retardation in growth and deformation of epidermal cells because of exposure to copper oxide (CuO)NPs has also been reported to be associated with changes in levels of indole acetic acid in wheat (Adams et al. 2017). Additionally, in another study on *Arabidopsis*, deleterious response of ZnO NPs on growth was attributed to the induction of oxidative stresses (Wang et al. 2016). In other studies, laboratory-synthesized NPs (0.1%, w/v) showed substantial growth promontory effects on chickpea seed germination, seedling length, and fresh and dry weight. NP-treated seedlings also showed a remarkable increase of chlorophyll in chickpea (Sathiyanarayanan and Banu, 2016). It has been reported that NP treatment improves the photosynthetic rate by improving the synthesis of photosynthetic pigments (Xie et al. 2012; Priyadarshini et al. 2012; Siddiqui and Al-Whaibi, 2014). The reason for increased photosynthetic pigment might also be because of enhanced electron transport rates, as suggested by Giraldo et al. (2014). Roots of *Brassica nigra* showed an 80% increase in 2,2-diphenyl-1picrylhydrazyl radical scavenging activity, total antioxidant, and reducing power potential in the presence of 10 mg/L CuO NPs in the media. Furthermore, though non-enzymatic antioxidant molecules, phenols, and flavonoids were also increased, it was concentration dependent (Zafar et al. 2016). Moreover, engineered carbon nanotubes increase the germination of seeds and the growth and development of plants (Lahiani et al. 2013; Siddiqui and Al-Whaibi, 2014). However, most of the studies on NPs are concerned with toxicity, while some studies that have been executed on NPs are of benefit to herbs. Study in the area of nanotechnology is needed to discover new utilizations to target the special delivery of chemical molecules, proteins, and nucleotides for genetic variations of crop plants (Torney et al. 2007; Scrinis and Lyons, 2007). Ma et al. (2010b) examined the toxicity of some rare earth oxide NPs(CeO_2, La_2O_3, nano-Gd_2O_3, and Yb_2O_3) on the root growth of some important crops of higher plants, namely, *Brassica napus*, *Raphanus sativus*, *Triticum aestivum*, *Lactuca sativa*, *Brassica oleracea*, *Lycopersicon esculentum*, as well as *Cucumis sativus*. The impacts of NPs examined on root elongation differed significantly among various NPs and plant types. Results showed that application of 2,000 mg/L nano-CeO_2 had no impact on the growth of roots of all studied plants, except lettuce. However, nano-La_2O_3, nano-Gd_2O_3, and nano-Yb_2O_3, in a concentration of 2,000 mg/L, prevented root growth of all studied taxa. Inhibitory efficacy of nano-La_2O_3, nano-Gd_2O_3, and nano-Yb_2O_3 also varied in several growth activities of taxa. For wheat, the deterrence chiefly happens within the seed incubation procedure, but lettuce and rape were prevented on both seed soaking and incubation proceedings. The 50% inhibitory doping differed greatly between the studied plants, for example, the concentration for rape was nearly 40 mg/L of nano-La_2O_3, 20 mg/L of nano-Gd_2O_3, and 70 mg/L of nano-Yb_2O_3, respectively. In the concentration domains, the RE^{3+} ion liberated from the NPs had poor traces on the growth of roots.The lower concentration of silicon dioxide NPs increased seed germination in *Lycopersicum esculentum*. The concentration of 8 g/L of silicon dioxide NPs enhanced seed germination by 22.16%, mean time of germination by 3.98%, seedling vigor index by 507.82%, and seed germination index by 22.15% compared to control plants. Concentration of 8 g/L of silicon dioxide NPs increased the fresh weight of seedlings by 116.58% and seedling dry weight by 117.46% compared to control seedlings, and also silicon dioxide NPs played a very important role in root and shoot growth (Siddiqui and Al-Whaibi, 2014). Wang et al. (2014) studied the effect of quantum dots and silica-coated quantum dot treatment on rice plants. They found that silica-coated quantum dots significantly promoted root growth of this plant, compared to control samples. Nano-SiO_2 developed the growth of the plant by enhancing exchange of gases, net rate of photosynthesis, level of transpiration, conductance of stomata, activities of photosystem II potential, rate of electron transportation, and photochemical quell (Xie et al. 2011; Siddiqui et al. 2014).

Prakash et al. (2017) treated *Arabidopsis thaliana* seedlings with five concentrations of ZnO NPs (0, 20, 50, 100, and 200 mg/L) and analyzed morphological changes. The resulting data confirmed that the fresh weight, as well as length of the primary root, was decreased after exposure to most of the concentrations (except for 20 mg/L) of ZnO NPs. However, treatment with 20 mg/L of ZnO NPs triggered a 9% increase in the formation of the lateral root. No root-to-shoot translocation of Zn was seen in seed lines and accumulation was mainly located in the tips of the root, primary-lateral root junctions, and root-shoot junctions. The effects of ZnO NPs were examined on *Gossypium hirsutum* L. plants. The results confirmed that the shoot and root lengths and total biomass significantly maximized with an increment of NP concentrations compared to control. The maximum growth rate observed was 130.0% and 115.2% for shoots and roots, respectively, in the presence of 200 mg/L NPs. It is very important to know that the growth rate of shoots is significantly higher than root tissues following ZnO NP exposure with respect to control plants (Venkatachalam et al. 2017).

Manganese (Mn) oxide NPs significantly increased elongation of roots in *Lactuca sativa* seedlings by more than 50%. It confirmed that manganese oxide NPs could be applied as Mn micronutrient fertilizer or enhancer of plant growth. Although these NPs improved agronomic productions, they had no or little phytotoxicity at levels greater than 50 ppm (Liu et al. 2016). Sergey et al. (2015) examined the various effects of iron oxide nanoparticles on *A. thaliana* physiology at two different concentrations: 3 and 25 mg/L. In the treated samples with a concentration of 3 mg/L, no effect was seen on seedlings or root length, while 25 mg/L-treated plants had reduced seedlings and root length. In the treated samples, no observable phenotypic variations were reported in the size of the plant or its total structure.

Askary et al. (2016a) examined the effects of iron oxide NPs on stem and leaf trichomes in *Mentha piperita* (Labiatae) under saline stress. The results proved that these NPs had a negative effect on the density of trichomes. It seems that iron NPs had significant effects on plant morphological and anatomical traits. The effects of various concentrations of iron NPs were examined on salt-treated plants of *Lallemantia royleana* (Askary et al. 2016b). Significant negative and positive correlations were observed between the densities of trichomes with concentrations of salt and iron oxide NPs. Significant negative correlations ($P \leq 0.05$) were recorded between concentrations of iron NPs with simple four-celled and sessile hairs, while these NPs had significant positive correlations ($P \leq 0.05$) with pelate and simple one-celled types. Pariona et al. (2017) studied the effect of citrate-coated magnetite NPs on the germination as well as early growth of *Quercus macdougallii*. They used two types of magnetite NPs, namely, NP1 and NP2. The morphology of NP1 was quasispherical, with sizes of 6–10 nm, but the NP2 had sizes between 65 and 160 nm. Castiglionea et al. (2016) employed two kinds of TiO_2 NPs to assess if various responses in function of different particle size and features were elicited. Furthermore, as corresponding bulk material considered for decades an inert and safe material, has been recently classified as possibly carcinogenic to humans (Group 2B carcinogen, IARC 2010), they also tested the effects of the same concentration of this form. To evaluate the hypothesized effects of these materials on seeds of *Vicia faba*, they considered various physiological, histochemical, cytological, and biochemical end points. Hafeez et al. (2015) examined the effects of Cu NPs for enhancing growth and yield of wheat cultivar Millat-2011. Their findings showed that germination of seed was not affected with 0.2–0.8 ppm of Cu NPs, while it decreased significantly at 1.0 ppm. In solution culture, concentration of Cu NPs of more than 2 ppm was detrimental to this plant. However, application of low concentrations of Cu NPs (0.2, 0.4, 0.6, 0.8, and 1.0 ppm) conspicuously amplified leaf area, chlorophyll content, and leaf freshness, as well as dry weight and also arid weight of the root in comparison to control samples.

3.2.3 GENETIC TOOL

Genetic engineering is a powerful tool to create genetic modifications for selected aspects and to allow integration and transfer of target gene(s) into the genome of target host. DNA delivery is the most important aspect of gene transfer in plants to develop the transgenes. There are several methods used for gene transfer, but all of them have some limitations. The small size of the NPs (1 nm = 10^{-9} m) allows them to bypass the cell barriers like the cell wall and plasma membrane and deliver genes into the cell of living systems, and they can also be used as a transgenic vehicle for nucleic acid (Rai et al. 2012). NP-mediated gene transfer methods are new and have the potential to directly transfer DNA into the cells, achieving stable integration and rapid expression of the transgene. NPs are particles at the atomic or molecular level with nanoscale dimension (Ball, 2002; Roco, 2003). NPs behave differently and exhibit different physiochemical properties as compared with their bulk counterpart (Nel et al. 2006). NPs are formed from a variety of materials, and their action depends upon their chemical composition, size, and shape (Brunner et al. 2006). On the basis of their origin, Roberto and Ruffini (2009) classified NPs into three major types as: (1) natural NPs; (2) incidental NPs, also called waste or anthropogenic particles, which are mainly formed due to the manmade activities; and (3) engineered NPs or nanomaterials.

The NPs which can be used as a vector for gene transfer include calcium phosphate, carbon based, silica, gold, magnetite, strontium phosphate, magnesium phosphate, and manganese phosphate (Potrykus, 1991). The NP mediated gene transfer mechanism is reported as follows: initially, NPs recognize and are adsorbed on the cell membrane, and later these NPs are internalized by endocytosis. During this process, if DNA escapes before the fusion of endosome with lysosome, it successfully enters into the cell cytoplasm, where its degradation by nucleases can occur. However, for an efficient transfer, it must be protected from nucleases and enter into the nucleus. Generally, the DNA enters into the nucleus with the help of nuclear pore complexes (NPCs) which are large numbers of proteins, forming a channel. Indeed, it is not clear how DNA alone or with NPs is transported into the nucleus. One of the reasons may be that DNA-loaded NPs may recognize the surface and become attached to the surface of the nuclear membrane, where the import in molecule can transfer the DNA into the nucleus. In this case, a NP protects DNA from nucleases until it enters the nucleus. On this basis, several NPs can be used to conjugate with nucleic acids for their proper transfer into the living cell (Ito et al. 1983), and various types of NPs have been used as nucleic acid carriers (Ruzin and McCarthy, 1986). The NPs have some advantages over other methods as they are not subject to microbial attack, can be easily synthesized, are less toxic, and exhibit good stability (Hoffmann-Tsay et al. 1994). Many scientists successfully adopted different nanoparticulate delivery carriers like polymeric NPs, metallic NPs, and quantum dots in gene therapy (Jun et al. 2008; Akhter et al. 2011). NPs play an important role as a gene carrier in the biotransformation process, as well as protection of DNA damage from ultrasound as reported by Liu et al. (2008). Many scientists used green fluorescent protein (GFP) tagged with NPs to demonstrate its effectiveness and reported that NPs alone enter into the cytoplasm as seen in the green cytoplasmic content (Jun et al. 2008). On the other hand, bioconjugation is the chemical strategy to conjugate two biomolecules. Understanding of the bioconjugation process has enabled applications of biomolecules to numerous fields like medicine, agriculture, and materials. Conjugation of DNA with NPs of calcium phosphate, carbon, silica, gold, magnetite, strontium phosphate, magnesium phosphate, and manganese phosphate has been used for DNA delivery; these have low toxicity and good storage capacity (Fukumori and Ichikawa, 2006; Sokolova and Epple, 2008). Among bionanoconjugates, gold nanoparticles coupled with biomolecules attracted great attention in the last decade due to their ease of preparation and conjugation, biocompatibility, good tenability, high stability, etc. Many diverse simple synthetic methods are available to synthesize NPs and conjugate with the biomolecules, which can be used effectively as smart delivery systems.

Genetic transformation allows the transfer of a foreign gene of interest (transgene) encoding the trait into the plant cell, in order to introduce a desired trait to a crop. Gene vectors play many important roles in genetic transformations which need to be constructed in order to transfer the gene of interest. A vector consists of a cassette which includes the genes of interest, flanked by the necessary controlling sequences, i.e., promoter, terminator, and marker genes, which allow efficient screening of the transformed and non-transformed. Quantum dots were also modified with streptavidin (available commercially) which can be covalently conjugated to different biotinylated peptides (Chen and Gerion, 2004; Lagerholm et al. 2004; Kim et al. 2008). Several approaches are available for the conjugation of biomolecules like proteins to NPs, including enzymes or antibodies. Initially only 'nonspecific' adsorption was employed: this can be achieved by incubating NPs with the protein, which allows the adsorption of NPs to the proteins by electrostatic attraction or by van der Waals forces, hydrogen bridges, thiol bonds (from cysteine residues), or amino groups modified by agents like bis-NHS for the coating on silica NPs (Jana et al. 2007). Josephson et al. (1999; 2001) have demonstrated the covalent conjugation of magnetic iron oxide NPs with peptides and oligonucleotides. There are several examples of NP-protein conjugation, which can be found in a number of reviews and research papers (Niemeyer, 2001; Ghadiali and Stevens, 2008). Yu et al. (2012) reported a new method to use zinc sulfide (ZnS) NPs as a gene carrier for plant gene transfer. The ZnS NPs with a diameter of around 3–5 nm were modified with positively charged poly-L-lysine (PLL) to bind negatively charged pBI121 plasmid DNA. The ZnS NP-mediated transformation of tobacco was carried out via the ultrasound-assisted method. The ZnS NPs mediated stable gene expression in tobacco plants was observed for the first time. Cui et al. (2012) used polyethyleneimine (PEI)-modified magnetic NPs as vectors to transfer genes to porcine somatic cells. The PEI-modified Fe_3O_4 magnetic NPs were mixed with pEGFP-N1 in a certain mass ratio and incubated for 30 min at room temperature; these modified NPs were employed to transfer the reporter gene into somatic cells, and the gene delivery efficiency was also examined (Hoshino et al. 2004).

Regarding mode of entry, transport, and effects of different NPs in plant tissues, the plant cell wall acts as a barrier, preventing the easy entry of any external agent including metal NPs into plant cells. The cell wall pore diameter ranging from 5 to 20 nm (Fleischer et al. 1999; Navarro et al. 2008) allows easy passage of the NPs, having the diameter less than the pore diameter of the cell wall, and enters the plasma membrane of the cell (Moore, 2006; Navarro et al. 2008). There is also a possibility of pore size enlargement or an induction of new pores in the cell wall upon interaction with engineered NPs which in turn enhance their uptake. NPs may also cross the membrane using membrane-embedded transporter proteins or through ion channels. On entering the cytoplasm, NPs may interact with different cytoplasmic organelles and biomolecules, affecting the metabolic processes in the cytoplasm (Jia et al. 2005). When NPs are applied on the leaf surface, they can enter through stomatal openings or through the bases of trichomes (Eichert et al. 2008; Fernandez and Eichert, 2009; Uzu et al. 2010). The NPs accumulated on photosynthetic surfaces can cause foliar heating resulting in the alterations to gas exchange. The stomatal obstruction caused by the NPs produces changes in various physiological and cellular functions of plants (Da Silva et al. 2006). McKnight et al. (2003, 2004) have reported the integration of plasmid DNA with surface-modified carbon nanofibers in viable cells for controlled biochemical manipulations. The effective integration and delivery of plasmid DNA was confirmed from the gene expression, similar to the micro-injection method of gene delivery (Neuhaus and Spangerberg, 1990; Bolik and Koop, 1991; Segura and Shea, 2001). The use of fluorescent-labeled starch NPs as plant transgenic vehicles was reported by Jun et al. (2008), in which the authors reported the gene transfer with the help of NP-biomolecule conjugate. The conjugate was designed in such a way that it binds and shuttles genes across the cell barrier like the cell wall, cell membrane, and nuclear membrane of plant cells by inducing instantaneous pore channels with the help of ultrasound waves. Different genes can be integrated on the fluorescence-labeled NP at the same time, and their imaging with a fluorescence microscope makes it possible to understand the movement and expression of the exterior genes transferred. Increasing the porosity of the cell wall and cell membrane by a suitable agent helps in NP-mediated DNA transfer in regenerative calli and soft tissues. The ability of NPs to penetrate plant cell walls also helps precise manipulation of gene expression at the single-cell level by delivering DNA and its activators in a controlled fashion (McKnight et al. 2003). Honeycomb mesoporous silica NPs(MSN) with 3 nm pores can transport DNA and chemicals into isolated plant cells and intact leaves. MSNs loaded with genes and their chemical inducers were capped with gold NPs at the ends to protect the molecules from leaching out. Removal of gold NP caps enabled the release of chemicals and triggered gene expression in plants under controlled conditions. Incubation of protoplast with fluo-rescently labeled MSNs revealed that surface modification of MSNs with triethylene glycol was necessary to penetrate the cells. The modification of MSNs with triethylene glycol also allowed adsorption of plasmid DNA on MSN surfaces. Plasmid DNA was released from the MSN upon its entry into the protoplasts; the plasmid DNA released allows expression of a marker gene like GFP in the cell, which can be detected by fluorescence microscopy. This method can detect marker expres-sion of 1,000 times less than that required for the conventional delivery method. Efficiency of this delivery method has pronounced applications in various protoplast-based gene expression studies. Nowadays, particle bombardment or gene gun is one of the popular methods to transfer DNA into intact plant cells (Klein et al. 1989; Deng et al. 2001).

Some of the approaches, which can be used to improve NP-based gene delivery into plant cells, are given below:

1. The surface charge in protoplast and plasma membranes varies from 10 to 30 mV (Reid et al. 2002). Modifications to this surface charge can be done to enhance the adhesion of bioconjugate DNA, by introducing a positive charge over their surface. Wiesman et al. (2007) have demonstrated a similar concept.
2. The porosity of protoplast is 0.1–5,000 nm (Berestovsky et al. 2001), and in plasma mem-brane, it varies between 1 and 50 nm (Berestovsky et al. 2001).
3. Stearylamine, triethylamine, and natural biodegradable polymers like chitosan can impart a positive charge on DNA conjugate nanomaterial surface (Li et al. 2011).
4. The advancement in nanotechnology may open the way for further development of effi-cient plant transformation systems.

Although *Agrobacterium*-mediated gene transfer is a widely and commonly used method, it has sev-eral drawbacks (Tzfira and Citovsky, 2003; Shrawat and Good, 2011). These include (1) limitation to carry genes <50 kb in size, (2) chances of transgene silencing, (3) poor gene transfer efficiency, and (4) effectiveness against dicot plants. To improve the efficiency of *Agrobacterium*-mediated gene transfer, newer modifications in indirect methods, e.g. application of acetosyringone, were applied to increase the host range (Verma and Mathur, 2011).

Raising the temperature enhances gene transfer that is successfully adopted and confirmed to the animal cell transfection process (Kenel et al. 2010). It seems from the literature that at a temperature above 37°C, the gene transfer tendency increases and a further increase up to 43°C within a certain period of time provides greater transient transfection. This was confirmed by the work on interleu-kin-2 and swine growth hormone expressions using indirect ELISA (Dillen et al. 1997; Baron et al. 2001). Dillen et al. (1997) demonstrated the effect of temperature (15–29°C) on *Agrobacterium*-mediated gene transfer in plants. Baron et al. (2001) reported the effect of temperature in the bio-transformation process with *A. tumefaciens* involved in co-cultivation of *Phaseolus acutifolius* and further in *Nicotiana tabacum* biotransformation. In both the situations, the level of transient uidA expression decreased notably when the temperature was raised above 22°C, decreased further at 27°C, and was undetectable at 29°C (Baron et al. 2001; Iba, 2002; Rakoczy-Trojanowska, 2002).

Gene loaded liposomes provide stability to genes, diminish or reduce the deletion in the DNA, while used along with the physical techniques control the release pattern of gene delivery, surface modification of liposome (positively charged) potentiate the penetration within a cell, show a higher degree of reproducibility applicable to a wide range of cell types, and free from cellular toxicity. In this technique, the mechanism of DNA entry through the protoplast seems to be mediated by endocytosis process of liposome by the cells (Shirazi et al. 2011), which may involve the following steps: (a) adhesion of the protoplast to the lysosomal surface, (b) fusion of liposomes and protoplast after adhesion, (c) formation of endosome within the cytoplasm, and (d) release of plasmids inside the cell. The liposomal technique has been successfully used to deliver DNA into the protoplasts of a number of plant species (e.g. tobacco, petunia, carrot, etc.). Deshayes et al. (1985) developed positively charged liposomes bearing *E. coli* plasmid (pLGV23neo) having a kanamycin resistance gene, which were used in the development of transgenic tobacco. Protoplasts isolated from the leaves of transgenic tobacco were resistant to 100 mg/ml kanamycin.

3.2.4 NANO-ENHANCED BIOLOGICAL TREATMENT OF AGRICULTURAL WASTEWATER

Providing clean and abundant fresh water for human use and industry applications, including agricultural and farming uses, is one of the most daunting challenges facing the world (Vorosmarty et al. 2010). Agriculture requires a considerable amount of fresh water, and in turn, often contributes substantially to pollution of ground water through the use of pesticides, fertilizers, and other agricultural chemicals. Effective technologies for remediation and purification will be needed to manage the volume of wastewater produced by farms on a continual basis and be cost-effective for all. Technical challenges include water quality and quantity, treatment and reuse, safety due to chemical and biological hazards, monitoring, and sensors. Nanotechnology R&D has shown great promise in providing novel and economically feasible solutions. Several aspects of nanotechnology solutions are briefly discussed below (Chen and Yada, 2011).

In industrialized nations, chemical and physical based (chlorine dioxide, ozone, and ultraviolet) microbial disinfection systems are commonly used. However, much of the world still does not have the industrial infrastructure necessary to support chemical-based disinfection of water. Hence, alternative technologies that require less intensive infrastructure and more cost-effective approaches such as nanoscale oligodynamic metallic particles which may exhibit a toxic effect on living cells even in relatively low concentrations, may be worthy of attention. Among the oligodynamic metallic nanoparticles, silver is considered the most promising nanomaterial with bactericidal and viricidal properties owing to its wide-range effectiveness, low toxicity, ease of use, its charge capacity, high surface to volume ratios, crystallographic structure, and adaptability to various substrates (Nangmenyi and Economy, 2009). Its antimicrobial mechanism is due to the production of reactive oxygen species (ROS) that cleaves DNA. Another nanoscale technological development for microbial disinfection is visible light photocatalysts of transition metal oxides made into NPs, nanoporous fibers, and nanoporous foams (Li et al. 2009). In addition to its effectiveness in disinfecting microorganisms, it can also remove organic contaminants such as pharmaceuticals and personal care products (PPCPs) and endocrine disrupting compounds (EDCs) (Chen and Yada, 2011).

Given the limited fresh water supplies both above and underground, it is likely that the desalination of sea water will become a major source of fresh water. Conventional desalination technology is reverse osmosis (RO) membranes which generally require large amounts of energy. A number of nanotechnologies have been attempted to develop low energy alternatives. Among them, the three most promising technologies appear to be protein polymer biomimetic membranes, aligned-carbon nanotube membranes, and thin film nanocomposite membranes (Hoek and Ghosh, 2009). Some of the prototypes have demonstrated up to 100 times better water permeability with nearly perfect salt rejection than RO. Carbon nanotube membranes, owing to their extremely high water permeability compared to other materials of similar size, have desalination efficiencies in the order of 1,000 times better than the current technology. Some of these membranes can also integrate other functionalities such as disinfection, de-odoring, de-fouling, and self-cleaning. Some of the above-mentioned technologies are currently in the commercial development stage, and may be introduced to the marketplace in the near future (Chen and Yada, 2011).

Functionalization of ligand-based nanocoating which is bonded to the surface of high surface and low-cost filtration substrate can effectively absorb high concentrations of heavy metal contaminants. The system can be regenerated in situ by treatment with bifunctional self-assembling ligand of the previously used nanocoating media. A startup company (Crystal Clear Technologies) has demonstrated that multiple layers of metal can be bonded to the same substrate (Farmen, 2009). Such a water treatment unit should be available in the near future for the removal of various heavy metals in water. Another approach to remove heavy metals and ions is the use of dendrimer enhanced filtration (DEF) (Diallo, 2009). Functionalized dendrimer scan binds cations and anions according to acidity.

3.2.5 NANOTECHNOLOGY IN ANIMAL PRODUCTION AND ANIMAL HEALTH

The diversity of animal production associated with agriculture (livestock, poultry, and aquaculture) is a very important source and integral part of human diets (meat, fish, egg, milk, and their processed products). Therefore, nanotechnologies aim to maintain and increase animal production and offer effective, sometimes novel, solutions to these challenges including production efficiency, animal health and feed, nutritional efficiency, diseases including zoonoses, product quality and value, byproducts and waste, and environmental footprints (Kuzma, 2010).The integrity of inputs, the quality of outputs and the abundance of byproducts of animal production for human, animal, and environmental health are crucial and optimal farm-scale economics of application. Nanotools can be used to enhance the safety of animal feed and waste (Kuzma, 2010). A critical element of sustainable agricultural production is the minimization of production input while maximizing output. One of the most significant inputs in animal production is feedstock. Low feeding efficiency results in high demand for feed, high discharges of waste, heavy environmental burden, high production cost, and competing with other uses of the grains, biomass, and other feed materials. Nanotechnology may significantly improve the nutrient profiles and efficiency of minor nutrient delivery of feeds. Most animal feeds are not nutritionally optimal, especially in developing countries (Chen and Yada, 2011). On the other hand, many animal diseases cause substantial losses in agricultural animal production including bovine mastitis, tuberculosis, respiratory disease complex, Johne's disease, avian influenza, and porcine reproductive and respiratory syndrome (PRRS). Zoonotic diseases not only cause devastating economic losses to animal producers, but also impose serious threats to human health, e.g. Variant Creutzfeldt-Jakob disease (VCJD). Nanotechnology has the potential to enable revolutionary changes in this area, and some specific technologies may be feasible in the near future given the current state of research and development (Emerich and Thanos, 2006; Scott, 2007). Nanotechnology offers numerous features in detection and diagnostics including high specificity and sensitivity, simultaneous detection of multiple targets, rapid, robust, onboard signal processing, communication, automation, convenience, and low cost. The uses of portable, implantable, or wearable devices are particularly welcome in agricultural field applications. Early detection is imperative so that quick, simple, and inexpensive treatment strategies can be taken to remedy the situation. Nanotechnology-based drugs and vaccines can be more effective in treating/preventing the diseases than current technologies, thus reducing cost. Precise delivery and controlled release of nanotechnology-enabled drugs leave little footprint in the animal waste and the environment, which alleviates the increasing concern of antibiotic resistance, and decreases health and environmental risks associated with the use of antibiotics. The targeted delivery and active NPs may enable new drug administration that is convenient, fast, non-intrusive to animals, and cost-effective (Chen and Yada, 2011).

Animal reproduction remains a challenge not only in developing countries but also in developed nations. Low fertility results in a low reproduction rate, increased financial input, and reduced efficiency of livestock operations (Narducci, 2007). For centuries, animals have been bred for important characteristics such as disease resistance, improved performance and growth, and product quality. In the past decade, genetic engineering of several livestock species has been achieved. Genetic engineering can speed the process of introducing desirable traits into livestock and allows for the introduction of entirely new ones (NRC, 2002). Genetic engineering of livestock is faced with technical challenges that nanotechnology might be able to overcome (Kuzma, 2010). Several technological fronts have also been explored in order to improve animal reproduction. Microfluidic technology has matured over the last two decades and has been integrated into many nanoscale processing and monitoring technologies including food and water quality, animal health, and environmental contaminations. Nanoscale delivery vehicles are sought to substantially improve bioavailability and better control the of release kinetics, reduce labor intensity, and minimize waste and discharge to the environment (Emerich and Thanos, 2006; Narducci, 2007). Another strategy that may be explored is to monitor animal hormone levels using implanted nanotechnology-enabled sensing devices with wireless transmission capability, thus the information on the optimal fertility period can become available in real time to assist the livestock operators for reproduction decision making (Afrasiabi, 2010). Modification of animal feeds has been effectively used to improve animal production as well as product quality and value. The regulation of nutrient utilization can be used to enhance the efficiency of animal production and to design animal-derived foods consistent with health recommendations and consumer perceptions. For example, the concepts of nutrient regulation have been used to redesign foods, such as milk fatty acids, cis-9, trans-11 conjugated linoleic acid (CLA) and vaccinic acid (VA), that may have a potential role in the prevention of chronic human diseases such as cancer and atherogenesis (Bauman et al. 2008). Animal waste is a serious concern in the animal production industry. Stricter environmental policies will prevent irresponsible discharge of animal waste. The unpleasant odors that emanate from intensive animal production facilities adversely affect air quality, and in turn, living conditions and the real estate value of the adjacent area. However, bioconversion of animal waste into energy and electricity can result in new revenue, renewable energy, high quality organic fertilizer, and improved environmental quality while adding value (Scott, 2002). Nanotechnology-enabled catalysts will play a critical role in efficient and cost-effective bioconversion and fuel cells for electricity production as well as enabling efficient energy storage which will greatly facilitate and benefit the development of distributed energy supplies, especially in rural communities where infrastructure is lacking (Davis et al. 2009; Soghomonian and Heremans, 2009). Such an approach may result in the elimination of the need for system-wide electricity grids, hence accelerating rural development and improving productivity and the business and living environments, and will be especially beneficial to developing countries (Chen and Yada, 2011) (Figure 3.1).

3.2.6 IMPROVE THE QUALITY OF THE SOIL

The detrimental effect of NPs on the microbial world could easily be manifested by inhibition of growth, wall formation, and cell morphological damage, which eventually exerts negative impacts on the microbial community (Thul et al. 2013) (Table 3.1). Akhavan and Ghaderi (2010) demonstrated that both graphene and graphene oxide nanowalls may lead to damage of the cell membrane of gram-negative bacteria (*Escherichia coli*) and gram-positive bacteria (*Staphylococcus aureus*), which may ultimately result in leakage of intracellular substances and thus cell death. Similarly, nano-TiO_2 and nano-ZnO also reduce the quantity and diversity of microorganisms in the soil and change compositions of the soil microbial community (Ge et al. 2011). After entering the microbial cell, NPs change the metabolism of microbes. Thul et al. (2013) demonstrated the effect of NPs on bacteria in soil, and found that bacteria were responsible for enhancing productivity in the soil environment. Holden et al. (2012) showed the combined effects of NP hazards in different populations, communities, and ecosystems, and reported that bacteria can also assist the mobilization of NPs by adsorbing and accumulating other forms of NPs via the food chain, and can change many populations (Figure 3.1). Some NPs including silver (Ag) NPs exert an antimicrobial effect at the ecosystem level.

Tilston et al. (2013) showed that soil microbes are a sufficient and skilled catalyst, which might either absorb or disperse the accumulated engineered NPs. Choi et al. (2008) reported that Ag NPs after attachment to microbial cells can cause detrimental effects because of cell wall pitting. After 28 days of treatment with nano zerovalent iron (nZVI), aroclor-1242 can change bacterial aroclor

FIGURE 3.1 Applications of nanotechnology in the agricultural sector.

congener profiles together with changes in the physicochemical properties of soil such as pH. Yang et al. (2013) demonstrated the effect of 35 nm carbon coated Ag NPs or silver ions (Ag^+) on nitrifier *Nitrosomonas* species and nitrogen fixer *Azotobacter vinelandii*. The results showed that exposure of a sublethal dose of Ag NPs or Ag^+ upregulated many nitrifying genes (*amoA1* and *amoC2*) in *Nitrosomonas europaea* (2.1 to 3.3-fold). NPs or nanomaterials can cause lipid peroxidation, which leads to damage in the cell membrane by accomplishing the generation of reactive oxygen species (ROS) (Manke et al. 2013; Huang et al. 2015). Cabiscol et al. (2000) illustrated the consequences of ROS on bacteria that peroxidize the lipid bilayer and lead to changes in the fluidity and permeability of the membrane, thus making cells more vulnerable to failure of nutrient uptake and osmotic stress. In the cell membrane, more peroxidized fatty acids result in DNA damage by releasing free radicals (Thul et al. 2013). Adams et al. (2006) compared the ecotoxicity of TiO_2, SiO_2, and ZnO NPs, which exert toxicity on bacteria in the presence of light. The interaction of microbes and NPs for comparing the physicochemical properties and biological response of engineered and metal oxide NPs has been illustrated (Adams et al. 2006). It has been observed that toxicity of NPs is species-specific and depends on the size and shape of metallic NPs (Adams et al. 2006; Thul et al. 2013). Additionally, Zuverza-Mena et al. (2017) also reported that the behavior of NPs on plants also depends on the exposure conditions and characteristics of NPs. Moreover, the toxic behavior of NPs also depends on environmental effects. Du et al. (2017) reported that at the ambient concentration of CO_2 (370 μmol/mol), TiO_2 NPs possess toxicity in seedlings of *Oryza sativa*, while at a high concentration of CO_2 (570 μmol/mol), TiO_2 NPs induced visible signs of toxicity by declining the overall biomass of the plant.

Besides this, surface coating of NPs is also capable of enhancing microbial toxicity (Adams et al. 2006; Thul et al. 2013). However, Medina-Velo et al. (2017) reported that coated ZnO NPs, at concentrations of 250 and 500 mg/kg, showed a beneficial role, i.e. increased leaf length (~13%) and root length (~44%), compared to the control. Moreover, Ge et al. (2012) suggested that engineered

NPs characteristically could change the bacterial colonies in a dose-dependent manner. Similarly, Priester et al. (2012) stated that engineered NPs such as cerium oxide (CeO_2) can eliminate N_2 fixation and impair growth in soybean. Aken (2015) reported the antimicrobial effect of Ag NPs by transcriptional analysis of some bacteria. Transcriptional analysis was performed to show the pattern of gene expressions (Aken, 2015). Nagy et al. (2011) and McQuillan and Shaw (2014) studied the impact of Ag NPs on model bacterium *E. coli* involving its whole-genome microarray and focused on patches of gene expressions in the regulation of metabolism of sulfur, oxidative balance, and homeostasis of Ag, Cu, and Fe. McQuillan et al. (2012) reviewed that exposure of Ag NPs on *E. coli* leads to stress. Radzig et al. (2013) found that Ag NPs can also result in oxidative damage of DNA, comparatively less resistance to porin synthesis, and disturb water homeostasis and redox balance. Du et al. (2011) showed adverse effects of TiO_2 and ZnO NPs on microbial communities in the soil. Yang et al. (2013) studied the impact of Ag NPs and Ag^+ ions on nitrogen-cycling bacteria. The authors showed that a sublethal dose of 35 nm carbon-coated Ag NPs or Ag^+ ions did not affect the expression of nitrogen-fixing genes *nifD*, *nifH*, *vnfD*, and *anfD* in *A. vinelandii* and denitrifying genes *narG*, *napB*, *nirH*, and *norB* in *Pseudomonas stutzeri*. Pelletier et al. (2010) reported that exposure of CeO_2 NPs results in a disturbance in cellular respiration, oxidative stress, and iron deficiency in *E. coli*. Similarly, Xie et al. (2011) also showed the effect of ZnO NPs on *Campylobacter jejuni* and reported upregulation of stress-responsive genes. Yang et al. (2012) reported the negative impact of quantum dots and other dissolved NPs in *Pseudomonas aeruginosa* by characterization of transporter genes and other stress-responsive genes. They reported that NPs induced the generation of ROS in spite of induced activities of catalases, peroxidases, etc. Sytar et al. (2013) showed that hydrogen peroxide, superoxide anion, and hydroxyl radical are examples of ROS that take oxygen from metal and cause cellular damage (Figure 3.1). Hossain and Mukherjee (2012, 2013) stated that during the mitotic cycle in *E. coli*, CdS and CdO NPs inhibit the septum formation and downregulate important cell division proteins. Cui et al. (2012) showed that 4,6-diaminopyrimidine sulfhydryl-modified Au NPs affect ribosomal protein S10 (proteosynthesis), peroxidase, hydroperoxide reductase, and F-type ATP synthase. In *E. coli* and *S. aureus*, graphene and graphene oxide nanowalls lead to rupture of the cell membrane, which results in leakage of intracellular substances and thus cell death (Huang et al. 2007; Akhavan and Ghaderi, 2010).

Hydrogels, nanoclays, and nanozeolites have been reported to enhance the water-holding capacity of soil (Sekhon, 2014), hence acting as a slow release source of water, reducing the hydric shortage periods during crop season. Applications of such systems are favorable for both agricultural purposes and reforestation of degraded areas. Organic e.g., polymer and carbon nanotubes, and inorganic, e.g. nano metals and metal oxide nanomaterials have also been used to absorb environmental contaminants (Khin et al. 2012), increasing soil remediation capacity and reducing times and costs of the treatments (Table 3.4).

3.3 APPLICATIONS OF NANOTECHNOLOGY IN THE FOOD SECTOR (AT THE MARKETING SITE)

The importance of nanotechnology to the food sector is due to its entry into many areas of food security and food systems, such as the detection of pathogens, saving food, disease treatment delivery methods, protection of the environment, reduction of pollution, and new tools for molecular and cellular biology. Furthermore, nanotechnology tools are used to increase the security of the food manufacturing industry by the shipping and processing of food products through sensors designed to detect pathogens and contaminants, and also encapsulation and delivery systems to active ingredients to specific targets to protect, carry, and deliver these materials (Sekhon, 2010; Srividhya and Chellaram, 2012). Food nanotechnology has become an area of emerging interest and has opened the way for food technologies to improve both the quality and quantity of food produced while also facilitating the evaluation of food safety, especially in terms of food additives (nanotechnology

TABLE 3.4

Effect of Nanoparticles on Soil Microorganisms

Nanoparticles Type	Target Microorganisms	Mode of action	References
Ag^+	*Escherichia coli, Staphylococcus aureus*	Adsorption to cells and intracellular invasion and reaction of free Ag^+ with thiol groups of proteins and inhibition of enzymes	Feng et al. (2000)
Ag^+	*Staphylococcus aureus*	Adsorption to cells and intracellular invasion, generation of free radicals and associated cell membrane damage	Kim et al. (2011)
Ag^+	*Pseudomonas putida, Salmonella typhimurium,* and *Vibrio cholera*	Binding to nucleic acids and impairment of DNA replication	Morones et al. (2005)
Ag^+	*Candida albicans* and *Saccharomyces cerevisiae*	Dissipation of membrane potential	Nasrollahi et al. (2011)
Ag^+	*E. coli*	Apoptosis or programmed cell death	Bao et al. (2015)
CeO_2	*E. coli*	Cell membrane damage and disruption of cell wall	Thill et al. (2006)
CeO_2	*S. typhimurium, Enterococcus faecalis,* and *Bacillus subtilis*	Cell membrane damage and disruption of cell wall	Krishnamoorthy et al. (2014)
CeO_2	*E. coli* and *B. subtilis*	Growth inhibition in response to the size of particles	Pelletier et al. (2010)
Cu^{2+}/CuO	*E. coli*	Action of metal ions at the cell surface leading to membrane damage	Karlsson et al. (2013)
Cu^{2+}/CuO	*Brevibacillus laterosporus, Chryseobacterium indoltheticum,* and *Pantoea ananatis*	Action of metal ions at the cell surface leading to membrane damage	Palza (2015) Concha-Guerrero et al. (2014)
Cu^{2+}/CuO	*E. coli*	Reaction with amino acids and protein interference	Karlsson et al. (2013) Palza (2015)
Cu^{2+}/CuO	*E. coli*	Dissipation of cell membrane potential leading to filamentous cell growth and loss of viability	Chatterjee et al. (2014)
Cu^{2+}/CuO	*E. coli* and *Lactobacillus brevis*	DNA degradation	Chatterjee et al. (2014) Deryabin et al. (2013) Kaweeteerawat et al. (2015)

(Continued)

TABLE 3.4 (CONTINUED)
Effect of Nanoparticles on Soil Microorganisms

Nanoparticles Type	Target Microorganisms	Mode of action	References
Cu^{2+}/CuO	B. laterosporus, C. indoltheticum, P. ananatis, and Vibrio fischeri	ROS generation and oxide reduction reactions	Concha-Guerrero et al. (2014); Heinlaan et al. (2008)
TiO_2	E. coli	Photocatalytic production of reactive oxygen species (ROS) leading to oxidative stress, cell membrane disruption, and DNA strand breaks	Barnes et al. (2013)
TiO_2	B. subtilis		Tsuang et al. (2008)
TiO_2	S. aureus		Heinlaan et al. (2008)
TiO_2	Pseudomonas aeruginosa		Adams et al. (2006)
TiO_2	V. fischeri		Kim et al. (2013)
TiO_2	Bacteroides fragilis, B. subtilis, Enterococcus hirae and S. typhimurium		Gogniat and Dukan (2007)
TiO_2	E. coli	Lipid modification and disruption of lipopolysaccharide wall layer	Sunada et al. (1998); Maness et al. (1999)
TiO_2	Rhodopseudomonas acidophila	Structural deformation of proteins	Chen et al. (2005)
Zn^{2+}	Various bacteria	Shape-dependent cellular uptake and toxicity	Sirelkhatim et al. (2015)
Zn^{2+}	E. coli	ROS generation and oxidative toxicity	Adams et al. (2006)
Zn^{2+}	V. fischeri		Heinlaan et al. (2008)
Zn^{2+}	B. subtilis		Arakha et al. (2015)
Zn^{2+}	Campylobacter jejuni		Xie et al. (2011)
Zn^{2+}	S. aureus		Premanathan et al. (2011)
Zn^{2+}	Streptococcus agalactiae	Cell membrane disorganization and increased membrane permeability	Huang et al. (2008)
Zn^{2+}	S. aureus		Reddy et al. (2007)
Zn^{2+}	E. coli		Brayner et al. (2006)
Zn^{2+}	C. jejuni		Xie et al. (2011)
Zn^{2+}	E. coli	Membrane lipid modification	Jiang et al. (2010)
Zn^{2+}	Klebsiella pneumoniae		Reddy et al. (2014)
Zn^{2+}	Bacillus licheniformis	Protein interference and inhibition of protein function	Sinha and Khare (2014)
Zn^{2+}	V. cholera		Chatterjee et al. (2010)
Zn^{2+}	S. aureus	Apoptosis	Premanathan et al. (2011)

(Continued)

TABLE 3.4 (CONTINUED)

Effect of Nanoparticles on Soil Microorganisms

Nanoparticles Type	Target Microorganisms	Mode of action	References
Carbon-based nanomaterials (C60/fullerene/carbon nanotubes)	*E. coli*	Physical characteristics such as size and composition	Kang et al. (2008)
Carbon-based nanomaterials (C60/fullerene/carbon nanotubes)	*Pseudomonas fluorescens*		Riding et al. (2012)
Carbon-based nanomaterials (C60/fullerene/carbon nanotubes)	*E. coli*	Mechanical stress on the cell wall or membrane and cytotoxicity	Fortner et al. (2005)
Carbon-based nanomaterials (C60/fullerene/carbon nanotubes)	*Shewanella oneidensis*		Tang et al. (2007) De Windt et al. (2006)
Carbon-based nanomaterials (C60/fullerene/carbon nanotubes)	*Pseudomonas putida*	Oxidative stress by ROS production, membrane lipid modification, and DNA damage	Fang et al. (2007)
Carbon-based nanomaterials (C60/fullerene/carbon nanotubes)	*B. subtilis* and *P. aeruginosa*		Olivi et al. (2013)
Carbon-based nanomaterials (C60/fullerene/carbon nanotubes)	*S. typhimurium*		Sera et al. (1996)
Silica/SiO$_2$	*E. coli* and *B. subtilis*	Oxidative stress and associated membrane disintegration	Adams et al. (2006)
Silica/SiO$_2$	*E. coli*	Interaction with bacterial lipopolysaccharides and destabilization of outer membrane	Capeletti et al. (2014)

within food materials), and packaging (nanotechnology used to wrap and protect food materials). Nanoscale food additives have the potential to extend a product's shelf life and improve its texture, flavor, and nutrient composition. Additives can also be used to detect food pathogens and provide functions like food quality indicators (Alfadul and Elneshwy, 2010). In terms of food packaging, nanotechnology generally increases a food's shelf life and quality. Among nanomaterials, nanoclays are used to maintain food quality and the formation of pathogens and toxins in packaged foods are prevented using antimicrobial agents such as silver, titanium oxide, zinc oxide, and other bio-NPs. The progression of nanotechnology in food engineering is currently following two major pathways, i.e., food nanosensing and food nanostructured ingredients (Chaudhry et al. 2008; Nickols-Richardson and Piehowski, 2008). Recently, nanotechnology and its related industries have been developed into a broad range of nutritional research linked to dairy and food processing, preservation, and the packaging and development of functional foods. Furthermore, food and dairy manufacturers, agricultural producers, and consumers are those who are mainly affected by the progression of nanotechnology (Charych et al. 1996; Duncan, 2011). In the food packaging industry, nanotechnology plays an important role not only in packaging materials but also in improving food safety, alerting consumers to concerns over the presence of contamination and bacteria within foods, producing stronger flavors and color quality, repairing tears in packaging, and releasing preservatives to extend package shelf life. To benefit from the advantages of nanotechnology, attention is given to the relationship between food material morphology and bulk physicochemical properties –highlighting the importance of biopolymers in solutions, gels, and films. Furthermore, functional nanostructures can be incorporated into individual biological molecules to develop biosensors that target natural sugars or proteins (Golding and Sein, 2004; Flanagan et al. 2006; Weiss et al. 2006; Augustin et al. 2009; Bouwmeester et al. 2009). Briefly, the applications of nanotechnology in the food industry include: security of manufacturing, processing, and shipping of food products through the use of nanosensors for pathogens and contaminants; devices that maintain historical environmental records for a particular product; tracking systems for individual shipments that integrate sensing, localization, reporting, and remotely controlling food products (smart/intelligent systems) that can increase the efficacy and security of food processing and transportation; and encapsulation and delivery systems that carry, protect, and deliver functional food ingredients to their specific destinations (Rhim, 2004; Gupta and Gupta, 2005). Currently available nano-based materials, such as nanoscale dispersions and nanocapsules that represent the main constituents of drugs, vitamins, antimicrobials, antioxidants, flavorings, colorants, and preservatives, have resulted in increased benefits from drugs and foods to humankind. Other important examples of nanotechnology employed in the food and drug industries are association colloids and surfactant micelles, vesicles, bilayers, reverse micelles, and liquid crystals that encapsulate and deliver polar, nonpolar, and/or amphiphilic functional ingredients (Sundarraj et al. 2014). Following the penetration of nanotechnology into the food industry and related technologies, preparation and controlling of the employed nanomaterials with specific characteristics and properties are among the significant sections using easy processing operations like dipping and washing. In fact, the composition, thickness, and structure of nanomaterials can control by changing the type of the order that the object is introduced into various dipping solutions, the adsorbing substances in the dipping solutions, the total number of dipping steps used, and finally the environmental and solution conditions used, e.g. pH, temperature, dielectric constant, and ionic strength. Nanotechnology is used for applications in food via both bottom-up and top-down methods. The top-down approach is based on physically processing food materials; the bottom-up approach is concerned with building and growing larger structures from atoms and molecules, such as the organization of casein micelles or starch and protein folding (Kralchevsky et al. 2005; Khot et al. 2012; Komaiko et al. 2016). Similar to other branches of science and technology, nanotechnology, as a modern approach to science, has found a very specific position in nutritional technology and its related industries. The use of nanofiltration in the food industry, to detect and control metabolites and pathogenic agents, has played a fundamental role in food storage methods. Improving the stability and lifetime of nutrients, especially in connection to reactive compounds, such as odors and colors that interact with other nutritional content, is one further possible use of nanotechnology. In another sector, this modern technology can effectively be employed to conserve nutrients against oxidation, deformation, and degradation reactions (Mahmoud and Berekaa, 2015; Lee et al. 2015).

3.3.1 FOOD PROCESSING

Nanosensors help farmers in maintaining farms with precise control, and report timely needs of plants (Mousavi and Rezaei, 2011). Nanosensors are devices consisting of an electronic data processing part and a sensing layer or part, which can translate a signal such as light or the presence of an organic substance or gas into an electronic signal structured at the nanometer scale. Thus, the development of nanosensors is mandatory to aid decision-making in crop monitoring, accurate analysis of nutrients and pesticides in soil, or for maximizing the efficiency of water use for smart agriculture. Indeed, nanosensors could demonstrate their potential in managing all the phases of the food supply chain, from crop cultivation and harvesting to food processing, transportation, packaging, and distribution (Scognamiglio et al. 2014). Dynamic measurement of soil properties (pH and nutrients, residual pesticides in crop and soil, and soil humidity) for the detection of pathogens and prediction of nitrogen uptake by nanosensor applications are only a few examples to enhance sustainable farming and agricultural operations (Bellingham, 2011). In particular, nanosensor systems can be developed to monitor the presence of insect pests, pathogens, or residual pesticides in order to better tune the amount of chemical pesticides to be employed for crop productivity management and the duration of their stay in the soil and environment, since they show higher sensitivity and specificity compared to the other traditional sensors, including controlled release mechanisms via nanoscale carriers monitored by nanosensors integrated in platforms employing wireless signals. This will avoid overdose of agricultural chemicals and pesticides during the course of cultivation and minimize inputs of fertilizers, reducing waste and improving productivity. Networks of nanosensors located throughout cultivated fields will assure a real time and comprehensive monitoring of the crop growth, furnishing effective high-quality data for best management practices (El Beyrouthya and El Azzi, 2014).

In view of the increasing scarcity of water resources in relation to the demand for different uses, the use of sensor technology was a prerequisite for intelligent agriculture to maximize the efficiency of water use. Nanosensor systems estimating soil water tension in real time may be coupled with autonomous irrigation controllers. The feature of irrigation control is difficult and highly complex for farmers to apply because they need to be fully familiar with all production inputs (e.g. climate, soil structure, and aquatic plant needs) (de Medeiros et al. 2001). Furthermore, nanosensors also have applications in fast, sensitive, and cost-effective detection of different targets to ensure food quality, safety, freshness, authenticity, and traceability along the entire food supply chain. Nanosensors represent one of the emerging technologies challenging the assessment of food quality and safety, being able to provide smart monitoring of food components (e.g. sugars, amino acid, alcohol, vitamins, and minerals) and contaminants (e.g. pesticides, heavy metals, toxins, and food additives). Food quality and food safety control represent a crucial effort not only to obtain healthy food, but also to avoid huge waste of food products. The potential of a nanosensor can also be demonstrated by the latest trends in intelligent or smart packaging to monitor the freshness properties of food, and check the integrity of the packages during transport, storage, and display in markets (Vanderroost et al. 2014). Much intelligent packaging involves nanosensors as monitoring systems to measure physical parameters (humidity, pH, temperature, light exposure), to reveal gas mixtures (e.g. oxygen and carbon dioxide), to detect pathogens and toxins, or to control freshness (e.g. ethanol, lactic acid, acetic acid) and decomposition (e.g. putrescine, cadaverine).

Nanosensors are used in the detection of adulterants, pathogens, toxins, toxic compounds, and harmful artificial colorings and flavors or ingredients in the food products. Nanosensors may be used for pathogen detection in food and food products and reduce the detection time. Such nanosensors could be incorporated in packaging material and would serve as an 'electronic tongue' or 'nose' by detecting chemicals released during food spoilage (Garcia et al. 2006). Microfluidic devices are also known to detect pathogens in real time and with high sensitivity. Microfluidic sensors are used to detect compounds of interest rapidly in only microliters of required sample volumes. Microfluidic sensors have widespread applications in medical, biological, and chemical analysis.

3.3.2 FOOD PACKAGING

Nano-enabled food contact materials (FCMs) and food packaging make up the major share of the current and short-term predicted market for applications in the food segment (Chaudhry et al. 2008). Most applications of nanotechnology in food and agriculture using these developments include lightweight but strong packaging materials and prolonged shelf life of packaged foodstuffs, and the likely low risk to the consumer attributable to the fixed or embedded nature of engineered nanomaterials in plastic polymers. Many nanotechnology derived FCMs are currently available worldwide. Nano-packaging applications as FCMs are anticipated to grow from a USD 66 million business in 2003 to over USD 360 million by 2008 (Scrinis and Lyons, 2007). The main areas of nanotechnology application fall into the following broad categories:

- FCMs incorporating nanomaterial to improve packaging properties such as flexibility, gas barrier properties, temperature/moisture stability, light and flame resistance, transparency, and mechanical stability.
- 'Active' FCMs that incorporate NPs with antimicrobial or oxygen scavenging properties.
- 'Intelligent' or 'smart' food packaging incorporating nanosensors for sensing and signaling of microbial and biochemical changes, release of antimicrobials, antioxidants, enzymes, flavors, and nutraceuticals to extend shelf life.
- Biodegradable polymer-nanomaterial composites by the introduction of inorganic particles, such as clay into the biopolymeric matrix and can also be controlled with surfactants that are used for the modification of layered silicate (Alfadul and Elneshwy, 2010; Sozer and Kokini, 2009; Chaudhry et al. 2008; Miller and Senjen, 2008; Brody, 2007; Doyle, 2006; Joseph and Morrison, 2006; Lopez-Rubio et al. 2006).
- Edible nanocoated materials that improve the storage life and quality of food products.

NP-reinforced materials, termed as nanocomposites, are polymers reinforced with small quantities (up to 5% by weight) of nanosized particles. These nanocomposites have high aspect ratios and are able to improve the properties and performance of the polymer. Polymer composites with nanoclay are among the first nanocomposites available on the market as improved materials for food packaging. Nanoclay with natural nanoscaled layer structure restricts the permeation of gases. Nanoclay-polymer composites have been made from a thermoplastic polymer reinforced with NPs of clay which include polyamides (PA), nylons, polyolefins, polystyrene (PS), ethylene-vinyl acetate (EVA) copolymer, epoxy resins, polyurethane, polyimides, and polyethylene terephthalate (PET). Commercially, a number of nanoclay-polymer composites are available. Known applications of nanoclay in multilayer film packaging include bottles for beer, edible oils, carbonated drinks, and films (Chaudhry et al. 2008; Brody, 2007). Examples of available nanoclay composites include Imperm® (from Nanocor® Inc.), Aegis® X (Honeywell), Durethan® KU2-2601 (Bayer AG), plastic beer bottles, Miller Brewing Co. (USA), and Hite Brewery Co. NP composites include DuPont Light Stabilizer 210, Durethan®, and Bayer's shaving (Scrinis and Lyons, 2007; Chaudhry et al. 2008; Sozer and Kokini, 2009; Miller and Senjen, 2008). The use of nanocomposites in food contact material has been approved by the United States Food and Drug Administration (USFDA) (Sozer and Kokini, 2009).

The commercially important nanomaterials are nanosilver and nanozinc oxide for antimicrobial action, nano-titanium dioxide for UV protection in transparent plastics, nano-titanium nitride for mechanical strength and as a processing aid, and nanosilica for surface coating in food packaging (Doyle, 2006; Miller and Senjen, 2014; Momin and Joshi, 2008; Chaudhry et al. 2008). Kodak company is developing antimicrobial packaging for food products as 'active packaging' which would absorb oxygen (Asadi and Mousavi, 2006). Other companies include Fresher Longer™ Miracle Food Storage Containers and FresherLonger™ Plastic Storage Bags from Sharper Image® (USA), Nano Silver Food Containers from A-DO Korea, and Nano Silver Baby Milk Bottle from Baby Dream® Co. Ltd. (South Korea). Oxygen scavenging packaging using enzymes between polyethylene films have also been developed (Lopez-Rubio et al. 2006). An active packaging application could also be designed to stop microbial growth once the package is opened by the consumer and rewrapped with an active-film portion of the package (Brody, 2007). ZnO quantum dots were utilized as a powder, bound in a polystyrene film (ZnO-PS), or suspended in a polyvinylpyrrolidone gel (ZnO-PVP) as antimicrobial packaging against *Listeria monocytogenes*, *Salmonella enteritidis*, and *E. coli* O157: H7 (Sun et al. 2009).

A multi-detection test – FoodExpertID – has been developed by bioMerieux for nano-surveillance response to food scares. Nanoscale radio frequency identification tags (RFID) have been developed to track containers or individual food items (Baruah and Dutta, 2009) and are being used in retailing chains (Joseph and Morrison, 2006; Asadi and Mousavi, 2006). The nanotech company pSiNutria is also developing nano-based tracking technologies, including an ingestible biosilicon which could be placed in foods for monitoring purposes and pathogen detection but could also be eaten by consumers (Scrinis and Lyons, 2007; Miller and Senjen, 2008).

Food packaging is used to maintain the safety and the quality of food products from production to consumption and to extend their shelf life by maintaining an adequate environment, avoiding spoilage by microorganisms, and keeping chemical contaminants, oxygen, moisture, light, etc., separated from food. In order to achieve that, packaging materials contribute both to creating the physical protection and the physicochemical conditions appropriate for obtaining an adequate shelf life and maintaining food quality and safety. The packaging should prevent microbial contamination, keep the adequate moisture content depending on the type of food, and act as a barrier against permeation of gases like CO_2, O_2, H_2O, and other volatile compounds (flavors, taints, etc.). At the same time, the basic properties of packaging materials such as thermal, mechanical, and optical properties should be maintained (Singh and Singh, 2005; Marsh and Bugusu, 2007). Food packaging differs from other durable goods such as furniture, electronics, or home appliances because of its short shelf life and the necessary safety aspects. Different materials are used for food packaging applications like plastic, paper, and paperboard, metal, glass, or a combination of those materials, with different physical and chemical characteristics depending on the type of food. Lately, there has been a lot of effort in the development of new packaging materials that are able to maintain food quality, increase the shelf life, and at the same time have good mechanical properties and are easy to process. One way to improve the packaging is by the addition of NPs to improve characteristics such as the barrier properties to different gases, antimicrobial properties, biodegradability, and to incorporate sensors that can inform of the quality of the food, etc. Currently, the largest category of nanotechnology applications for the food sector is in food packaging materials: the global food and beverage packaging market with nanomaterials was USD 4.13 billion in 2008 and has been predicted to grow to USD 7.3 billion by 2014 (IR&P Inc. 2009), which represents an annual growth rate of 11.65%. A critical issue for a food packaging material is permeability; no material is completely impermeable to water vapor, atmospheric gases, or different natural substances that can be contained within the food or in the packaging material itself (Robertson, 2012). Different foods need different permeabilities through the packaging material; thus, fresh fruits and vegetables need a continual supply of oxygen for sustained cellular respiration and to avoid the spread of anaerobic bacteria. Plastic containers for carbonated beverages, on the other hand, must have high carbon dioxide and oxygen barriers to avoid decarbonation and oxidation of the beverage. For other products the issue is water vapor permeability. As we have seen, food products require sophisticated and different packaging functions. In addition, as the distance between producers and consumers is getting longer with the global market, the demands on the packaging industry are likely to rise. Among the different materials used for food packaging, organic polymeric materials [polypropylene (PP), polyethylene (PE), polyethylene terephthalate (PET), polystyrene (PS), and polyvinyl chloride (PVC)] due to their low cost, ease of processing, light weight, and formability are attractive alternatives for the food packaging industry. The major problem is their inherent permeability to gases and other small molecules. It depends on their polarity and the position of the polymeric side chains, hydrogen bonding, molecular weight, cross-linking, crystallinity, and synthesis and processing methodology. No pure polymer possesses all the barrier and mechanical properties needed for every food packaging application, so usually polymer blends or complex multilayer films are used. The option of using polymer nanocomposites as materials for food packaging is the last one proposed for solving the mentioned problems. They are created by dispersing inert nanoscale fillers throughout the polymeric matrix. Many different nanomaterials are used as fillers like silica NPs (Ke et al. 2010; Bracho et al. 2012), clay nanoplatelets (Ku et al. 2004; Schuetz et al. 2011), organoclay (Ham et al. 2013; Gokkurt et al. 2013), graphene (Lee et al. 2013; Yousefi et al. 2013), polysaccharide nanocrystals (Lin et al. 2012), carbon nanotubes (Swain et al. 2013; Prashantha et al. 2009), cellulose-based nanomaterials (Floros et al. 2012; Sandquist, 2013), chitosan NPs(Chang et al. 2010), and other metal NPs like ZnO_2 (Esthappan et al. 2013), colloidal Cu (Cárdenas et al. 2009), or Ti (Li et al. 2011). The inclusion of nanofillers in the polymer matrix affects the barrier properties in two ways: by creating a tortuous path for gas diffusion and by causing changes to the polymer matrix itself at the interfacial regions (Choudalakis and Gotsis, 2009). In the first way, due to the impermeability of the inorganic nanofillers, gas molecules cannot follow a straight-line path (perpendicular to the film surface) but must diffuse around the NPs. The result is a longer mean path for gas diffusion through the film in the presence of the nanofillers. It allows the manufacturer to accomplish larger effective film thickness with lower amounts of polymer. In the second way, if the interactions between NPs and polymer are favorable, the polymer strands close to each NP would be partially immobilized. As a result, the gas molecules have attenuated hopping rates between free volume holes or altered density and/or size of holes in the interfacial zones. This has been directly observed using positron annihilation lifetime spectroscopy (PALS) (Wang et al. 2007). Also, the presence of surfactants and other additives used to incorporate the nanofiller efficiently into the matrix can modify the solubility or diffusivity of the gases. These effects in the interfacial regions are particularly important in polyolefins (Picard et al. 2007), which are polymer matrices that possess very high native gas permeability. These mechanisms are the reasons why nanomaterials have been more successful than micromaterials as fillers for polymer composites. They have much higher aspect ratios, and the interfacial volume element is significantly greater in a polymer nanocomposite than in a polymer microcomposite from the same materials. Active packaging relates to the incorporation of additives to the packaging systems with the purpose of maintaining or extending the shelf life and product quality. The active additives can be incorporated directly into the packaging matrix, attached to the interior of the packaging material, or introduced inside the package in separate containers such as sachets (Restuccia et al. 2010). It provides dynamic, rather than the conventional passive, protection to the food inside and has an active role in food preservation, different from just providing an inert barrier from external conditions (Lim, 2011). The additives release or absorb substances into or from the food and the surrounding environment (Brody, 2001), thus promoting food preservation. Most active packaging materials used include substances that absorb ethylene, oxygen, carbon dioxide, moisture, flavors, or odors and other materials that release antioxidants, carbon dioxide, antimicrobial agents, or flavors (Vermeiren et al. 1999) (Figure 3.2).

Oxygen is directly or indirectly responsible for the degradation of many foods. Direct oxidation reactions are responsible, for example, for the rancidity of vegetable oils and browning of fruits. Food deterioration can also be produced by indirect action of O_2 due to food spoilage by aerobic microorganisms. Very low O_2 levels inside the packaging can be maintained by incorporating O_2 scavengers, which are useful for several applications. Xiao-e et al. (2004) described the photocatalytic activity of nanocrystalline titania: UV illumination of nanocrystalline TiO_2/polymer films in

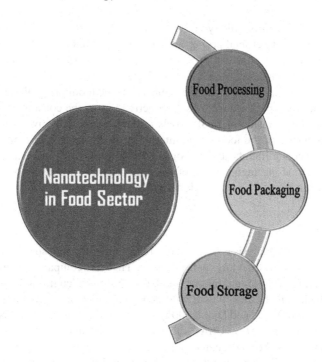

FIGURE 3.2   Applications of nanotechnology in the food sector.

the presence of excess organic hole scavengers resulted in the oxygen removal in a closed environment. According to the authors, it can be used for packaging many oxygen-sensitive products, but a major flaw is the UVA light requirement. In another approach, different thermoplastic polymers (PET, PP, FEP, LLDPE, and nylon) have been infused with metal and metal oxide NPs, and it has been discovered that films containing less than 1% wt Pd and Pt NPs were active as oxygen scavengers and reduced the oxygen flux by two orders of magnitude (Yu et al. 2004).

Different kinds of NPs have been used to create antimicrobial food packaging. Among them, nanosilver is the most widespread (Rai et al. 2009), even with commercialized items already in the market. The antimicrobial properties of silver have been known since ancient times where wine and water were stored in silver vessels. Silver has many advantages over other antimicrobial agents. It has a broad spectrum and is toxic to numerous strains of fungi, bacteria, algae, and some viruses, with varying degrees of toxicity. In its elemental form, it is shelf-stable for long periods of time. Silver can also be easily incorporated into many different materials such as plastics and textiles, making it especially useful for food packaging applications. Despite the known antimicrobial properties of bulk silver, the mechanism of its activity is still unknown. There is also some controversy about the manner in which silver nanoparticles (Ag NPs) are toxic to bacterial cells: some postulate that the activity comes from the Ag^+ ions detached from the surfaces of Ag NPs which will act by the same mechanisms as conventional silver antimicrobials (Lok et al. 2007). However, some research shows that Ag NPs are more toxic than the equivalent amount of dissociated Ag^+ions (Limbach et al. 2007). Morones et al. (2005) showed that the amount of Ag^+ions released from Ag NPs in their experiment was too low to account completely for their toxicity. Inorganic NPs can be easily incorporated into the polymers to create antimicrobial nanocomposites (Althues et al. 2007), and that is one of the biggest advantages over molecular antimicrobials. Ag NPs can be engineered to remain potent antimicrobial agents for long periods of time (Roe et al. 2008) due to their controlled release properties. And this makes nanosilver/polymer composites very attractive materials for food packaging. The antimicrobial activity of the silver nanocomposites is dependent on different factors that affect the Ag+ release rate, such as the degree of polymer crystallinity, filler type (silver zeolites or Ag NPs), hydrophilicity of the matrix, and particle size. Other nanomaterials have also been found to have antimicrobial properties, like TiO_2 (Hamal et al. 2010), MgO (Huang et al. 2005), Cu and CuO (Cárdenas et al. 2009), ZnO (Bajpai et al. 2010), Cd (Xie et al. 2011), chitosan (Lu et al. 2010), and carbon nanotubes (Kang et al. 2009).

Intelligent packaging refers to those packages that allow the monitoring of the conditions and the quality of the content of the package from the production line to the consumer. Intelligent packaging can be related to the inclusion of smart labels in the package that give information about the physicochemical properties of the food or the interior of the package, like temperature, pH or different gases, chemical contaminants, and pathogen concentrations. Although a very promising area of research related to food packaging, most of the research on nanosensors or assays for analytes related to food is still in the early stages of development.

3.3.3 FOOD STORAGE

Ensuring food safety and its analysis is the need of time for any food industry. Thus, nanosensors are used for spoilage detection in food and their products. The U.S. company Oxonica Inc. has developed nano-barcodes (20–500 nm in diameter and 0.04–15 mm in length) to be used for individual items or pallets, which must be read with a modified microscope for anti-counterfeiting purposes (Miller and Senjen, 2008; Warad and Dutta, 2005). Engineered nanosensors are being developed by Kraft along with Rutgers University (USA) within packages to change color to warn the consumer if a food is beginning to spoil or has been contaminated by pathogens using electronic 'noses' and 'tongues' to 'taste' or 'smell' scents and flavors (Scrinis and Lyons, 2007; Sozer and Kokini, 2009; Joseph and Morrison, 2006; Asadi and Mousavi, 2006). Nestlé, British Airways, and Monoprix supermarkets are using chemical nanosensors that can detect color change (Pehanich, 2006).

A nanolaminate consists of two or more layers of material with nanometer dimensions that are physically or chemically bonded to each other (1–100 nm per layer, usually 5 nm). These could be used to encapsulate various hydrophilic, amphiphilic, or lipophilic substances, active functional agents such as antimicrobials, antibrowning agents, antioxidants, enzymes, flavors, and colors with enhanced moisture and gas barrier properties. These are developed from food-grade ingredients (proteins, polysaccharides, lipids). These functional agents would improve the storage life and quality of coated foods such as meats, cheese, fruit and vegetables, confectionery, bakery goods, and fast food (Weiss et al. 2006). The U.S. company Sono-Tek Corp. announced in early 2007 that it had developed an edible antibacterial nanocoating which can be applied directly to bakery goods and is currently under testing with its clients (Miller and Senjen, 2008).

Edible films are defined as a thin layer of material that can be ingested and act as a barrier to oxygen, solute movement, and moisture for the food. This material can be employed as a continuous layer between food components or can coat the whole food (Guilbert, 1986). Edible films and coatings have many advantages over synthetic films, and for this reason there has been much research in this area in recent years. Their main advantage over traditional synthetics is that they can be ingested along with the packaged products. If the films are eaten, the packaging disposal problem is avoided, and even if the films are not eaten, as they are produced exclusively from renewable, edible ingredients, can still contribute to the reduction of environmental pollution because they are supposed to degrade faster than polymeric materials (Bourtoom, 2008). The purposes of using coatings and films, depending on the food, are to hinder the migration of different components of the food or the ambient like moisture, oil, carbon dioxide, oxygen, and aromas, make the food more attractive, enhance its mechanical properties and integrity, and carry food additives or nutritional supplements (Kester and Fennema, 1986; Krochta and Mulder-Johnston, 1997). An edible film coating, acting as a barrier to oxygen, moisture, or aromas, can also help cut packaging requirements and, therefore, waste. For example, a multilayer plastic package can be reduced to a single-component recyclable package if the barrier characteristics of an edible film permit it to do so (Miller and Krochta, 1997). The sensory characteristics of an edible film, which are of importance to its functionality, such as transparency, gloss, color sticking, and roughness, can be selected depending on the purposes (Debeaufort et al. 1998). Edible films and coatings can also contain food additives to improve the general coating performance (i.e., strength, flexibility, or adherence); enhance the food texture, flavor, and color; and control microbial growth (Cuppet, 1994). In this area of food packaging, nanotechnology can be applied in two ways: on the one hand, the edible film can be a nanolaminate, manufactured as various nanolayers of diverse materials, added one at a time using the electrostatic attractions between the different layers. On the other hand, the edible film can be a nanocomposite, incorporating NPs or emulsified nanodroplets containing active ingredients to enhance texture, appearance, or taste. A nanolaminate is a very thin food grade film composed of two or more layers with nanometer dimensions (1–100 nm per layer), where the individual laminates are physically or chemically bonded with each other. A very good control of the properties of the laminate can be obtained by a layer-by-layer preparation of the coating (Weiss et al. 2006). This is the main advantage over conventional edible coatings used as barrier layers to increase shelf life (gums and waxes). The process of coating foods with nanolaminates consists of spraying or dipping layer by layer a series of solutions containing target compounds to the surface of the food (McClements et al. 2005). The electrodeposition technique can be used to coat very hydrophilic food systems such as precut vegetables and fruits adding antimicrobial agents or more vitamins (Vargas et al. 2008). Nowadays, edible nanolaminates are fabricated from lipids, polysaccharides, and proteins. They are poor at protecting against moisture, although polysaccharide and protein-based films are good barriers against carbon dioxide and oxygen. On the other hand, lipid-based nanolaminates, although good at protecting food from moisture, are a poor barrier to other gases and have limited mechanical strength. Research is being conducted in identifying additives as polyols to improve them because neither lipids nor polysaccharides or proteins deliver all of the desired properties in an edible coating. At present, coating foods that include nanolaminates involves either dipping the food into different solutions containing target substances or spraying the substance onto the food surface. Frequently, the formation of the nanolaminate is a result of the electrostatic attraction between compounds with opposite charges, and there are also various methods which could cause adsorption. As a result, different nanolaminates might include various functional agents as antioxidants, antimicrobials, enzymes, anti-browning, colors, and flavors. The second approach is the use of edible films with nano-sized fillers. In this case, clay nanoparticles are the most important group of nanoparticles used to enhance the properties of edible films. The physical and mechanical properties of the pure polymer or conventional composites can be improved on a large scale when using a nanometer-sized dispersion of clay to create a polymer-clay nanocomposite (Rhim and Ng, 2007). Nanoclay particles have been combined with both proteins (Shotornvit et al. 2009) and polysaccharides (Casariego et al. 2009; Tang et al. 2008b) to create enhanced edible films. Films have also been created by adding other NPs, such as tripolyphosphate-chitosan (De Moura et al. 2009), microcrystalline cellulose (Bilbao-Sáinz et al. 2010), or silicon dioxide (Tang et al. 2009), to biopolymers. Using these NPs, the moisture barrier properties have been improved and the microbial growth restricted. Rhim et al. (2006) studied different NPs to improve the physical properties of chitosan-based films that also showed antimicrobial activity to a certain extent. The optical properties are affected to a greater or lesser extent depending on the type of nanoclay used; this has been observed in isolated whey protein-based films (Shotornvit et al. 2009) and gelatin-clay nanocomposites (Farahnaky et al. 2014) where light transmittance of the gelatin decreased with the inclusion of the nanoclay. The incorporation of other NPs like porous silica-coated titania dramatically changed the appearance of whey protein-isolated films from a transparent appearance to opaque (Kadam et al. 2013). Recently, the addition of bacterial cellulose nanocrystals to create a gelatin nanocomposite film (George and Siddaramaiah, 2012) reduced the moisture affinity of gelatin, a very interesting property for edible packaging applications. The dynamic mechanical properties and degradation temperature of gelatin were also improved. Other nanomaterials like carbon nanotubes have also been used as nanofillers in gelatin films.

Nanoencapsulation is the incorporation of ingredients in small vesicles or walled material with nano (or submicron) sizes (Surassmo et al. 2009). Nanoencapsulation in the form of micelles, liposomes, or biopolymer-based carrier systems has been used to develop delivery systems for additives and supplements for use in food and beverage products. Nanoencapsulation is the extension of microencapsulation, which has been used by the food industry for food ingredients and additives for many years. These nanomaterials offer several advantages such as preserving the ingredients and additives during processing and storage, masking unpleasant tastes and flavors, controlling the release of additives, better dispersion of water-insoluble food ingredients and additives, and improved uptake of the encapsulated nutrients and supplements (Chen et al. 2006a, b; Weiss et al. 2006; Momin et al. 2013). The protection of bioactive compounds, such as vitamins, antioxidants, proteins, and lipids as well as carbohydrates, may be achieved using this technique for the production of functional foods with enhanced functionality and stability. The improved uptake and bioavailability alone have opened up a vast area of application in food products that incorporate nanosized vitamins, nutraceuticals, antimicrobials, antioxidants, functional ingredients, etc. After food packaging, nanoencapsulation is currently the largest area of nanotechnology application in the food industry, and a number of products based on nanocarrier technology are already available on the market. Nanoencapsulation can reduce the amount of active ingredients needed in formulation and therefore the cost (Huang et al. 2009).

REFERENCES

Abreu FO, Oliveira EF, Paula HC, de Paula RC (2012) Chitosan/cashew gum nanogels for essential oil encapsulation. *Carbohydrate Polymers*, 89(4), 1277–1282.

Adak T, Kumar J, Shakil NA, Walia S (2012) Development of controlled release formulations of Imidacloprid employing novel nano-ranged amphiphilic polymers. *Journal of Environmental Science and Health, Part B*, 47(3), 217–225.

Adams LK, Lyon DY, Alvarez PJ (2006) Comparative eco-toxicity of nanoscale TiO_2, SiO_2, and ZnO water suspensions. *Water Research*, 40(19), 3527–3532.

Adams J, Wright M, Wagner H, Valiente J, Britt D, Anderson A (2017) Cu from dissolution of CuO nanoparticles signals changes in root morphology. *Plant Physiology and Biochemistry*, 110, 108–117.

Adel MM, Atwa WA, Hassan ML, Salem NY, Farghaly DS, Ibrahim SS (2014) Biological activity and field persistence of Pelargonium graveolens (Geraniales: Geraniaceae) loaded solid lipid nanoparticles (SLNs) on *Phthorimaea operculella* (Zeller)(PTM)(Lepidoptera: Gelechiidae). *International Journal of Scientific and Engineering and Research*, 4(11), 514–520.

Afrasiabi ZA (2010) New nanoscale biosensor for detection of luteinizing hormone in smallruminants to determine optimum breeding time. A USDS/NIFA Evans-Allen project. http://cris.csrees.usda.gov/cgi-bin/starfinder/17118/crisassist.txt As of 12-April-2011.

Agrawal S, Rathore P (2014) Nanotechnology pros and cons to agriculture: A review. *International Journal of Current Microbiology and Applied Sciences*, 3(3), 43–55.

Aken VB (2015) Gene expression changes in plants and microorganisms exposed to nanomaterials. *Current Opinion in Biotechnology*, 33, 206–219.

Akhavan O, Ghaderi E (2010) Toxicity of graphene and graphene oxide nanowalls against bacteria. *ACS Nano*, 4(10), 5731–5736.

Akhter S, Ahmad M, Singh A, Ahmad I, Rahman M, Anwar M, Jain G, Ahmad F, Khar R (2011) Cancer targeted metallic nanoparticle: Targeting overview, recent advancement and toxicity concern. *Current Pharmaceutical Design*, 17(18), 1834–1850.

Alfadul SM, Elneshwy AA (2010) Use of nanotechnology in food processing, packaging and safety–review. *African Journal of Food, Agriculture, Nutrition and Development*, 10(6), 2719–2739.

Alghuthaymi MA, Almoammar H, Rai M, Said-Galiev E, Abd-Elsalam KA (2015) Myconanoparticles: Synthesis and their role in phytopathogens management. *Biotechnology and Biotechnological Equipment*, 29(2), 221–236.

Althues H, Henle J, Kaskel S (2007) Functional inorganic nanofillers for transparent polymers. *Chemical Society Reviews*, 36(9), 1454–1465.

Anjali CH, Sharma Y, Mukherjee A, Chandrasekaran N (2012) Neem oil (*Azadirachta indica*) nanoemulsion— A potent larvicidal agent against *Culex quinquefasciatus*. *Pest Management Science*, 68(2), 158–163.

Anusuya S, Banu KN (2016) Silver-chitosan nanoparticles induced biochemical variations of chickpea (*Cicer arietinum* L.). *Biocatalysis and Agricultural Biotechnology*, 8, 39–44.

Arora S, Sharma P, Kumar S, Nayan R, Khanna PK, Zaidi MGH (2012) Gold-nanoparticle induced enhancement in growth and seed yield of Brassica juncea. *Plant Growth Regulation*, 66(3), 303–310.

Arthurs SP, Lacey LA, Behle RW (2006) Evaluation of spray-dried lignin-based formulations and adjuvants as solar protectants for the granulovirus of the codling moth, *Cydia pomonella* (L). *Journal of Invertebrate Pathology*, 93(2), 88–95.

Asadi G, Mousavi M (2006) Application of nanotechnology in food packaging. http://iufost.edpsciences.org.

AshaRani PV, Low Kah Mun G, Hande MP, Valiyaveettil S (2008) Cytotoxicity and genotoxicity of silver nanoparticles in human cells. *ACS Nano*, 3(2), 279–290.

Askary M, Amirjani MR, Saberi T (2014) Evaluation of the effects of iron nanofertilizer on leaf growth, antioxidants and carbohydrate contents of *Catharanthus roseus*. *JPPF*, 3(7), 43–56.

Askary M, Talebi SM, Amini F, Bangan ADB (2016). Effects of stress on foliar trichomes plasticity in *Mentha piperita*. *Nusantara Bioscience*, 8(1), 32–38.

Asli S, Neumann PM (2009) Colloidal suspensions of clay or titanium dioxide nanoparticles can inhibit leaf growth and transpiration via physical effects on root water transport. *Plant, Cell and Environment*, 32(5), 577–584.

Atha DH, Wang H, Petersen EJ, Cleveland D, Holbrook RD, Jaruga P, Dizdaroglu M, Xing B, Nelson BC (2012) Copper oxide nanoparticle mediated DNA damage in terrestrial plant models. *Environmental Science and Technology*, 46(3), 1819–1827.

Augustin MA, Hemar Y (2009) Nano-and micro-structured assemblies for encapsulation of food ingredients. *Chemical Society Reviews*, 38(4), 902–912.

Bajpai SK, Chand N, Chaurasia V (2010) Investigation of water vapor permeability and antimicrobial property of zinc oxide nanoparticles-loaded chitosan-based edible film. *Journal of Applied Polymer Science*, 115(2), 674–683.

Balaji APB, Mishra P, Kumar RS, Mukherjee A, Chandrasekaran N (2015) Nanoformulation of poly (ethylene glycol) polymerized organic insect repellent by PIT emulsification method and its application for Japanese encephalitis vector control. *Colloids and Surfaces, Part B: Biointerfaces*, 128, 370–378.

Ball P (2002) Natural strategies for the molecular engineer. *Nanotechnology*, 13(5), 15–28.

Baron C, Domke N, Beinhofer M, Hapfelmeier S (2001) Elevated temperature differentially affects virulence, VirB protein accumulation, and T-pilus formation in different Agrobacterium tumefaciens and Agrobacterium vitis strains. *Journal of Bacteriology*, 183(23), 6852–6861.

Baruah S, Dutta J (2009) Nanotechnology applications in pollution sensing and degradation in agriculture: A review. *Environmental Chemistry Letters*, 7(3), 191–204.

Bauman DE, Perfield JW II, Harvatine KJ, Baumgard LH (2008) Regulation of fat synthesis by conjugated linoleic acid: Lactation and the ruminant model. *Journal of Nutrition*, 138(2), 403–409.

Bellingham BK (2011) Proximal soil sensing.*Vadose Zone Journal*, 10, 1342–1342.

Berekaa MM (2015) Nanotechnology in food industry; advances in food processing, packaging and food safety. *International Journal of Microbiology and Applied Sciences*, 4(5), 345–357.

Berestovsky G, Ternovsky V, Kataev A (2001) Through pore diameter in the cell wall of Chara corallina. *Journal of Experimental Botany*, 52(359), 1173–1177.

Bilbao-Sáinz C, Avena-Bustillos RJ, Wood DF, Williams TG, McHugh TH (2010) Composite edible films based on hydroxypropyl methylcellulose reinforced with microcrystalline cellulose nanoparticles. *Journal of Agriculture and Food Chemistry*, 58(6), 3753–3760.

Bolik M, Koop H (1991) Identification of embryogenic microspores of barley (Hordeum vulgare) by individual selection and culture and their potential for transformation by microinjection. *Protoplasma*, 162(1), 61–68.

Bourtoom T (2008) Edible films and coatings: Characteristics and properties. *International Food Research Journal*, 15, 237–248.

Bouwmeester H, Dekkers S, Noordam MY, Hagens WI, Bulder AS, De Heer C, Voorde SECJ, Wijnhoven SWP, Marvin HJP, Sips AJ (2009) Review of health safety aspects of nanotechnologies in food production. *Regulatory Toxicology and Pharmacology*, 53(1), 52–62.

Bracho D, Dougnac VN, Palza H, Quijada R (2012) Functionalization of silica nanoparticles for polypropylene nanocomposite applications. *Journal of Nanomaterials*. 2012, 1–8. doi:10.1155/2012/263915.

Brody AL (2001) What's active in active packaging. *Food Technology*, 55, 104–106.

Brody AL (2007) Case studies on nanotechnologies for food packaging. *Food Technology*, 07, 102–107.

Brunel F, El Gueddari NE, Moerschbacher BM (2013) Complexation of copper (II) with chitosan nanogels: Toward control of microbial growth. *Carbohydrate Polymers*, 92(2), 1348–1356.

Brunner T, Wick P, Manser P, Spohn P, Grass R, Limbach L, Bruinink A, Stark W (2006) In vitro cytotoxicity of oxide nanoparticles: Comparison to asbestos, silica, and effect of particle solubility. *Environmental Science and Technology*, 40(14), 4374–4381.

Burman U, Saini M, Kumar P (2013) Effect of zinc oxide nanoparticles on growth and antioxidant system of chickpea seedlings. *Toxicological and Environmental Chemistry*, 95(4), 605–612.

Buzea C, Pacheco II, Robbie K (2007) Nanomaterials and nanoparticles: Sources and toxicity. *Biointerphases*, 2(4), MR17–MR71.

Cabiscol Català E, Tamarit Sumalla J, Ros Salvador J (2000) Oxidative stress in bacteria and protein damage by reactive oxygen species. *International Microbiology*, 3(1), 3–8.

Campos EVR, De Oliveira JL, Da Silva CMG, Pascoli M, Pasquoto T, Lima R, Abhilash PC, Fraceto LF (2015) Polymeric and solid lipid nanoparticles for sustained release of carbendazim and tebuconazole in agricultural applications. *Scientific Reports*, 5, 13809.

Cañas JE, Long M, Nations S, Vadan R, Dai L, Luo M, Ambikapathi R, Lee EH, Olszyk D (2008) Effects of functionalized and nonfunctionalized single-walled carbon nanotubes on root elongation of select crop species. *Environmental Toxicology and Chemistry: Anais an International Journal*, 27(9), 1922–1931.

Cárdenas G, Meléndrez M, Cruzat C, Cancino AG (2009). Colloidal Cu Nanoparticles/ chitosan composite film obtained by microwave heating for food package applications. *Polym Bull*, 62(4), 511–524.

Casariego A, Souza BWS, Cerqueira MA, Teixeira JA, Cruz L, Díaz R, Vicente AA (2009) Chitosan/ clay films' properties as affected by biopolymer and clay micro/nanoparticles' concentrations. *Food Hydrocolloids*, 23, 1895–1902.

Castiglione MR, Giorgetti L, Bellani L, Muccifora S, Bottega S, Spanò C (2016) Root responses to different types of TiO2 nanoparticles and bulk counterpart in plant model system Vicia faba L. *Environmental and Experimental Botany*, 130, 11–21.

Chang PR, Jian RJ, Yu JG, Ma XF (2010) Starch-based composites reinforced with novel chitin nanoparticles. *Carbohydrate Polymers*, 80(2), 420–425.

Charych D, Cheng Q, Reichert A, Kuziemko G, Stroh M, Nagy JO, Spevak W, Stevens RC (1996) A 'litmus test'for molecular recognition using artificial membranes. *Chemistry and Biology*, 3(2), 113–120.

Chaudhry Q, Scotter M, Blackburn J, Ross B, Boxall A, Castle L, Aitken R, Watkins R (2008) Applications and implications of nanotechnologies for the food sector. *Food Additives and Contaminants Part A, Chemistry, Analysis, Control, Exposure and Risk Assessment*, 25(3), 241–258.

Chen F, Gerion D (2004) Fluorescent CdSe/ZnS nanocrystal-peptide conjugates for longterm, nontoxic imaging and nuclear targeting in living cells. *Nanoletters*, 4(10), 1827–1832.

Chen H, Weiss J, Shahidi F (2006) Nanotechnology in nutraceuticals and functional foods. *Food Technology*, 3, 30–36.

Chen H, Yada R (2011) Nanotechnologies in Agriculture: New tools for sustainable developments. *Trends in Food Science & Technology*, 22(11), 585–594.

Chen L, Remondetto GE, Subirade M (2006) Food protein-based materials as nutraceutical delivery systems. *Trends in Food Science and Technology*, 17(5), 272–283.

Chithrani BD, Chan WC (2007) Elucidating the mechanism of cellular uptake and removal of protein-coated gold nanoparticles of different sizes and shapes. *Nano Letters*, 7(6), 1542–1550.

Choi O, Deng KK, Kim NJ, Ross L Jr, Surampalli RY, Hu Z (2008) The inhibitory effects of silver nanoparticles, silver ions, and silver chloride colloids on microbial growth. *Water Research*, 42(12), 3066–3074.

Choudalakis G, Gotsis AD (2009) Permeability of polymer/clay nanocomposites: A review. *European Polymer Journal*, 45(4), 967–984.

Choudhury SR, Pradhan S, Goswami A (2012) Preparation and characterisation of acephate nano-encapsulated complex. *Nanoscience Methods*, 1(1), 9–15.

Chowdappa P, Gowda S (2013) Nanotechnology in crop protection: Status and scope. *Pest Management in Horticultural Ecosystems*, 19(2), 131–151.

Cientifica Report (2006) Nanotechnologies in the food industry, published August 2006. www.cientifica.com/ www/details.php?id¼447. Accessed 24 Oct 2010.

Cui Y, Zhao Y, Tian Y, Zhang W, Lü X, Jiang X (2012) The molecular mechanism of action of bactericidal gold nanoparticles on Escherichia coli. *Biomaterials*, 33(7), 2327–2333.

Cuppet SL (1994) Edible coatings as carriers of food additives, fungicides and naturals antagonists. In: Krochta JM, Baldwin EA, Nisperos-Carriedo M (eds) *Edible Coatings and Films to Improve Food Quality*, pp 121–137. Lancaster: Technomic Pub. Co.

Cushen M, Kerry J, Morris M, Cruz-Romero M, Cummins E (2012) Nanotechnologies in the food industrye-Recent developments, risks and regulation. *Trends in Food Science and Technology*, 24(1), 30–46.

Da Costa JT, Forim MR, Costa ES, De Souza JR, Mondego JM, Junior ALB (2014) Effects of different formulations of neem oil-based products on control Zabrotes subfasciatus (Boheman, 1833)(Coleoptera: Bruchidae) on beans. *Journal of Stored Products Research*, 56, 49–53.

Da Silva L, Oliva M, Azevedo A, De Araujo J (2006) Responses of resting plant species to pollution from an iron pelletization factory. *Water, Air and Soil Pollution*, 175(1–4), 241–256.

Dasgupta N, Ranjan S (2018) *Nanotechnology in Food Sector: An Introduction to Food Grade Nanoemulsions*. Springer –Nature, Singapore, pp 1–18.

Davis R, Guliants VV, Huber G, Lobo RF, Miller JT, Neurock M (2009) An international assessment of research in catalysis by nanostructure and materials. Baltimore, MD: World Technology Evaluation Enter. http://www.wtec.org/catalysis/WTEC-CatalysisReport-6Feb2009- color-hi-res.pdf.

de Medeiros GA, Arruda FB, Sakai E, Fujiwara M (2001) The influence of crop canopy on evapotranspiration and crop coefficient of beans (Phaseolus vulgaris L.). *Agricultural Water Management*, 49(3), 211–224.

De Moura MR, Aouada FA, Avena-Bustillos RJ, McHugh TH, Krochta JM, Mattoso LHC (2009) Improved barrier and mechanical properties of novel hydroxypropyl methylcellulose edible films with chitosan/tripolyphosphate nanoparticles. *Journal of Food Engineering*, 92(4), 448–453.

de Oliveira JL, Campos EVR, Bakshi M, Abhilash PC, Fraceto LF (2014) Application of nanotechnology for the encapsulation of botanical insecticides for sustainable agriculture: Prospects and promises. *Biotechnology Advances*, 32(8), 1550–1561.

Debeaufort F, Quezada-Gallo JA, Voilley A (1998) Edible films and coatings: Tomorrow's packagings: A review. *Critical Reviews in Food Science and Nutrition*, 38(4), 299–313.

Dehkourdi EH, Mosavi M (2013) Effect of anatase nanoparticles (TiO 2) on parsley seed germination (Petroselinum crispum) in vitro. *Biological Trace Element Research*, 155(2), 283–286.

Deng X, Wei Z, ANH (2001) Transgenic peanut plants obtained by particle bombardment via somatic embryogenesis regeneration system. *Cell Research*, 11(2), 156–160.

Deshayes A, Herrera-Estrella L, Caboche M (1985) Liposome-mediated transformation of tobacco mesophyll protoplasts by an Escherichia coli plasmid. *EMBO Journal*, 4(11), 2731–2737.

Diallo M (2009) Water treatment by dendrimer-enhanced filtration: Principles and applications. In: Savage NF et al (eds) *Nanotechnology Applications for Clean Water*, pp 143–155. Norwich, NY: William Andrew.

Dillen W, De Clercq J, Kapila J, Zambre M, Montagu M (1997) The effect of temperature on Agrobacterium tumefaciens-mediated gene transfer to plants. *Plant Journal*, 12(6), 1459–1463.

Doyle ME (2006) Nanotechnology: A brief literature review. http://www.wisc.edu/fri/briefs/ FRIBrief_Nanotech_Lit_Rev.pdf.

Du W, Gardea-Torresdey JL, Xie Y, Yin Y, Zhu J, Zhang X, Ji R, Gu K, Peralta-Videa JR, Guo H (2017) Elevated CO2 levels modify TiO2 nanoparticle effects on rice and soil microbial communities. *Science of the Total Environment*, 578, 408–416.

Du W, Sun Y, Ji R, Zhu J, Wu J, Guo H (2011) TiO 2 and ZnO nanoparticles negatively affect wheat growth and soil enzyme activities in agricultural soil. *Journal of Environmental Monitoring*, 13(4), 822–828.

Duncan TV (2011) Applications of nanotechnology in food packaging and food safety: Barrier materials, antimicrobials and sensors. *Journal of Colloid and Interface Science*, 363(1), 1–24.

Eichert T, Kurtz A, Steiner U, Goldbach H (2008) Size exclusion limits and lateral heterogeneity of the stomatal foliar uptake pathway for aqueous solutes and water-suspended nanoparticles. *Physiologia Plantarum*, 134(1), 151–160.

El Beyrouthya M, El Azzi D (2014) Nanotechnologies: Novel solutions for sustainable agriculture. *Advances in Crop Science and Technology*, 2(3), e118.

Emerich DF, Thanos CG (2006) The pinpoint promise of nanoparticle-based drug delivery and molecular diagnosis. *Biomolecular Engineering*, 23(4), 171–184.

Esmaeili A, Saremnia B (2012) Preparation of extract-loaded nanocapsules from Onopordon Leptolepis DC. *Industrial Crops and Products*, 37(1), 259–263.

Esthappan SK, Sinha MK, Katiyar P, Srivastav A, Joseph R (2013) Polypropylene/zinc oxide nanocomposite fibers: Morphology and thermal analysis. *Journal of Polymer Materials*, 30, 79–89.

Fan R, Huang YC, Grusak MA, Huang CP, Sherrier DJ (2014) Effects of nano-TiO2 on the agronomically-relevant Rhizobium–legume symbiosis. *Science of the Total Environment*, 466, 503–512.

Farahnaky A, Dadfar SMM, Shahbazi M (2014) Physical and mechanical properties of gelatinclay nanocomposite. *Journal of Food Engineering*, 122, 78–83.

Farmen L (2009) Commercialization of nanotechnology for removal of heavy metals in drinking water. In: Savage NF et al (eds) *Nanotechnology Applications for Clean Water*, pp 115–130. Norwich, NY: William Andrew.

Feizi H, Moghaddam PR, Shahtahmassebi N, Fotovat A (2012) Impact of bulk and nanosized titanium dioxide (TiO 2) on wheat seed germination and seedling growth. *Biological Trace Element Research*, 146(1), 101–106.

Feng BH, Peng LF (2012) Synthesis and characterization of carboxymethyl chitosan carrying ricinoleic functions as an emulsifier for azadirachtin. *Carbohydrate Polymers*, 88(2), 576–582.

Fernandez V, Eichert T (2009) Uptake of hydrophilic solutes through plant leaves: Current state of knowledge and perspectives of foliar fertilization. *Critical Reviews in Plant Sciences*, 28(1–2), 36–68.

Flanagan J, Singh H (2006) Microemulsions: A potential delivery system for bioactives in food. *Critical Reviews in Food Science and Nutrition*, 46(3), 221–237.

Fleischer M, O'Neill R, Ehwald (1999) The pore size of non-graminaceous plant cell wall is rapidly decreased by borate ester cross-linking of the pectic polysaccharide rhamnogalacturonan II. *Plant Physiology*, 121(3), 829–838.

Floros M, Hojabri L, Abraham E, Jose J, Thomas S, Pothan L, Leao AL, Narine S (2012) Enhancement of thermal stability, strength and extensibility of lipid-based polyurethanes with cellulose-based nanofibers. *Polymer Degradation and Stability*, 97(10), 1970–1978.

Forim MR, Costa ES, da Silva MFDGF, Fernandes JB, Mondego JM, Boiça Junior AL (2013) Development of a new method to prepare nano-/microparticles loaded with extracts of *Azadirachta indica*, their characterization and use in controlling *Plutella xylostella*. *Journal of Agricultural and Food Chemistry*, 61(38), 9131–9139.

Fraceto LF, Maruyama CR, Guilger M, Mishra S, Keswani C, Singh HB, deLima R (2018) *Trichoderma harzianum* based novel formulations: Potential Applications for Management of Next-Gen agricultural challenges. *Journal of Chemical Technology and Biotechnology*, 93(8), 2056–2063.

Frandsen MV, Pedersen MS, Zellweger M, Gouin S, Roorda SD, Phan TQC (2015) U.S. Patent Application No. 14/562, p 905.

Fukumori Y, Ichikawa H (2006) Nanoparticles for cancer therapy and diagnosis. *Advanced Powder Technology*, 17(1), 1–28.

Gao H, Shi W, Freund LB (2005) Mechanics of receptor-mediated endocytosis. *Proceedings of the National Academy of Sciences of the United States of America*, 102(27), 9469–9474.

Garcia M, Aleixandre M, Gutierrez J, Horrillo M (2006) Electronic nose for wine discrimination. *Sensors and Actuators Part B*, 113(2), 911–916.

Garcia M, Forbe T, Gonzales E (2010) Potential applications of nanotechnology in the agro-food sector. *Ciencia y Tecnologia de los Alimentos Campinas*, 30(3), 573–581.

Ge Y, Schimel JP, Holden PA (2011) Evidence for negative effects of TiO2 and ZnO nanoparticles on soil bacterial communities. *Environmental Science and Technology*, 45(4), 1659–1664.

Ge Y, Schimel JP, Holden PA (2012) Identification of soil bacteria susceptible to TiO2 and ZnO nanoparticles. *Applied and Environmental Microbiology*, 78(18), 6749–6758.

George J, Siddaramaiah (2012) High performance edible nanocomposite films containing bacterial cellulose nanocrystals. *Carbohydrate Polymers*, 87(3), 2031–2037.

Ghadiali J, Stevens M (2008) Enzyme-responsive nanoparticle systems. *Advanced Materials*, 20, 1–5.

Giraldo JP, Landry MP, Faltermeier SM, McNicholas TP, Iverson NM, Boghossian AA, Reuel NF, Hilmer AJ, Sen F, Brew JA, Strano MS (2014) Plant nanobionics approach to augment photosynthesis and biochemical sensing. *Nature Materials*, 13(4), 400.

Gokkurt T, Durmus A, Sariboga V, Oksuzomer MAF (2013) Investigation of thermal, rheological, and physical properties of amorphous poly(ethylene terephthalate)/organoclay nanocomposite films. *Journal of Applied Polymer Science*, 129(5), 2490–2501.

Golding M, Sein A (2004) Surface rheology of aqueous casein–monoglyceride dispersions. *Food Hydrocolloids*, 18(3), 451–461.

González-Melendi P, Fernández-Pacheco R, Coronado MJ, Corredo E, Testillano PS, Risueño MC, Marquina C, Ibarra MR, Rubiales D, Pérez-de-Luque A (2007) Nanoparticles as smart treatment-delivery systems in plants: Assessment of different techniques of microscopy for their visualization in plant tissues. *Annals of Botany*, 101(1), 187–195.

Gopal M, Kumar R, Goswami A (2012) Nano-pesticides—A recent approach for pest control. *Plant Protection Science*, 4, 1–7.

Grillo R, dos Santos NZP, Maruyama CR, Rosa AH, de Lima R, Fraceto LF (2012) Poly (ε-caprolactone) nanocapsules as carrier systems for herbicides: Physico-chemical characterization and genotoxicity evaluation. *Journal of Hazardous Materials*, 231, 1–9.

Guan H, Chi D, Yu J, Li X (2008) A novel photodegradable insecticide: Preparation, characterization and properties evaluation of nano Imidacloprid. *Pesticide Biochemistry and Physiology*, 92(2), 83–91.

Guilbert S (1986) Technology and application of edible protective films. In: Mathlouthi M (ed) *Food Packaging and Preservation*, pp 371–394. London: Elsevier Applied Science.

Gupta AK, Gupta M (2005) Synthesis and surface engineering of iron oxide nanoparticles for biomedical applications. *Biomaterials*, 26(18), 3995–4021.

Hafeez A, Razzaq A, Mahmood T, Jhanzab HM (2015) Potential of copper nanoparticles to increase growth and yield of wheat. *Journal of Nanoscience with Advanced Technology*, 1(1), 6–11.

Hagens WI, Oomen AG, de Jong WH, Cassee FR, Sips AJ (2007) What do we (need to) know about the kinetic properties of nanoparticles in the body?*Regulatory Toxicology and Pharmacology*, 49(3), 217–229.

Ham M, Kim JC, Chang JH (2013) Thermal property, morphology, optical transparency, and gas permeability of PVA/SPT nanocomposite films and equi-biaxial stretching films. *Polymer (Korea)*, 37(5), 579–586.

Hamal DB, Haggstrom JA, Marchin GL, Ikenberry MA, Hohn K, Klabunde KJ (2010) A multifunctional biocide/sporocide and photocatalyst based on titanium dioxide (TiO2) co-doped with silver, carbon, and sulfur. *Langmuir*, 26(4), 2805–2810.

Hawthorne J, Musante C, Sinha SK, White JC (2012) Accumulation and phytotoxicity of engineered nanoparticles to Cucurbita pepo. *International Journal of Phytoremediation*, 14(4), 429–442.

Hellmann C, Greiner A, Wendorff JH (2011) Design of pheromone releasing nanofibers for plant protection. *Polymers for Advanced Technologies*, 22(4), 407–413.

Hoek EMV, Ghosh AK (2009) Nanotechnology-based membranes for water purification. In: Savage NF et al (eds) *Nanotechnology Applications for Clean Water*, pp 47–58. Norwich, NY: William Andrew.

Hoffmann-Tsay S, Ernst R, Hoffmann F (1994) Design, synthesis and application of surface-active chemicals for the promotion of electrofusion of plant protoplasts. *Bioelectrochemistry and Bioenergetics*, 34(2), 115–122.

Holden PA, Nisbet RM, Lenihan HS, Miller RJ, Cherr GN, Schimel JP, Gardea-Torresdey JL (2012) Ecological nanotoxicology: Integrating nanomaterial hazard considerations across the subcellular, population, community, and ecosystems levels. *Accounts of Chemical Research*, 46(3), 813–822.

Hoshino A, Fujioka K, Oku T, Suga M, Sasaki Y, Ohta T, Yasuhara M, Suzuki K, Yamamoto K (2004) Physicochemical properties and cellular toxicity of nanocrystal quantum dots depend on their surface modification. *Nanoletters*, 4(11), 2163–2169.

Hossain ST, Mukherjee SK (2012) CdO nanoparticle toxicity on growth, morphology, and cell division in Escherichia coli. *Langmuir*, 28(48), 16614–16622.

Hossain ST, Mukherjee SK (2013) Toxicity of cadmium sulfide (CdS) nanoparticles against Escherichia coli and HeLa cells. *Journal of Hazardous Materials*, 260, 1073–1082.

Huang L, Li D-Q, Lin Y-J, Wei M, Evans DG, Duan X (2005) Controllable preparation of NanoMgO and investigation of its bactericidal properties. *Journal of Inorganic Biochemistry*, 99(4), 986–993.

Huang QR, Yu HL, Ru QM (2009) Bioavailability and delivery of nutraceuticals using nanotechnology. *Journal of Food Science*. Epub. Online 10 Dec 2009.

Huang S, Wang L, Liu L, Hou Y, Li L (2015) Nanotechnology in agriculture, livestock, and aquaculture in China: A review. *Agronomy for Sustainable Development*, 35(2), 369–400.

Huang WC, Tsai PJ, Chen YC (2007) Functional gold nanoparticles as photothermal agents for selective-killing of pathogenic bacteria.

Hussain SM, Hess KL, Gearhart JM, Geiss KT, Schlager JJ (2005) In vitro toxicity of nanoparticles in BRL 3A rat liver cells. *Toxicology in Vitro*, 19(7), 975–983.

Hwang I, Kim T, Bang S, Kim K, Kwon H, Seo M, Youn Y, Park H, Yasunga-Aoki C, Yu Y (2011) Insecticidal effect of controlled release formulations of etofenprox based on nano-bio technique. *Journal of the Faculty of Agriculture, Kyushu University*, 56(1), 33–40.

Iba K (2002) Acclimative response to temperature stress in higher plants: Approaches of gene engineering for temperature tolerance. *Annual Review of Plant Biology*, 53, 225–2245.

Ito H, Fukuda Y, Murata K, Kimura A (1983) Transformation of intact yeast cells treated with alkali cations. *Journal of Bacteriology*, 153(1), 163–168.

Jamal M, Moharramipour S, Zandi M, Negahban M (2013) Efficacy of nanoencapsulated formulation of essential oil from Carum copticum seeds on feeding behavior of plutella xylostella (Lep.: Plutellidae). Journal of Entomological Society of Iran, 33, 23–31.

Jampílek J, Kráľová K (2015) Application of nanotechnology in agriculture and food industry, its prospects and risks. *Ecological Chemistry and Engineering S*, 22(3), 321–361.

Jampílek J, Kráľová K (2017) Nanopesticides: Preparation, targeting, and controlled release. In: Alexandru Grumezescu *New Pesticides and Soil Sensors*, pp. 81–127. Academic Press. Cambridge, Massachusetts, USA

Jana N, Earhart C, Ying J (2007) Synthesis of water-soluble and functionalized nanoparticles by silica coating. *Chemistry of Materials*, 19(21), 5074–5082.

Jasim B, Thomas R, Mathew J, Radhakrishnan EK (2017) Plant growth and diosgenin enhancement effect of silver nanoparticles in Fenugreek (Trigonella foenum-Graecum L.). *Saudi Pharmaceutical Journal,* 25(3), 443–447.

Jia G, Wang H, Yan L, Wang X, Pei R, Yan T, Zhao Y, Xinbiao G (2005) Cytotoxicity of carbon nanomaterials: Single-wall nanotube, multi-wall nanotube, and fullerene. *Environmental Science and Technology,* 39(5), 1378–1383.

Jo YK, Kim BH, Jung G (2009) Antifungal activity of silver ions and nanoparticles on phytopathogenic fungi. *Plant Disease,* 93(10), 1037–1043.

Joseph T, Morrison M (2006) Nanotechnology in agriculture food: A nanoforum report published in 2006. www.nanoforum.org.

Josephson L, Perez J, Weissleder R (2001) Magnetic nanosensors for the detection of oligonucleotide sequences. *Angewandte Chemie,* 113(17), 3304–3306.

Josephson L, Tung C, Moore A, Weissleder R (1999) High-efficiency intracellular magnetic labeling with novel superparamagnetic-Tat peptide conjugates. *Bioconjugate Chemistry,* 10(2), 186–191.

Jun L, Xuan-Ming L, Su-Yao X, Chuen-Yi T, Dong-Ying T, Xuanming L, Liu X (2008) Preparation of fluorescence starch-nanoparticle and its application as plant transgenic vehicle. *Journal of Central South University of Technology,* 15(6), 768–773.

Kadam DM, Thunga M, Wang T, Kessler MR, Grewell D, Lamsai B, Yu C (2013) Preparation and characterization of whey protein isolate films reinforced with porous silica coated titania nanoparticles. *Journal of Food Engineering,* 117(1), 133–140.

Kah M, Beulke S, Tiede K, Hofmann T (2013) Nanopesticides: State of knowledge, environmental fate, and exposure modeling. *Critical Reviews in Environmental Science and Technology.* 43(16):1823–1867.

Kang S, Mauter MS, Elimelech M (2009) Microbial cytotoxicity of carbon-based nanomaterials: Implications for river water and wastewater effluent. *Environmental Science and Technology,* 43(7), 2648–2653.

Kavitha P, Manjunath M, Huey-min H (2018) Nanotechnology applications for environmental industry. In: Chaudhery Mustansar Hussain Mustansar HC (ed) *Handbook of Nanomaterials for Industrial Applications.* Elsevier. Cambridge, MA 02139, United State

Ke YC, Wang YG, Yang L (2010) Improving the hydrophobic, water barrier and crystallization properties of poly(ethylene terephthalate) by incorporating monodisperse SiO2 particles. *Polymer International,* 59(10), 1350–1359.

Keck CM, Müller RH (2013) Nanotoxicological classification system (NCS)–a guide for the risk-benefit assessment of nanoparticulate drug delivery systems. *European Journal of Pharmaceutics and Biopharmaceutics,* 84(3), 445–448.

Kenel F, Eady C, Brinch S (2010) Efficient Agrobacterium tumefaciens mediated transformation and regeneration of garlic (Allium sativum) immature leaf tissue. *Plant Cell Reports,* 29(3), 223–230.

Kester SL, Fennema OR (1986) Edible films and coatings: A review. *Food Technology,* 48, 47–59.

Khin MM, Nair AS, Babu VJ, Murugan R, Ramakrishna S (2012) A review on nanomaterials for environmental remediation. *Energy and Environmental Science,* 5(8), 8075–8109. doi: 10.1039/c2ee21818f.

Khodakovskaya M, Dervishi E, Mahmood M, Xu Y, Li Z, Watanabe F, Biris AS (2009) Carbon nanotubes are able to penetrate plant seed coat and dramatically affect seed germination and plant growth. *ACS Nano,* 3(10), 3221–3227.

Khot LR, Sankaran S, Maja JM, Ehsani R, Schuster EW (2012) Applications of nanomaterials in agricultural production and crop protection: A review. *Crop Protection,* 35, 64–70.

Kim S, Lee S, Lee I (2012) Alteration of phytotoxicity and oxidant stress potential by metal oxide nanoparticles in Cucumis sativus. *Water, Air and Soil Pollution,* 223(5), 2799–2806.

Kim SW, Kim KS, Lamsal K, Kim YJ, Kim SB, Jung M, Sim SJ, Kim HS, Chang SJ, Kim JK, Lee YS (2009) An in vitro study of the antifungal effect of silver nanoparticles on oak wilt pathogen Raffaelea sp. *Journal of Microbiology and Biotechnology,* 19(8), 760–764.

Kim Y, Oh Y, Oh E, Ko S, Han M, Kim H (2008) Energy transfer-based multiplexed assay of proteases by using gold nanoparticle and quantum dot conjugates on a surface. *Analytical Chemistry,* 80(12), 4634–4641.

Klein T, Kornstein L, Stanford J, Fromm M (1989) Genetic transformation of maize cells by particle bombardment. *Plant Physiology,* 91(1), 440–444.

Komaiko JS, McClements DJ (2016) Formation of food-grade nanoemulsions using low-energy preparation methods: A review of available methods. *Comprehensive Reviews in Food Science and Food Safety,* 15(2), 331–352.

Konotop YO, Kovalenko MS, Ulynets VZ, Meleshko AO, Batsmanova LM, Taran NY (2014) Phytotoxicity of colloidal solutions of metal-containing nanoparticles. *Cytology and Genetics,* 48(2), 99–102.

Kralchevsky PA, Ivanov IB, Ananthapadmanabhan KP, Lips A (2005) On the thermodynamics of particle-stabilized emulsions: Curvature effects and catastrophic phase inversion. *Langmuir*, 21(1), 50–63.

Krishnaraj C, Jagan EG, Ramachandran R, Abirami SM, Mohan N, Kalaichelvan PT (2012) Effect of biologically synthesized silver nanoparticles on Bacopa monnieri (Linn.) Wettst.plant growth metabolism. *Process Biochemistry*, 47(4), 651–658.

Krishnaraj C, Ramachandran R, Mohan K, Kalaichelvan PT (2012) Optimization for rapid synthesis of silver nanoparticles and its effect on phytopathogenic fungi. *Spectrochimica Acta, Part A – Molecular and Biomolecular Spectroscopy*, 93, 95–99.

Krochta JM, Mulder-Johnston C (1997) Edible and biodegradable polymer films: Challenges and opportunities. *Food Technology*, 51, 61–74.

Ku BC, Froio D, Steeves D, Kim DW, Ahn H, Ratto JA, Blumstein A, Kumar J, Samuelson LA (2004) Cross-linked multilayer polymer-clay nanocomposites and permeability properties. *Journal of Macromolecular Science – Pure and Applied Chemistry A*, 41(12), 1401–1410.

Kumar P, Burman U, Santra P (2015) Effect of nano-zinc oxide on nitrogenase activity in legumes: An interplay of concentration and exposure time. *International Nano Letters*, 5(4), 191–198.

Kumar S, Bhanjana G, Sharma A, Sidhu MC, Dilbaghi N (2014) Synthesis, characterization and on field evaluation of pesticide loaded sodium alginate nanoparticles. *Carbohydrate Polymers*, 101, 1061–1067.

Kumar S, Chauhan N, Gopal M, Kumar R, Dilbaghi N (2015) Development and evaluation of alginate–chitosan nanocapsules for controlled release of acetamiprid. *International Journal of Biological Macromolecules*, 81, 631–637.

Kumar V, Guleria P, Kumar V, Yadav SK (2013) Gold nanoparticle exposure induces growth and yield enhancement in Arabidopsis thaliana. *Science of the Total Environment*, 461, 462–468.

Kuzma J (2010) Nanotechnology in animal productione upstream assessment of applications. *Livestock Science*, 130(1–3), 14–24.

Lagerholm B, Wang M, Ernst L, Ly D, Liu H, Bruchez M, Waggoner A (2004) Multicolor coding of cells with cationic peptide coated quantum dots. *Nanoletters*, 4(10), 2019–2022.

Lahiani MH, Dervishi E, Chen J, Nima Z, Gaume A, Biris AS, Khodakovskaya MV (2013) Impact of carbon nanotube exposure to seeds of valuable crops. *ACS Applied Materials and Interfaces*, 5(16), 7965–7973.

Lai F, Wissing SA, Müller RH, Fadda AM (2006) Artemisia arborescens L essential oil-loaded solid lipid nanoparticles for potential agricultural application: Preparation and characterization. *AAPS PharmSciTech*, 7(1), E10.

Lambert K, Bekal S (2002) Introduction to plant-parasitic nematodes. *Plant Health Instructor*, 10, 1094–1218.

Lao SB, Zhang ZX, Xu HH, Jiang GB (2010) Novel amphiphilic chitosan derivatives: Synthesis, characterization and micellar solubilization of rotenone. *Carbohydrate Polymers*, 82(4), 1136–1142.

Lee KJ, Park SH, Govarthanan M, Hwang PH, Seo YS, Cho M, Leed WH, Lee JY, Kamala-Kannana S, Oh BT (2013) Synthesis of silver nanoparticles using cow milk and their antifungal activity against phytopathogens. *Materials Letters*, 105, 128–131.

Lee S, Chung H, Kim S, Lee I (2013) The genotoxic effect of ZnO and CuO nanoparticles on early growth of buckwheat, Fagopyrum esculentum. *Water, Air and Soil Pollution*, 224(9), 1668.

Lee S, Kim S, Kim S, Lee I (2013) Assessment of phytotoxicity of ZnO NPs on a medicinal plant, Fagopyrum esculentum. *Environmental Science and Pollution Research International*, 20(2), 848–854.

Lee SY, Lee SJ, Choi DS, Hur SJ (2015) Current topics in active and intelligent food packaging for preservation of fresh foods. *Journal of the Science of Food and Agriculture*, 95(14), 2799–2810.

Lee Y, Kim D, Seo J, Han H, Khan SB (2013) Preparation and characterization of poly(propylene carbonate)/exfoliated graphite nanocomposite films with improved thermal stability, mechanical properties and barrier properties. *Polymer International*, 62(9), 1386–1394.

Li C, Guo T, Zhou D, Hu Y, Zhou H, Wang S, Chen J, Zhang Z (2011) A novel glutathione modified chitosan conjugate for efficient gene delivery. *Journal of Controlled Release*, 154(2), 177–188.

Li M, Huang Q, Wu Y (2011) A novel chitosan-poly (lactide) copolymer and its submicron particles as Imidacloprid carriers. *Pest Management Science*, 67(7), 831–836.

Li Q, Wu P, Shang JK (2009) Nanostructured visible-lightphotocatalysts for water purification. In: Savage NF et al (eds) *Nanotechnology Applications for Clean Water*, pp 17–37. Norwich, NY: William Andrew.

Li R, Liu CH, Ma J, Yang YJ, Wu HX (2011) Effect of org-titanium phosphonate on the properties of chitosan films. *Polymer Bulletin*, 67(1), 77–89.

Lim CJ, Basri M, Omar D, Abdul Rahman MB, Salleh AB, Raja Abdul Rahman RNZ (2013) Green nanoemulsion-laden glyphosate isopropylamine formulation in suppressing creeping foxglove (A. gangetica), slender button weed (D. ocimifolia) and buffalo grass (P. conjugatum). *Pest Management Science*, 69(1), 104–111.

Lim L-T (2011) Active and intelligent packaging materials. In: Moo-Young M, Butler M, Webb C, Moreira A, Grodzinski B, Cui ZF, Agathos S (eds) *Comprehension Biotechnology*, vol. 4, pp 629–644. Amsterdam: Elsevier.

Limbach LK, Wick P, Manser P, Grass RN, Bruinink A, Stark WJ (2007) Exposure of engineered nanoparticles to human lung epithelial cells: Influence of chemical composition and catalytic activity on oxidative stress. *Environmental Science and Technology*, 41(11), 4158–4163.

Lin N, Huang J, Dufresne A (2012) Preparation, properties and applications of polysaccharide nanocrystals in advanced functional nanomaterials: A review. *Nanoscale*, 4(11), 3274–3294.

Lin D, Xing B (2007) Phytotoxicity of nanoparticles: Inhibition of seed germination and root growth. *Environmental Pollution*, 150(2), 243–250.

Lin D, Xing B (2008) Root uptake and phytotoxicity of ZnO nanoparticles. *Environmental Science and Technology*, 42(15), 5580–5585.

Liu J, Wang F, Wang L, Xiao S, Tong C, Tang D, Liu X (2008) Preparation of fluorescence starchnanoparticle and its application as plant transgenic vehicle. *Journal of Central South University of Technology*, 15(6), 768–773.

Liu R, Zhang H, Lal R (2016) Effects of stabilized nanoparticles of copper, zinc, manganese, and iron oxides in low concentrations on lettuce (Lactuca sativa) seed germination: Nanotoxicants or nanonutrients?*Water, Air and Soil Pollution*, 227(1), 42.

Liu Y, Ai K, Cheng X, Huo L, Lu L (2010) Gold-nanocluster-based fluorescent sensors for highly sensitive and selective detection of cyanide in water. *Advanced Functional Materials*, 20(6), 951–956.

Liu Y, Tong Z, Prud'homme RK (2008). Stabilized polymeric nanoparticles for controlled and efficient release of bifenthrin. *Pest Management Science: Formerly Pesticide Science*, 64(8), 808–812.

Loha KM, Shakil NA, Kumar J, Singh MK, Adak T, Jain S (2011) Release kinetics of β-cyfluthrin from its encapsulated formulations in water. *Journal of Environmental Science and Health, Part B*, 46(3), 201–206.

Loha KM, Shakil NA, Kumar J, Singh MK, Srivastava C (2012) Bio-efficacy evaluation of nanoformulations of β-cyfluthrin against Callosobruchus maculatus (Coleoptera: Bruchidae). *Journal of Environmental Science and Health, Part B*, 47(7), 687–691.

Lok C-N, Ho C-M, Chen R, He Q-Y, Yu W-Y, Sun H, Tam PK-H, Chiu J-F, Che C-M (2007) Silver nanoparticles: Partial oxidation and antibacterial activities. *Journal of Biological Inorganic Chemistry*, 12(4), 527–534.

López-Moreno ML, de la Rosa G, Hernández-Viezcas JA, Peralta-Videa JR, Gardea-Torresdey JL (2010) XAS corroboration of the uptake and storage of CeO_2 nanoparticles and assessment of their differential toxicity in four edible plant species. *Journal of Agricultural and Food Chemistry*, 58(6), 3689.

Lopez-Rubio A, Gavara R, Lagaron JM (2006) Bioactive packaging: Turning foods into healthier foods through biomaterials. *Trends in Food Science and Technology*, 17(10), 567–575.

Ma Y, Kuang L, He X, Bai W, Ding Y, Zhang Z, Zhao Y, Chai Z (2010) Effects of rare earth oxide nanoparticles on root elongation of plants. *Chemosphere*, 78(3), 273–279.

Madhuban G, Rajesh K, Arunava G (2012) Nano-pesticides-A recent approach for pest control. *Journal of Plant Protection Sciences*, 4(2), 1–7.

Manke A, Wang L, Rojanasakul Y (2013) Mechanisms of nanoparticle-induced oxidative stress and toxicity. *BioMed Research International*, 2013, 1-15. /doi.org/10.1155/2013/942916.

Marrani D (2013) Nanotechnologies and novel foods in European law. *NanoEthics*, 7(3), 177–188.

Marsh K, Bugusu B (2007) Food packaging – Roles, materials, and environmental issues. *Journal of Food Science*, 72(3), R38–R55.

McClements DJ, Decker EA, Weiss J (2005) Novel procedure for creating nano-laminated edible films and coatings. US Patent Application UMA, pp 05–27.

McGrath MT (2004) The plant health instructor: What are fungicides. http://www.apsnet.org/edcenter/intr opp/topics/Pages/fungicides.aspx.

McKnight T, Melechko A, Griffin G, Guillorn M, Merkulov V, Serna F, Hensley D, Doktycz M, Lowndes D, Simpson M (2003) Intracellular integration of synthetic nanostructures with viable cells for controlled biochemical manipulation. *Nanotechnology*, 14(5), 551–556.

McKnight T, Melechko A, Hensley D, Mann D, Griffin G, Simpson M (2004) Tracking gene expression after DNA delivery using spatially indexed nanofiber arrays. *Nano Letters*, 4(7), 1213– 1219.

McQuillan JS, Groenaga Infante H, Stokes E, Shaw AM (2012) Silver nanoparticle enhanced silver ion stress response in Escherichia coli K12. *Nanotoxicology*, 6(8), 857–866.

McQuillan JS, Shaw AM (2014) Differential gene regulation in the Ag nanoparticle and Ag+-induced silver stress response in Escherichia coli: A full transcriptomic profile. *Nanotoxicology*, 8(Suppl1), 177–184.

Medina-Velo IA, Barrios AC, Zuverza-Mena N, Hernandez-Viezcas JA, Chang CH, Ji Z, Zink JI, PeraltaVidea JR, Gardea-Torresdey JL (2017) Comparison of the effects of commercial coated and uncoated ZnO nanomaterials and Zn compounds in kidney bean (Phaseolus vulgaris) plants. *Journal of Hazardous Materials*, 332, 214–222.

Memarizadeh N, Ghadamyari M, Adeli M, Talebi K (2014) Preparation, characterization and efficiency of nanoencapsulated Imidacloprid under laboratory conditions. *Ecotoxicology and Environmental Safety*, 107, 77–83.

Miller KS, Krochta JM (1997) Oxygen and aroma barrier properties of edible films: A review. *Trends in Food Science and Technology*, 8(7), 228–237.

Miller G, Senjen R (2008) Out of the laboratory and on to our plates - Nanotechnology in food and agriculture. http://www.foeeurope.org/activities/nanotechnology/Documents/Nano_food_report.pdf.

Mishra S, Keswani C, Abhilash PC, Fraceto LF, Singh HB (2017) Integrated approach of Agri-nanotechnology: Challenges and future trends. *Frontiers in Plant Sciences*. doi:10.3389/fpls.2017.00471.

Mishra S, Keswani C, Singh A, Singh BR, Singh SP, Singh HB (2016) Microbial nanoformulation: Exploring potential for coherent nano-farming. In: VK Gupta, GD Sharma, MG Tuohy, R Gaur (eds) *The Handbook of Microbial Bioresources*, pp 107–120. CABI-UK, Wallingford OX10 8DE, United Kingdom.

Mishra S, Singh A, Keswani C, Singh HB (2014b) Nanotechnology: Exploring potential application in agriculture and its opportunities and Constraints. *Biotech Today*, 4(1), 9–14.

Mishra S, Singh BR, Singh A, Keswani C, Naqvi AH, Singh HB (2014a) Biofabricated silver nanoparticles act as a strong fungicide against *Bipolaris sorokiniana* causing spot blotch disease in wheat. *PLoS One*, 9(5), e97881.

Momin JK, Chitra J, Prajapati JB (2013) Potential of nanotechnology in functional foods. *Emirates Journal of Food and Agriculture*, 25(1), 10–19.

Momin JK, Joshi BH (2015) Nanotechnology in foods. In: Rai, M., Ribeiro, C., Mattoso, L., Duran, N. (Eds.)*Nanotechnologies in Food and Agriculture*, pp. 3–24. Cham: Springer.

Moore M (2006) Do nanoparticles present ecotoxicological risks for the health of the aquatic environment?*Environment International*, 32(8), 967–976.

Morales MI, Rico CM, Hernandez-Viezcas JA, Nunez JE, Barrios AC, Flores JP, Peralta-Videa JR, Gardea-Torresdey JL (2013) Toxicity assessment of cerium oxide nanoparticles in cilantro (*Coriandrum sativum* L.) plants grown in organic soil. *Journal of Agricultural and Food Chemistry*, 61(26), 6224–6230.

Moraru CI, Panchapakesan CP, Huang Q, Takhistov P, Liu S, Kokini JL (2003) Nanotechnology: A new frontier in food science. *Food Technology*, 57(12), 24–29.

Morones JR, Elechiguerra JL, Camacho A, Holt K, Kouri JB, Ramı́rez JT, Yacaman MJ (2005) The bactericidal effect of silver nanoparticles. *Nanotechnology*, 16(10), 2346–2353.

Mousavi SR, Rezaei M (2011) Nanotechnology in agriculture and food production. *Journal of Applied Environmental and Biological Sciences*, 1, 414–419.

Moustafa HZ, Mohamad TG, Torkey H (2015) Effect of formulated nanoemulsion of eucalyptus oil on the cotton bollworms. Journal of Biological and Chemical Research, 33, 478–484.

Musante C, White JC (2012) Toxicity of silver and copper to Cucurbita pepo: Differential effects of nano and bulk-size particles. *Environmental Toxicology*, 27(9), 510–517.

Nagy A, Harrison A, Sabbani S, Munson RS Jr, Dutta PK, Waldman WJ (2011) Silver nanoparticles embedded in zeolite membranes: Release of silver ions and mechanism of antibacterial action. *International Journal of Nanomedicine*, 6, 1833.

Nangmenyi G, Economy J (2009) Nonmetallic particles for oligodynamic microbial disinfection. In: Savage NF et al (eds) *Nanotechnology Applications for Cleans Water*, pp 3–15. Norwich, NY: William Andrew.

Narducci D (2007) An introduction to nanotechnologies: What's in it for us?*Veterinary Research Communications*, 31(Suppl 1), 131–137.

Navarro E, Baun A, Behra R, Hartmann N, Filser J, Miao A, Quigg A, Santschi P, Sigg L (2008) Environmental behaviour and ecotoxicity of engineered nanoparticles to algae, plants and fungi. *Ecotoxicology*, 17(5), 372–386.

Negahban M, Moharramipour S, Zandi M, Hashemi SA, Ziayee F (2012) Nano-insecticidal activity of essential oil from Cuminum cyminum on Tribolium castaneum. In: *Proc 9th. Int. Conf. on Controlled Atmosphere and Fumigation in Stored Products, Antalya, Turkey*. pp 63–6.

Nehoff H, Parayath NN, Domanovitch L, Taurin S, Greish K (2014). Nanomedicine for drug targeting: Strategies beyond the enhanced permeability and retention effect. *International Journal of Nanomedicine*, 9, 2539.

Nekrasova GF, Ushakova OS, Ermakov AE, Uimin MA, Byzov IV (2011) Effects of copper (II) ions and copper oxide nanoparticles on Elodea densa Planch.*Russian Journal of Ecology*, 42(6), 458.

Nel A, Xia T, Madler L, Li N (2006) Toxic potential of materials at the nanolevel. *Science*, 311(5761), 622–627.

Neuhaus G, Spangerberg G (1990) Plant transformation by microinjection techniques. *Physiologia Plantarum*, 79, 213–217.

Nguyen HM, Hwang IC, Park JW, Park HJ (2012) Photoprotection for deltamethrin using chitosan-coated beeswax solid lipid nanoparticles. *Pest Management Science*, 68(7), 1062–1068.

Nickols-Richardson SM, Piehowski KE (2008) Nanotechnology in nutritional sciences. *Minerva Biotecnologica*, 20(3), 117.

Niemeyer C (2001) Nanoparticles, proteins, and nucleic acids: Biotechnology meets materials science. *Angewandte Chemie International Edition*, 40(22), 4128–4158.

NRC (2002) *National Research Council Animal Biotechnology: Science-Based Concerns*. Washington, DC: National Academy Press.

Owolade OF, Ogunleti DO, Adenekan MO (2008) Titanium dioxide affects disease development and yield of edible cowpea. *Agricultural and Food Chemistry*, 7(50), 2942–2947.

Pariona N, Martínez AI, Hernandez-Flores H, Clark-Tapia R (2017) Effect of magnetite nanoparticles on the germination and early growth of Quercus macdougallii. *Science of the Total Environment*, 575, 869–875.

Paulkumar K, Gnanajobitha G, Vanaja M, Rajeshkumar S, Malarkodi C, Pandian K, Annadurai G (2014) Piper nigrum leaf and stem assisted green synthesis of silver nanoparticles and evaluation of its antibacterial activity against agricultural plant pathogens. *The Scientific World Journal*, 2014. 1-9. doi.org /10.1155/2014/829894.

Pehanich M (2006) Small gains in processing and packaging. *Food Processing*, 11, 46–48.

Pelletier DA, Suresh AK, Holton GA, McKeown CK, Wang W, Gu B, Mortensen NP, Allison DP, Joy DC, Allison MR, Brown SD (2010) Effects of engineered cerium oxide nanoparticles on bacterial growth and viability. *Applied and Environmental Microbiology*, 76(24), 7981–7989.

Pereira AE, Grillo R, Mello NF, Rosa AH, Fraceto LF (2014) Application of poly (epsilon-caprolactone) nanoparticles containing atrazine herbicide as an alternative technique to control weeds and reduce damage to the environment. *Journal of Hazardous Materials*, 268, 207–215.

Perlatti B, de Souza Bergo PL, Fernandes JB, Forim MR (2013) Polymeric nanoparticle-based insecticides: A controlled release purpose for agrochemicals. In: S. Trdan. *Insecticides-Development of Safer and More Effective Technologies*. InTech, Rijeka, Croatia 523–550. doi: 10.5772/53355.

Picard E, Gauthier H, Gérard JF, Espuche E (2007) Influence of the intercalated cations on the surface energy of montmorillonites: Consequences for the morphology and gas barrier properties of polyethylene/ montmorillonites nanocomposites. *Journal of Colloid and Interface Science*, 307(2), 364–376.

Pokhrel LR, Dubey B (2013) Evaluation of developmental responses of two crop plants exposed to silver and zinc oxide nanoparticles. *Science of the Total Environment*, 452, 321–332.

Popat A, Liu J, Hu Q, Kennedy M, Peters B, Lu GQM, Qiao SZ (2012) Adsorption and release of biocides with mesoporous silica nanoparticles. *Nanoscale*, 4(3), 970–975.

Potrykus I (1991) Gene transfer to plants: Assessment of published approaches and results. *Annual Review of Plant Physiology and Plant Molecular Biology*, 42(1), 205–225.

Prakash M, Nair G, Chung M (2017) Regulation of morphological, molecular and nutrient status in Arabidopsis thaliana seedlings in response to ZnO nanoparticles and Zn ion exposure. *Science of the Total Environment*, 575, 187–198.

Prasad R, Bagde US, Varma A (2012) An overview of intellectual property rights in relation to agricultural biotechnology. *African Journal of Biotechnology*, 11(73), 13476–13752.

Prasad TNVKV, Sudhakar P, Sreenivasulu Y, Latha P, Munaswamy V, Reddy KR, Sreeprasad TS, Sajanlal PR, Pradeep T (2012) Effect of nanoscale zinc oxide particles on the germination, growth and yield of peanut. *Journal of Plant Nutrition*, 35(6), 905–927.

Prashantha K, Soulestin J, Lacrampe MF, Krawczak P, Dupin G, Claes M (2009) Masterbatchbased multi-walled carbon nanotube filled polypropylene nanocomposites: Assessment of rheological and mechanical properties. *Composites Science and Technology*, 69(11–12), 1756–1763.

Priester JH, Ge Y, Mielke RE, Horst AM, Moritz SC, Espinosa K, Gelb J, Walker SL, Nisbet RM, An YJ, Schimel JP (2012) Soybean susceptibility to manufactured nanomaterials with evidence for food quality and soil fertility interruption. *Proceedings of the National Academy of Sciences of the United States of America*, 109(37), E2451–E2456.

Radzig MA, Nadtochenko VA, Koksharova OA, Kiwi J, Lipasova VA, Khmel IA (2013) Antibacterial effects of silver nanoparticles on gram-negative bacteria: Influence on the growth and biofilms formation, mechanisms of action. *Colloids and Surfaces, Part B: Biointerfaces*, 102, 300–306.

Ragaei M, Sabry AKH (2014) Nanotechnology for insect pest control. *International Journal of Science, Environment and Technology*, 3(2), 528–545.

Rai M, Deshmukh S, Gade A, Kamel-Abd-Elsalam (2012) Strategic nanoparticle-mediated gene transfer in plants and animals – A novel approach. *Current Nanoscience*, 8(1), 170–179.

Rai M, Ingle A (2012) Role of nanotechnology in agriculture with special reference to management of insect pests. *Applied Microbiology and Biotechnology*, 94(2), 287–293.

Rai M, Ribeiro C, Mattoso L, Duran N (2015) *Nanotechnologies in Food and Agriculture*, p 347. Cham/Heidelberg/New York/Dordrecht/London: Springer.

Rai M, Yadav A, Gade A (2009) Silver nanoparticles as a new generation of antimicrobials. *Biotechnology Advances*, 27(1), 76–83.

Rajesh S, Raja DP, Rathi JM, Sahayaraj K (2012) Biosynthesis of silver nanoparticles using Ulva fasciata (Delile) ethyl acetate extract and its activity against Xanthomonas campestris pv. malvacearum. *Journal of Biopesticides*, 5, 119.

Rakoczy-trojanowska M (2002) Alternative methods of plant transformation – A short review. *Cellular and Molecular Biology Letters*, 7(3), 849–858.

Rees P, Brown MR, Summers HD, Holton MD, Errington RJ, Chappell SC, Smith PJ (2011) A transfer function approach to measuring cell inheritance. *BMC Systems Biology*, 5(1), 31.

Reid W, Zhang Q, Sekimoto H (2002) Influence of membrane surface charge on nutrient uptake by plants. *Developments in Plant and Soil Sciences*, 92, 198–199.

Restuccia D, Spizziri UG, Parisi OI, Cirillo G, Curio M, Iemma F, Puoci F, Vinci, Picci N (2010) New EU regulation aspects and global market of active and intelligent packaging for food industry applications. *Food Control*, 21(11), 1425–1435.

Rhim JW (2004) Increase in water vapor barrier property of biopolymer-based edible films and coatings by compositing with lipid materials. *Food Science and Biotechnology*: 13(4), 528–535.

Rhim JW, Hong SI, Park HM, Ng PKW (2006) Preparation and characterization of chitosan-based nanocomposite films with antimicrobial activity. *Journal of Agriculture and Food Chemistry*, 54(16), 5814–5822.

Roberto C, Ruffini C (2009) Nanoparticles and higher plants. *Cryologia*, 62(2), 161–165.

Robertson GL (2012) *Food Packaging: Principles and Practice*. Boca Raton: CRC Press.

Roco M (2003) Broader societal issue of nanotechnology. *Journal of Nanoparticle Research*, 5(3/4), 181–189.

Roco MC, Mirkin CA, Hersam MC (2011) *Nanotechnology Research Directions for Societal Needs in 2020: Retrospective and Outlook*. Springer Science & Business Media, Dordrecht, Netherlands.

Roe D, Karandikar B, Bonn-Savage N, Gibbins B, Roullet JB (2008) Antimicrobial surface functionalization of plastic catheters by silver nanoparticles. *Journal of Antimicrobial Chemotherapy*, 61(4), 869–876.

Ruzin S, McCarthy S (1986) The effect of chemical facilitators on the frequency of electrofusion of tobacco mesophyll protoplast. *Plant Cell Reports*, 5(5), 342–345.

Saharan V, Mehrotra A, Khatik R, Rawal P, Sharma SS, Pal A (2013) Synthesis of chitosan based nanoparticles and their in vitro evaluation against phytopathogenic fungi. *International Journal of Biological Macromolecules*, 62, 677–683.

Saharan V, Sharma G, Yadav M, Choudhary MK, Sharma SS, Pal A, Biswas P (2015) Synthesis and in vitro antifungal efficacy of Cu–chitosan nanoparticles against pathogenic fungi of tomato. *International Journal of Biological Macromolecules*, 75, 346–353.

Saini P, Gopal M, Kumar R, Srivastava C (2014) Development of pyridalyl nanocapsule suspension for efficient management of tomato fruit and shoot borer (Helicoverpa armigera). *Journal of Environmental Science and Health, Part B*, 49(5), 344–351.

Sandquist D (2013) New horizons for microfibrillated cellulose. *Appita Journal*, 66(156), 162.

Sarkar DJ, Kumar J, Shakil NA, Walia S (2012) Release kinetics of controlled release formulations of thiamethoxam employing nano-ranged amphiphilic PEG and diacid based block polymers in soil. *Journal of Environmental Science and Health, Part A*, 47(11), 1701–1712.

Savithramma N, Ankanna S, Bhumi G (2012) Effect of nanoparticles on seed germination and seedling growth of Boswellia ovalifoliolata an endemic and endangered medicinal tree taxon. *Nano Vision*, 2(1), 2.

Schuetz MR, Kalo H, Linkebein T, Groschel AH, Muller AHE, Wilkie CA, Breu J (2011) Shear stiff, surface modified, mica-like nanoplatelets: A novel filler for polymer nanocomposites. *Journal of Materials Chemistry*, 21(32), 12110–12116.

Scognamiglio V, Arduini F, Palleschi G, Rea G (2014) Biosensing technology for sustainable food safety. *TrAC Trends in Analytical Chemistry*, 62, 1–10.

Scott NR (2002) Rethink, redesign, reengineer. *Resource*, 9(9), 8–10.

Scott NR (2007) Nanoscience in veterinary medicine. *Veterinary Research Communications*, 31(1), 139–141.

Scrinis G, Lyons K (2007) The emerging nanocorporate paradigm: Nanotechnology and the transformation of nature, food and agri-food systems. *International Journal of Sociology of Food and Agriculture*, 15(2), 22–44.

Segura T, Shea L (2001) Materials for non viral gene delivery. *Annual Review of Materials Research*, 31(1), 25–46.

Sekhon BS (2010) Food nanotechnology–an overview. *Nanotechnology, Science and Applications*, 3, 1.

Sekhon BS (2014) Nanotechnology in agri-food production: An overview. *Nanotechnology, Science and Applications*, 7, 31.

Sergey B, LeFebvre M, Sherwood J, Xu Y, Bao Y, Ramonell KM (2015) Developmental and reproductive effects of iron oxide nanoparticles in Arabidopsis thaliana. *International Journal of Molecular Sciences*, 16(10), 24174–24193.

Shakil NA, Singh MK, Pandey A, Kumar J, Parmar VS, Singh MK, Pandey RP, Watterson AC (2010) Development of poly (ethylene glycol) based amphiphilic copolymers for controlled release delivery of carbofuran. *Journal of Macromolecular Science, Part A. Pure and Applied Chemistry*, 47(3), 241–247.

Sharma P, Bhatt D, Zaidi MGH, Saradhi PP, Khanna PK, Arora S (2012) Silver nanoparticle-mediated enhancement in growth and antioxidant status of Brassica juncea. *Applied Biochemistry and Biotechnology*, 167(8), 2225–2233.

Shenashen M, Derbalah A, Hamza A, Mohamed A, El Safty S (2017) Antifungal activity of fabricated mesoporous alumina nanoparticles against root rot disease of tomato caused by Fusarium oxysporium. *Pest Management Science*, 73(6), 1121–1126.

Shirazi R, Ewert K, Leal C, Majzoub R, Bouxsein N, Safinya C (2011) Synthesis and characterization of degradable multivalent cationic lipids with disulfide-bond spacers for gene delivery. *Biochimica et Biophysica Acta*, 1808(9), 2156–2166.

Shotornvit R, Rhim J, Hong S (2009) Effect of nano-clay type on the physical and antimicrobial properties of whey protein isolate/clay composite films. *Journal of Food Engineering*, 91(3), 468–473.

Shrawat A, Good A (2011) Agrobacterium tumefaciens-mediated genetic transformation of cereals using immature embryos. *Methods in Molecular Biology*, 710, 355–372.

Siddiqui MH, Al-Whaibi MH (2014). Role of nano-SiO2 in germination of tomato (Lycopersicum esculentum seeds Mill.). *Saudi Journal of Biological Sciences*, 21(1), 13–17.

Siddiqui MH, Al-Whaibi MH, Faisal M, Al Sahli AA (2014) Nano-silicon dioxide mitigates the adverse effects of salt stress on Cucurbita pepo L.*Environmental Toxicology and Chemistry*, 33(11), 2429–2437.

Singh RK, Singh N (2005) Quality of packaged food. In: Han JH (ed) *Innovations in Food Packaging*, pp 24–44. San Diego: Elsevier Academic Press.

Sioutas C, Delfino RJ, Singh M (2005) Exposure assessment for atmospheric ultrafine particles (UFPs) and implications in epidemiologic research. *Environmental Health Perspectives*, 113(8), 947–955.

Soghomonian V, Heremans JJ (2009) Characterization of electrical conductivity in a zeolite like material. *Applied Physics Letters*, 95(15), 152112.

Sokolova V, Epple M (2008) Inorganic nanoparticles as a carrier for nucleic acid into cells. *Angewandte Chemie International Edition*, 47(8), 1382–1395.

Song S, Liu X, Jiang J, Qian Y, Zhang N, Wu Q (2009) Stability of triazophos in self-nanoemulsifying pesticide delivery system. *Colloids and Surfaces A: Physicochemical and Engineering Aspects*, 350(1–3), 57–62.

Sozer N, Kokini JL (2009) Nanotechnology and its applications in the food sector. *Trends in Biotechnology*, 27(2), 82–89.

Srinivasan C, Saraswathi R (2010) Nano-agriculture——Carbon nanotubes enhance tomato seed germination and plant growth. *Current Science*, 99(3), 274–275.

Srividhya S, Chellaram C (2012) Role of marine life in nanomedicine. *Ind J Innov Develop*, 1(Suppl8), 31–33.

Sun C, Shu K, Wang W, Ye Z, Liu T, Gao Y, Zheng H, Yin Y, Yin Y (2014) Encapsulation and controlled release of hydrophilic pesticide in shell cross-linked nanocapsules containing aqueous core. *International Journal of Pharmaceutics*, 463(1), 108–114.

Sun D, Jin T, Su JY, Zhang H, Sue HJ (2009) Antimicrobial efficacy of zinc oxide Quantum Dots against Listeria monocytogenes, Salmonella enteritidis, and Escherichia coli O157:H7. *Journal of Food Science*, 74(1), M46–M52.

Sundarraj A, GuhanNath S, Aaron S, Ranganathan T (2014) Recent innovations in nanotechnology in food processing and its various applications—a review. *International Journal of Pharmaceutical Sciences Review and Research*, 29(2), 116–124.

Surassmo S, Min SG, Bejrapha P, Choi MJ (2009) Effects of surfactants on the physical properties of capsicum-oleoresin-loaded nanocapsules formulated through the emulsion-diffusion method. *Food Research International*, 43(1), 8–17. doi:10.1016/j.foodres.2009.07.008.

Swain SK, Pradhan AK, Sahu HS (2013) Synthesis of gas barrier starch by dispersion of functionalized multiwalled carbon nanotubes. *Carbohydrate Polymers*, 94(1), 663–668.

Sytar O, Kumar A, Latowski D, Kuczynska P, Strzałka K, Prasad MNV (2013) Heavy metal-induced oxidative damage, defense reactions, and detoxification mechanisms in plants. *Acta Physiologiae Plantarum*, 35(4), 985–999.

Tang H, Xiong H, Tang S, Zou P (2009) A starch-based biodegradable film modified by nano silicon dioxide. *Journal of Applied Polymer Science*, 113(1), 34–40.

Tang X, Alavi S, Herald TJ (2008) Effect of plasticizers on the structure and properties of starchclay nanocomposite films. *Carbohydrate Polymers*, 74(3), 552–555.

Thorek DL, Tsourka A (2008) Size, charge and concentration dependent uptake of iron oxide particles by nonphagocytic cells. *Biomaterials*, 29(26), 3583–3590.

Thul ST, Sarangi BK, Pandey RA (2013) Nanotechnology in agroecosystem: Implications on plant productivity and its soil environment. *Expert Opinion on Environmental Biology*, 2, 2Ā7.

Tilston EL, Collins CD, Mitchell GR, Princivalle J, Shaw LJ (2013) Nanoscale zerovalent iron alters soil bacterial community structure and inhibits chloroaromatic biodegradation potential in aroclor 1242-contaminated soil. *Environmental Pollution*, 173, 38–46.

Torney F, Trewyn BG, Lin VSY, Wang K (2007) Mesoporous silica nanoparticles deliver DNA and chemicals into plants. *Nature Nanotechnology*, 2(5), 295.

Tzfira T, Citovsky V (2003) The Agrobacterium-plant cell interaction.taking biology lessons from a bug. *Plant Physiology*, 133(3), 943–947.

Unsworth JB, Corsi C, Van Emon JM, Farenhorst A, Hamilton DJ, Howard CJ, Hunter R, Jenkins JJ, Kleter GA, Kookana RS, Lalah JO, Leggett M, Miglioranza KS, Miyagawa H, Peranginangin N, Rubin B, Saha B, Shakil NA, Lalah JO (2015) Developing global leaders for research, regulation, and stewardship of crop protection chemistry in the 21st century. *Journal of Agricultural and Food Chemistry*, 64(1), 52–60.

US DOA (2003) *Nanoscale Science and Engineering for Agriculture and Food Systems: A Report Submitted to Cooperative State Research, Education and Extension Service.* Washington, DC: Department of Agriculture; the United States Department of Agriculture: National Planning Workshop.

Uzu G, Sobanska S, Sarret G, Munoz M, Dumat C (2010) Foliar lead uptake by lettuce exposed to atmospheric pollution. *Environmental Science and Technology*, 44(3), 1036–1042.

Vanderroost M, Ragaert P, Devlieghere F, De Meulenaer B (2014) Intelligent food packaging: The next generation. *Trends in Food Science and Technology*, 39(1), 47–62.

Vargas M, Pastor C, Chiralt A, McClements DJ, González-Martínez C (2008) Recent advances in edible coatings for fresh and minimally processed fruits Gonza ´lez-Martı ´nez C. *Critical Reviews in Food Science and Nutrition*, 48(6), 496–511.

Venkatachalam P, Priyanka N, Manikandan K, Ganeshbabu I, Indiraarulselvi P, Geetha N, Muralikrishna K, Bhattacharya RC, Tiwari M, Sharma N, Sahi SV (2017) Enhanced plant growth promoting role of phycomolecules coated zinc oxide nanoparticles with P supplementation in cotton (Gossypium hirsutum L.). *Plant Physiology and Biochemistry*, 110, 118–127.

Verma P, Mathur A (2011) Agrobacterium tumefaciens-mediated transgenic plant production via direct shoot bud organogenesis from pre-plasmolyzed leaf explants of Catharanthus roseus. *Biotechnology Letters*, 33(5), 1053–1060.

Vermeiren L, Devlieghere G, van Beest M, de Kruijf N, Debevere J (1999) Developments in the active packaging of foods. *Trends in Food Science and Technology*, 10(3), 77–86.

Villagarcia H, Dervishi E, de Silva K, Biris AS, Khodakovskaya MV (2012) Surface chemistry of carbon nanotubes impacts the growth and expression of water channel protein in tomato plants. *Small*, 8(15), 2328–2334.

Vorosmarty CJ, Mc Intyre PB, Gessner MO, Dudgeon D, Prusevich A, Green P (2010) Global threats to human water security and river biodiversity. *Nature*, 467(7315), 55–561.

Wang A, Zheng Y, Peng F (2014) Thickness-controllable silica coating of CdTe QDs by reverse microemulsion method for the application in the growth of rice. *Journal of Spectroscopy*, 2014: 1-5. dx.doi.org/10.1155/2014/169245.

Wang L, Li X, Zhang G, Dong J, Eastoe J (2007) Oil-in-water nanoemulsions for pesticide formulations. *Journal of Colloid and Interface Science*, 314(1), 230–235.

Wang SJ, Liu LM, Fang PF, Chen Z, Wang HM, Zhang SP (2007) Microstructure of polymer–clay nanocomposites studied by positrons. *Radiation Physics and Chemistry*, 76(2), 106–111.

Wang X, Yang X, Chen S, Li Q, Wang W, Hou C, Gao X, Wang L, Wang S (2016) Zinc oxide nanoparticles affect biomass accumulation and photosynthesis in Arabidopsis. *Frontiers in Plant Science*, 6, 1243.

Warad HC, Dutta J (2005) Nanotechnology for Agriculture and Food Systems: A View. In *Proceedings of the 2nd International Conference on Innovations in Food Processing Technology and Engineering, ed. by A Noomhorn and VK Jindal. Asian Institute of Technology, Bangkok (2005).* Also available online: http://www. nano.ait.ac.th/Emerging/Supply/suppl13-Agriculture%20And%20Food%20Systems%20_%20A%20View.pdf

Weiss J, Takhistov P, Mcclements DJ (2006) Functional materials in food nanotechnology. *Journal of Food Science,* 71(9), R107–R116.

Wibowo D, Zhao CX, Peters BC, Middelberg AP (2014) Sustained release of fipronil insecticide in vitro and in vivo from biocompatible silica nanocapsules. *Journal of Agricultural and Food Chemistry,* 62(52), 12504–12511.

Wiesman Z, Dom N, Sharvit E, Grinberg S, Linder C, Heldman E, Zaccai M (2007) Novel cationic vesicle platform derived from Vernonia oil for efficient delivery of DNA through plant cuticle membranes. *Journal of Biotechnology,* 130(1), 85–94.

Xiao-e L, Green ANM, Haque SA, Mills A, Durrant JR (2004) Light-driven oxygen scavenging by titania/polymer nanocomposite films. *Journal of Photochemistry and Photobiology, Part A,* 162(2–3), 253–259.

Xie Y, He Y, Irwin PI, Jin T, Shi X (2011) Antibacterial activity and mechanism of action of zinc oxide nanoparticles against Campylobacter jejuni. *Applied and Environmental Microbiology,* 77(7), 2325–2331.

Xie Y, Li B, Tao G, Zhang Q, Zhang C (2012) Effects of nano-silicon dioxide on photosynthetic fluorescence characteristics of Indocalamus barbatus McClure. *Journal of Nanjing Forestry University (Natural Sciences Edition),* 36(2), 59–63.

Xie Y, Li B, Zhang Q, Zhang C, Lu K, Tao G (2011) Effects of nano-TiO2 on photosynthetic characteristics of Indocalamus barbatus. *Journal of Northeast Forestry University,* 39(3), 22–25.

Yang FL, Li XG, Zhu F, Lei CL (2009) Structural characterization of nanoparticles loaded with garlic essential oil and their insecticidal activity against Tribolium castaneum (Herbst)(Coleoptera: Tenebrionidae). *Journal of Agricultural and Food Chemistry,* 57(21), 10156–10162.

Yang H, Fan W, Vaneski A, Susha AS, Teoh WY, Rogach AL (2012) Heterojunction engineering of CdTe and CdSe quantum dots on TiO2 nanotube arrays: Intricate effects of size-dependency and interfacial contact on photoconversion efficiencies. *Advanced Functional Materials,* 22(13), 2821–2829.

Yang Y, Wang J, Xiu Z, Alvarez PJ (2013) Impacts of silver nanoparticles on cellular and transcriptional activity of nitrogen-cycling bacteria. *Environmental Toxicology and Chemistry,* 32(7), 1488–1494.

Yin YH, Guo QM, Yun HAN, Wang LJ, Wan SQ (2012) Preparation, characterization and nematicidal activity of lansiumamide B nano-capsules. *Journal of Integrative Agriculture,* 11(7), 1151–1158.

Yousefi N, Gudarzi MM, Zheng QB, Lin XY, Shen X, Jia JJ, Sharif F, Kim JK (2013) Highly aligned, ultralarge-size reduced graphene oxide/polyurethane nanocomposites: Mechanical properties and moisture permeability. *Composites – Part A,* 49, 42–50.

Yu J, Liu RYF, Poon B, Nazarenko S, Koloski T, Vargo T, Hiltner A, Vaer E (2004) Polymers with palladium nanoparticles as active membrane materials. *Journal of Applied Polymer Science,* 92(2), 749–756.

Yu Q, Huang H, Chen R (2012) Synthesis of CuO nanowalnuts and nanoribbons from aqueous solution and their catalytic and electrochemical properties. *Nanoscale,* 4(8), 2613–2620.

Yuan H, Li J, Bao G, Zhang S (2010) Variable nanoparticle-cell adhesion strength regulates cellular uptake. *Physical Review Letters,* 105(13), 138101.

Zafar H, Ali A, Zia M (2017) CuO nanoparticles inhibited root growth from *Brassica nigra* seedlings but induced root from stem and leaf explants. *Applied Biochemistry and Biotechnology,* 181(1), 365–378.

Zhang D, Hua T, Xiao F, Chen C, Gersberg RM, Liu Y, Stuckey D, Ng WJ (2015) Phytotoxicity and bioaccumulation of ZnO nanoparticles in Schoenoplectus tabernaemontani. *Chemosphere,* 120, 211–219.

Zhang J, Li M, Fan T, Xu Q, Wu Y, Chen C, Huang Q (2013) Construction of novel amphiphilic chitosan copolymer nanoparticles for chlorpyrifos delivery. *Journal of Polymer Research,* 20(3), 107.

Ziaee M, Moharramipour S, Mohsenifar A (2014) Toxicity of Carum copticum essential oil-loaded nanogel against Sitophilus granarius and Tribolium confusum. *Journal of Applied Entomology,* 138(10), 763–771.

4 Trichoderma-Based Nanopesticides
Next-Generation Solution for Disease Management

Yashoda Nandan Tripathi, Chetan Keswani, Walia Zahra,
Kumari Divyanshu, and Ram Sanmukh Upadhyay

CONTENTS

4.1   INTRODUCTION

Chemical pesticides have been commonly used in agriculture for enhancing the production and yield of food and fibers and for checking vector-borne diseases. These pesticides include insecticides, fungicides, herbicides, nematicides, etc. Although these pesticides have proven to be a boon in terms of elimination of pests, they can simultaneously also be thought of as a curse when their impact on unsustainable management of soil resources is assessed. The pesticides have been found to cause detrimental effects on all aspects of the environment, viz. soil, air, water, biota, etc. Hence, the need for the development of eco-friendly pesticides, also referred to as biopesticides, has become inevitable (Keswani et al. 2019a,b). Among all the synthetic pesticides, fungicides are most popular because of the fact that fungal diseases are the major cause of crop loss worldwide (Gawai 2015). Looking at the drawbacks of using chemical pesticides, use of biofungicides has gained importance in controlling targeted diseases of fungal origin (Bhattacharya et al. 2016). Biofungicides are essentially derived from biological organisms like bacteria, fungi, animals, or plants and they suppress insect pests in an ecofriendly way. Depending upon the microorganisms involved, there can be several mechanisms of biocidal activity of the biofungicides like rhizosphere competence, parasitism, antibiosis, inducing metabolic changes, stimulating plant growth, etc. (Saraf et al. 2014; Shrivastava et al. 2014).

4.1.1   *Trichoderma*

The history of fungus *Trichoderma* dates back to 1794 when it was first reported and described by Persoon (1794) and later on it was suggested to have a link with the sexual state of a *Hypocrea* species. Earlier, it was difficult to assign the genus *Trichoderma/Hypocrea* morphologically and was proposed to have only a single species, that is *Trichoderma viride*. With the advent of modern molecular technologies, now more than 200 phylogenetically defined species of *Trichoderma* have been discovered (Atanasova et al. 2013). The sexual teleomorph stage has been observed in many *Trichoderma* species which colonize woody as well as herbaceous plants. Nonetheless, there are still many strains in which the sexual stage has not been discovered.

Trichoderma can grow in all types of soil and in all kinds of climatic zones because it can grow very fast, prolifically produce spore, and can also produce antibiotics even under highly a competitive environment for space, nutrients, and light (Montero-Barrientos et al. 2011; Mukherjee et al. 2013). It is a ubiquitous fungus and can be found wherever decaying plant material is available because of its inherent property of colonizing cellulosic material (Jaklitsch 2009). Species of this genus are characterized by rapid growth, mostly bright green conidia and a repetitively branched conidiophore structure. Since the 1920s this fungus has been known to function as a biocontrol agent (BCA) against plant pathogens (Waghunde et al. 2016). The different species of this genus have the capability to enhance plant growth and development, they have elevated reproductive ability, can modify the rhizosphere, are capable of growing in adverse conditions, are competent enough to use the nutrients, and they also show strong aggressiveness against phytopathogenic fungi (Keswani et al. 2014). *Trichoderma* can inhibit the growth of various plant pathogens because of its ability to secrete various hydrolytic enzymes. Due to its multi-enzymatic system, it is considered to be among the best biocontrol agents (Ahluwalia et al. 2014).

Trichoderma-based products are considered as relatively novel in the biocontrol market and have become very popular. Currently, more than 50 different formulations of *Trichoderma*-based biofungicides are available in the international market and this constitutes 60% of all the fungal based BCAs registered worldwide (Waghunde et al. 2016).

In order to successfully colonize a habitat and to thrive in it, the organism needs to defend its ecological niche despite competition for nutrients, space, and light. *Trichoderma* appears to be master of this game and acts as an efficient mycoparasite, antagonist, or biocontrol agent by utilizing its enzymes or chemicals as weapons (Vinale et al. 2008). There was a surge in research on the antagonistic properties of *Trichoderma* spp. only after Weindling found *T. atroviride* to be acting as a parasite on other fungi in 1932. In the current scenario, along with *T. atroviride*, several other species like *T. harzianum*, *T. virens*, and *T. asperellum* are the focus of research in the field of biocontrol. This fungus can not only control growth of different fungi but also has a growth inhibitory effect on nematodes (Goswami et al. 2008). In order to suppress the growth of pathogens, the *Trichoderma* spp. makes use of proteolytic enzymes, ATP-Binding Cassette (ABC) transporter membrane pumps, diffusible or volatile metabolites, and other secondary metabolites as active measures.

4.1.2 NANOTECHNOLOGY

The term 'nanotechnology' is derived from the Greek word 'nanos' meaning 'dwarf'. It is a branch of science that utilizes technology to control matter at the molecular level. Matter behaves entirely differently at the nanoscale level when compared to its macroscopic form. According to the National Nanotechnology Initiative, nanotechnology is the manipulation of matter with at least one dimension sized from 1 to 100 nanometers and these materials are then referred to as nanoparticles (NPs). The NPs not only have a strong affinity for proteins but they also have very flexible physical properties such as a large surface area to volume ratio (Mishra et al. 2016). Nanotechnology has become a thrust area of research in the past decade because of the fascinating properties of the NPs. Several developed economies are spending billions of dollars in this field. Up to 2012 the United States had invested USD 3.7 billion on its National Nanotechnology Initiative, which is a program for science, engineering, and technology research and development for nanoscale projects.

Looking at the enormous success of nanotechnology in various fields, agricultural scientists have now turned their attention towards agri-nanotechnology for improving agriculture (Mishra et al. 2016). Nanotechnology offers an ecofriendly alternative to plant disease management over conventional chemical methods that cause problems of toxicity. Various traditional practices in agriculture like integrated pest management are not only insufficient and cause deterioration of the soil but they also have detrimental effects on animals and human beings. Therefore, substitutes for these chemicals are being explored, and the utilization of nanotechnology in agriculture is proving beneficial in conserving nature (Ragaei and Sabry 2014). Several industries such as the pharmaceutical, agrochemical, and biotechnological industries are now shifting their attention to NP synthesis (Vijaykumar et al. 2013). Synthesis of NPs is a growing trend in the agrochemical industry because of the physical, chemical, and biological properties of the NPs. These nanomaterials help in the elimination of sprayed chemical products by smartly delivering active ingredients (Gogos et al. 2012).

4.2 APPLICATION OF NANOTECHNOLOGY IN AGRICULTURE

The field of agriculture forms the basis of the livelihood of more than 60% of the population of most developing countries. Meanwhile, the sector is widely challenged by the change in climatic conditions, use of large amounts of chemical fertilizers, and inappropriate use of resources (Raliya et al. 2017). Different, less-toxic chemical agents and newer delivery systems are being used to improve crop productivity in order to decrease the bulk use of chemicals in the field of agriculture and hence involved in providing more profound solutions to the problems in the field. Nanotechnology deals with working with the smallest possible particles and enhancing agricultural productivity. The use of this technology can thus enhance crop productivity in the agricultural sector as NPs possess special physiochemical characteristics such as tunable pore size, high surface area, and reactivity, and their applications include plant genome improvement, study of the mechanisms underlying the plant disease pathogenesis, and the delivery of various agrochemicals. Veterinary care, nutrient deficiency detection, and nanosensors are some other examples of the direct applications of nanotechnology in agriculture. Due to the exclusive functional properties of NPs such as higher solubility in suspension, particle size, higher surface area which efficiently penetrates the seed coats and hence the emerging roots, and enhancement of the bioavailability of the molecules to the seed radicals (Dehner et al. 2010).

Other applications of nanotechnology include the improvement of crop productivity, e.g. zinc nanofertilizers used in the enhancement of crop production of *Pennisetum americanum* (Tarafdar et al. 2014), soil and water management, plant disease detection and enhancement of the fertilizers and pesticides (Park et al. 2006; Corradini et al. 2010). Thus, nanotechnology plays a prominent role in strengthening the sustainability of agriculture by acting as the carrier of important nutrients and allowing their controlled release.

Nanotechnology is being widely implemented in the field of agriculture and the nutrient elements are either delivered in the form of particles or emulsions, and it is being investigated whether or not these NPs can be used to replace the traditional fertilizers at different growth stages of the crop (Millan et al. 2008). Moreover, the NPs which are being engineered are programmed so that when they are stimulated by environmental factors or as the function of time, they release important nutrients in a controlled manner according to the crop's requirements (Narware et al. 2019). The release of nutrients depends upon the pH, moisture, or the magnetic or ultrasonic pulses, etc. (Corradini et al. 2010).

Due to their less toxic nature, unique optical activities, and biocompatibility, NPs have gained prominent attention in biological studies. When NPs are incorporated in biological tests for measuring, the activity or presence of selected analytics in a sample become faster, flexible, and sensitive (Vidotti et al. 2011; Kandasamy and Prema 2015). Thus, application of NPs can be advantageous over the traditional procedures.

4.2.1 NANOFERTILIZERS

Nanofertilizers are being widely used nowadays to reduce the pollution of soil and water and as an alternative to reduce the use of bulk fertilizers in agriculture for balanced crop nutrition. They facilitate the slower and steadier release of nutrients and hence reduce the loss of important nutrients, and facilitate its usage efficiency, thereby reducing the cost of environmental protection (Lu et al. 2002; Raliya et al. 2015; Janmohammadi et al. 2016).

Although nanofertilizers are easily found in markets, agricultural fertilizers are still not manufactured by major marketing companies in particular. Nanosilica, iron, titanium oxide, iron Mn/ZnSe quantum dots (QDs), ZnCdSe/ZnS core shell QDs, gold nanorods, etc. are used in manufacturing nanofertilizers in many cases. In the present decade, metal oxide nanoparticles such as FeO, ZnO, TiO_2, etc. have been extensively tested for their biological fate and toxicity to be used in the agricultural sector (Dimkpa 2014; Zhang et al. 2016). The effectiveness of nanofertilizers has not been investigated much yet, however some of them have been found to be effective as fertilizers (Prasad et al. 2018).

Radiolabeling of the metal oxide NPs is generally done either by direct proton bombardment or they are enriched with ^{18}O in order to generate ^{18}F during their synthesis (Llop et al. 2014). In the presence of protein and cell media, the degree of aggregation, size, and zeta potential of the metal oxide NPs are investigated, and their uptake and intracellular fate are studied by using transmission electron microscopy, confocal laser scanning microscopy, Raman chemical imaging spectroscopy, and ion beam microscopy (Llop et al. 2014).

4.2.2 NANOPESTICIDES

As insect pests are predominantly involved in harming agricultural crops NP-based pesticides can be used to control these pests for better crop productivity. The nanopesticides being engineered these days have the property to slowly release the components with greater specificity, permeability, and solubility (Mishra et al. 2014b; Mishra et al. 2017). This is effectively achieved by either protecting the active constituents of the pesticide from degradation or by enhancing their pest control efficacy for longer period. The nanopesticide formulation also helps in reducing the exposure of hazardous chemicals to human beings which is environmentally friendly for crop protection

(Nuruzzaman et al. 2016). Thus, the establishment of a favorable pesticide delivery system will be quite helpful in enhancing food production on a global scale and hence will be helpful in decreasing the negative impact on the environment (Bhattacharyya et al. 2016).

Nanoencapsulation is therefore promoted to enhance the quality of nanopesticides to be used at agricultural scale. A few chemical companies have also openly promoted the sale of these nanopesticides and the examples of the same include Subdue MAXX, Penncap-M, Ospray's Chyella, and Karate ZEON (Gouin 2004).

For the successful implication of nanotechnology in agriculture, the importance of checking the toxicity of manufactured NPs is very important. The toxicity of the metal NPs generally depends on their binding specificity to their biological binding site, solubility, or other physiochemical features (Du et al. 2017). Metal NPs are known to exhibit anticandidal, antifungal, and antibacterial activities. The cytotoxicity of these NPs depends upon the charge on the membrane of the target organism and also on the structure of the targeted cell wall. Thus, molds are found to be more sensitive than yeast followed by gram-negative bacteria and gram-positive bacteria. The toxicity of the NPs is attributed to the electrostatic interaction in between them and also due to their accumulation in the cytoplasm (Rana and Kalaichelvan 2013).

Eco-toxicological research is necessary to understand the environmental consequence of NPs and their effect on natural systems. Moreover, the delayed impacts of the toxic effects of these NPs on the environment needs extensive research so that nanotechnology can be implemented without having any hazardous effects (Singh et al. 2017). Other aspects to be investigated include the bioaccumulation of these NPs in the food chain and their interaction with other pollutants present in the environment. In plants, NPs enter the cellular system and are translocated into the shoot and other aerial parts. The rate of transpiration can be affected, the rate of photosynthesis can be altered, and the translocation of the food material can be hindered due to accumulation of these NPs (Du et al. 2017). Thus, the study of the toxic effects of these NPs is important to understand as it can be helpful in reducing the hazardous effects at various trophic levels (Tripathi et al. 2016).

4.3 BIOSYNTHESIS OF NANOPARTICLES

Generally, NP synthesis involves three methods, viz. physical, chemical, and biological and usually it involves two approaches: bottom-up and top-down. Bottom-up includes biological samples like whole organism, tissue, or chemically formed NPs, while top-down includes physical machining or templating of sample size to nanometer scale (Singh et al. 2015) (Figure 4.1). NP biosynthesis through the biological method is an appealing and trending proposal because NP formation occurs with fewer defects as they are formed from the reduction and oxidation of small entities. The enzymes, proteins, sugars, and phytochemicals like flavonoids, phenolics, terpenoids, cofactors, etc. present in the biological entity help in the reduction, capping, and stabilization process (Singh et al. 2015).

4.3.1 GREEN SYNTHESIS OF NANOPARTICLES

Green synthesis of metal NPs has been studied in different microbes, algae, and plants like alfalfa, *Aloe vera*, *Cinnamonum camphora*, *Capsicum annum*, *Medicago sativa*, *Brassica juncea*, *Brassica chicory*, *Azadiracta indica*, and *Cymbopogon flexuosus* (Rai et al. 2018).

Microorganisms have the ability to absorb and accumulate metals along with the secretion of large amounts of enzymes involved in the enzymatic reduction of metal ions (Mishra et al. 2016). Fungi are the preferred choice of the nanotechnologists because of the variety of benefits which they offer over other microorganisms. They are compatible with bioreactors and fermenters, have higher flow pressure and agitation resistance, and produce extracellular reductive proteins in large amounts (Narayanan and Sakthivel 2010). Apart from that, they are easy to handle, require simple nutrients, have high wall binding capacity and intracellular metal uptake capabilities, high tolerance, and the ability to bio-accumulate metals make them the most desirable candidate for green synthesis of NPs

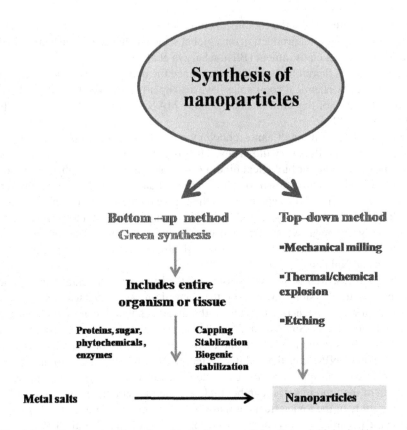

FIGURE 4.1   Schematic representation of synthesis of nanoparticles through biological/green methods.

(Ahluwalia et al. 2014). Fungal species, which are most often used for synthesis of NPs, include *Trichoderma reseei, T. viride, Phytophthora infestans, Aspergillus niger, A. flavus, A. clavatus, Fusarium oxysporum, Verticillum* sp., *Penicillium* sp., and *Pleurotus sajor-caju* (Devi et al. 2013).

Extracellular synthesis of metal NPs is advantageous over the intracellular process in high scale synthesis because of reduced handling, and sophisticated isolation of NPs can be done from fungal mycelium. In addition, the recovery of NPs from filamentous fungi is cheaper when compared to intracellular extraction from other microbes. Hence, filamentous fungi like *Trichoderma* which produces NPs extracellularly are a better choice for high scale synthesis of metal NPs (Vahabi and Dorcheh 2014).

4.3.2   BIOSYNTHESIS OF METAL NANOPARTICLES USING *TRICHODERMA* SP.

Metallic NPs generally include gold, silver, alloys, and other metal NPs. Around 70% of the fungi that are being used to biosynthesise metal NPs are either human or plant pathogenic. *Trichoderma* sp. are the only non-pathogenic fungi which are reported to have compatibility with genetic manipulation, are ecofriendly, and have a high scale production process. Different *Trichoderma* sp. are considered as an efficient bio-synthesiser of silver nanoparticles (AgNPs). Extracellular biosynthesis of AgNPs has been done by different *Trichoderma* species which include *T. reesei* (Vahabi 2011), *T. viride* (Fayaz 2010), *T. asperellum* (Mukherjee et al. 2008).

4.3.3   BIOSYNTHESIS OF SILVER NANOPARTICLES

AgNPs have been shown to have a growth effect on various microorganisms like bacteria, fungi, and even on viruses (Elamawi 2018). Apart from their unique properties, the production cost of AgNPs is very low and they also are ecofriendly, reliable, and biocompatible. The AgNPs synthesized using fungi have several advantages over the other methods/organisms employed in synthesizing these NPs. The advantages include tolerance of fungi towards a high concentration of metal NP in the medium, large scale production of NPs can be easily managed, good dispersion of NPs, and higher protein expressions. Since the investment in synthesizing AgNPs using fungi is also very low, the large scale production of AgNP, fungi are the preferred organism (Vahabi 2011; Fraceto 2018). The general procedure for the characterization of biosynthesized AgNPs is shown in Figure 4.2.

Several filamentous pathogenic and non-pathogenic fungi have been reported for their ability to extra-cellular synthesize biomass free AgNPs. Pathogenic filamentous fungi include *Aspergillus fumigatus* and among non-pathogenic filamentous fungi are the species of genus *Trichoderma* – *T. asperellum* and *T. reesei;* the latter is most popular for producing extracellular enzymes at the industrial scale (Vahabi et al. 2011). Among all the available biocontrol agents, *Trichoderma* spp. are the most well-known because of their ability to produce a number of hydrolytic enzymes such as chitinases, glucanases, and proteases, which help them in performing their mycoparasitic activity (Ahluwalia et al. 2014).

4.3.4 THROUGH *TRICHODERMA REESEI*

T. reesei is a well-known biological agent at industrial level because it produces more extracel-lular enzymes and at higher scale compared to other fungi. The extracellular reductase enzymes produced by these fungi reduce toxic Ag^+ ions to nontoxic AgNP extra-cellularly. Industrial scale AgNP biosynthesis using *T. reesei* is advantageous because this fungus has high metal NP concen-tration tolerance in the medium, the filtration process is easy, dispersion of NPs is ideal, and protein expression is maximum.

4.3.5 THROUGH *TRICHODERMA ATROVIRIDE*

Several researchers have reported biosynthesis of AgNPs from *T. atroviride* by using simple meth-ods. Saravanakumar and Wang, (2018) and Ponmurugan (2016) have followed a similar strategy to biosynthesize AgNPs from *T. atroviride*. They grew the fungal culture in suitable broth, followed

Characterization of silver nanoparticles

Primary characterization

UV-visible spectral study through

Color change of the fungal filtrate supplemented with silver nitrate. AgNPs were characterized in dual beam UV-visible spectrophotometer.

TEM analysis

A drop of AgNPS suspension on carbon-coated copper grids and allowing water to evaporate as described by El-Sonbaty

FIGURE 4.2 General steps for the characterization of silver nanoparticles.

by shaking it in an orbital shaker with agitation of 180 rpm at 28°C for 4 days. This was followed by fungal biomass harvesting and several rounds of washing, and finally mixing with silver nitrate. Finally, characterization of the biosynthesized AgNP was done by using techniques such as UV-visible spectroscopy, scanning electron microscopy (SEM), energy dispersive X-ray (EDX), X-ray diffraction (XRD), and Fourier transform infrared (FTIR) spectroscopy.

4.3.6 THROUGH *TRICHODERMA KONINGII*

Tripathi et al. (2013) reported that *T. koningii* have the property to synthesize proteins and enzymes that act as capping and reducing agents which enhances the AgNPs synthesis. During this process the Ag^+ was reduced to $Ag^°$, and further, its quality was characterized by techniques such as dynamic light scattering (DLS), X-ray diffraction (XRD), and transmission electron microscopy (TEM). Similarly, Devi et al. (2013) worked on *T. asperellum, T. harzianum, T. longibrachiatum, T. pseudokoningii,* and *T. virens* for the biosynthesis of AgNPs and evaluated that *T. virens* have maximum capability of NP synthesis. This was confirmed by UV-vis study and high resolution transmission electron microscopy (HRTEM) used for the analysis of morphology of the NPs.

4.3.7 THROUGH *TRICHODERMA HARZIANUM*

The mechanism for mycosynthesis of AgNPs through a catalytic reaction and reduction process of *T. harzianum* is shown in Figure 4.3. Sundaravadivelan and Padmanabhan, (2014) have found that AgNPs produced using *T. harzianum* showed larvicidal and pupicidal potentials that can help in control of mosquito *Aedes aegypti*. Similarly, Shelar and Chavan, (2015) synthesized AgNPs using *T. harzianum* and found that it significantly increased the germination index in sunflower and soybean.

4.4   BIOSYNTHESIS OF GOLD NANOPARTICLES THROUGH FUNGUS

Gold nanoparticles (AuNPs) were synthesized for the first time in the Roman period for decorating glasses. Beveridge and Murray, (1980) were the first to synthesize AuNPs using microbes. They showed the synthesis of 5–25 nm octahedral NPs in the cell wall of *Bacillus subtilis* 168. Several physical and chemical methods are available for synthesis of AuNPs but nowadays focus has shifted towards a greener approach because of the ease of biosynthesis and elimination of harsh chemical procedures

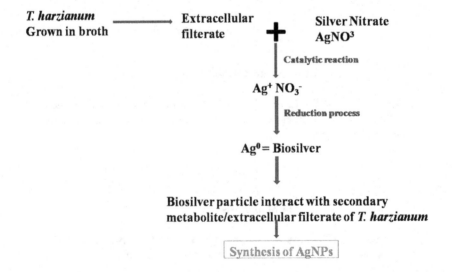

FIGURE 4.3   Mechanisms of mycosynthesis of silver nanoparticles from *T. harzianum* extracellular filtrate.

(Das et al. 2012). In recent years, synthesis of AuNPs has been an area of great interest among researchers but mostly prokaryotes are employed for this purpose. For industrial scale production of NPs, fungi are the preferred organisms because of their ability to synthesize a large amount of extracellular proteins. Therefore, scientists are now using such fungi as *Fusarium oxysporum*, *Colletotrichum*, and *Trichoderma* spp. for extracellular biosynthesis of AuNPs (Sathiskumar et al. 2009).

4.4.1 GOLD NANOPARTICLE SYNTHESIS BY *TRICHODERMA* SP.

A serious drawback of NP synthesis through the green approach is the reaction time which can extend to several days when compared to physical and chemical approaches. However, rapid synthesis of AuNPs has been reported by some workers but the drawback is that they used pathogenic microbes which can cause serious problems in later stages. AuNPs were synthesized within a minute from a cell-free extract of *T. viride* and this NP served as an efficient biocatalyst which reduced 4-nitrophenol to 4-aminophenol in the presence of $NaBH_4$ and had antimicrobial activity against pathogenic bacteria (*E. coli*, *S. sonnei*, and *P. syringae*). A flowchart for the quick synthesis of AuNPs from *T. viride* is shown in Figure 4.4.

Similarly, Mukherjee et al. (2013) demonstrated green synthesis of uniform pseudospherical optically flat triangular nanoprisms of AuNPs from a non-pathogenic biocontrol agent *T. asperellum* for reduction of $HAuCl_4$ at room temperature. This finding was particularly interesting because in similar conditions, $AgNO_3$ was reduced by cell-free extract of *T. asperellum* to produce spherical shaped NPs. Figure 4.5 shows the mechanism of biosynthesis of AuNP from *Trichoderma*.

4.4.2 THROUGH *TRICHODERMA HARZIANUM*

Tripathi et al. (2018) reported that AuNPs synthesized using biomass of *T. harzianum* had a size in the range 32–44 nm which suggests that fungal biomass was supporting the synthesis of a narrow

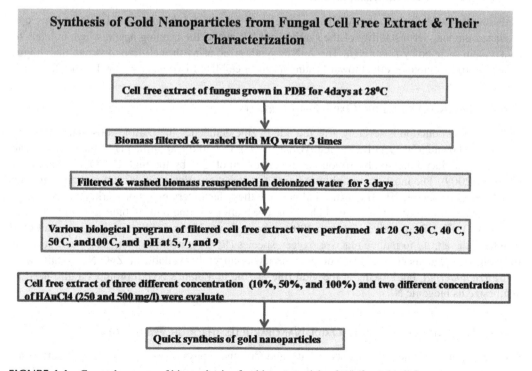

FIGURE 4.4   General process of biosynthesis of gold nanoparticles from fungal cell free extract.

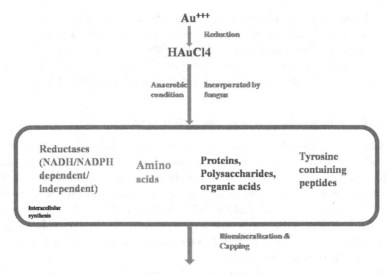

Extracellular Synthesis of gold nanoparticles

FIGURE 4.5   Mechanism behind biosynthesis of gold nanoparticles.

size range of nanoparticles. The synthesized AuNPs showed strong antibacterial activity and the Minimum Inhibitory Concentration (MIC) of AuNPs was found to be 20 µg/ml against *E. coli* MTCC 1305. These biogenic AuNPs also showed effective catalytic reduction of methylene blue as 39% reduction was observed in 30 min.

4.4.3 THROUGH *TRICHODERMA KONINGII*

Maliszewska (2013) synthesized AuNPs using cell-free filtrate of *T. koningii* and FTIR spectroscopy predicted the long term stability of the synthesized NP due to the capping agent, which is likely to be a protein secreted by *T. koningii*. The AuNPs synthesized using the cell-free filtrate of this fungus showed a strong cytotoxic effect towards human cancer cell line LoVo and sub-line LoVo/DX.

4.4.4 BIOSYNTHESIS OF ZINC OXIDE NANOPARTICLES

Agricultural production is severely affected due to zinc deficiency in alkaline soils that are rich in calcium carbonate ($CaCO_3$) (Takkar and Walker 1993). Due to the alkaline pH of the soil, the zinc solubility is reduced due to absorption and precipitation of zinc by the high $CaCO_3$ concentration (Alloway 2009). The most commonly used zinc fertilizers are zinc oxide (ZnO) and zinc sulfates ($ZnSO_4.H_2O$ or $ZnSO_4.7H_2O$) but in alkaline soils these fertilizers may not work. ZnO NPs may help in overcoming this problem by helping in dissolution of zinc and its bioavailability in soils with $CaCO_3$. ZnNPs of a size less than 100 nm have shown better antibacterial activity and they also have the ability to induce reactive oxygen species (ROS) generation that causes cell death of the pathogens (Duhan et al. 2017). Not much review is currently available on ZnO NPs synthesized using *Trichoderma*, but looking at the vital role of zinc in a plant's health prompts a dire need to synthesize its biogenic NPs.

4.4.5 BIOSYNTHESIS OF COPPER OXIDE NANOPARTICLES THROUGH *TRICHODERMA*

Copper is a microelement that exists as Cu^{2+} and Cu^+ under physiological conditions. It acts as an important component of various biochemical and physiological regulations in plants. At the cellular level, it plays an important role in oxidative phosphorylation, signal trafficking machinery, and iron mobilization. It is an important component of many enzymes such as polyphenol oxidase, amino oxidase, plastocyanin, laccase, and super oxide dismutase (Harold et al. 2007). The minimum concentration of copper that is required by the plants is 10^{-14} to 10^{-16} M. Concentration higher than this leads to toxicity, growth inhibition, photosynthesis interferences, photo respiration, and increased oxidative stress in plants. Deficiency of copper is asymptomatic, and it causes severe yield loss. So, in order to increase the copper uptake by the plants, it can be delivered in the form of NPs.

Copper nanoparticle (CuNP) synthesis by fungi has proven not only beneficial in improving crop yield because of its antimicrobial activity, but it also helps in bioremediation. The small size and high surface area to volume ratio of CuNPs allows them to interact closely with microbial membranes and higher temperature and low pH helps in better solubilization of copper in bacterial membrane and thus causes toxicity (Shobha et al. 2014). Salvadori et al. (2014) reported that dead biomass of *Trichoderma koningiopsis* rapidly produced CuNPs extra-cellularly in aqueous solution. *T. koningiopsis* exhibited high tolerance to copper (1057 mg/L^{-1}) and it efficiently synthesized CuNPs with an average size of 87.5 nm. This experiment concluded that *T. koningiopsis* is an ideal candidate for mycoremediation of CuNPs from contaminated water systems. Various types of metal NPs that are synthesized using different *Trichoderma* species are listed in Table 4.1.

4.5 ANTIMICROBIAL ACTION OF *TRICHODERMA* NANOPARTICLES AGAINST PLANT PATHOGENS

As chemical pesticides are being used indiscriminately, leading to the development of different races of the pathogens, an alternative strategy is being looked at to fight the pathogens without polluting the environment and inhibiting the chances for the development of newer races of pathogens. Thus, depending upon the theme of nanotechnology, synthesis of biological origin pesticides is now under consideration. Some microorganisms are found to be efficient in reducing the toxicity of the metals by reducing them and thus rendering them non-hazardous to the biological systems. Due to their advantageous physicochemical properties over yeast, plants, bacteria, etc., fungi have become the most preferred choice for the nanotechnologists (Sanghi and Verma 2009). They offer easy handling, require simple nutrients, are capable of intracellular metal uptake, and have high wall-binding capacity. Because of the properties of production of a variety of proteins, and rapid mycelial growth and economic viability, fungi have gained greater interest to be used as stabilizers and reducing agents for the synthesis of NPs (Mukherjee et al. 2008). The examples of the fungi being used for the synthesis

TABLE 4.1

Various *Trichoderma* spp. Involved in Metal Nanoparticle Biosynthesis

Trichoderma species	Nanoparticle synthesized	References
T. virens VN-11	Silver NPs	Devi et al. (2013)
T. asperellum	Silver NPs	Devi et al. (2013)
	Copper oxide NPs	Saravanakumar et al. (2019)
T. harzianum	Silver NPs	Shelar and Chavan (2015)
	Gold NPs	Tripathi et al. (2018)
T. longibranchiatum	Silver NPs	Omran et al. (2018)
T. pseudokoningii	Silver NPs	Tripathi et al. (2013)
T. viride ATCC36838	Silver NPs	Othman et al. (2017)
T. hamatum	Silver NPs	Hussein (2016)
T. reesei	Silver NPs	Vahabi et al. (2011)
T. koningiopsis	Copper NPs	Salvadori et al. (2014)
T. atroviride	Silver NPs	Saravanakumar and Wang (2018)

of NPs include *T. reseei* (Vahabi et al. 2011), *T. viride* (Fayaz et al. 2010), *T. asperellum* (Mukherjee et al. 2008), *A. niger* (Jaidev and Narasimha 2010), *Phytophthora infestans* (Thirumurugan et al. 2009), *Fusarium oxysporum* (Durán et al. 2005), etc. Large numbers of microorganisms are also seen to form the composites of NPs either intra- or extra-cellularly. Fungi, among all these microorganisms, have the primary focus because of their properties of tolerance and bioaccumulation. These are also efficient in secreting the extracellular enzymes and hence can be employed in the production of enzymes on a larger scale. The economic viability and ease in the handling of biomass also makes the fungi advantageous for the synthesis of metallic NPs (Devi et al. 2013).

Trichoderma has been widely reported to have antagonistic ability against various plant pathogens (Lorito et al. 2010; Machowicz-Matejko and Zalewska 2015). It shows anti-pathogenic activity because of the production of the hydrolytic enzymes that cleave the cell wall, mycoparasitism, and acting as a growth promoter for the crops with bioremediation abilities. Species of *Trichoderma* are found to colonize the apoplast of the roots of plants (Tripathi et al. 2013).

In this genus, molecular interaction is observed among biocontrol agents, antagonistic fungus, and plants (Prabhu et al. 2017). They are also known for producing a variety of secondary metabolites including pyrones, non-ribosomal peptides, terpenoids, and indolic-derived compounds (Contreras-Cornejo et al. 2016). The secondary metabolite enzyme, ergosterol, is known to play a prominent role in the formation of the fungal cell wall and is also helpful in enhancing the defense mechanism against the invading pathogens (Abdel-Fattah et al. 2007). The fungus has also been used for the biological control of other fungi in case of plant pathogens (Shibazaki and Gonoi 2014).

T. asperellum (Mukherjee et al. 2008) and *T. reesei* (Vahabi et al. 2011) are the non-pathogenic fungi which have been reported to synthesize the AgNPs. Another mycoparasitic filamentous fungus *T. harzianum* is known to secrete a variety of hydrolytic enzymes and has the potential to inhibit the growth of plant pathogens and hence, due to its property of biomass conversion it has obtained greater attention from the industrial and research sectors. It is also considered as the best biological control agent because it is composed of the multi-enzymatic system consisting of proteases, chitinases, and glucanases. Other glycosyl hydrolases are also secreted by *T. harzianum* and consist of xylanases, secretome, cellulases, and mannanases (Do Vale et al. 2012) under suitable culture conditions. Furthermore, some compounds consisting of good reducing properties have also been found in *T. harzianum*, such as anthraquinones and naphthoquinones (Liu et al. 2007). These properties of *T. harzianum* have encouraged researchers to produce NPs and a group of researchers have produced AgNPs using this fungus (Ahluwalia et al. 2014). The ergosterol-producing *T. harzianum* also produces silver ions that can be applied to gain defense against other pathogenic fungi for plants. Silver salts are found to have antiseptic activity and the property has been reported in Greek, Indian and Roman texts for 5,000 years (Castro-Longoria et al. 2011).

The biosynthesis of AgNPs is preferred because the synthesis procedure is quite safe, and the procedure is ecofriendly. Because of the tolerance level of fungi and its capability of bio-accumulating the metals, the synthesis of NPs from fungi is considered an important branch. Since fungi are capable of producing the AgNPs biogenically and their biomass is easy to handle, they are favored over other microorganisms (El-Shanshoury et al. 2011). *T. harzianum, T. longibrachiatum, T. virens, T. pseudokoningii,* and *T. asperellum* are widely used in the synthesis of AgNPs (Devi et al. 2013; Elgorban 2016).

Mishra et al. (2014a) have reported on the viability of AgNPs against plant pathogens by using bacterium as a reducing agent. These NPs have successfully restricted the pathogenic fungus *Bipolaris sorokiniana* which is known to negatively affect the growth of wheat crops. Do Vale et al. (2012) reported on the anti-fungal potential of AgNPs, which significantly reduced the growth of different species of phytopathogenic fungi and their inhibitory effect, is dependent on the species of pathogen and type and concentration of AgNPs.

Researchers have shown the synthesis of AgNPs using the enterobacteria culture supernatant and AuNPs by using the supernatant of *Penicillium*. However, as these microbes involved in the rapid synthesis of NPs are also the well-known pathogen for plants, *Trichoderma* is often considered for the synthesis of Ag and AuNPs (Du et al. 2011).

4.5.1 MECHANISM OF ACTION OF NANOPARTICLES

Phytopathogenic fungi are killed by five mechanisms, i.e. antibiosis, induced systemic resistance, parasitism, cross protection mediated by virus and competition for the nutrients, and space (Ghorbanpour et al. 2018) (Figure 4.6). Biocontrol agents generally compete with the pathogens for nutrients such as nitrogen, oxygen, and carbohydrates by inhabiting the shared habitat such as photosphere, rhizosphere, and other plant tissues (Spadaro et al. 2010). The species of *Trichoderma* act as important biocontrol agents that are resistant to antimicrobial compounds produced by plants and compete for space and nutrients (Błaszczyk et al. 2014). Depending upon their properties and adapting to the environmental conditions, the fungal biocontrol agents exert almost the same mode of action. Parasitism includes the killing of plant pathogens by biocontrol agents through the destruction of propagules or by the development of organs such as haustoria, and by secretion of secondary metabolites and enzymes capable of destroying pathogenic fungal structures (Daguerre et al. 2014). Antibiosis, on the other hand, refers to the secretion of antimicrobial compounds which help in suppressing the growth of phytopathogenic fungi by biocontrol agents. *Trichoderma* produces a variety of antibiotics and wall-degrading enzymes and helps in protecting the plants from pathogenic fungi (Hermosa et al. 2014; Strakowska et al. 2014). The mechanisms behind the antimicrobial activity include the binding of the toxic metal ions such as Cd^{2+}, Cu^{2+}, Zn^{2+}, and Ag^{2+} released from their respective NPs to the –SH (sulfhydryl group) of sulfur-containing proteins, thereby affecting the proteins important for the membrane and affecting cell-permeability. These toxic ions are capable of destroying the DNA, altering the membrane potential, inducing protein oxidation, and hindering electron transport. The level of ROS increases, causing damage to the cell by inducing the metal-catalyzed oxidation reactions (El-Argawy et al. 2017).

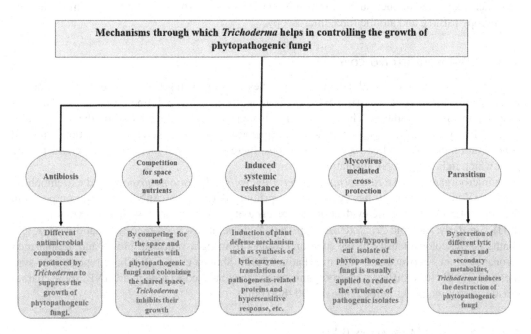

FIGURE 4.6 Mechanisms through which *Trichoderma* inhibits the invasion of pathogenic fungi.

AgNPs are found to be effective against a large number of microbes, viruses, and fungi. These NPs have served as a promising alternative to antibiotic therapy and have higher potential to resolve the problem of multidrug resistance, hence they are regarded as next-generation antibiotics (Rai and Ingle 2012; Stiufiuc et al. 2013). The antimicrobial activity of AgNPs is shown by binding to the cell wall or membrane thereby penetrating the cell and damaging the organelles like vacuoles, mitochondria, and ribosomes thus inducing the oxidative stress. Moreover, by managing the inflammatory response and modulating the AgNPs, they also help in modulating the immune system and managing the inflammatory response thereby helping in killing off the pathogenic microorganisms (Tian et al. 2007).

Cellular dysfunctions occur due to the interaction of AgNPs with the microbial membrane. The NPs alter the structure of the membrane and change its permeability and transport activity (Singh et al. 2015). Kvitek et al. (2008) have suggested that the antimicrobial activity of AgNPs can be enhanced by using anionic detergents such as sodium dodecyl sulfate (SDS) and other surfactants. AgNPs are found to damage the organelles, particularly with ribosomes, resulting in their degeneration and thus inhibition of protein synthesis through translation (Morones et al. 2005; Rai and Ingle 2012). Ag^+ on the other hand can bind with the functional groups present on the protein, leading to their deactivation (Klueh et al. 2000; Rai and Ingle 2012). The 3D structure of the proteins is altered by AgNPs and Ag^+ which can lead to functional defects in the protein of microorganisms (Lok et al. 2006). These NPs also interact with the nucleosides of the nucleic acid. The silver ion intercalates in between the nitrogenous bases and disrupts the hydrogen bonding between the base pairs (Klueh et al. 2000). Thus, by binding of the AgNPs, the transcription of genes in microorganisms is blocked (Morones et al. 2005). These NPs also change the state of DNA from relaxed to condensed form, therefore inhibiting its replication ability (Feng et al. 2000).

4.6 DIFFERENT TECHNIQUES AND INSTRUMENTATION USED FOR THE CHARACTERIZATION OF NANOPARTICLES

Preliminary characterization of synthesized nanomaterial is done using techniques like UV–visible absorption spectroscopy (UV-vis), Fourier transform infrared spectrum analysis, X-ray diffraction, and microscopic techniques such as transmission electron microscopy [TEM], scanning electron microscopy [SEM], and atomic force microscopy.

4.6.1 UV-VISIBLE SPECTROSCOPY

The primary detection of metal nanoparticle synthesis is done using UV-vis spectroscopy. For different metals, the UV-vis spectrum exhibits plasmon resonance at a defined wavelength that indicates the metal NP synthesis. The preliminary characterization is done based on the visible color change. Every metal NP has a characteristic absorption peak. On increase in reaction time and concentration of biological extracts with salt ions, there is a gradual increase in the characteristic peak which indicates the NP formation.

In microorganisms like *R. capsulata*, *Sclerotium rolfsii*, *Thermomonospora*, and extremophilic yeasts, UV-vis spectroscopy has been used for primary detection of AuNPs. Absorption peaks in the range of 500–550 nm are evident of the presence of AuNPs in these organisms (Mourato et al. 2011). Similarly, in some *Trichoderma* spp. and *A. flavus* it has been reported that during AgNP synthesis, the color intensity of the solution changes from light yellow to brown and the plasmon resonance ranges between 390–400 nm (Fatima 2016). Shobha et al. (2014) have reported the synthesis of CuNPs from different microorganisms and they have concluded that CuNPs show characteristic absorption peaks at the range of 200–800 nm.

4.6.2 ELECTRON MICROSCOPY (EM)

It is obligatory to traceably calibrate the size and form of individual particles by their direct imaging when characterization of NPs is conducted (Buhr et al. 2009). The high resolution and high imaging speed of scanning electron microscopy (SEM) and transmission electron microscopy (TEM) makes them the preferred methods for direct imaging and dimensional measurements of micro- and nanostructures. Das et al. (2009) reported that in order to confirm the crystalline nature of AuNPs, high resonance transmission electron microscopy (HRTEM) can be used which shows the lattice fringes of AuNPs synthesized from *R. oryzae*. Similarly, other microscopy techniques such as field emission scanning electron microscopy (FESEM) and atomic force microscopy (AFM) have been used for the characterization of AuNPs (Das et al. 2009; Sathishkumar et al. 2011). Sarvanakumar and Wang, (2018) reported synthesis of AgNPs from *T. atroviride* and performed TEM to find that the structure of biosynthesized AgNPs was anisotropic with an average size of 15–25 nm.

4.6.3 FOURIER TRANSFORM INFRARED SPECTROSCOPY (FTIR)

The structural features and the nature of associated functional groups of biological extracts with nanoparticles is characterized using Fourier transform infrared spectroscopy (FTIR). It measures infrared intensity vs. wavelength (wave number) of light that reflects the dependence of nanoparticle optical properties, viz. the resonance wavelength, the extinction cross-section, and the ratio of scattering to absorption, on the NP dimensions (Shobha et al. 2014). Using the FTIR spectrum several organic and inorganic molecules can be assigned to the metal NPs at a particular band. In FTIR spectrum many alkaline, amine, proteins, and aromatic peptides at the bands of 1,115.4 and 3,450 cm^{-1} were assigned to the metallic and O stretching vibrations of the metallic oxides respectively. Anand (2015) observed similar results with the FTIR spectrum for AgNPs with the 3,457.16 cm^{-1} synthesized by *Trichoderma gamsii*.

4.6.4 X-RAY DIFFRACTION

X-ray diffraction (XRD) helps in establishing the metallic nature of particles and provides information on translational symmetry, size, and shape of the unit cell from peak positions and information on electron density inside the unit cell (Prema 2010). XRD analysis is useful in the confirmation of synthesis of AgNPs through an approximate 20 nm particle size calculated using the Debye–Scherer equation (Shah et al. 2015). Phase composition along with the mean size of NPs is analyzed through XRD. Crystal structure and phase composition, as well as the mean size of the AuNPs, are analyzed by XRD spectroscopy (Bhambure et al. 2009; Binupriya et al. 2010).

4.6.5 ENERGY DISPERSIVE SPECTRUM (EDS), ENERGY DISPERSIVE X-RAY (EDX), AND X-RAY PHOTOELECTRON SPECTRAL (XPS) TECHNIQUES

Energy dispersive spectrum (EDS), energy dispersive X-ray (EDX), and X-ray photoelectron spectral (XPS) techniques can be used for the elemental analysis, electronic state, and chemical characterization (Das et al. 2009; He et al. 2007). EDS, EDX, and XPS have been used in the characterization of AuNPs synthesized by *E. coli*, *Y. lipolytica*, *Rhizopus oryzae*, and *R. capsulate* (Das et al. 2009; He et al. 2007; Shedbalkar et al. 2014). EDX has been used to display the elemental signal at the high percentage, indicating the affluent synthesis of AgNPs by *T. atroviride* KNUP001 (Saravanakumar and Wang 2018).

4.6.6 DYNAMIC LIGHT SCATTERING (DLS)

The size distribution profile of small NPs in suspension can be analyzed using DLS. It works on the principle of photon correlation spectroscopy or quasi-elastic light scattering. DLS quantifies the changes and fluctuations in the scattered light intensity due to the diffusion of NPs. The different techniques used in the characterization of nanoparticles are shown in Figure 4.7.

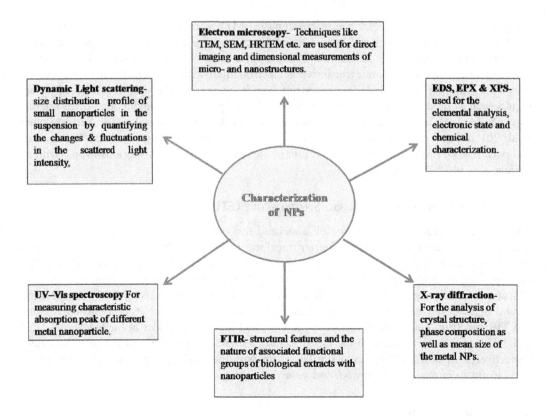

FIGURE 4.7   Different techniques used for characterization of nanoparticles.

REFERENCES

Abdel-Fattah, G.M., Shabana, Y.M., and Ismail, A.E. (2007). *Trichoderma harzianum*: A biocontrol agent against *Bipolaris oryzae. Mycopathologia* 164(2): 81–89.

Ahluwalia, V., Kumar, J., Sisodia, R., Shakil, N.A., and Walia, S. (2014). Green synthesis of silver nanoparticles by *Trichoderma harzianum* and their bio-efficacy evaluation against *Staphylococcus aureus* and *Klebsiella pneumonia. Ind Crop Prod* 55: 202–206.

Alloway, B.J. (2009). Soil factors associated with zinc deficiency in crops and humans. *Environ Geochem Health* 31(5): 537–548.

Anand, B.G. (2015). Biosynthesis of silver nano-particles by marine sediment fungi for a dose dependent cytotoxicity against HEp2 cell lines. *Biocatal Agric Biotechnol* 4(2): 150–157.

Atanasova, L., Druzhinina, I.S., and Jaklitsch, W.M. (2013). Two hundred *Trichoderma* species recognized on the basis of molecular phylogeny. In: *Trichoderma: Biology and Applications*, eds. P.K. Mukherjee, U.S. Singh,B.A.Horwitz, M. Schmoll, and M. Mukherjee, 10–42. Wallingford: CABI-UK.

Beveridge, T.J., and Murray, R.G.E. (1980). Sites of metal deposition in the cell wall of *Bacillus subtilis. J Bacteriol* 141(2): 876.

Bhambure, R., Bule, M., Shaligram, N., Kamat, M., and Singhal, R. (2009). Extracellular biosynthesis of gold nanoparticles using *Aspergillus niger*—Its characterization and stability. *Chem Eng Technol* 32(7): 1036–1041.

Bhattacharyya, A., Duraisamy, P., and Govindarajan, M. (2016). Nano-biofungicides: Emerging trend in insect pest control. In: *Advances and Applications through Fungal Nanobiotechnology*, ed. R. Prasad, 307–319. Cham: Springer International Publishing.

Binupriya, A.R., Sathishkumar, M., and Yun, S.I. (2010). Biocrystallization of silver and gold ions by inactive cell filtrate of Rhizopus stolonifer. *Colloids Surf B Biointerfaces* 79(2): 531–534.

Błaszczyk, L., Siwulski, M., Sobieralski, K., Lisiecka, J., and Jędryczka, M. (2014). *Trichoderma* spp.: Application and prospects for use in organic farming and industry. *J Plant Protect Res* 54(4): 309–317.

Buhr, E., Senftleben, N., Klein, T., Bergmann, D., Gnieser, D., Frase, C.G., and Bosse, H. (2009). Characterization of nanoparticles by scanning electron microscopy in transmission mode. *Meas Sci Technol* 20(8): 084025.

Castro-Longoria, E., Vilchis-Nestor, A.R., and Avalos-Borja, M. (2011). Biosynthesis of silver, gold and bimetallic nanoparticles using the filamentous fungus *Neurospora crassa*. *Colloids Surf B Biointerfaces* 83(1): 42–48.

Contreras-Cornejo, H.A., Macías-Rodríguez, L., del-Val, E., and Larsen, J. (2016). Ecological functions of *Trichoderma* spp. and their secondary metabolites in the rhizosphere: Interactions with plants. *FEMS Microbiol Ecol* 92(4): fiw036.

Corradini, E., de Moura, M.R., and Mattoso, L.H.C. (2010). A preliminary study of the incorparation of NPK fertilizer into chitosan nanoparticles. *eXPRESS Polym Lett* 4(8): 509–515.

Daguerre, Y., Siege, K., and Edel-Hermann, V. (2014). Fungal proteins and genes associated with biocontrol mechanisms of soil-borne pathogens: A review. *Fungal Biol Rev* 28(4): 97–125.

Das, S.K., Das, A.R., and Guha, A.K. (2009). Gold nanoparticles: Microbial synthesis and application in water hygiene management. *Langmuir* 25(14): 8192–8199.

Das, S.K., Khan, M.M.R., Guha, A.K., Das, A.R., and Mandal, A.B. (2012). Silver-nano biohybride material: Synthesis, characterization and application in water purification. *Bioresour Technol* 124: 495–499.

Dehner, C.A., Barton, L., and Maurice, P.A. (2010). Size-dependent bioavailability of hematite (α-Fe2O3) nanoparticles to a common aerobic bacterium. *Environ Sci Technol* 45(3): 977–983.

Devi, T.P., Kulanthaivel, S., and Kamil, D. (2013). Biosynthesis of silver nanoparticles from *Trichoderma* species. *Indian J Exp Biol* 51(7): 543–547.

Dimkpa, C.O. (2014). Can nanotechnology deliver the promised benefits without negatively impacting soil microbial life? *J Basic Microbiol* 54(9): 889–904.

Do Vale, L.H., Gómez-Mendoza, D.P., and Kim, M.S. (2012). Secretome analysis of the fungus *Trichoderma harzianum* grown on cellulose. *Proteomics* 17(17): 2716–2728.

Du, W., Sun, Y., and Ji, R. (2011). TiO$_2$ and ZnO nanoparticles negatively affect wheat growth and soil enzyme activities in agricultural soil. *J Environ Monit* 13(4): 822–828.

Du, W., Tan, W., and Peralta-Videa, J.R. (2017). Interaction of metal oxide nanoparticles with higher terrestrial plants: Physiological and biochemical aspects. *Plant Physiol Biochem* 110: 210–225.

Duhan, J.S., Kumar, R., Kumar, N., Kaur, P., Nehra, K., and Duhan, S. (2017). Nanotechnology: The new perspective in precision agriculture. *Biotechnol Rep* 15: 11–23.

Durán, N., Marcato, P.D., and Alves, O.L. (2005). Mechanistic aspects of biosynthesis of silver nanoparticles by several *Fusarium oxysporum* strains. *J Nanobiotechnology* 3(1): 8.

Elamawi, R.M., Al-Harbi, R.E., and Hendi, A.A. (2018). Biosynthesis and characterization of silver nanoparticles using *Trichoderma longibrachiatum* and their effect on phytopathogenic fungi. *Egypt J Biol Pest Control* 28(1): 28.

El-Argawy, E., Rahhal, M.M., and El-Korany, A. (2017). Efficacy of some nanoparticles to control damping-off and root rot of sugar beet in El-Behiera Governorate. *Asian J Plant Pathol* 11: 35–47.

Elgorban, A.M., Al-Rahmah, A.N., and Sayed, S. (2016). Antimicrobial activity and green synthesis of silver nanoparticles using *Trichoderma viride*. *Biotechnol Biotechnol Equip* 30(2): 299–304.

El-Shanshoury, A.E., ElSilk, S.E., and Ebeid, M.E. (2011). Extracellular biosynthesis of silver nanoparticles using *Escherichia coli* ATCC 8739, *Bacillus subtilis* ATCC 6633, and *Streptococcus thermophilus* ESh1 and their antimicrobial activities. *ISRN Nanotechnol* 2011: 1–7.

Fatima, F. (2016). Extracellular mycosynthesis of silver nanoparticles and their microbicidal activity. *J Glob Antimicrob Resist* 7(Supplement C): 88–92.

Fayaz, A.M., Balaji, K., and Girilal, M. (2010). Biogenic synthesis of silver nanoparticles and their synergistic effect with antibiotics: A study against gram-positive and gram-negative bacteria. *Nanomedicine* 6(1): 103.

Feng, Q.L., Wu, J., and Chen, G.Q. (2000). A mechanistic study of the antibacterial effect of silver ions on *Escherichia coli* and *Staphylococcus aureus*. *J Biomed Mater Res B Appl Biomater* 52(4): 662–668.

Fraceto, L.F., Maruyama, C.R., Guilger, M., Mishra, S., Keswani, C., Singh, H.B., and deLima, R. (2018). *Trichoderma harzianum* based novel formulations: Potential Applications for Management of Next-Gen agricultural challenges. *J Chem Technol Biotechnol* 93(8): 2056–2063.

Gawai, D.U. (2015). Antifungal activity of essential oil of *cymbopogoncitratus*stapf against different *Fusarium* species. *Bionano Front* 8(2): 186–189.

Ghorbanpour, M., Omidvari, M., and Abbaszadeh, D.P. (2018). Mechanisms underlying the protective effects of beneficial fungi against plant diseases. *Biol Control* 117: 147–157.

Gogos, A., Knauer, K., and Bucheli, T.D. (2012). Nanomaterials in plant protection and fertilization: Current state foreseen applications and research priorities. *J Agric Food Chem* 60(39): 9781–9792.

Goswami, J., Pandey, R.K., Tewari, J.P., and Goswami, B.K. (2008). Management of root knot nematode on tomato through application of fungal antagonists, Acremonium strictum and *Trichoderma harzianum*. *J Environ Sci Health B* 43(3): 237–240.

Gouin, S. (2004). Microencapsulation: Industrial appraisal of existing technologies and trends. *Trends Food Sci Technol* 15(7–8): 330–347.

Harold, C., Passam, Ioannis., C., Karapanos, Penelope, J.B., and Dimitrios, S. (2007). A review of recent research on tomato nutrition, breeding and post-harvest technology with reference to fruit quality, the European. *J Plant Sci Biotechnol* 1(1): 1–21.

He, S., Guo, Z., Zhang, Y., Zhang, S., Wang, J., and Gu, N. (2007). Biosynthesis of gold nanoparticles using the bacteria *Rhodopseudomonas capsulata*. *Mater Lett* 61(18): 3984–3987.

Hermosa, R., Cardoza, M.B., and Rubio, M.E. (2014). Secondary metabolism and antimicrobial metabolites of *Trichoderma*. In: *Biotechnology and Biology of Trichoderma*, eds. V.K. Gupta, M. Schmoll, A. Herrera-Estrella, R.S. Upadhyay, I. Druzhinina, and M. Tuohy, 125–137. Amsterdam: Elsevier.

Hussein, M. (2016). Silver tolerance and silver nanoparticle biosynthesis by Neoscytalidiumnovaehollandae and Trichoderma inhamatum. *Eur J Biol Res* 6(1): 28–35.

Jaidev, L.R., and Narasimha, G. (2010). Fungal mediated biosynthesis of silver nanoparticles, characterization and antimicrobial activity. *Colloids Surf B* 81(2): 430.

Jaklitsch, W.M. (2009). European species of *Hypocrea* Part I. The green-spored species. *Stud Mycol* 63: 1–91.

Janmohammadi, M., Amanzadeh, T., and Sabaghnia, N. (2016). Impact of foliar application of nano micronutrient fertilizers and titanium dioxide nanoparticles on the growth and yield components of barley under supplemental irrigation. *Acta Agric Slov* 107(2): 265–276.

Kandasamy, S., and Prema, R.S. (2015). Methods of synthesis of nano particles and its applications. *J Chem Pharm* 7: 278–285.

Keswani, C., Mishra, S., Sarma, B.K., Singh, S.P., and Singh, H.B. (2014). Unraveling the efficient applications of secondary metabolites of various *Trichoderma* spp. *Appl Microbiol Biotechnol* 98(2): 533–544.

Keswani, C., Prakash, O., Bharti, N., Vílchez, J.I., Sansinenea, E., Lally, R.D., Borriss, R., Singh, S.P., Gupta, V.K., Fraceto, L.F. and de Lima, R. (2019a). Re-addressing the biosafety issues of plant growth promoting rhizobacteria. *Sci Total Environ* 690: 841–852.

Keswani, C., Dilnashin, H., Birla, H., and Singh, S.P. (2019b). Re-addressing the commercialization and regulatory hurdles for biopesticides in India. *Rhizosphere* 11: 100155.

Klueh, U., Wagner, V., and Kelly, S. (2000). Efficacy of silver-coated fabric to prevent bacterial colonization and subsequent device-based biofilm formation. *J Biomed Mater Res B Appl Biomater* 53(6): 621–631. The Japanese Society for Biomaterials, and The Australian Society for Biomaterials and the Korean Society for Biomaterials.

Kvitek, L., Panáček, A., and Soukupova, J. (2008). Effect of surfactants and polymers on stability and antibacterial activity of silver nanoparticles (NPs). *J Phys Chem Biophys C* 112(15): 5825–5834.

Liu, S.Y., Lo, C.T., and Chen, C. (2007). Efficient isolation of anthraquinone-derivatives from *Trichoderma harzianum* ETS 323. *J Biochem Biophys Methods* 70(3): 391–395.

Llop, J., Estrela,-Lopis, I., and Ziolo, R.F. (2014). Uptake, biological fate, and toxicity of metal oxide nanoparticles. *Part Part Syst Charact* 31(1): 24–35.

Lok, C.N., Ho, C.M., and Chen, R. (2006). Proteomic analysis of the mode of antibacterial action of silver nanoparticles. *J Proteome Res* 5(4): 916–924.

Lorito, M., Woo, S.L., and Harman, G.E. (2010). Translational research on *Trichoderma*: From'omics to the field. *Annu Rev Phytopathol* 48: 395–417.

Lu, C., Zhang, C., and Wen, J. (2002). Research of the effect of nanometer materials on germination and growth enhancement of Glycine max and its mechanism. *Soybean Sci* 21(3): 168–171.

Machowicz-Matejko, E., and Zalewska, E.D. (2015). Pharmacological substances in vitro in limiting growth and development of fungi Colletotrichum genera. *J Ocul Pharmacol Ther* 31(5): 303–309.

Maliszewska, I. (2013). Microbial mediated synthesis of gold nanoparticles: Preparation, characterization and cytotoxicity studies. *Dig J Nanomater Bios* 8(3): 1123–1131.

Millan, G., Agosto, F., and Vazquez, M. (2008). Use of clinoptilolite as a carrier for nitrogen fertilizers in soils of the Pampean regions of Argentina. *Cienc Investig Agrar* 35(3): 293–302.

Mishra, S., Keswani, C., Abhilash, P.C., Fraceto, L.F., and Singh, H.B. (2017). Integrated approach of Agri-nanotechnology: Challenges and future trends. *Front Plant Sci.* doi:10.3389/fpls.2017.00471.

Mishra, S., Keswani, C., Singh, A., Singh, B.R., Singh, S.P., and Singh, H.B. (2016). Microbial nanoformulation: Exploring potential for coherent nano-farming. In: *The Handbook of Microbial Bioresources*, eds. V.K. Gupta, G.D. Sharma, M.G. Tuohy, and R. Gaur, 107–120. Wallingford, Oxfordshire, UK: CABI-UK.

Mishra, S., Singh, B.R., Singh, A., Keswani, C., Naqvi, A.H., and Singh, H.B. (2014a). Biofabricated silver nanoparticles act as a strong fungicide against *Bipolaris sorokiniana* causing spot blotch disease in wheat. *PloS One* 9(5): e97881.

Mishra, S., Singh, A., Keswani, C., and Singh, H.B. (2014b). Nanotechnology: Exploring potential application in agriculture and its opportunities and Constraints. *Biotech Today* 4(1): 9–14.

Montero-Barrientos, M., Hermosa, R., Cardoza, R.E., Gutiérrez, S., and Monte, E. (2011). functional analysis of the *Trichoderma harzianum nox1* gene, encoding an NADPH oxidase, relates production of reactive oxygen species to specific biocontrol activity against *Pythium ultimum*. *Appl Environ Microbiol* 77(9): 3009–3016.

Morones, J.R., Elechiguerra, J.L., and Camacho, A. (2005). The bactericidal effect of silver nanoparticles. *Nanotechnology* 16(10): 2346.

Mourato, A., Gadanho, M., Lino, A.R., and Tenreiro, R. (2011). Biosynthesis of crystalline silver and gold nanoparticles by extremophilic yeasts. *Bioinorg Chem Appl* 2011: 1–8.

Mukherjee, P., Roy, M., and Mandal, B.P. (2008). Green synthesis of highly stabilized nanocrystalline silver particles by a nonpathogenic and agriculturally important fungus *T. asperellum*. *Nanotechnology* 19: 103.

Mukherjee, P.K., Horwitz, B.A., Singh, U.S., Mukherjee, M., and Schmoll, M. (2013). *Trichoderma* in agriculture, industry and medicine: An overview. In: *Trichoderma: Biology and Applications*, eds. P.K. Mukherjee, B.A. Horwitz, U.S. Singh, M. Mukherjee, and M. Schmoll, 1–9. Nosworthy, Way, Wallingford, Oxon, UK: CABI.

Narayanan, K.B., and Sakthivel, N. (2010). Biological synthesis of metal nanoparticles by microbes. *Adv Colloid Interface Sci* 156(1–2): 1–13.

Narware, J., Yadav, R.N., Keswani, C., Singh, S.P., and Singh, H.B. (2019). Silver nanoparticle-based biopesticides for phytopathogens: Scope and potential in agriculture. In: *Nano-Biopesticides Today and Future Perspectives*, ed. O. Koul, 303–314. Cambridge, MA: Academic Press.

Nuruzzaman, M.D., Rahman, M.M., and Liu, Y. (2016). Nanoencapsulation, nano-guard for pesticides: A new window for safe application. *J Agric Food Chem* 64(7): 1447–1483.

Omran, B.A., Nassar, H.N., Younis, S.A., Fatthallah, N.A., Hamdy, A., El-Shatoury, E.H., and El-Gendy, N.S. (2018). Physiochemical properties of *Trichoderma longibrachiatum* DSMZ 16517-synthesized silver nanoparticles for the mitigation of halotolerant sulphate-reducing bacteria. *J Appl Microbiol* 126(1): 138–154.

Othman, A.M., Elsayed, M.A., Elshafei, A.M., and Hassan, M.M. (2017). Application of response surface methodology to optimize the extracellular fungal mediated nanosilver green synthesis. *Genet Eng Biotechnol J* 15(2): 497–504.

Park, H.J., Kim, S.H., and Kim, H.J. (2006). A new composition of nanosized silica-silver for control of various plant diseases. *Plant Pathol J* 22(3): 295–302.

Persoon, C.H. (1794). NeuerVersucheinersystematischenEintheilung der Schwämme. *NeuesMagazinfür Bot* 1: 63–128.

Prabhu, D., Vidhyavathi, R., and Jeyakanthan, J. (2017). Computational identification of potent inhibitors for streptomycin 3″-adenylyltransferase of *Serratia marcescens*. *Microb Pathog* 103: 94–106.

Prasad, R., Bhattacharyya, A., and Nguyen, Q.D. (2018). Nanotechnology in sustainable agriculture: Recent developments, challenges, and perspectives. *Front Microbiol* 8: 1014.

Prema, P. (2010). Chemical mediated synthesis of silver nanoparticles and its potential antibacterial application. *Analysis and Modeling to Technol Appl* 6: 151–166.

Ragaei, M., and Sabry, A.K.H. (2014). Nanotechnology for insect pest control. *Int J Sci Environ Technol* 3(2): 528–545.

Rai, M., and Ingle, A. (2012). Role of nanotechnology in agriculture with special reference to management of insect pests. *Appl Microbiol Biotechnol* 94(2): 287–293.

Rai, M., Ingle, A.P., Paralikar, P., Anasane, N., Gade, R., and Ingle, P. (2018). Effective management of soft rot of ginger caused by Pythium spp. and Fusarium spp.: Emerging role of nanotechnology. *Appl Microbiol Biotechnol* 102(16): 6827–6839.

Raliya, R., Biswas, P., and Tarafdar, J.C. (2015). TiO2 nanoparticle biosynthesis and its physiological effect on mung bean (*Vigna radiata* L.). *Biotechnol Rep* 5: 22–26.

Raliya, R., Saharan, V., and Dimkpa, C. (2017). Nanofertilizer for precision and sustainable agriculture: Current state and future perspectives. *J Agr Food Chem* 66(26): 6487–6503.

Rana, S., and Kalaichelvan, P.T. (2013). Ecotoxicity of nanoparticles. *ISRN Toxicol* 2013: 1–11.

Salvadori, M.R., Ando, R.A., Oller, Do., Nascimento, C.A., and Correa, B. (2014). Bioremediation from wastewater and extracellular synthesis of copper nanoparticles by the fungus *Trichoderma koningiopsis*. *J Environ Sci Health A* 49(11): 1286–1295.

Sanghi, R., and Verma, P. (2009). Biomimetic synthesis and characterisation of protein capped silver nanoparticles. *Bioresour Technol* 100(1): 501–504.

Saraf, M., Pandya, U., and Thakkar, A. (2014). Role of allelochemicals in plant growth promoting rhizobacteria for biocontrol of phytopathogens. *Microbiol Res* 169(1): 18–29.

Saravanakumar, K., Shanmugam, S., Varukattu, N.B., MubarakAli, D., Kathiresan, K., and Wang, M.H. (2019). Biosynthesis and characterization of copper oxide nanoparticles from indigenous fungi and its effect of photothermolysis on human lung carcinoma. *J Photochem Photobiol B Biol* 190: 103–109.

Saravanakumar, K., and Wang, M.H. (2018). Trichoderma based synthesis of anti-pathogenic silver nanoparticles and their characterization, antioxidant and cytotoxicity properties. *Microb Pathog* 114: 269–273.

Sathishkumar, K., Amutha, R., Arumugam, P., and Berchmans, S. (2011). *J Appl Mater Interfaces* 3: 1418.

sathiskumar, M., Sneha, K., Won, S.W., Cho, C.W., and Kim, S. (2009). Cinnamon zeylanicum bark extract and powder mediated green synthesis of nanocrystalline silver particles and its bactericidal activity. *Colloids Surf B Biointerfaces* 73(1): 332–338.

Shah, A.T., Din, M.I., Bashir, S., Qadir, M.A., and Rashid, F. (2015). Green synthesis and characterization of silver nanoparticles using Ferocactusechidne extract as a reducing agent. *Anal Lett* 48(7): 1180–1189.

Shedbalkar, U., Singh, R., Wadhwani, S., Gaidhani, S., and Chopade, B.A. (2014). Microbial synthesis of gold nanoparticles: Current status and future prospects. *Adv Colloid Interface Sci* 209: 40–48.

Shelar, G.B., and Chavan, A.M. (2015). Myco-synthesis of silver nanoparticles from *Trichoderma harzianum* and its impact on germination status of oil seed. *Biolife* 3(1): 109–113.

Shibazaki, A., and Gonoi, T. (2014). Lectin-microarray technique for glycomic profiling of fungal cell surfaces. *Methods Mol Biol* 1200: 287–294.

Shobha, G., Vinutha, M., and Ananda, S. (2014). Biological Synthesis of Copper Nanoparticles and its impact. *Int J Pharm Sci Invent* 3(8): 28–38.

Shrivastava, S., Prasad, R., and Varma, A. (2014). Anatomy of root from eyes of a microbiologist. In: *Root Engineering*, eds. A. Morte, and A. Varma, vol. 40, 3–22. Berlin/Heidelberg: Springer.

Singh, B.R., Singh, B.N., and Singh, A. (2015). Mycofabricated biosilver nanoparticles interrupt *Pseudomonas aeruginosa* quorum sensing systems. *Sci Rep* 5: 13719.

Singh, S., Vishwakarma, K., and Singh, S. (2017). Understanding the plant and nanoparticle interface at transcriptomic and proteomic level: A concentric overview. *Plant Gene* 11: 265–272.

Spadaro, D., Ciavorella, A., and Dianpeng, Z. (2010). Effect of culture media and pH on the biomass production and biocontrol efficacy of a *Metschnikowia pulcherrima* strain to be used as a biofungicide for postharvest disease control. *Can J Microbiol* 56(2): 128–137.

Stiufiuc, R., Iacovita, C., and Lucaciu, C.M. (2013). SERS-active silver colloids prepared by reduction of silver nitrate with short-chain polyethylene glycol. *Nanoscale Res Lett* 8(1): 47.

Strakowska, J., Błaszczyk, L., and Chełkowski, J. (2014). The significance of cellulolytic enzymes produced by *Trichoderma* in opportunistic lifestyle of this fungus. *J Basic Microbiol* 54(Supplement1): S2–13.

Sundaravadivelan, C., and Padmanabhan, M.N. (2014). Effect of mycosynthesized silver nanoparticles from filtrate of *Trichoderma harzianum* against larvae and pupa of dengue vector Aedesaegypti L. *Environ Sci Pollut R* 21(6): 4624–4633.

Takkar, P.N., and Walker, C.D. (1993). The distribution and correction of zinc deficiency. In: *Zinc in Soils and Plants, Development in Plant and Soil Science*, ed. A.D. Robson, 151–165. Boston: Kluwer Academic Publishers.

Tarafdar, J.C., Raliya, R., and Mahawar, H. (2014). Development of zinc nanofertilizer to enhance crop production in pearl millet (*Pennisetum americanum*). *Agric Res* 3(3): 257–262.

Thirumurugan, G., Shaheedha, S.M., and Dhanaraju, M.D. (2009). In vitro evaluation of antibacterial activity of silver nanoparticles synthesised by using *Phytophthora infestans*. *Int J ChemTech Res* 1(3): 714–716.

Tian, J., Wong, K.K., and Ho, C.M. (2007). Topical delivery of silver nanoparticles promotes wound healing. *ChemMedChem Chem Enabling Drug Discov* 2(1): 129–136.

Tripathi, D.K., Singh, S., and Singh, S. (2016). Impact of nanoparticles on photosynthesis: Challenges and opportunities. *Materials Focus* 5(5): 405–411.

Tripathi, R.M., Gupta, R.K., Shrivastav, A., Singh, M.P., Shrivastav, B.R., and Singh, P. (2013). *Trichoderma koningii* assisted biogenic synthesis of silver nanoparticles and evaluation of their antibacterial activity. *Adv Nat Sci Nanosci Nanotechnol* 4(3): 32–35.

Tripathi, R.M., Shrivastav, B.R., and Shrivastav, A. (2018). Antibacterial and catalytic activity of biogenic gold nanoparticles synthesised by *Trichoderma harzianum*. *IET Nanobiotechnol* 12(4): 509–513.

Vahabi, K., and Dorcheh, S.K. (2014). Biosynthesis of silver nano-particles by Trichoderma and its medical applications. In: *Biotechnology and Biology of Trichoderma*, eds. V.K. Gupta, M. Schmoll, A. Herrera-Estrella, R.S. Upadhyay, I. Druzhinina, and M.G. Tuohy, 393–404. Amsterdam: Elsevier.

Vahabi, K., Mansoori, G.A., and Karimi, S. (2011). Biosynthesis of silver nanoparticles by fungus *Trichoderma reesei* (a route for large-scale production of AgNPs). *Insciences J* 1(1): 65–79.

Vidotti, M., Carvalhal, R.F., and Mendes, R.K. (2011). Biosensors based on gold nanostructures. *J Brazil Chem Soc* 22(1): 3–20.

Vijayakumar, M., Priya, K., Nancy, F.T., Noorlidah, A., and Ahmed, A.B.A. (2013). BioSyn-thesis, characterisation and anti-bacterial effect of plant-mediated silvernanoparticles using *Artemisia nilagirica*. *Ind Crops Prod* 41: 235–240.

Vinale, F., Sivasithamparam, K., Ghisalberti, E.L., Marra, R., Woo, S.L., and Lorito, M. (2008). Trichoderma–plant–pathogen interactions. *Soil Biol Biochem* 40(1): 1–10.

Waghunde, R.R., Shelake, R.M., and Sabalpara, A.N. (2016). Trichoderma: A significant fungus for agriculture and environment. *Afr J Agric Res* 11(22): 1952–1965.

Weindling, R. (1932). Trichoderma lignorum as a parasite of other soil fungi. *Phytopathology* 22: 837–845.

Zhang, X.F., Liu, Z.G., Shen, W., and Gurunathan, S. (2016). Silver nanoparticles: Synthesis, characterization, properties, applications, and therapeutic approaches. *Int J Mol Sci* 17(9): 1–34.

5 Role of Nanopesticides in Agricultural Development

Al-Kazafy Hassan Sabry

CONTENTS

5.1 INTRODUCTION

It is expected that world population will reach 9.8 billion by 2050 (WPF 2016). This increase in world population means an increasing demand for food which will need an increase in crop, vegetable, and fruit production. Therefore, food production needs a revolution in all fields of agricultural development. Pesticides are considered the main component in agricultural development. By using pesticides, every dollar spent in agriculture production yields four dollars (Pimente et al. 1992). This means that using pesticides increases agricultural production by 300%. Peshin and Dhawan (2009) estimated that agricultural pests (harmful insects, fungi, viruses, etc.) caused 2,000 billion dollars of damage to agricultural production each year (Fareed et al. 2013). Despite the advantages of pesticide use, there are also some disadvantages. These disadvantages include pesticide poisoning. Each year there are approximately one million human pesticide poisonings and 20,000 deaths (WHO/UNEP 1989).

Using nanotechnology in pesticide formulation can reduce the amount of pesticide used against pests, and also reduce the costs of pesticides by the use of nanopesticide technology (Mishra et al. 2014a; Mishra et al. 2014b). Using nanopesticides improved the physical traits of pesticide formulation such as stability and slowed the release of active ingredients (Bang et al. 2009).

This review aims to throw some light on the role of nanopesticide formulations in increasing agricultural production.

5.2 WHAT ARE PESTICIDES?

According to the Federal Insecticides, Fungicides and Rodenticides Act (FIFRA) pesticides are defined as any substance or mixture of substances intended for preventing, destroying antifeedant or mitigating pests (FAO 1989). Therefore, pesticides can be natural and artificial formulations; they include insecticides, fungicides, rodenticides, etc.

5.2.1 What Are Nanopesticides?

Nanopesticides are defined as any pesticide formulation including nanometer-size structures (Kah et al. 2013). This nanosize structure ranges from 1 to less than 1,000 nm. On the other hand, nanopesticides can be defined as small engineered structures which provide the pesticides with new properties not found in the original pesticides, such as slow degradation and release of active ingredients and increased efficacy (Hemraj 2017; Mishra et al. 2017).

5.2.2 Advantages of Nanopesticides

There are many advantages to nanopesticides compared with the normal formulation. These advantages include:

1. **High stability:** nanopesticides are more stable than the normal formulation. This allows them to be more effective against pests,
2. **High solubility:** the higher solubility of nanopesticides means they are more effective.
3. **Decrease in the active ingredient:** the nanoemulsion of nanopesticides has less active ingredients than the normal formulation. This means there is a decrease in toxicity to non-target organisms and a decrease in pesticide contamination.
4. **Decrease in adjuvants (additives, surfactant, emulsion, dispersion, etc.)** This means a decrease in production costs.
5. **Nanoemulsion pesticides have small particles compared with the normal formulation:** These nanoparticles allow easy penetration of the plant cuticle or insect cuticle, making the toxic action faster. Small particles also mean more surface area.

5.2.3 Types of Nanopesticide Formulations

There are four main types of nanopesticide formulations:

5.2.3.1 Nanoemulsion Pesticides

Nanoemulsion pesticides mean that a kinetically stable clear dispersion consists of two immiscible phases, the oil phase and the water phase, and other additives such as surfactant molecules. This nanoemulsion has small particles or droplets, with a size range between 5 and 200 nm and usually comprises the dispersed phase (Shah and Bhalodia 2010). Jaiswal et al. (2015) defined nanoemulsion as a colloidal particulate system with a particle size range of 10 to 1,000 nm.

Nanopesticides consist of two main parts; the first one is the active ingredient which is responsible for the toxic action. The second part is the emulsion, which allows the mixing of the water and oil (Trotta 1999). The main role of emulsion is to reduce the surface tension to under 10 dynes/cm. Due to the reduction of the surface tension, the oil and water will mix perfectly. On the other hand, Tadros et al. (2004) recorded that nanoemulsion consists of a lipophilic phase and hydrophilic surfactant.

5.2.3.1.1 Role of Nanoemulsion Insecticide in Agricultural Development

Many nanoemulsion insecticide formulations have been used against insect pests. Louni et al. (2018) used a botanical extract from *Mentha longifolia* in a nanoemulsion formulation against Mediterranean flour moth, *Ephestia khehniella* larvae. The size of the nanoemulsion particles range was 14–36 nm. The results obtained showed that the nanoemulsion was more effective than the normal formulation. Pant et al. (2017) prepared a nanoemulsion formulation for permethrin (pyrethroid pesticides) and tested it against American cockroaches. Zeng et al. (2019) evaluated norcantharidin nanoemulsion against the diamond-back moth, *Plutella xylostella*. The results were clear that the nanoemulsion formulation of norcantharidin was more stable and soluble than the normal formulation. The effectiveness of norcantharidin nanoemulsion (the LC_{50} 60.414 mg/l) was increased against *P. xylostella* larvae compared with the normal formulation (LC_{50} 175.602 mg/l). The mortality percentage for the 3rd instar larvae reached 90% after 48 hours of nanoemulsion norcantharidin treatment. Zhao et al. (2017) prepared positive charge nanoemulsion of lambda-cyhalothrin to increase the binding affinity between the plant leaves and treated insecticide. The results obtained showed a high penetration or absorption of positive charged nanoemulsion through the plant leaves. The stability of lambda-cyhalothrin nanoemulsion was increased compared with normal emulsion. Abouelkassem et al. (2015) used nanoemulsion and normal formulation of jojoba extract against the rice weevil, *Sitophilus oryzae*. The results showed that the nanoemulsion of jojoba extract was more effective against *S. oryzae* than the normal formulation. The LC_{50}s were 0.31 and 3.12 ml/kg wheat. Fernandes et al. (2014) prepared an insecticide nanoemulsion extracted from *Manilkara subsericea* fruits. This nanoemulsion consisted of 5% of *M. subsericea* extract, 85% water, 5% oil and 5% sorbitanmonooleate/polysorbate 80 (as a surfactant). The results showed that this nanoemulsion was effective against *Dysdercus peruvianus* (cotton pest). The primary target site of this nanoemulsion is acetylcholine esterase. The results showed this nanoemulsion was also safe for non-target organisms. Mossa et al. (2017) used a nanoemulsion of camphor essential oil against *Sitophilus granarius* (wheat weevil). The results showed that nanoemulsion of camphor essential oil was highly effective against *S. granarius*. The LC_{50} of nanoemulsion was 181.49 $\mu g\ g^{-1}$ compared with 282.01 $\mu g\ g^{-1}$ with the normal emulsion. Choupanian and Omar (2018) tested the nanoemulsion of neem oil against the adults of both *Sitophilus oryzae* and *Tribolium castaneum*. The neem oil nanoemulsion caused 100% mortality for *S. oryzae* adults after only 24 hours of treatment. The results also showed that the adult *S. oryzae* were more susceptible to the nanoemulsion than *T. castaneum*. On the other hand, camphor essential oil was extracted from *Eucalyptus globules* leaves, converted to nanoemulsion formulation by ultrasound and used against wheat weevil, *S. granaries* (Mossa et al. 2017). The nanoemulsion droplet size of 99.0 nm was reached after 40 minutes of sonication. The LC_{50} of this nanoemulsion formulation was 181.49 $\mu g/g$ compared to 282.01 $\mu g/g$ in normal emulsion formulation.

5.2.3.1.2 Role of Nanoemulsion Herbicides in Agricultural Development

Herbicides are very important in controlling harmful weeds. Most pesticides used in integrated pest management programs are herbicides. Due to less than 0.1% of the active ingredient reaching the target site of treated weeds, especially with foliage treatments, the nanoemulsion herbicide formulation is the best solution for this problem. Lim et al. (2013) prepared a nanoemulsion of glyphosate isopropylamine to control creeping foxglove, slender button weed, and buffalo grass. This nanoemulsion was prepared by mixing short-chain alkyl polyglucosides (SAPGs) as a surfactant with organosilicone. The second step was mixing this mixture to the oil phase (the short-chain fatty acid methyl esters, SFAMEs). The mixture was stirred until the emulsion formed and then water was added. The results obtained showed the nanoemulsion herbicide formulation increased the herbicide penetration through plant cuticles and therefore increased the herbicide activity. Hazrati et al. (2017) used green nanoemulsion against harmful weeds. They tested the activity of nanoemulsion containing garden savory (*Satureja hortensis*) against *Amaranthus retroflexus* and *Chenopodium album*. The diameter of the nanoemulsion droplets was less than 130 nm. The results showed that the green nanoemulsion was effective against both weeds at 2–4 true leaf stages and therefore a promising natural herbicide, especially under greenhouse conditions. Balah, and Abd El Azim (2016) prepared natural nanoemulsion from *Thymus capitatus* and *Majorana hortensis* against the seeds and seedling growth of both *Convolvulus arvensis* and *Setaria viridis*. Two types of formulation were prepared: nanoemulsion and microemulsion. The size of nanoemulsion particles ranged between 5 and 22 nm. The results showed that the nanoemulsion formulation exhibited high activity as a herbicide, especially against the *C. arvensis* at 5–7 leaf stage in the greenhouse.

5.2.3.1.3   Role of Nanoemulsion Fungicides in Agricultural Development

Any toxic substance that kills or inhibits the growth of fungi is called a fungicide. Fungicide is an important part of an integrated pest management program. Nanotechnology trends use nanoemulsion formulation to control pathogenic fungi and reduce the environmental contamination of the traditional formulation of fungicides. da Silva et al (2019) used nanoemulsion of mancozeb and eugenol against the *Glomerella cingulated* fungus which is responsible for the rotting of ripe grapes. Eugenol nanoemulsion was the safest formulation for use in agriculture and least toxic to mammalian cells. On the other hand, Díaz-Blancas et al. (2016) used nanoemulsion formulation for tebuconazole (effective fungicide used for preventive and curative actions). By using this fungicide two problems occurred; the first one was low solubility, and the second was high stability in the soil. To overcome these problems tebuconazole nanoemulsion was prepared with a particle size range of 9 to 250 nm. Abd-Elsalam and Khokhlov (2015) used eugenol oil as a nanoemulsion formulation to control *Fusarium oxysporum*. The eugenol nanoemulsion was stable and the particle size ranged from 50 to 110 nm. These nanoparticles allowed penetration of the plant cuticle and acted on the infected fungi.

5.2.3.1.4   Role of Nanoemulsion Miticides in Agricultural Development

Mites are considered one of the most serious pests to agricultural crops. Mites need a suitable formulation that has a high surface area. Mossa et al. (2018) prepared garlic oil nanoemulsion formulation and used it against *Aceria oleae* (Nalepa) and *Tegolophushassani* (Keifer). The results found that the nanoemulsion formulation was an effective and safe acaricide. The nanoemulsion was prepared by the ultrasonic method for 35 minutes. The droplet size was 93.4 nm. Garlic nanoemulsion has high activity as an acaricide against eriophyid mites compared with normal emulsion. The LC_{50}s were 298.225 and 09.634 µg/ml, respectively. Increasing nanoemulsion toxicity may be due to increasing the surface area of emulsion droplets. Heng-ke et al. (2016) evaluated two commercial compounds: fenpropathrin (insecticide) and cyflumetofen (acaricide) against *Tetranychus cinnabarinus* female adults. A nanoemulsion formulation of two compounds was used. After 24 hours of pesticide treatment, the LC_{50}s were 711.62 and 4.32 mg/l respectively. The results showed that the nanoemulsion of both pesticides used had good efficacy against *T. cinnabarinus* compared with the normal emulsion. The authors found that the nanoemulsion formulation had good solubility and dispersibility compared with nanoemulsion.

5.2.3.2   Nanosuspension Pesticides

Generally, suspension can be defined as a heterogeneous mixture consisting of tiny solid particles not dissolving in a liquid or gaseous substance. This means that the small particles of the active pesticide ingredient are not dissolved in water like in wettable powder pesticides. This is the main problem facing the use of suspension pesticides. To overcome this problem, nanosuspension pesticides were used as an alternative formulation to suspension concentrate pesticides (Mishra et al. 2016). Nanosuspension pesticides not only increase the pesticide solubility but also increase the activity of active ingredients (Patel and Agrawal 2011). Nanosuspension is defined as a submicron colloidal dispersion of nanoscale particles stabilized by surfactants (Barret 2004).

Nanosuspension pesticides consist of two main parts:

1. Poor water-soluble active ingredient
2. Dispersion phase

The nanoparticles of active ingredients in suspension pesticides make possible the intravenous administration of poorly soluble active ingredients without any blockade in target organisms. The water dispersibility of the nanoformulation particle size can therefore improve the efficacy of the pesticide active ingredient compared with conventional formulations (Zhang et al. 2016).

5.2.3.2.1 Advantages of Nanosuspension Pesticides

1. Increase in pesticide solubility
2. Suitable for use with hydrophilic pesticides (pesticides dissolved in water)
3. Decrease of pesticide concentrations or doses. Subsequent decrease in pesticide contamination
4. Increase in the persistence of pesticides
5. Increase in the efficacy of pesticides against target organisms
6. Nanosuspension of pesticides means a larger surface area of nanoparticles which can improve the coverage, adhesion, and penetration of particles through the surface of crop leaves and targeted organisms (Fu et al. 2013)
7. The main advantage of nanosuspension is avoiding the use of organic solvents

5.2.3.2.2 Preparation of nanosuspension

There are two methods for nanosuspension preparation:

1. Top-down method: this means decreasing the particle size to nanoscale using such methods as wet-milling and high pressure homogenization (HPH) (Karadag et al. 2014)
2. Bottom-up method: this means increasing the particle size to nanoscale using microprecipitation and supercritical fluid (Lévai et al. 2015)

5.2.3.2.3 Role of Nanosuspension Insecticides in Agricultural Development

The main disadvantage of most insecticides used in insect control was the poor solubility of these insecticides in water. Nanosuspension insecticides solved this problem by increasing the insecticides' solubility and efficacy. Wang et al. (2019) prepared lambda-cyhalothrin nanosuspension with a 12.0 nm particle size. The particle size increased from 12.0 to 24.8 nm after storage at 0°C for seven days. The suspensibility was increased to 99.6%. The activity of lambda-cyhalothrin nanosuspension was evaluated against mustard aphid. The results showed that the LC_{50} lambda-cyhalothrin nanosuspension was 1.7 times less than the conventional formulation. Increasing the surface area of nanosuspension formulation increased the coverage of insecticides on plant leaves.

Cui et al. (2018) prepared avermectin nanosuspension with a particle size of 188 nm and used it against diamond-back moths. The suspensibility of avermectin nanosuspension was increased to 98.8%. The wetting time of the solid nanodispersion in water also decreased to 13 seconds. The results showed that by increasing the suspensibility and decreasing the wetting time of the avermectin active ingredient the efficacy of this nanosuspension increased 1.5 times that of the normal suspension. Using nanoparticles in insecticides decreased environmental pollution and increased crop production. Cui et al. (2016) prepared chlorantraniliprole nanosuspension and used it against diamond-back moth larvae in rice fields. The nanoparticle size was 29 nm. The suspensibility and wetting time of chlorantraniliprole nanosuspension were 97.32% and 13 seconds respectively. The rate of nanosuspension droplet stability was 1.5 and 3 times compared to the conventional chlorantraniliprole formulation. Due to the increase in suspensibility and decrease in particle size, the toxicity of chlorantraniliprole nanosuspension increased 3.3 times compared with the conventional chlorantraniliprole formulation. Pan et al. (2015) converted the conventional formulation of lambda-cyhalothrin into the nanosuspension formulation to increase the solubility of the lambda-cyhalothrin formulation. The solubility of the lambda-cyhalothrin nanosuspension not only increased but the stability and efficacy were also increased. The mean particle size of the lambda-cyhalothrin nanosuspension was 16.0 nm. Saini et al. (2014) compared the activity of pyridalyl (a new synthetic insecticide) nanosuspension and conventional formulation against *Helicoverpa armigera* larvae. The nanosuspension particle size was 100 nm. The results obtained showed that the LC_{50} of pyridalyl nanosuspension was 40 µg/mL^{-1} compared with 205 µg/mL^{-1}. This means that the nanosuspension formulation was 6.25 times more effective than conventional formulation.

5.2.3.2.4   Role of Nanosuspension Fungicides in Agricultural Development

It is known that most fungicides used in agriculture have poor solubility, so the conversion of these fungicides to nanosuspension formulation increases the fungicide solubility (Fraceto et al. 2018). Yao et al. (2018) prepared a nanosuspension formulation from azoxystrobin. This fungicide was very insoluble in the conventional formulation (Taylor and Zhang 2016). The particle size of the azoxystrobin nanosuspension was 238.1 ± 1.5 nm. Decreasing the nanoparticle size increased the efficacy of the new formulation *Fusarium oxysporum* by 1.7 times compared with the conventional formulation. Increasing the suspension efficacy is not only the main advantage of nanosuspension but it also increases the formulation stability and decreases the contact angle between the droplets and plant surface. Improvement of fungicide characteristics can protect the treated crops from fungi attack and increase crop production. Kumar et al. (2015) prepared a nanosuspension formulation of hexaconazole (a fungicide used worldwide) for protecting crops from damage by several pathogens of apples, vines, coffee, peanuts, bananas, cucurbits, peppers, rice, groundnut, and mango. The particle size of the nanosuspension formulation was 100 nm. This new formulation had thermal stability compared with the conventional formulation. The efficacy of the nanosuspension formulation was double the conventional one. Griffin et al. (2017) extracted an antifungal compound from desert thumb plants, *Cynomorium coccineum* L. This extraction was converted to nanosuspension formulation and used against the pathogenic fungus *Candida albicans*. This nanosuspension formulation was very effective against *C. albicans*. The size of nanoparticles ranged between 300 and 600 nm.

5.2.3.3   Nanocapsule Pesticides

Nanocapsule pesticide refers to the active ingredient of the pesticides found in the nanoscale shell. It is known that the conventional pesticide formulation releases quickly into the environment, so it has short time persistence. The main function of the nanoscale shell is to control the release of the active ingredient, so the nanocapsule pesticide stays effective against the target pest for a long time.

5.2.3.3.1   The Components of Nanocapsule Pesticide

1. The shell of nanopesticides: this shell is manufactured in a biodegradable agent. If this shell is introduced to the environment it would degrade and release the active ingredient. The shell can be called the polymer; it can be made of an artificial or natural substance. Now, most of the nanocapsule shell is made of natural materials such as chitosan, starch, or alginates to reduce the toxic action of artificial substances. There are some conditions which must exist in the shell material, such as:
   - It must be non-toxic
   - It must be compatible with the environment
   - It must be relatively cheap
   - It must have good mechanical strength
2. The active pesticide ingredient responsible for the toxic action of the pesticide: this active ingredient exists in the shell to release slowly, so the must have affinity between the polymer and the active ingredient.

5.2.3.3.2   The Advantages of Nanocapsule Pesticides

There are many benefits for nanocapsule formulation. These benefits include:

1. Slow release of the active ingredient
2. Reducing the amount of the active ingredient
3. Reducing the cost of formulation production
4. Increasing the pesticide's persistence
5. It can be used against the target pests
6. Reducing of pesticides load

7. Reducing the underground water contamination
8. Reducing resistance of the target pests toward the pesticides

5.2.3.3.3 Role of Nanocapsule Insecticides in Agricultural Development

The main threat facing insecticides in the environment is the effect of environmental factors such as light, the sun, and UV light on the treated insecticides; nanocapsule insecticides can overcome these factors. Caballero et al. (2019) prepared nanocapsules of deltamethrin (insecticide belonging to the pyrethroids group). The results showed that the efficacy of deltamethrin nanocapsules was increased compared with the conventional formulation. When the deltamethrin nanocapsule formulation was mixed with the normal formulation of indoxacarb (oxadiazine group) the efficacy of indoxacarb increased more than the indoxacarb mixed with the piperonyl butoxide. This means that the nanocapsule formulation has active insecticides. Pasquoto-Stigliani et al. (2017) prepared neem nanocapsules by using neem oil coated with poly(ϵ-caprolactone). The size of the nanocapsules was 400 nm. The results also showed the nanocapsules containing neem oil did not affect the soil microbiota for 300 days compared with the normal formulation. The efficacy of the nanocapsules was increased compared with the traditional formulation. The phytotoxicity of the nanocapsules against maize plants was also reduced.Wibowo et al. (2014) prepared silica nanocapsules and mixed them with fipronil (phenyl pyrezoles group). The loading of fipronil was carried out by direct dissolution in the oil core before biomimetic growth of a layer of silica shell surrounding the core, with encapsulation efficiency as high as 73%. The efficacy of the nanocapsule formulation was increased against termites. Nanocapsule oil also was used against the adult red flour beetle, *Tribolium castaneum*. Khoobdel et al. (2017) prepared nanocapsule oil from *Rosmarinus officinalis* essential oil and used it against *T. castaneum*. The diameter of the prepared nanocapsules was 145 ± 15 nm. The results showed that the nanocapsule oil was very effective against the *T.castaneum* adults and it can control the oil release. Sun et al. (2014) prepared methomyl-loaded (carbamates group) nanocapsules and assessed them against armyworm larvae. The percentage of larval mortality reached 100% after two days of treatment and stayed effective against infested larvae during seven days of treatment. This persistence may be due to the slow release of methomyl nanocapsules. Khanahmadi et al. (2017) used nanocapsules of *Artemisia haussknechtii* essential oil as a fumigant against *S. oryzae* and *T. castaneum*. The results showed that the mortality percentage reached 100% with a concentration of 166 ppm. The efficacy of nanocapsule formulation and phosphine (conventional fumigant) were compared. The LC_{50}s of the nanocapsule formulation were 30.29 and 45.82 ppm, against *S. oryzae* and *T. castaneum* respectively, while the corresponding results with phosphine were 550.11 and 713.24 ppm.

5.2.3.3.4 Role of Nanocapsule Fungicides in Agricultural Development

Chauhan et al. (2017) prepared hexaconazole (a fungicide used worldwide) nanocapsules coated with chitosan to control fungicide release. The hexaconazole nanocapsule formulation was released continually for 14 days compared to five days with the commercial formulation. This meant that the stability of the nanocapsule formulation was greater than for the commercial formulation. The antifungal activity of hexaconazole nanocapsules was compared with the normal formulation against *Rhizoctonia solani*. The growth of mycelia in tested fungus was evaluated. The results showed that the length of mycelia in *R. solani* was 18 mm after 72 hours of hexaconazole nanocapsule treatment compared to 45 mm with the commercial formulation. Grillo et al. (2014) attributed this difference to the fact that nanocapsules have a high surface area, which allows greater interaction of the fungus with hexaconazole over a greater area.

5.2.3.3.5 Role of Nanocapsules Herbicides in Agricultural Development

Bombo et al. (2019) prepared atrazine (one of the most used herbicides) nanocapsules to overcome the leaching of the conventional formulation to underground water and enhance the activity of this herbicide. The atrazine nanocapsule was transported directly through the vascular tissue of the weed leaves and into the cells where it degraded the chloroplasts resulting in herbicidal activity. The results showed that the nanoformulation was stable and effective compared with the traditional formulation. Oliveira et al. (2015) prepared atrazine nanocapsules and used them against mustard, *Brassica juncea*. The results showed that when the mustard was treated with atrazine nanocapsules net photosynthesis and PSII maximum quantum yield decreased, and leaf lipid peroxidation increased, leading to shoot growth inhibition and the development of severe symptoms. The results also showed an increase of atrazine nanocapsules compared with the normal formulation. Clemente et al. (2014) determined the side effect of simazine, atrazine and ametryn nanocapsules coated by poly(ε-caprolactone) on aquatic organisms (the alga *Pseudokirchneriella subcapitata* and the microcrustacean *Daphnia similis*). The results showed that the nanocapsule formulation had low toxicity against the algae and high toxicity to the microcrustacean. The cytogenetic evaluation showed that the nanocapsule formulations of tested herbicides were less toxic on aquatic organisms than the commercial formulations. This result can help to reduce the amount of herbicide used in soil treatment.

5.2.3.4 Nanoparticle Pesticides

The main object of using nanoparticles is to increase the surface area of the pesticide formulation and decrease underground water contamination. Generally, nanoparticle size ranges between 1 and 100 nm (Abou El-Nour Kholoud et al. 2010). Recently, many pesticides have been converted from the conventional formulation to nanoparticles to improve their physical and chemical characteristics.

Nanoparticles consist of three layers:

1. Surface layer which contacts with polymer, surfactants, and/or metal ion
2. Shell layer
3. The core

5.2.3.4.1 Advantages of Nanoparticle Formulation

1. Increase in pesticide stability
2. Increase in pesticide efficacy
3. Good pesticide distribution in treated fields
4. Decrease in the amount of formulation
5. Decrease in underground water contamination
6. Decrease in the cost of pesticide formulation
7. Fast detection of pesticide residues

5.2.3.4.2 Preparation of Nanoparticles

There are two main methods to prepare the nanoparticle formulation:

1. Top-down method: The top-down method is still one of the important methods for synthesizing nanomaterials (Teli et al. 2010)
2. Bottom-up method: The synthesis of large polymer molecules is a typical example of the bottom-up approach, where individual building blocks (monomers) are assembled into a large molecule (or polymerized into bulk material)

5.2.3.4.3 Role of Nanoparticles Insecticides in Agricultural Development

The main problem facing insecticide application is the stability of insecticides under climatic factors and reducing the hazards of insecticide stability. Nanoparticle formulations may limit this problem. Zhao et al. (2018) used a new insecticide formulation to improve insecticide persistence and reduce the side effect of this stability to the environment by using spirotetramat (lipid synthesis inhibitors) – loading mesoporous silica nanoparticles (MSNs) in cucumber fields. The use of silica nanoparticles reduces spirotetramate degradation in the plant fluid and also reduces the spirotetramate residues. Wang et al. (2018) compared the efficacy of a normal formulation of aver-mectin (worldwide insecticides) with avermectin – loaded avermectin/polysuccinimide with glycine methyl ester nanoparticles (AVM-PGA) against the diamond-back moth, *Plutella xylostella*. The results found that the nanoparticle formulation was more effective than the normal formulation. The percentages of mortalities were 96.3 and 51.5% respectively. Rouhani et al. (2013) used silica and silver nanoparticles against the larvae and adults of *Callosobruchus maculates*. The results showed that the LC_{50}s of silica and silver nanoparticles against adults were 0.68 and 2.06 g/kg of cowpeas respectively. On the other hand, the LC_{50}s against larvae were 1.03 and 1.00 g/kg. The results also showed silica nanoparticles were more effective than silver nanoparticles against both *C. maculates* adults and larvae. Khooshe-Bast et al. (2016) evaluated zinc oxide nanoparticles against greenhouse whitefly, *Trialeurodes vaporariorum* adults. The LC_{50} was 7.35 mg/l.

5.2.3.4.4 Role of Nanoparticles Fungicides in Agricultural Development

The small size of nanoparticles allows control of a wide range of agricultural pathogenic fungi. Malandrakis et al. (2019) used copper, silver, and zinc nanoparticles against foliar and soil-borne plant pathogens as an alternative to conventional fungicides. The final results showed that the nanoparticle formulations were 10–100 times more effective than traditional formulations. Jamdagni et al. (2018) used green and chemical silver nanoparticles against some common agricultural fungi-cides. The size of the nanoparticles was 31.86 and 41.91 nm, respectively. The silver nanoparticles were treated alone and in combination with conventional fungicides such as carbendazim, manco-zeb, and thiram. The results showed that the green nanosilver was more effective than the chemical nanosilver. The results also showed the nanosilver particles have synergistic action on commonly used fungicides against *F. oxysporum*. Elmer et al. (2018) used different nanoparticles of B, CuO, MnO, SiO, TiO, and ZnO against *F. oxysporum* in watermelon fields. The total yield was increased to 39% with the use of CuO nanoparticles compared with untreated fields. Gallardo et al. (2016) used silver nanoparticles against toxigenic fungi affecting cocoa (*Theobroma cacao*). The results showed that the use of silver nanoparticles (80 ppm) completely inhibits *Aspergillus flavus* growth.

5.2.3.4.5 Role of Nanoparticles Herbicides in Agricultural Development

The main problem in herbicide application is the leaching of herbicides and accumulation in under-ground water. Nanotechnology is developing new herbicide formulations to overcome herbicide leaching. Nanoparticle formulation is one of the most common formulations used as a herbicide. Nanoherbicides are one of the new strategies for combating the problems of conventional herbicides (Narware et al. 2019). Li et al. (2018) found that the silver nanoparticles have a synergistic action of diclofop-methyl (herbicides) against *Arabidopsis thaliana*. When the silver nanoparticles were combined with diclofop-methyl plant growth was significantly inhibited, and chlorophyll synthesis was found in *A. thaliana*.

5.2.3.4.6 Role of Nanoparticles Nematicides in Agricultural Development

Plant pathogenic nematodes are the most destructive pest. This pest has acquired resistance against many conventional pesticides so new pesticide formulations such as nanoparticle pesticides may have reduced pest resistance. Cromwell et al. (2014) evaluated the nematicidal activity of silver nanopar-ticles against root-knot nematodes (*Meloidogyne* spp.). The concentrations of silver nanoparticles were used at 30–150 mg/ml. With all concentrations greater than 99% *M. incognita* became inactive after 6 hours from application. Silver nanoparticles at 150 mg/l reduced the *M. graminis* in the soil samples by 82% and 92% after two and four days of exposure, respectively.

5.3 CONCLUSION

Greenpeace reported that more than 70% of pesticide formulations used in China were not taken up by plants, but it leached into the soil and dissolved in groundwater (Fan 2017). Development of new pesticide formulations to reduce the consumption amount of pesticides is badly needed. These new trends are not only to reduce pesticide use but also to improve the efficacy, safety, stability, and solubility of pesticide formulation. Development of new pesticide formulations mainly depends on changes in the physicochemical properties of the active ingredients. This can be done by using new pesticide formulations such as nanopesticides.

Nanotechnology has developed new pesticide formulations to overcome the disadvantages of the old formulations. These new formulations include nanoemulsion, nanosuspension, nanocapsule, and nanoparticles (Sabry and Ragaei 2018). These formulations can be used at a rate of 50 ml of nanoemulsion per hectare or 8 g of nanoparticles per hectare. These new formulations have many advantages:

- High stability
- Easy removal
- High solubility
- Avoiding the use of organic solvent
- Good control in active ingredient release
- Easy handling
- Easy storage
- Low residue
- Highly specific on pests
- Highly effective

REFERENCES

Abd-Elsalam KA, Khokhlov AR (2015) Eugenol oil nanoemulsion: Antifungal activity against *Fusarium oxysporum* f.sp. *vasinfectum*and phytotoxicity on cottonseeds. *ApplNanosci* 5(2): 255–265.

Abouelkassem SH, Abdelrazeik AB, Rakha OM (2015) Nanoemulsion of Jojoba oil, preparation, character- ization and insecticidal activity against *Sitophilus oryzae* (Coleoptera: Curculionidae) on wheat. *Int J Agricinnov Res* 4(1): 2319–1473.

Abou El-NourKholoud M. M., Eftaiha A, Al-Warthan A, Ammar AAR (2010) Synthesis and applications of silver nanoparticles. *Arab J Chem* 3(3): 135–114.

Balah MA, Abd El Azim WM (2016) Emulsions and nanoemulsion formation from wild and cultivate d thyme and marjoram essential oils for weeds control. *J Plant Prot Path Mansoura Univ* 7(10): 641–648.

Bang SH, Yu YM, Hwang IC, Park HJ (2009) Formation of size-controlled nano carrier systems by self- assembly. *J Microencap* 26(8): 722–733.

Barret ER (2004) Nanosuspensions in drug delivery. *Nat Rev* 3: 785–796.

Bombo AB, Pereira AES, Lusa MG, de Medeiros OE, de Oliveira JL, Campos EVR, de Jesus MB, Oliveira HC, Fraceto LF, Mayer JLS (2019) A mechanistic view of interactions of a nanoherbicide with target organism. *J Agric Food Chem* 67(16): 4453–4462.

Caballero JP, Murillo L, List O, Bastiat G, Flochlay-Sigognault A, Guerino F, Lefrançois C, Lautram N, Lapied B, Apaire-Marchais V (2019) Nanoencapsulated deltamethrin as synergistic agent potentiates insecticide effect of indoxacarb through an unusual neuronal calcium-dependent mechanism. *Pest Biochem Physiol* 157: 1–12.

Chauhan N, Dilbaghi N, Gopal M, Kumar R, Kim K, Kumar S (2017) Development of chitosan nanocapsules for the controlled release of hexaconazole. *Int J BiolMacromol* 97: 616–624.

Choupanian M, Omar D (2018) Formulation and physicochemical characterization of neem oil nanoemulsions for control of *Sitophilusoryzae* (L., 1763) (Coleoptera: Curculionidae) and *Triboliumcastaneum* (Herbst, 1797) (Coleoptera: Tenebrionidae). *Türk.entomolderg* 42(2): 127–139.

Clemente Z, Grillo R, Jonsson M, Santos NZ, Feitosa LO, Lima R, Fraceto LF (2014) Ecotoxicological evalu- ation of poly(epsilon-caprolactone) nanocapsules containing triazine herbicides. *J Nanosci Nanotechnol* 14(7): 4911–4917.

Cromwell WA, Yang J, Starr JL, Jo, YK (2014) Nematicidal effects of silver nanoparticles on root-knot nema- tode in bermudagrass. *J Nematol* 46(3): 261–266.

Cui B, Feng L, Wang C, Yang D, Yu M, Zeng Z (2016) Stability and biological activity evaluation of chloran- traniliprole solid nanodispersions prepared by high pressure homogenization. *PloS One* 11(8): e0160877.

Cui B, Wang C, Zhao X, Yao J, Zeng Z, Wang Y (2018) Characterization and evaluation of avermectin solid nanodispersion prepared by microprecipitation and lyophilisation techniques. *PloS One* 13(1): e0191742.

Da Silva GS, Dos RTR, Copetti PM, Favarin FR, Sagrillo MR, da Silva AS, Segat JC, Baretta D, Ourique AF (2019) Evaluation of cytotoxicity, genotoxicity and ecotoxicity of nanoemulsions containing mancozeb and eugenol. *Ecotoxicol Environ Saf* 169: 207–215.

Díaz-Blancas V, Medina DI, Padilla-Ortega E, Bortolini-Zavala R, Olvera-Romero M, Luna-Bárcenas G (2016) Nanoemulsion formulations of fungicide tebuconazole for agricultural applications. *Molecules* 26(10): 1–12.

Elmer W, De La Torre-Roche R, Pagano L, Majumdar S, Zuverza-Mena N, Dimkpa C, Gardea-Torresdey J, White JC (2018) Effect of metalloid and metal oxide nanoparticles on fusarium wilt of watermelon. *Plant Dis* 102(7): 1394–1401.

Fan LY (2017) China founds pesticide office to combat pollution, Overuse. http://www.sixthtone.com/news/1000987/china-founds-pesticide-office-to-combat-pollution%2C-overuse#. Accessed Oct 12, 2017.

FAO (1989) *International Code of Conduct on the Distribution and Use of Pesticides*, Rome, Italy: FAO, p. 35.

Fareed M, Pathak MK, Bihari V, Kamal R, Srivastava AK, Kesavachandran CN (2013) Adverse respiratory health and hematological alterations amongagricultural workers occupationally exposed to organophosphatepesticides. A crosssect stud North India. PLoS One 8(7): e69755.

Fernandes CP, Almeida FB, Silveira AN, Gonzalez MS, Mello CB, Feder D, Apolinário R, Santos MG, Carvalho JCT, Tietbohl LAC, Rocha L, Falcão DQ (2014) Development of an insecticidal nanoemulsion with *Manilkarasubsericea*(Sapotaceae) extract. *Nanobiotech* 12(22): 1–9.

Fraceto LF, Maruyama CR, Guilger M, Mishra S, Keswani C, Singh HB, deLima R (2018) *Trichoderma harzianum* based novel formulations: Potential Applications for Management of Next-Gen agricultural challenges. *J Chem Technol Biotechnol* 93(8): 2056–2063.

Fu Q, Sun J, Zhang D, Li M, Wang Y, Ling G (2013) Nimodipine nanocrystals for oral bioavailability improvement: Preparation, characterization and pharmacokinetic studies. *Colloids Surf B* 109: 161–166.

Griffin S, Alkhayer R, Mirzoyan S, Turabyan A, Zucca P, Sarfraz M, Nasim MJ, Trchounian A, Rescigno A, Keck CM, Jacob C (2017) Nanosizingcynomorium: Thumbs up for potential antifungal applications. *Inven* 2(24): 1–10.

Hazrati H, Saharkhiz MJ, Niakousari M, Moein M (2017) Natural herbicide activity of *Saturejahortensis*L. essential oil nanoemulsionon the seed germination and morphophysiological features of two important weed species. *Ecotoxicol Environ Saf* 142 : 423–430.

Hemraj C (2017) Nanopesticide: Current status and future possibilities 5(1): 1–4.

Heng-ke Zhao, Yue Lan, Can Nan (2016) Preparation of 8% fenpropathrin·cyflumetofennano-emulsion and its performance. *J Sciengricu Sin* 49(14): 2700–2710.

Gallardo RV, Cruz JFO, Ortíz-Rodriguez OO (2016) Fungicidal effect of silver nanoparticles on toxigenic fungi in cocoa. *Pesq Agropec Bras Brasília* 51(12): 1929–1936.

Grillo R, Pereira AE, Nishisaka CS, de Lima R, Oehlke K, Greiner R, Fraceto LF (2014) Chitosan/tripolyphosphate nanoparticles loaded with paraquat herbicide: An environmentally safer alternative for weed control. *J Hazard Mater* 278: 163–171.

Jaiswal M, Dudhe R, Sharma PK (2015) Nanoemulsion: An advanced mode of drug delivery system. *3 Biotech* 5(2): 123–127. https://doi.org/10.1007/s13205-014-0214-0.

Jamdagni P, Rana JS, Khatri P (2018) Comparative study of antifungal effect of green and chemically synthesised silver nanoparticles in combination with carbendazim, mancozeb, and thiram. *IET Nanobiotechnol* 12(8): 1102–1107.

Kah M, Beulke S, Tiede K, Hofmann T (2013) Nanopesticides: State of knowledge, environmental fate, and exposure modeling. *Crit Rev Environ Sci Technol* 43(16): 1823–1867.

Karadag A, Ozcelik B, Huang Q (2014) Quercetin nanosuspensions produced by high-pressure homogenization. *J Agr Food Chem* 62(8): 1852–1859.

Khanahmadi M, Pakravan P, Azandaryani AH, Negahban M, Ghamari E (2017) Fumigant toxicity of*Artemisia haussknechtii* essential oil and its nanoencapsulated form. *J Entomol Zool Stud* 5(2): 1776–1783.

Khoobdel M, Ahsaei SM, Farzaneh M (2017) Insecticidal activity of polycaprolactone nanocapsules loaded with *Rosmarinus officinalis* essential oil in *Triboliumcastaneum*(Herbst). *Entomol Res* 47(3): 175–184.

Khooshe-Bast Z, Sahebzadeh N, Ghaffari-Moghaddam M, Mirshekar A (2016) Insecticidal effects of zinc oxide nanoparticles and *Beauveria bassiana* TS11 on *Trialeurodesvaporariorum* (Westwood, 1856) (Hemiptera: Aleyrodidae). *Actaagric Slov* 107(2): 299–309.

Kumar R, Nair KK, Alam I, Gogoi R, Singh PK, Srivastava C, Gopal M, Goswami A (2015) Development and quality control ofnanohexaconazole as an effective fungicideand its biosafety studies on soil nitifiers. *J NanosciNanotechnol* 15(2): 1350–1356.

Lévai G, Martín Á, de Paz E, Rodríguez-Rojo S, Cocero MJ (2015) Production of stabilized quercetin aqueoussuspensions by supercritical fluid extraction of emulsions. *J Supercrit Fluid* 100: 34–45.

Li X, Ke M, Zhang M, Peijnenb WJGM, Fan X, Xu J, Zhang Z, Lu T, Fu Z, Qian H (2018) The interactive effects of diclofop-methyl and silver nanoparticles on *Arabidopsis thaliana*: Growth, photosynthesis and antioxidant system. *Environ Pollut* 232: 212–219.

Lim CJ, Basri M, Omar D, Abdul Rahman MB, Salleh AB, Raja ARN (2013) Green nanoemulsion-laden glyphosate isopropylamine formulation in suppressing creeping foxglove (*A.gangetica*), slender button weed (*D. ocimifolia*) and buffalo grass (*P. conjugatum*). *Pest Managsci* 69: 104–111.

Louni M, Shakarami J, Negahban M (2018) Insecticidal efficacyof nanoemulsion containing *Menthalongifolia* essential oil against *Ephestiakuehniella* (Lepidoptera: Pyralidae). *J Crop Prot* 7(2): 171–182.

Malandrakis AA, Kavroulakis N, Chrysikopoulos CV (2019) Use of copper, silver and zinc nanoparticles against foliar and soil-borne plant pathogens. *Sci Total Environ* 670: 292–299.

Mishra S, Keswani C, Abhilash PC, Fraceto LF, Singh HB (2017) Integrated approach of Agri-nanotechnology: Challenges and future trends. *Front Plant Sci* doi: 10.3389/fpls.2017.00471.

Mishra S, Keswani C, Singh A, Singh BR, Singh SP, Singh HB (2016) Microbial nanoformulation: Exploring potential for coherent nano-farming. In: *The Handbook of Microbial Bioresources*, Eds. Gupta VK, Sharma GD, Tuohy MG, Gaur R, Wallingford: CABI-UK, pp. 107–120.

Mishra S, Singh A, Keswani C, Singh HB (2014b) Nanotechnology: Exploring potential application in agriculture and its opportunities and Constraints. *Biotech Today* 4(1): 9–14.

Mishra S, Singh BR, Singh A, Keswani C, Naqvi AH, Singh HB (2014a) Biofabricated silver nanoparticles act as a strong fungicide against *Bipolaris sorokiniana* causing spot blotch disease in wheat. *PloS One* 9(5): e97881

Mossa AH, Abdelfattah NAH, Mohafrash SMM (2017) Nanoemulsion of camphor (*Eucalyptus globulus*) essential oil, formulation, characterization and insecticidal activity against wheat weevil, *Sitophilusgranarius*. *Asian J Crop Sci* 9: 50–62.

Mossa AH, Afia SI, Mohafrash SMM, Abou-Awad BA (2018) Formulation and characterization of garlic (*Allium sativum* L.) essential oil nanoemulsion and its acaricidal activity on *Eriophyid olive* mites (Acari: Eriophyidae). *Environ Sci Pollut Res Int* 25(11): 10526–10537.

Narware J, Yadav RN, Keswani C, Sing SP, Singh HB (2019) Silver nanoparticle-based biopesticides for phytopathogens: Scope and potential in agriculture. In: *Nano-Biopesticides Today and Future Perspectives*, Ed. Koul O, San Diego, CA: Academic Press, pp. 303–314.

Oliveira HC, Stolf-Moreira R, Martinez CBR, Grillo R, de Jesus MB, Fraceto LF (2015) Nanoencapsulation enhances the post-emergence herbicidal activity of atrazine against mustard plants. *PloS One* 10(7): e0132971.

Pan Z, Cui B, Zeng Z, Feng L, Liu G, Cui H, Pan H (2015) Lambda-cyhalothrin nanosuspension prepared by the meltemulsification-high pressure homogenization method. *J Nanomat* 2015: 1–8.

Pant M, Kumar N, Ompal Dubey S, Patanjali PK (2017) Biodiesel waste based new generation formulation of permethrin for cockroach control. *J Scient Indus Res* 76(3): 184–186

Pasquoto-Stigliani T, Campo EVRs, Oliveira JL, Silva CMG, Bilesky-José N, Guilger M, Troost J, Oliveira HC, Stolf-Moreira R, Fraceto LF, de Lima R (2017) Nanocapsules containing neem (*Azadirachtaindica*) oil: Development, characterization, and toxicity evaluation. *Sci Rep* 7(1): 5929.

Patel VR, Agrawal YK (2011) Nanosuspension: An approach to enhance solubility of drugs. *J Adv Pharm Tech Res* 2(2): 81–87.

Peshin R, Dhawan AK (2009) Pimentel D pesticides and pest controls. In: *Integrated Pest Management: Innov Developproc*, vol. 1, pp. 83–87.

Pimentel D, Acquay H, Biltonen M, Rice P, Silva M, Nelson J, Lipner V, Giordano S, Horowitz A, D'Amore M (1992) Environmental and economic costs of pesticides. *BioScience* 42(10): 750–760.

Rouhani M, Samih MA, Kalantari S (2013) Insecticidal effect of silica and silver nanoparticles on the cowpea seed beetle, *Callosobruchusmaculatus* F. (Col.: Bruchidae). *J Entomol Res* 4(4): 297–305.

Sabry KH, Ragaei M (2018) Nanotechnology and their applications in insect's pest control. In *Nanobiotechnology Applications in Plant Protection*, Eds. Abd-Elsalam, Kamel A, Prasad, Ram, Basel, Switzerland: Springer International Publishing, pp. 1–28.

Saini P, Gopal M, Kumar R, Srivastava C (2014) Development of pyridalyl nanocapsule suspension for efficient management of tomato fruit and shoot borer (*Helicoverpaarmigera*). *J EnvironSci Health Part B* 49: 344–351.

Shah P, Bhalodia D (2010) Nanoemulsion: A pharmaceutical review. *Sys Rev Pharm* 1(1): 24–32.

Sun C, Shu K, Wang W, Ye Z, Liu T, Gao Y, Zheng H, He G, Yin Y (2014) Encapsulation and controlled release of hydrophilic pesticide in shellcross-linked nanocapsules containing aqueous core. *Int J Pharmac* 463(1): 108–114.

Tadros TF, Izquierdo P, Esquena J, Solans C (2004) Formation and stability of nanoemulsions. *Adv Colloid Interface Sci* 108–109: 303–318.

Taylor LS, Zhang GG (2016) Physical chemistry of supersaturated solutions and implications for oral absorption. *Adv Drug Deliv Rev* 101: 122–142.

Teli KM, Mutalik S, Rajanikant GK (2010) Nanotechnology and nanomedicine: Going small means aiming big. *Curr Pharm Des* 16(16): 1882–1892.

Trotta M (1999) Influence of phase transformation on indomethacin release from microemulsions. *J Control Release* 60(2–3): 399–405.

Wang C, Cui B, Guo L, Wang A, Zhao X, Wang Y, Sun C, Zeng Z, Zhi H, Chen H, Liu G, Cui H (2019) Fabrication and evaluation of lambda-cyhalothrin nanosuspension by one-step melt emulsification technique. *Nanomat* 9(145): 1–13.

Wang G, Xiao Y, Xu H, Hu P, Liang W, Xie L, Jia J (2018) Development of multifunctional avermectin poly(succinimide) nanoparticles to improve bioactivity and transportation in rice. *J Agric Food Chem* 66(43): 11244–11253.

Wibowo D, Zhao CX, Peters BC, Middelberg AP (2014) Sustained release of fipronil insecticide in vitro and in vivo from biocompatible silica nanocapsules. *J Agric Food Chem* 62(52): 12504–12511.

Yao J, Cui B, Zhao X, Wang Y, Zeng Z, Sun C, Yang D, Liu G, Gao J, Cui H (2018) Preparation, characterization, and evaluation of azoxystrobin nanosuspension produced by wet media milling. *Appl Nanosci* 8(3): 297–307.

Zhang Y, Li Y, Zhao X, Zu Y, Wang W, Wu W (2016) Preparation, characterization and bioavailability of oral puerarin nanoparticles by emulsion solvent evaporation method. *Rsc Adv* 6(74): 69889–69901.

Zhao P, Yuan W, Xu C, Li F, Cao L, Huang Q (2018) enhancement of spirotetramat transfer in cucumber plant using mesoporous silica nanoparticles as carriers. *J Agric Food Chem* 66(44): 11592–11600.

Zhao X, Zhu Y, Zhang C, Lei J, Ma Y, Du F (2017) Positive charge pesticide nanoemulsions prepared by the phase inversion composition method with ionic liquids. *RSC Adv* 7(77): 48586–48596.

6 Applications and Practices of Myconanoparticles in Food and Agriculture

Evrim Özkale

CONTENTS

6.1 INTRODUCTION

Nanoparticles (NPs) have several applications in different fields such as medicine, environment, drug design, drug delivery, cosmetics, textiles, food industry, optics, and optical devices. NPs also exhibit significant antimicrobial activity towards a diverse range of pathogens, and lead to considerable interest by researchers due to the increasing resistance of microbes against the available antibiotics and development of resistant microbial strains.

NP synthesis has several applications in different fields such as research of DNA sequencing and pharmaceuticals, biosensors, biomolecular detection and diagnostics, engineering applications like tissue engineering, genetic engineering, optics, and optoelectronics. Among the synthesized NPs, silver NPs have a broad range of applications, such as their use in water filtering apparatus, catalysis, optical receptors for bio-labeling, antimicrobial activities, anticancer activities, antioxidants, anti-dermatophytic, anti-inflammatory, antitumor, hepatoprotective, cytotoxic, and immunomodulatory hypotensive activities, anti-HIV (human immunodeficiency virus), anti-diabetic, wound dressing, surgical masks, food packaging, paints, and additives in bone cement, etc. (Ahmad et al. 2010).

Since the chemical methods for the synthesis of NPs have some drawbacks such as involving a large amount of heat and energy, and leave hazardous residues, employing innocuous biological sources is therefore is a better option. The natural mechanism of microorganisms for detoxification of metal ions intra- or extracellularly through reduction by bioaccumulation, precipitation, biomineralization, and biosorption have roles in the production of NPs (Narware et al. 2019).

6.2 BIOSYNTHESIS OF NANOPARTICLES BY FUNGI

In industrial nanobiotechnology different biological systems such as bacteria, fungi, algae, or their original materials are involved. Among these, fungi, which secrete extracellular proteins, have been found to be the most favorable for the synthesis of metal NPs (MNPs) (Gade et al. 2010). As compared to bacteria, they are easy to cultivate, reliable, ecofriendly, and the methodology for extracellular synthesis is extremely cost-effective and less time consuming (Mishra et al. 2016).

In fungal-based MNP biosynthesis, the biomineralization process is done via recruiting different metal ions by intracellular and extracellular enzymes and other biomolecules secreted from fungi (Ahmad et al. 2003; Karimi and Vahabi, 2017; Mishra et al. 2017). The mycosynthesis of MNPs, or myconanotechnology, is at the interface between mycology and nanotechnology and includes an exciting new applied interdisciplinary science with considerable potential due to the wide range and diversity of fungi (Rai et al. 2009; Gade et al. 2010). MNPs have a number of applications in the medicine, agriculture and textile industries (Gade et al. 2010). Extracellular synthesis of MNPs is conducted by the trapping of the metal ions on the surface of the cells and reducing them in the presence of the enzymes while intracellular synthesis occurs into the fungal cell in the presence of enzymes. Since they have greater tolerance and uptake capacity for metals, particularly in terms of the binding capacity of metal salts, fungal biomass is an advantage for large-scale production of NPs. Including this, their downstream processing is quite easy so that it is easy to cultivate for the efficient low-cost synthesis of MNPs (Yadav et al. 2015).

Various generation systems and methodologies for the fabrication and optimization of several MNPs have been reported for the myconanofactories. When aqueous silver ions, exposed to the *F. oxysporum*ions were reduced by a nitrate dependent reductase quinone process, it led to the formation of 10–30 nm silver NPs (Duran et al. 2005). A rapid photobiological approach to generate silver NPs from several strains of *F. oxysporum* in the presence of a conventional halogen-tungsten lamp was reported by Mohammadian et al. (2007). Using this method, nanoparticles that were 10–60 nm in size were produced in less than 1 hour.

Several protocols have also been developed to synthesize myconanoparticles of varying size (0–100 nm) and shape (hexagons, pentagons, circles, squares, rectangles) at both intra- and extracellular levels by physical and chemical methods (Riddin et al. 2006; Rai et al. 2009).

Much of the research on fungal extracellular synthesis of NPs through fungal growth has been undertaken with several metals but mostly with silver. These silver myconanoparticles have been reported from *Aspergillus fumigatus* (Bhainsa and D'Souza, 2006), *Phoma* (Chen et al. 2003), *Fusarium oxysporum* (Ahmad et al. 2003; Duran et al. 2005; Bhansal et al. 2004, 2005; Karbasian et al. 2008; Mohammadian et al. 2007), *F. semitectum* (Basavaraja et al. 2007), *F. acuminatum* (Ingle et al. 2008), *F. solani* (Ingle et al. 2009), *Trichoderma asperellum* (Mukherjee et al. 2008), *Trichoderma reesei* (Vahabi et al. 2011), *Trichoderma harzianum* (Fraceto et al. 2018), *Aspergillus niger* (Gade et al. 2008), *Alternaria alternata* (Gajbiye et al. 2009), and *Cladosporium cladosporioides* (Balaji et al. 2009). The potential of silver MNPs as antibacterial agents has been widely investigated. In these studies, the sizes of NPs have ranged from 5 to 70 nm. The size and formation of these particles are important for good dispersion. Gade et al. (2008) considered the potential bioactivity of silver NPs of around 20 nm and in spherical shape produced by *Aspergillus niger* against a range of bacteria. Ingle et al. (2008) have also tested the NPs produced where spherical size distribution in the range of 5–40 nm with an average diameter of 13 nm by the fungus *F. Acuminatum* Ell. and Ev. (USM-3793) against different human pathogens, and have reported that efficient antibacterial activity was found in multidrug-resistant and highly pathogenic bacteria, *E. coli*, *Salmonella typhi*, *Staphylococcus epidermidis*, and *S. aureus*. The silver NPs proved to be toxic to each of the above species and the effect is 1.4–1.9× stronger than that of pure silver ions (Ingle et al. 2008).

In extra and intracellular production of various NPs by fungi, one of the main advantages of biosynthesis of MNPs is their major role in the protection of the environment which is also the ultimate target of other green technologies. Recently mycogenesis of myconanoparticles has attracted much attention among researchers (Mishra et al. 2014a; Mishra et al. 2014b). Compared to bacteria, fungi produce a large quantity of NPs as they secrete a large number of enzymes which are directly involved in the bioreduction and stabilization process of NPs. Enzymes secreted by these organisms, especially in the case of fungi, the reductases enzyme might be involved in the mechanisms of NP production and stabilization (Khandel and Kumar Shahi, 2018).

It has been found that synthesis of NPs using extracellular cell filtrate is more beneficial over intracellular metabolites. Certain fungal species such as *F. oxysporum* can readily synthesize MNPs extracellularly using high levels of secreted proteins and/or enzymes that not only stabilize the particles but allow for an improved yield over an intracellular one (Riddin et al. 2006). An extracellular synthesis has obvious advantages over an intracellular process when it comes to downstream processing, since there would be very little handling of the fungal biomass (Gade et al. 2008).

6.3 APPLICATIONS OF FUNGAL MNPS IN FOOD AND AGRICULTURE

A remarkable array of new technologies, biotechnology, information technology, nanotechnology, etc., now exists that can be integrated into agricultural research and development to impact agricultural productivity, food and nutritional security, and economic growth. Agricultural productivity, soil health, water security, and food quality in storage and distribution are identified as the primary determinants of food security that can be impacted by developments in nanotechnology (Sastry et al. 2011). They identified areas of nanoresearch determinants of food security, these are productivity, soil health, water security and storage, and distribution of food. Some applications for these areas are a) natural biopolymer-based nanocomposite films and incorporated into food packaging for safe storage, b) nanowire immunosensors array for the quick detection of food-borne pathogens and c) nanoscale titanium dioxide particles as a blocking agent of UV light in plastic packaging.

Agriculture is the backbone of most developing countries. Rickman et al. (2009) reported on the application of nanotechnology in precision farming to increase crop yields (outputs) while minimizing input (fertilizers, pesticides, herbicides) (Gade et al. 2010). The field of nanotechnology is now an integral part of research and development for the large-scale manufacturing of agricultural products and processed foods and drinks, as well as for food packaging, worldwide (Moriarty, 2001).

6.3.1 VEGETABLE AND FRUIT PRESERVATION

In many agriculture-based countries, the main problem is to keep fruit and vegetables safe and fresh until they reach consumers. Even though fruit and vegetables have a natural waxy coating, it is not adequate to offer protection against water loss and a high respiration rate, thus resulting in weight and protein losses during storage. The U.S. Centers for Disease Control and Prevention estimates around 76 million cases of foodborne illness in the United States each year, resulting in about 5,000 deaths (Fayaz et al. 2009, Mead et al. 1999). The use of protective nanocoating and suitable packaging has become a topic of great interest in the field of food nanotechnology because of its potential for the increased shelf life of many food products (Ahvenairen, 2003). Because chemical synthesis methods produce a toxic substance as a byproduct, there is a big challenge to develop a new protocol that is a reliable, green chemistry process for the synthesis of NPs that does not use toxic chemicals in the synthesis protocols. Recently, scientists have looked to microorganisms as possible ecofriendly nanofactories for the synthesis of NPs (Bolander et al. 2006), such as Cadmium sulphide (CdS) (Ahmad et al. 2002), gold (Mukherjee et al. 2002), and silver (Ahmad et al. 2003). The first report of biogenic silver NPs incorporated with polysaccharide film used for vegetable and food preservation was conducted by Fayaz et al. (2009). The study on the extracellular biogenic synthesis of silver NPs by *T. viride* was carried out and demonstrated that silver NPs incorporated into sodium alginate films along with glycerol could be used for vegetable and fruit preservation.

Aguilar-Mendez et al. (2011) found that in the presence of silver NPs, *Colletotrichum gloesporioides* showed significantly delayed growth. Hence, it could be concluded that nanosilver could be used as an alternative to chemical fungicide to manage plant diseases.

Kim et al. (2008) reported the antifungal potential of double capsulated silver NP (1.5 nm) solution against rose powdery mildew which is a common plant disease caused by *Sphaerotheca pannosa* var. *rosae*. They sprayed the nanosilver (1.5 nm) solution, diluted up to 10 ppm, on the infected area of the plant. They found that after two days of spraying, around 95% of the rose powdery mildew disappeared and did not recur for a week. Verma et al. (2010) reported that nanoparticles synthesized from fungus *Aspergillus clavatus* exhibited good antibacterial activity against some disease-causing pathogenic bacteria, viz., *E. coli* and *Pseudomonas flurorescens*.

Nanobiosensors can be used effectively for sensing a wide variety of fertilizers, herbicide, pesticide, insecticide, pathogens, moisture, and soil pH (Rai and Ingle, 2012; Raiv et al. 2012). According to Mousavi and Rezaei (2011), the concept of precision farming includes a system controller for each growth factor such as nutrition, light, temperature, etc. These systems should also have information for planting and harvest times, which can be controlled by satellite systems. The systems allow the farmer to know the best time for planting and harvesting to avoid encountering bad weather conditions (Hamzah et al. 2018).

Nanomaterials are also useful for the development of, and used for the preparation of, devices which can be applied in precision farming. Precision farming generally involves the use of devices made of biosensors which help in agriculture. A nanobiosensor is a modified version of a biosensor, which may be defined as a compact analytical device or unit incorporating a biological or biologically derived sensitized element linked to a physico-chemical transducer (Turner, 2000).

6.4 CONCLUSION

The rapid development of nanotechnology has transformed many domains of food science, especially those that involve the processing, packaging, storage, transportation, functionality, and other safety aspects of food. A wide range of nanostructured materials from inorganic metal, metal oxides, and their nanocomposites, to nanoorganic materials with bioactive agents, has been applied in the food and agriculture industry. Since there is no need for toxic agents in either biosynthesis of NPs by fungi or fungal based material, or recovery and purification, fungi have been extensively used in the production of NPs. Industrial biotechnology also takes advantage of all-powerful tools in biotechnology to manipulate the protein structure, gene regulation, and metabolic pathway for enhancing and improving NP production by fungi (Karimi Dorcheh and Vahabi, 2017).

Despite the huge benefits nanotechnology has to offer, there are emerging concerns regarding the use of nanotechnology, as the accumulation of nanostructured materials (NSMs) in human bodies and in the environment can cause several health and safety hazards. Therefore, safety and health concerns, as well as regulatory policies, must be considered while manufacturing, processing intelligently, and actively packaging and consuming nanoprocessed food products (Bajpai et al. 2018).

REFERENCES

Aguilar-Méndez MA, San Martín-Martínez E, Ortega-Arroyo L, Cobián-Portillo G, Sánchez-Espíndola E (2011) Synthesis and characterization of silver nanoparticles: Effect on phytopathogen *Colletotrichum gloesporioides*. *J Nanopart Res* 13(6): 2525–2532.

Ahmad A, Mukherjee P, Mandal D, Senapati S, Islam Khan M, Rajiv Kumar R, Sastry M (2002) Enzyme mediated extracellular synthesis of CdS particles by the fungus *Fusarium oxysporum*. *J Am Chem Soc* 124: 12108–12109.

Ahmad A, Mukherjee P, Senapati S, Mandal D, Khan MI, Kumar R, Sastry M (2003) Extracellular biosynthesis of silver nanoparticles using the fungus *Fusarium oxysporum*. *Colloids Surf B* 28(4): 313–318.

Ahmad A, Senapati S, Khan MI, Kumar R, Sastry M (2003) Extracellular biosynthesis of monodisperse gold nanoparticles by a novel extremophilic actinomycete, *Thermomonospora*sp. *Langmuir* 19: 3550–3553.

Ahmad N, Sharma S, Alam MK, Singh VN, Shamsi SF, Mehta BR, Fatma A (2010) Rapid synthesis of silver nanoparticles using dried medicinal plant of basil. *Coll Surf Biointerfaces* 81(1): 81–86.

Ahvenaien R (2003) *Novel Packaging Techniques*. CRC Press, Boca Raton, FL.

Bajpai VK, Kamle M, Shukle S et al. (2018) Prospects of using nanotechnology for food preservation, safety and security. *J Food Drug Anal* 26(4): 1201–1214.

Balaji DS, Basavaraja S, Deshpande R, Mahesh DB, Prabhakar BK, Venkataraman A (2009) Extracellular biosynthesis of functionalized silver nanoparticles by strains of *Cladosporiumcladosporioides* fungus. *Coll Surf B Biointerfaces* 68: 88–92.

Bansal V, Rautray D, Ahmad A, Sastry M (2004) Biosynthesis of zirconia nanoparticles using the fungus *Fusarium oxysporum*. *J Mater Chem* 14(22): 3303–3305.

Bansal V, Rautray D, Bharde A, Ahire K, Sanyal A, Ahmad A, Sastry M (2005) Fungus mediated biosynthesis of silica and titania particles. *J Mater Chem* 15(26): 2583–2589.

Basavaraja V, Balaji SD, Lagashetty A, Rajasab AH, Venkataraman A (2007) Extracellular biosynthesis of silver nanoparticles using the fungus *Fusarium semitectum*. *Mater Res Bull* 43(5): 1164–1170.

Bhainsa KC, D'Souza SF (2006) Extracellular biosynthesis of silver nanoparticle using the fungus *Aspergillus fumigatus*. *Colloids Surf B* 47(2): 160–164.

Bolander ME, Mukhopadhyay D, Saikar G, Mukherjee P (2006) The use of microorganisms for the formation of metal nanoparticles and their application. *Appl Microbiol Biotechnol* 69(5): 485–449.

Chen JC, Lin ZH, Ma XX (2003) Evidence of the production of silver nanoparticles via pretreatment of *Phoma* sp. 32883 with silver nitrate. *Lett App Microbiol* 37(2): 105–108.

Duran N, Marcato PL, Alves OL et al. (2005) Mechanistic aspects of biosynthesis of nanoparticles by several *Fusarium oxysporum* strains. *J Nanobiotechnol* 3: 1–8.

Fayaz AM, Balaji K, Girilal M, Kalaichelvan PT, Venkatesan R (2009) Mycobased synthesis of silver nanoparticles and their incorporation into sodium alginate films for vegetable and fruit preservation. *J Agric Food Chem* 57(14): 6246–6252.

Fraceto LF, Maruyama CR, Guilger M, Mishra S, Keswani C, Singh HB, deLima R (2018) *Trichoderma harzianum* based novel formulations: Potential Applications for Management of Next-Gen agricultural challenges. *J Chem Technol Biotechnol* 93(8): 2056–2063.

Gade AK, Bonde P, Ingle AP et al. (2008) Exploitation of *Aspergillus niger* for the synthesis of silver nanoparticles. *J Biobased Mater Bioener* 2: 243–247.

Gade AK, Ingle A, Whiteley C, Rai M (2010) Mycogenic metal nanoparticles: Progress and applications. *Biotechnol Lett* 32(5): 593–600.

Gajbhiye MB, Keshewani JG, Ingle AP et al. (2009) Fungus mediated synthesis of silver nanoparticles against pathogenic fungi in combination of fluconazole. *Nanomed NBM* 5: 282–286.

Hamzah HM, Salah RF, Maroof MN (2018) *Fusarium mangiferae* as new cell factories for producing silver nanoparticles. *J Microbiol Biotechnol* 28(10): 1654–1663.

Ingle A, Gade A, Pierrat S, Sonnichsen C, Rai M (2008) Mycosynthesis of silver nanoparticles using the fungus *Fusarium acuminatum* and its activity against some human pathogenic bacteria. *Curr Nanosci* 4(2): 141–144.

Ingle A, Rai MK, Gade A, Bawashar M (2009) Fusarium solani: A novel biological agent for the extracellular synthesis of silver nanoparticles. *J Nanopart Res* 11(8): 2079–2085.

Karbasian M, Atyabi SM, Siadat SD, Momem SB, Norouzian D (2008) Optimizing nano-silver formation by *Fusarium oxysporum* (PTCC5115) employing response surface methodology. *Am J Agric Bio Sci* 3: 433–437.

Karimi Dorcheh S, Vahabi K (2017) Biosynthesis of nanoparticles by fungi: Large-scale production. In Merillion J-M, Ramawat KG (eds) *Fungal Metabolites*. Springer Inter Publ, Cham, Switzerland, pp. 395–414.

Khandel P, Kumar Shahi S (2018) Mycogenic nanoparticles and their bio-prospective applications. *J Nanostruc Chem*. doi:10.1007/s40097-018-0285-2.

Kim KJ, Sung WS, Moon SK, Choi JS, Kim JG, Dong GL (2008) Antifungal effect of silver nanoparticles on dermatophytes. *J Microbiol Biotechnol* 18(8): 1482–1484.

Mead PS, Slutsker L, Dietz V et al. (1999) Food related illness and death in United States. *Emerg Infect Dis* 5(5): 607–625.

Mishra S, Keswani C, Abhilash PC, Fraceto LF, Singh HB (2017) Integrated approach of Agri-nanotechnology: Challenges and future trends. *Front Plant Sci*. doi:10.3389/fpls.2017.00471.

Mishra S, Keswani C, Singh A, Singh BR, Singh SP, Singh HB (2016) Microbial nanoformulation: Exploring potential for coherent nano-farming. In Gupta VK, Sharma GD, Tuohy MG, Gaur R (eds) *The Handbook of Microbial Bioresources*. CABI-UK, Wallingford, pp. 107–120.

Mishra S, Singh A, Keswani C, Singh HB (2014b) Nanotechnology: Exploring potential application in agriculture and its opportunities and Constraints. *Biotech Today* 4(1): 9–14.

Mishra S, Singh BR, Singh A, Keswani C, Naqvi AH, Singh HB (2014a) Biofabricated silver nanoparticles act as a strong fungicide against *Bipolaris sorokiniana* causing spot blotch disease in wheat. *PLoS One* 9(5): e97881.

Mohammadian A, Shaojaosadati SA, Reeze MH (2007) *Fusarium oxysporum* mediates photo-generation of silver nanoparticles. *Sci Iran* 14: 323–326.

Moriarty P (2001) Nanostructured materials. *Rep Prog Phys* 64(3): 297–381.

Mousavi SR, Rezaei, M (2011) Nanotechnology in agriculture and food production. *J Appl Environ Biol Sci* 1(10): 414–419.

Mukherjee P, Roy M, Mandal BP et al. (2008) Green synthesis of highly stabilized nanocrystalline silver particles by a non-pathogenic and agriculturally important fungus *T. asprerellum*. *Nanotechnology* 19: 103–110.

Mukherjee P, Senapati S, Mandal D, Ahamd A, Khan MI, Kumar R, Sastry M (2002) Extracellular synthesis of gold nanoparticles by the fungus *Fusarium oxysporum*. *Chem Biochem* 3(5): 461–463.

Narware J, Yadav RN, Keswani C, Singh SP, Singh HB (2019) Silver nanoparticle-based biopesticides for phytopathogens: Scope and potential in agriculture. In Koul O (ed) *Nano-Biopesticides Today and Future Perspectives*. Academic Press, London, UK, pp. 303–314.

Rai M, Ingle A (2012) Role of nanotechnology in agriculture with special reference to managemnet of insect-pests. *Appl Microbiol Biotechnol* 94(2): 287–293.

Rai M, Yadav P, Bridge P, Gade A (2009) Myconanotechnology: A new emerging science. In Rai B (ed) *Applied Mycology*. CABI-UK, Wallingford, pp. 268–267.

Rai V, Acharya S, Dey N (2012) Implications of nanobiosensors in agriculture. *J Biomater Nanobiotechnol* 3(2): 315–324.

Rickman D, Luvall JC, Shaw J, Mask P, Kissel D, Sullivan, D (2009) Precision agriculture: Changing the face of farming. http://www.agiweb.org/geotimes/ nov03/feature_agric.html (Accessed on November 1, 2019).

Riddin TL, Gericke M, Whiteley CG (2006) Analysis of the inter and extracellular formation of platinum particles by *Fusarium oxysporum* f. sp. *lycopersici* using response surface methodology. *Nanotechnology* 17(14): 3482–3489.

Sastry RK, Rashmi HB, Rao NH (2011) Nanotechnology for enhancing food security in India *Food Policy* 36(3): 391–400.

Tumer AP (2000) Biosensors: Sense and sensitivity. *Science* 290(5495): 1315–1317.

Vahabi K, Mansoori GA, Karimi S (2011) Biosynthesis of silver nanoparticles by the fungus *Trichodermareesei*. *Insciences J* 1: 65–79.

Verma VC, Kharwar RN, Gange AC (2010) Biosynthesis of anti-microbial silver nanoparticles by the endophytic fungus *Aspergillus clavatus*. *Nanomedicine* 5(1): 33–40.

Yadav A, Kon K, Kratosova G, Dran N, Ingle AP, Rai M (2015) Fungi as an efficient mycosystem for the synthesis of metal ions: Progress and key aspects of research. *Biotechnol Lett* 37(11): 2099–2120.

Section 3

Pharmaceutical Nanotechnology

7 Core Competencies in Cancer Nanotechnology in Brazil
Contributions of Scientific and Technological Knowledge in the Health Sector

*Jorge Lima de Magalhães, Luc Quoniam,
Maria Simone de Menezes Alencar, Marcos Emiliano Lima
Alves Hir, Suzanne de Oliveira Rodrigues Schumacher,
and Adelaide Maria de Souza Antunes*

CONTENTS

7.1 BACKGROUND

In 2003, Brazil stated in its Industrial, Technological and Foreign Trade Policy (PITCE – Brazilian term), that nanotechnology would be considered as one of the future carrier technologies. The decision prioritized this technology in order to envision the ability to generate opportunities for science, technology, and industry over a longer-term horizon. This was based on the country's accumulated scientific competence in related areas (Grassi 2008).

Thus, nanotechnology has been touted as a highly innovative technology in the sense of profoundly altering the production of a wide range of industrial products, as they have a broad application in various sectors of the economy and society. The manipulation of nanoscale materials creates new possibilities to explore the functionality of molecular architecture through the control of optical, electronic, and reactivity properties (Alencar 2008).

Several sectors of the economy are likely to be impacted by nanoscience/nanotechnology, such as applications in healthcare, chemistry, petrochemicals, computing, energy, agribusiness, metallurgy, textiles, environmental protection, etc. It is noteworthy that the major areas of science integrate in a multidisciplinary way, involving skills from all sectors, such as pharmacy, chemistry, physics, biology, electronics, and others (Alencar 2008).

Nanotechnology is the creation of useful materials, devices, and synthesis used to manipulate matter at an incredibly small scale between 1 and 100 nm (Sinha et al. 2006). It concerns the study of devices that are themselves or have essential components in the 1–1,000 nm dimensional range – that is, from a few atoms to subcellular size.

According to Hahn and Weinberg (2002), human cancer is a complex disease caused by genetic instability and the accumulation of multiple molecular alterations. Cancer nanotechnology is emerging as a new field of interdisciplinary research, cutting across the disciplines of biology, chemistry, engineering, and medicine, and is expected to lead to major advances in cancer detection, diagnosis, and treatment (Nie et al. 2007).

Several programs have supported research on novel nanodevices capable of detecting cancer at its premalignant stage, locating cancerous tissue within the body, delivering anti-neoplastic drugs to the cancer cells, and determining if these cells are being killed by the drugs (Sinha et al. 2006). Nanotechnology is a new revolution for cancer therapy.

According to Ferrari (2005), cancer-related examples of nanotechnologies include injectable drug delivery nanovectors such as liposomes for the therapy of breast cancer, biologically targeted nanosized magnetic resonance imaging (MRI) contrast agents for intra-operative imaging in the context of neuro-oncological interventions, and novel, nanoparticle-based methods for high-specificity detection of DNA and protein (Ferrari 2005).

Nanotechnology more commonly refers to structures that are up to several hundred nanometers in size and are developed by top-down or bottom-up engineering of individual components. Nanomaterials have a large surface area to volume ratio and their physicochemical properties, such as friction and interaction with other molecules, are distinct from equivalent materials at a larger scale. The most common use of nanotechnology in medicine has been in the areas of developing novel therapeutic and imaging modalities that have the potential to outperform the current best technology in these areas (Alexis et al. 2008).

According to Sinha et al. (2006), nanotechnology is a fast-expanding area of science. This area of research is anticipated to lead to the development of novel, sophisticated, multifunctional applications which can recognize cancer cells, deliver drugs to target tissue, aid in reporting outcomes of therapy, provide real-time assessment of therapeutic and surgical efficacy, and most importantly, monitor intracellular changes to help prevent precancerous cells from becoming malignant (Sinha et al. 2006).

Given the importance of nanotechnology for the health sector, it is essential to reflect on industrial property for industrial advancement (Antunes and Magalhaes 2008b). Intellectual property is an important asset for businesses, and knowledge is becoming increasingly crucial for competitiveness, technology, and economic development. This is particularly true for technology-intensive sectors, where knowledge is regarded as a company's most important asset (Miller 2000).

Therefore, it is crucially important for organizations to invest in research, development, and innovation, if they are to remain active and competitive. Information Science has tools that can help organizations to produce, store, and manage data on any activities or processes, resulting in more effective management for innovation. With the increasingly turbulent, complex, and competitive conditions in the markets in which companies operate, the use of industrial/intellectual property has become a way of assuring the continuation of their activities in the future by protecting innovations and restricting how their competitors can act. The industrial property information contained in patents identifies the latest science and technology developments, making it a powerful competitive weapon (Pierret 2006).

Like this, the knowledge presented in each technology promotes the continuous scientific and technological development translated into various sources, such as patents. They characterize the state of the art of innovation and present themselves as opportunities for the strategic management of knowledge (Magalhães et al. 2001).

In this sense, thinking about new practices in applications for the value of information and knowledge management in industrial property with scientific core competencies in cancer is a great opportunity for industrial innovation (Lawrence and Giles 2000a). It is important to have a scenario about the cancer core competencies in the health sector.

7.2   KNOWLEDGE MANAGEMENT AND INDUSTRIAL PROPERTY

Knowledge management requires the involvement and support of all the company's stakeholders to preserve, transmit, and develop knowledge. Indeed, it is individuals who are at the center of the creation of value and who hold the key to the success of such a project. The management of the knowledge and knowhow of the company is therefore not universal, it depends strongly on the culture of the country in which it is practiced (Balmisse 2006).

Knowledge management is a field of management science theory, although the notion links an epistemological concept (knowledge), a concept that is at the center of many philosophical debates and that is linked to the perspective of efficient action in the company. The hybridization which results from the simultaneous use of the two terms induces the idea of an action situated in relation to teleological knowledge or, in other words, beliefs and certainties of a very particular nature. This hybridization is also described as one of the signs of the institutionalization of the company which is today supposed to create and manage knowledge (Cowan 2001; Pirró, Mastroianni, and Talia 2010).

In order to manage some knowledge, it is first necessary to have information and, therefore, to have data to be processed (Antunes et al. 2013; Jamil, Malheiro, and Ribeiro 2013; Lawrence and Giles 2000b). According to BALMISSE (2006), the different notions to the definitions of knowledge are: knowledge is the information assimilated to carry out an action; information is the data put into context; and, data are the gross element outside of context.

In the age of knowledge, intellectual capital plays an important role in economics and business, and therefore key to competitiveness and economic development in technology. In areas of high density, such as pharmaceuticals (nanotechnology), aerospace, and telecommunications, among others of equal weight and impact, knowledge becomes the most important asset and, therefore, the need to better manage this knowledge (Lastres and Sarita 1999; Singh et al. 2016; Keswani 2019).

The management of knowledge is presented as a major stake in the functioning of organizations and societies. It is also under this common denominator that the link between the knowledge-based enterprise and the knowledge-based society is established, the development of knowledge-based enterprises being in the direction of building a society of knowledge, the prosperity of the enterprises was assuring that of the companies, and global employment as the sum of the jobs of the enterprises. This makes it possible to assert that the understanding of the enterprise of knowledge is the only valid one, without really having to face the problem of the transition from the enterprise to the knowledge society (Pesqueux 2005).

Considering the evolution of the web in the 21st century, it is noted that the speed of information generation is unprecedented. The world's per capita technological capacity to store information has almost doubled every 40 months since the 1980s. The internet evolved from a purely static platform and assumed a dynamic and interactive role (Minelli, Chambers, and Dhiraj 2013), allowing users to exchange large amounts of information instantly. 'By the end of 2016, the amount of information created and replicated from sources of all types, posts on social networks, uploaded photos and videos, commercial transaction records, GPS signals and navigation traces (among other sources) will reach the order of zettabytes – 1 billion gigabytes' (Huyghe 2009; McKinsey Global Institute 2011).

Once the proliferation of data and information through the internet and the flow of information is established and there are no restrictions on distance and availability, the big question that arises is the ability to screen, interpret, and convert this volume of information into knowledge. The identification and analysis of the amount of information of a given area, whether scientific, technological or business, or their respective correlations is a mammoth task (Aleixo and Duarte 2015; Magalhães and Quoniam 2013).

A management analysis is therefore necessary to integrate actions and to provide effectiveness to generate technological innovation and consequently wealth of the country. Thus, a better management for information architecture, also known as business architecture, encourages new understanding (Ross, Weill, and Robertson 2006).

Certo and Peter (2010) regard analysis of the environment, which indicates that this context (business architecture requires information architecture) is monitoring the entire organizational environment to identify opportunities, challenges, and threats, regarding assessment to current and future risks (Certo and Peter 2010).

In this context, identifying the essential competencies in certain areas of knowledge and promoting their multidisciplinary integration is fundamental for the advancement and growth of science and technological development. In this scenario, industrial property is a strong indicator of technological innovation for the country (Magalhães et al. 2014; Possas et al. 2015).

Industrial property in the health sector is an important indicator and contributor to possible innovation in this area. Therefore, in the health sector the pharmaceutical industry plays a key role for humanity. Meanwhile, the world pharmaceutical industry is a highly concentrated and oligopolistic market. This market is characterized by being based on innovation, presenting great technical and scientific development, besides having a long maturation time until the introduction of new products into the market. Like any technology-based sector, it is strongly linked to the system of industrial property protection through the granting of patents* (Antunes and Magalhaes 2008c; 2008a).

It is worth mentioning the creation of the Patent Act. In 1994, it was signed by members of the World Trade Organization (WTO) Agreement on Trade-Related Aspects of Intellectual Property Rights (TRIPS) which is known as an international agreement. This agreement establishes that the entire production of the member country must grant patents for chemicals and pharmaceuticals. The patents in the field have become more intense and controlled in member countries of the WTO. This agreement establishes that the production and sale of drugs only can be made by the holder of the patent, for a period of at least 20 years. In this way, it prevents the production and marketing of pharmaceutical products by other pharmaceutical companies, encouraging investment in innovation. In Brazil, it is regulated by Law No. 9.279 of May 14, 1996.

Therefore, pharmaceutical patents have become very important; they contain innovation and reveal the details of the invention, be it process or product. On the other hand, researchers and/or managers consider patents as an important indicator of innovation. Studies show that patents of a company, countries, and owners of an invention can reveal the technological dynamism of an industry and provide information about the direction of technological change.

New trends influence the industrial development of a country, such as knowledge, like a major resource and learning, like a central process. Therefore, it is essential to broaden the base of expertise in human resources and hence increase the innovation potential (Lastres and Sarita 1999).

Formation of competencies for innovation has been previously defined as an intelligence cooperative, which translates as construction of knowledge in collaboration with peers at work. This mindset requires collaborative development processes capable of producing high quality information for scientific and technological knowledge. The experts have unrestricted access to information created by the scientific community, collaborative review of the contributions of members, governance based more on authority than on sanctions, and involvement in integrated levels and responsibilities (Ambrosi, Peugeot, and Pimenta 2005).

In 1977, Tidd et al. discussed innovation as a change in products and services offered by an organization, or even a change in the process or in a way to prepare the products or services. It also can be considered the way the organization delivers its products (Tidd, Pavitt, and Bessant 2005).

The activities of companies, research groups, government, and institutions of countries are effective when they attribute value and quality in their information is considered critical. These factors will lead the organization to succeed in their internal and external planning such as strategies for the

---

* Patent is a temporary title to an invention or utility model granted by the state to inventors or authors or other natural or legal persons having rights over creation. With this right, the inventor or the patent holder has the right to prevent third parties, without their consent, from producing, using, selling, or importing a product subject to their patent and/or process, or product obtained directly by process patented by him. In contrast, the inventor undertakes to disclose in detail all the technical content of the subject matter protected by the patent (MDIC. INPI – Instituto Nacional de Propriedade Industrial 2016).

short and long term. Thus, the capital of intellectual property plays an important role in enterprises and knowledge becomes increasingly 'key', to competitiveness, technology, and economic development. This happens mainly in the high-density technology sector where knowledge is considered as the most important asset of the company (Trigo et al. 2007).

In this sense, considering the information contained in the database of Lattes curricula of Brazil, it will be identified, extracted, and treated on the core competencies of the Brazilian leaders in terms of scientific competence (publications in periodicals) and technological competencies (patent applications).

7.3 SEARCH STRATEGY FOR IDENTIFYING CORE COMPETENCIES IN CANCER NANOTECHNOLOGY IN BRAZIL

Preliminarily, as a keyword search strategy, the data is already consolidated in the area of nanotechnology applied to health in the scientific papers of Alencar (2008), Alencar et al. (2017) and Amaral (2016). These data (key terms) were then compared with the Medical Subject Headings (MeSH) classification (https://www.ncbi.nlm.nih.gov/mesh).

The MeSH thesaurus is a controlled and hierarchically organized vocabulary produced by the National Library of Medicine. It is used for indexing, cataloging, and searching biomedical and health-related information.

In the MeSH database all the keywords and their 'roots' for nanotechnology and cancer are identified. Thus, a comparison was applied between the consolidated MeSH database terms and the terms described in the scientific papers. In this sense, the intersection between the terms was identified and a search strategy was generated that included all scientific works that are linked to the health sector in cancer nanotechnology.

The search strategy used to identify the core competencies in the area of nanotechnology for cancer in Brazil was (nano* OR quantum dots* OR fullerene*) AND (cancer OR oncologic*). The use of word search truncation, such as the asterisk, provided several other keywords correlated to the current theme.

To identify the core competencies in the area, the Brazilian Lattes platform was used. The Lattes platform is an information system (integrated database, web-based query interface, etc.) maintained by the Brazilian federal government to manage information on science, technology, and innovation related to individual researches and institutions working in Brazil (BRASIL, MCTI. 2019). In 2010 Lane (2010) highlighted the Lattes platform as exemplary (Lane 2010).

Similarly, it is important to point out that a database called the Research Group directory that aggregates institutionalized research groups and led by Brazilian researchers, was developed by the Brazilian government.

In this sense, the strategy of the keywords obtained earlier in this Lattes platform was applied. As there are more than 1,800,000 scientific curricula on the platform, the criterion was established to select only those specialists who meet the requirements 'Brazilian and foreign doctors with any level of productivity scholarship', and who have a presence in the Research Group directory in CNPq Group*.

Annually, CNPq selects Research Productivity Grants, which aims to value researchers who have outstanding scientific, technological and innovation production in their respective areas of knowledge, and encourages the increase in scientific, technological, and quality innovation production.

As shown in Table 7.1, after applying the search on the Lattes platform 751 specialists were initially identified in the universe of 1,800,000, in November 2019. In this chapter, it was chosen to restrict core competencies in cancer nanotechnology to the criterion 'Brazilian and foreign doctors with any level of productivity scholarship and Research Group directory presence' (total 231 specialists).

---

* The Brazilian National Council for Scientific and Technological Development is an organization of the Brazilian federal government under the Ministry of Science and Technology, dedicated to the promotion of scientific and technological research and to the formation of human resources for research in the country.

TABLE 7.1

Identification and Selection of Core Competencies in Cancer Nanotechnology

Search strategy: (nano OR quantum dots OR fullerene) AND (cancer OR oncologic)

Criteria in Lattes Platform	Specialists Numbers
All (doctors and other researchers, Brazilians and foreigners registered in Brazil	751
Brazilian and foreign doctors registered in Brazil	660
Brazilian doctors	599
Brazilian and foreign doctors with any level of productivity scholarship	239
Brazilian doctors with any level of productivity scholarship	227
Brazilian and foreign doctors with any level of productivity scholarship and Research Group directory presence	231

With the achievement of 231 core competencies in cancer nanotechnology, the search of this curriculum and respective treatment of existing data in each curriculum was applied. These constitute an excellent repository of senior scientific and technological productions in the area in question, since it covers the maximum and senior production in the area (based on all the previous criteria). For this, the identification, extraction, and data processing were applied to the open source software ScriptLattes (Mena-Chalco and Cesar Junior 2009).

7.4 CORE COMPETENCIES IN CANCER NANOTECHNOLOGY IN BRAZIL

The results are available at http://vlab4u.info/nanocorecompetenciebrazil/ after executing the script described in the methodology. It is important to note that ScriptLattes is an open-source software used for knowledge extraction and visualization from large data sets and is an important research topic in computer science with a strong potential impact in all scientific fields (Mena-Chalco and Cesar Junior 2009).

In this sense, 231 core competencies in cancer nanotechnology can be visualized with their respective scientific (journal articles) and technological (patent applications) productions. Similarly, the correlations and temporal evolutions of their research. On the main page, you can see the index summary for the productions of these specialists (Figure 7.1)*.

When the 'papers' field (Produção Bibliográfica – Brazilian term) is clicked, the temporal evolution of the scientific articles published over the years by these core competencies can be observed. It is noteworthy that until the 1990s, nanotechnology publication was very elementary, with slight growth from this point until the 2000s. From 2000 to 2012 growth was constant and reached around 2,000 papers per year from 231 specialists in the area of the research (Figure 7.2). This interest demonstrates the promising area in question.

It is noted that the number of awards obtained by these senior scientists was at the same rate of growth as the scientific publications. Of the total obtained (2,826), around 2,000 awards were granted in only the last 20 years (Figure 7.3).

As an example, some awards are highlighted, such as: Best oral presentation – Photo-Induced Electron Transfer from Hematite and Zinc Oxide Nanostructures to Cytochrome C: Systems Applicable to Spintronics; Best poster in cancer center – Next Frontiers to Cure Cancer; Best poster award – Symposium on Immunomodulation in Cancer Regeneration: Leishmanicidal Activity of Biogenic Fe_3O_4 Nanoparticles.

The core competencies are arranged by geolocation in Brazil – note greater concentration in the south and southeast. The most represented institutions are the University of São Paulo, Campinas State University, Oswaldo Cruz Foundation, Federal University of Rio de Janeiro, Fluminense

---

* It should be noted that the data may not be complete due to the lack of updating by specialists and also due to the proximity of dates (2018 and 2019).

Senior Intellectual Production in Brazil in Nanotechnology for Oncology

| Members | Bibliographic production | Technical production | Tutors (supervisions) | Projects | Awards | Events | Collaboration graph | Geolocation map | Metrics |

Bibliographic production
- Complete articles published in journals (28736)
- Organized published books or editions (337)
- Published book chapters (1814)
- Texts in magazine news newspapers (1954)
- Complete papers published in conference proceedings (6469)
- Expanded abstracts published in conference proceedings (2809)
- Abstracts published in conference proceedings (28273)
- Articles accepted for publication (210)
- Work Presentations (9139)
- Other types of bibliographic production (514)
- Total bibliographic production (80255)

Technical production
- Technological products (398)
- Processes or techniques (444)
- Technical works (2869)
- Other types of technical production (2875)
- Total technical production (6586)

FIGURE 7.1 Main display of the results of core competencies in cancer nanotechnology.

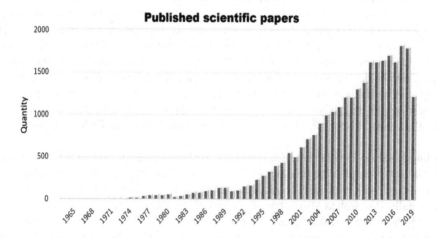

FIGURE 7.2 Papers about cancer nanotechnology of core competencies authors by Brazilian researches.

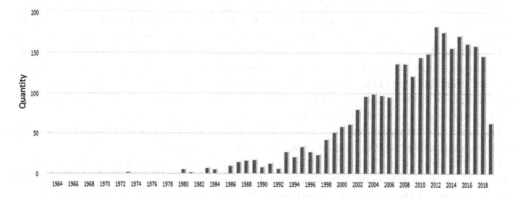

FIGURE 7.3 Awards earned by core competencies in cancer nanotechnology.

Federal University, Federal University of Rio Grande do Sul, Federal University of Minas Gerais, and Maringá State University. In general, universities and research excellence centers are located in the following states: São Paulo, Rio de Janeiro, Rio Grande do Sul, and Minas Gerais. It is noteworthy that there is always a link in health sector research with nanotechnology in cancer.

The core competencies in cancer nanotechnology are part of a body of expertise from different areas of science, such as pharmacy, chemistry, biology, physics, materials engineering, biochemistry, etc. These specialists have scientific and technological productions not only endogenously, but also externally to their institutions. This fact underscores the scope and multidisciplinary and factorial view of scientific and technological research. There is a Collaboration Rank listing (numerical value that indicates the impact of a member on the collaborations of their publications and patent applications – a measure like PageRank) – all the experts and in the most diverse collaborations. Both the collaboration ranking and the degree of feedback are detailed on the web page http://vla b4u.info/nanocorecompetenciebrazil/ – click on the 'Grafo de Colaborações' (Brazilian term).

In Table 7.2 shows the main researchers with publications and or patent applications. The core competencies in cancer nanotechnology are distributed in various areas of knowledge, such as exact and earth sciences, medicine, physics, chemistry, pharmacy, genetics, biomedical engineering, etc. Specialists Vanderlei Salvador Bagnato, Lydia Masako Ferreira, and Wanderley de Souza have the largest number of scientific publications: 784, 737, and 699 respectively. Concerning the patent applications, it is noted that the leading experts are Stanisçuaski Silvia Guterres, Adriana Pohlmann Raffin, and Nelson Eduardo Duran Caballero with 75, 75, and 69 respectively.

Figure 7.4 shows the Collaboration Rank from expert number 1 in terms of patent applications: Professor Dr Silvia Stanisçuaski Guterres. The research network is external to the research institution. The institution is Rio Grande do Sul and pharmacy is the predominant research department. Similarly, Professor Adriana Raffin Pohlmann (pharmacy department of the Federal University of Rio Grande do Sul) and Professor Oswaldo Novais de Oliveria Júnior (physical department of the University of São Paulo). Importantly, Professor Silvia has 426 scientific productions in networks.

Looking at industrial property emanating from patent applications whose technologies are promising for the industrial area, it is noted that Professor Silvia Guterres also ranks at number 1. The following are Professor Dr Nelson Eduardo Duran Caballero and Professor Dr Ruben Dario Sinisterra Millán.

Regarding collaborations for industrial property work (patent applications), it is noted that these senior specialists involved 2,766 other specialists in their scientific research. Prof Dr Silvia has co-authored scientific work with 1,012 other scientists. Prof Nelson Caball has already involved 1,166 scientists. Finally, Prof Ruben Millán has co-authored with 596 scientists.

Figure 7.5 shows the scientific productions and patents filed by the 231 core competencies in cancer nanotechnology. The cluster distribution shows the scientific and technological publications arranged by region in Brazil. Note that the southern and southeastern regions have the largest representation of patent deposits. Of the 1,564 patents filed, 70% are from this region. This fact denotes the strong concentration of scientific and technological knowledge in southern and southeastern Brazil.

Concerning the scientific productions (28,695), the concentration of technological knowledge in the southeast, south, northeast, and mid-west is observed. The northern region is more prominent. Note the strong interaction between the existing researchers in the southeast, south, and northeast. Most of these co-authors pass through the southern and southeastern regions; however, the presence of the northern and mid-western regions is noted.

It is noted that the patent deposits versus the scientific production made by the senior scientists are more robust and present in larger numbers when they are co-authored.

Given the above, it is observed that the greater the interaction of scientific research of senior cancer specialists in nanotechnology, the stronger is this interaction and the results are substantial in both scientific and technological production. Technology emanating from patents is a key area for the health sector and provides advancement in the subject matter. That is, the greater the investment

TABLE 7.2
Top Researchers with Scientific Publications and/or Patent Applications

Researcher	Institution	Region	State of Brazil	Number of Periodic	Patent Application	Main Area	Area
Vanderlei Salvador Bagnato	Universidade de São Paulo	Southeast	São Paulo	784	34	Exact and Earth Sciences	Physical
Lydia Masako Ferreira	Universidade Federal de São Paulo, UNIFESP	Southeast	São Paulo	737	12	Health Sciences	Medicine
Wanderley de Souza	Universidade Federal do Rio de Janeiro, Instituto de BioPhysical Carlos Chagas Filho, Laboratorio de Ultraestrutura Celular Hertha Meyer.	Southeast	Rio de Janeiro	699	1	Biological Sciences	Parasitology
Osvaldo Novais de Oliveira Junior	Universidade de São Paulo,	Southeast	São Paulo	551	9	Exact and Earth Sciences	Physical
Nelson Eduardo Duran Caballero	Universidade Estadual de Campinas, Instituto de Quimica-UNICAMP.	Southeast	São Paulo	542	69	Biological Sciences	Biochemistry
Paulo Cesar de Morais	UniversidadeCatólica de Brasília	Midwest	Distrito Federal	452	7	Exact and Earth Sciences	Physical
Vasco Ariston de Carvalho Azevedo	Universidade Federal de Minas Gerais	Southeast	Minas Gerais	443	12	Biological Sciences	Genetics
Jose Mauro Granjeiro	Instituto Nacional de Metrologia, Qualidade e Tecnologia, Diretoria de Metrologia Aplicada às Ciências da Vida – Dimav.	Southeast	Rio de Janeiro	385	2	Engineering	Biomedical engineering
Celso Vataru Nakamura	Universidade Estadual de Maringá	South	Paraná	368	10	Biological Sciences	Microbiology
Flavio Fernando Demarco	Universidade Federal de Pelotas, Faculdade de Odontologia, Departamento de Odontologia Restauradora	South	Rio Grande do Sul	364	not applicable	Health Sciences	Odontology
Sílvia Stanisçuaski Guterres	Universidade Federal do Rio Grande do South,	South	Rio Grande do Sul	310	75	Health Sciences	Pharmacy
Adriana Raffin Pohlmann	Universidade Federal do Rio Grande do South,	South	Rio Grande do Sul	275	75	Health Sciences	Pharmacy
Oswaldo Luiz Alves	Universidade Estadual de Campinas	Southeast	São Paulo	248	49	Exact and Earth Sciences	Chemistry

(Continued)

TABLE 7.2 (CONTINUED)
Top Researchers with Scientific Publications and/or Patent Applications

Researcher	Institution	Region	State of Brazil	Number of Periodic	Patent Application	Main Area	Area
Koiti Araki	Universidade de São Paulo	Southeast	São Paulo	214	42	Exact and Earth Sciences	Chemistry
Ruben Dario Sinisterra Millán	Universidade Federal de Minas Gerais, Instituto de Ciências Exatas, Departamento de Chemistry.	Southeast	Minas Gerais	142	66	Exact and Earth Sciences	Chemistry
Victor Dmitriev	Universidade Federal do Pará, Centro Tecnológico, Departamento de Electricalengineering.	North	Pará	124	36	Engineering	Electrical engineering
Fabiana Kömmling Seixas	Universidade Federal de Pelotas	South	Rio Grande do Sul	119	32	Biological Sciences	Genetics
Luiz Orlando Ladeira	Universidade Federal de Minas Gerais	Southeast	Minas Gerais	104	50	Exact and Earth Sciences	Physical

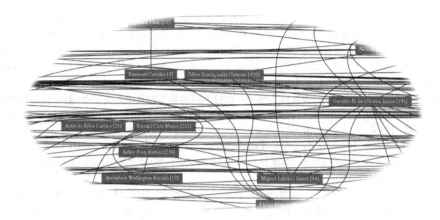

FIGURE 7.4 Part of core competencies in cancer nanotechnology collaborations graph – highlight Silvia Guterres.

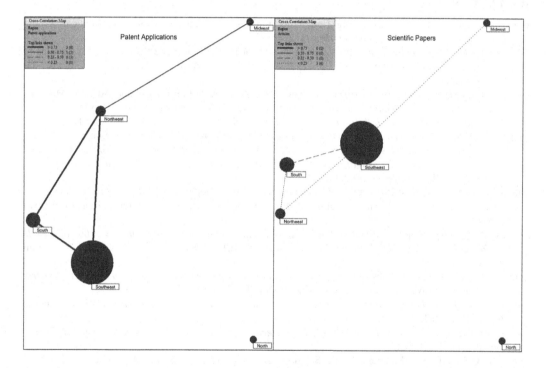

FIGURE 7.5 Cluster of scientific and technological productions by regions in Brazil.

in basic scientific research, the higher the chance that applied research progresses to the point of patent filing. Note that the 'square' represents the southeast region and is the largest size in 'space'. These point to the largest number of patent filings and the respective strength of the scientific and technological institutions in this state.

7.5 CONCLUSION

From the open source ScriptLattes the identification of 231 core cancer competencies for nano-technology in the country were examined. It is noted that the essential information is the great differential for decision makers in the 21st century. It translates into knowledge for innovation, when the trail of knowledge that is formed by the extraction of data that generates information and, therefore, knowledge is transferred. However, the management of this knowledge is fundamental for the scientific and technological advancement of organizations and, consequently, the wealth of countries and for the well-being of their populations.

The health sector, combined with future-bearing technology (nanotechnology), is promising in scientific and technological knowledge in the country. There are 28,695 scientific papers published, 337 books, 1,810 book chapters, and 2,824 awards granted in the area in question.

The prominent institutions in the area of these core competencies are located mainly in the south and southeast. There is strong interaction between them for both scientific and technological publications (patent applications).

In general, the scientific and technological knowledge in the field of nanotechnology applied to the health sector in Brazil is a strong contribution to the advancement of science and technology in humanity.

REFERENCES

Aleixo, J.A., and P. Duarte. 2015. "Big Data Opportunities in Healthcare. How Can Medical Affairs Contribute?" *Revista portuguesa de farmacoterapia* 7: 230–36.

Alencar, M.S.M. 2008. "Estudo de Futuro através da Aplicação de Técncias de Prospecção Tecnológica: o caso da nanotecnologia". UFRJ/EQ/EPQB. http://epqb.eq.ufrj.br/resumos/estudo-de-futuro-o-caso-da-nanotecnologia/.

Alencar, Maria Simone de Menezes, Rosany Bochner, Miriam Ferreira Freire Dias, and Adelaide Maria de Souza Antunes. 2017. "Análise da produção científica brasileira sobre nanotecnologia e saúde". *Revista Eletrônica de Comunicação, Informação e Inovação em Saúde* 11(1). doi:10.29397/reciis.v11i1.1199.

Alexis, Frank, June-Wha Rhee, Jerome P. Richie, Aleksandar F. Radovic-Moreno, Robert Langer, and Omid C. Farokhzad. 2008. "New Frontiers in Nanotechnology for Cancer Treatment". *Urologic Oncology: Seminars and Original Investigations, The Potential of Nanotechnology in Urologic Oncology* 26(1): 74–85. doi:10.1016/j.urolonc.2007.03.017.

Amaral, D.G. 2016. "Estudo Prospectivo do Patenteamento em Nanotecnologia no Brasil: uma análise de cadeia de valor". INPI – Instituto Nacional de Propriedade Industrial. http://biblioteca.inpi.gov.br/sophi a_web/.

Ambrosi, A., V. Peugeot, and D. Pimenta. 2005. *Enjeux de mots – Regards multiculturels sur les sociétés de l'information*. France: C&F Editions. http://www.decitre.fr/livres/enjeux-de-mots-9782915825039.html #table_of_content.

Antunes, A.M.S., J.L. Magalhaes, orgs. 2008a. *Oportunidades em medicamentos genéricos: A indústria farmacêutica Brasileira*. Rio de Janeiro, Brazil: INTERCIENCIA.

———, orgs. 2008b. *Patenteamento e Prospecação tecnológica no Setor Farmacêutico*. Rio de Janeiro, Brazil: INTERCIENCIA.

———, orgs. 2008c. *Patenteamento e Prospecação tecnológica no Setor Farmacêutico*. Rio de Janeiro, Brazil: INTERCIENCIA.

Antunes, A.M.S., F.M.L. Mendes, S.O.R. Schumacher, L. Quoniam, and J.L. Magalhães. 2013. "The Contribution of Information Science through Intellectual Property to Innovation in the Brazilian Health Sector: Library and Information Science Book Chapter I IGI Global". In *Rethinking the ConZipcodetual Base for New Practical Applications in Information Value and Quality*, 345. IGI Global. http://www. igi-global.com/chapter/the-contribution-of-information-science-through-intellectual-property-to-inn ovation-in-the-brazilian-health-sector/84214.

Balmisse, Gilles. 2006. *Guide des Outils du Knowledge Management: Panorama, choix et mise en oeuvre*. http://www.leslivresblancs.fr/informatique/applications-pro/knowledge-management/livre-blanc/ou tils-du-km-panorama-choix-et-mise-en-oeuvre-62.html.

BRASIL, MCTI. 2019. "Plataforma Lattes". http://lattes.cnpq.br/.

Certo, S.C., and J.P. Peter. 2010. *Administração Estratégica: Planejamento e implantação de estratégias*. 3° ed. São Paulo, Brazil: PRENTICE-HALL.

Cowan, Robin. 2001. "Expert Systems: Aspects of and Limitations to the Codifiability of Knowledge". *Research Policy* 30(9): 1355–72. doi:10.1016/S0048-7333(01)00156-1.

Ferrari, Mauro. 2005. "Cancer Nanotechnology: Opportunities and Challenges". *Nature Reviews Cancer* 5(3): 161–71. doi:10.1038/nrc1566.

Grassi, Robson Antonio. 2008. "Política industrial e compromissos críveis: Uma proposta de análise e de ação governamental". *Brazilian Journal of Political Economy* 28(4): 678–97. doi:10.1590/S0101-31572008000400009.

Hahn, William C., and Robert A. Weinberg. 2002. "Modelling the Molecular Circuitry of Cancer". *Nature Reviews Cancer* 2(5): 331–41. doi:10.1038/nrc795.

Huyghe, F.B. 2009. "Web 2.0: Influence, outils et réseaux". *Revue Internationale d'Intelligence Economique* 2009: 11.

Jamil, George Leal, Armando Malheiro, and Fernanda Ribeiro, orgs. 2013. *Rethinking the ConZipcodetual Base for New Practical Applications in Information Value and Quality*. IGI Global. http://www.igi-global.com/chapter/perZip codetion-of-the-information-value-for-public-health/84218.

Keswani, C. (Ed.). 2019. *Bioeconomy for Sustainable Development*. Singapore: Springer-Nature. 392 pages, ISBN- 978-981-13-9430-0.

Lane, Julia. 2010. "Let's Make Science Metrics More Scientific". *Nature* 464(7288): 488–89. doi:10.1038/464488a.

Lastres, H.M.M., and A. Sarita. 1999. *Informação e globalização na era do conhecimento*. Rio de Janeiro: Editora Campus Ltda.

Lawrence, Steve, and C. Lee Giles. 2000a. "Accessibility of Information on the Web". *Intelligence* 11(1): 32–9. doi:10.1145/333175.333181.

Lawrence, Steve, and C. Lee Giles. 2000b. "Accessibility of Information on the Web". *Intelligence* 11(1): 32–9. doi:10.1145/333175.333181.

Lima de Magalhães, Jorge, Flavia Maria Lins Mendes, Adelaide Maria de Souza Antunes, Zulmira Hartz, Jorge Lima de Magalhães, Flavia Maria Lins Mendes, Adelaide Maria de Souza Antunes, and Zulmira Hartz. 2001. "The Contribution of Information Science Through Scientific and Technological Knowledge in Intellectual Property". Chapter. http://Services.Igi-Global.Com/Resolvedoi/Resolve.Aspx?Doi=10.401 8/978-1-5225-6225-2.Ch013. 1o de janeiro de 2001. https://www.igi-global.com/gateway/chapter/20 8568.

Magalhães, J.L., and L. Quoniam. 2013. "PerZipcodetion of the Information Value for Public Health: A Case Study for Neglected Diseases: Library and Information Science Book Chapter | IGI Global". In *Rethinkin the ConZipcodetual Base for New Practical Applications in Information Value and Quality*, 345. IGI Global. http://www.igi-global.com/chapter/perZip codetion-information-value-public-health/84218.

Magalhães, Jorge L., Luc Quoniam, Jesús P. Mena-Chalco, and André Santos. 2014. "Extração e tratamento de dados na base lattes para identificação de core competencies em dengue". *Informação e Informação* 19(3): 30–54. doi:10.5433/1981-8920.2014v19n3p30.

McKinsey Global Institute. 2011. "Big data: The next frontier for innovation, competition, and productivity | McKinsey Global Institute | Technology & Innovation | McKinsey & Company". http://www.mckinsey.com/insights/mgi/research/technology_and_innovation/big_data_the_next_frontier_for_innovation.

MDIC. INPI – Instituto Nacional de Propriedade Industrial. 2016. "Consulta à Base de Dados do INPI". https://gru.inpi.gov.br/pePI/servlet/LoginController?action=login.

Mena-Chalco, Jesús Pascual, and Roberto Marcondes Cesar Junior. 2009. "ScriptLattes: An Open-Source Knowledge Extraction System from the Lattes Platform". *Journal of the Brazilian Computer Society* 15(4): 31–9. doi:10.1007/BF03194511.

Miller, Jerry P. 2000. *Millenium Intelligence: Understanding and Conducting Competitive Intelligence in the Digital Age*. Medford, NJ: Information Today, Inc.

Minelli, M., M. Chambers, and A. Dhiraj. 2013. *Big Data, Big Analytics*. EUA: John Wiey & Sons, Inc. https://books.google.com.br/books/about/Big_Data_Big_Analytics.html?hl=pt-BR&id=Mg3WvT8uHV4C.

Nie, Shuming, Yun Xing, Gloria J. Kim, and Jonathan W. Simons. 2007. "Nanotechnology Applications in Cancer". *Annual Review of Biomedical Engineering* 9(1): 257–88. doi:10.1146/annurev.bioeng.9.060906.152025.

Pesqueux, Yvon. 2005. "Management de la connaissance: un modèle organisationnel?" Management de la connaissance onun modele organisationnel r maio. https://halshs.archives-ouvertes.fr/halshs-0000400 5/document.

Pierret, Jean-Dominique. 2006. "Methodologie et structuration d'un outil de decouverte de connaissances base sur la litterture biomedicale : Une application basee sur le MeSH". Université du Sud Toulon Var. http://tel.archives-ouvertes.fr/tel-00011704.

Pirró, Giuseppe, Carlo Mastroianni, and Domenico Talia. 2010. "A Framework for Distributed Knowledge Management: Design and Implementation". *Future Generation Computer Systems* 26(1): 38–49. doi:10.1016/j.future.2009.06.004.

Possas, Cristina, Adelaide Maria de Souza Antunes, Flavia Maria Lins Mendes, Suzanne de Oliveira Rodrigues Schumacher, Reinaldo Menezes Martins, and Akira Homma. 2015. "Access to New Technologies in Multipatented Vaccines: Challenges for Brazil". *Nature Biotechnology* 33(6): 599–603. doi:10.1038/nbt.3244.

Ross, Jeanne W., Peter Weill, and David Robertson. 2006. *Enterprise Architecture as Strategy: Creating a Foundation for Business Execution*. Boston, MA: Harvard Business Review Press.

Singh, H.B., A. Jha, and C. Keswani (Eds.). 2016. *Intellectual Property Issues in Biotechnology*. Wallingford, UK: CABI. 304 pages. ISBN-13: 978-1780646534.

Sinha, R., G.J. Kim, S. Nie, and D.M. Shin. 2006. "Nanotechnology in Cancer Therapeutics: Bioconjugated Nanoparticles for Drug Delivery" *Molecular Cancer Therapeutics* 5(8). doi:10.1158/1535-7163. MCT-06-0141.

Tidd, J., K. Pavitt, and J. Bessant. 2005. Managing Innovation: Integrating Technological, Market and Organizational Change by Tidd, Joseph, 1960-, Pavitt, Keith, Bessant, John, 1952-. 3rd ed. John Wiley. http://prism.talis.com/mmu/items/1677998.

Trigo, Miguel Rombert, Luis Borges Gouveia, Luc Quoniam, and Edson Luiz Riccio. 2007. *Using Competitive Intelligence as a Strategic Tool in a Higher Education Context*. Organizado por D. Remenyi. Nr Reading: Academic Conferences Ltd.

8 Disruptive Nanotechnology Implications and Bio-Systems – Boon or Bane?

Lakshmipathy Muthukrishnan and Sathish Kumar Kamaraj

CONTENTS

8.1   INTRODUCTION

One of the most phenomenal forms of life on Earth, the microorganisms, have dominated the biosphere for more than three billion years. It is unbelievable that these tiny creatures survived where the possibility of life in deep-sea rock wedges, glacial ice, and hydrothermal vents seemed to be far from inevitable. Besides developing a prototype for metabolic pathways during their evolutionary lineage, some signature processes are still confined to prokaryotes alone. They are predominantly the early transformers of the Earth marking a magnificent transition from anaerobic to oxygenized atmospheric conditions engendering the evolution and sustenance of various forms of life on Earth (Fenchel et al. 2012). Indeed, they also have a key role in regulating and controlling global biogeochemical cycles to maintain equilibrium between terrestrial and marine ecosystems.

It would be astonishing if man-microbe interaction in all walks of life were unavoidable. Extraordinarily, the human womb serves to be the sterile environment where the sensational development of the fetus is pertinently uninfluenced by microbes. Interestingly, with the event of birth, humans encounter microorganisms that significantly mark the beginning of a lifelong mutualistic relationship. The indigenous microbiota of healthy humans and its enormous diversity play crucial roles in bridging the gap between abiotic and biotic components in the development, protection, and maintenance of *Homo sapiens*.

8.1.1   SURVIVAL STRATEGIES

In the present hostile world filled with a bewildering array of obnoxious agents, the need for survival and proficiency in propagating their 'selfish genes' inside a suitable biological system continues. There is a common bond that relies on the causes of deadly infections and their response from the host, and forms the basis for coexistence. The recent trend defines that the relationship between man and microbe has been altered from mutualism to antagonism, by which humans are victimized by varied dysfunction in their life by way of infections and diseases.

Bacterial infections are still a leading cause of death for millions of people worldwide. Although the advent of antibiotics and vaccines has certainly taken the dread out of many infectious diseases, the threat of infection is still a fact of life. It is also evident from history that they have harmed humans and disrupted society over millennia. Microbial diseases undoubtedly played a major role in the decline of the Roman Empire and the conquest of the New World, with the outbreak of plague or Black Death in 1347 that struck Europe with brutal force (McNeill 1976) killing one-third of the population within four years. There are records of 45 pandemics between 1500 and 1720. Though the disease was found quiescent in the 18th and 19th centuries, it was in Hong Kong that the last pandemic was reported in 1894 which spread throughout the world. India was one of the worst-hit Asian countries by this pandemic marking a death toll of 10 million by 1918. In continuation, cholera was the next widespread disease in the 19th century; it started from the Indian subcontinent and extended to China, Indonesia, and the Caspian Sea claiming around 40 million lives (Kelly Lee 2003). In a pandemic sequel, tuberculosis (TB) has been designated as the second most infectious disease after HIV/AIDS. According to World Health Organization (WHO) documentation, in 2013 nearly 9 million new cases were diagnosed and around 1.5 million deaths were due to TB having been reported mostly from developing countries.

8.1.2   HOW WERE INFECTIOUS AGENTS ENCOUNTERED?

The beginning of the third millennium has seen a dramatic increase in the incidence of infectious diseases, contributing to the increase in mortality worldwide. It was then a remarkable phenomenon but immunity was hypothesized by Thucydides in 430 BC that those who recovered from the plague could nurse the sick as the disease would not be contracted for a second time. The modern era of chemotherapy began with Paul Ehrlich's finding (a toxic dye molecule) that could work against sleeping sickness and syphilis, demonstrating the concept of selective toxicity. This was followed by long-term research and testing of numerous chemical agents for their therapeutic potential.

In 1928, Sir Alexander Fleming (1929) and Gerhard Domagk pioneered the modern chemotherapeutic era with their discovery of the first antibiotic penicillin (produced by *Penicilliumnotatum*) and prontosil, a major breakthrough in the medical field. Soon after the discovery of penicillin, Selman Waksman (1942) came out with new a antibiotic, streptomycin isolated from the actinomycete *Streptomyces griseus*. Subsequently, the clinically useful antimicrobial agents such as chloramphenicol, neomycin, terramycin, and tetracycline were isolated by 1953.

Although selective toxicity of therapeutic antimicrobials was considered, major concern over specific toxicity and the mechanism of action in the inhibition of microbial growth remained unanswered. At present, the known antimicrobial agents have been classified according to their mechanism of action. For instance, β-lactam and glycopeptides target bacterial cell wall synthesis; aminoglycosides and macrolide inhibit protein synthesis; quinolones interfere with DNA synthesis and sulphonamides with energy metabolism. But in the current scenario, those antimicrobials appraised as 'elixir,' 'magic bullets,' and 'wonder drug' have begun to fail in the treatment regime.

8.1.3 The Onset of Antimicrobial Resistance

There are a number of investigations reported on the origin, development, and challenges pertaining to the emergence of antibiotic resistance. Although several studies have shown horizontal gene transfer as the predominant mechanism leading to the emergence of multidrug-resistant strains of bacteria, the superbugs, and their strategy for coexistence. However, the difficulty in identifying the newly proposed crucial bacterial targets by antibiotics has culminated in the emergence of resistance in the context of antibiotic-mediated stress, an important criterion in bacterial evolution and pathogenesis (Vester et al. 2001).

The onset of antibiotic resistance in several bacterial strains has drawn the attention of scientists, medics, and pharmacologists across the globe to have this issue resolved. Despite numerous attempts in developing new antimicrobials and the maneuvering of existing drugs, antibiotic resistance has outpaced exploration of mysteries in the context of selection pressure associated with evolution and pathogenesis (Payne et al. 2007). This challenging and augmenting pattern of antibiotic-mediated resistance on a par with the rate of incidences of infectious diseases demands long-term technological solutions considering the novel formulation strategy, therapeutic reliability, bioavailability, biocompatibility, and healthcare costs.

8.2 CLASSIFICATION OF NANOMATERIALS

8.2.1 Nanoparticles

Nanoparticles are any atomic aggregates bonded together with any dimensions between 1 and 100 nm. A nanoparticle, though tiny, has a large surface area exhibiting increased surface reactivity and quantum effects, entirely different from that of its bulk counterpart (ASTM 2006; Buzea et al. 2007). Their characteristic quantum confinement, plasmon resonance, and super-paramagnetism determine the unusual properties exhibited by nanoparticles. They can be tailored for the desired size and shape with magnificent electrochemical properties using principles of radiation chemistry. Apart from size and shape, the hydrophobic nature of nanoparticles determines their application in the biomedical field as drug delivery agents. Alongside, the stability of nanoparticles *in vivo* relies greatly upon the surface charge (zeta potential), the composition of the particle, and the medium of dispersion.

8.2.2 Carbon-Based Materials

They are a class of allotropic forms of carbon taking the most common forms of hollow spheres or ellipsoids (usually fullerenes) or tubes (nanotubes). The versatility of hybridization offers unique properties to different allotropic forms. The high mechanical strength with efficient thermal and electric conductivity has been used in nanoelectronics, nanofiltration, photovoltaics, and biomedical applications. To accomplish biomedical applications, non-covalent and covalent approaches influencing surface functionalization need to be adopted. For instance, carbon nanotubes have been used effectively for the delivery of acetylcholine across the blood-brain barrier for the treatment of Alzheimer's disease in mice (Yang et al. 2010).

8.2.3 METAL-BASED MATERIALS

The members of this class constitute zero-dimensional nanomaterials such as nanocrystals (quantum dots). They may be metallic (nanogold, nanosilver, and metal oxides, such as titanium dioxide), ceramic or composite nanoparticles. They feature properties such as quantum confinement (semiconductor) and surface plasmon resonance and super-paramagnetism in metallic particles. Nanoparticles with optoelectric properties have been used in devices such as solar cells and those with opto-chemical and electromagnetic properties have been used in tissue engineering, target-specific drug delivery, and biomedical imaging techniques.

8.2.4 DENDRIMERS

They represent a class of polymeric nanomaterials in the form of branched units. The chain ends present on the surface can be easily tailored for different chemical reactions and used in catalysis. Dendrimers featuring monodispersity, solubility, encapsulation ability, and functionalizable surface units make them a potential candidate for drug delivery as a nanodrug carrier (Tekade et al. 2009).

8.2.5 COMPOSITES

It is the combination of two or more nanoparticles with different physico-chemical properties. The resulting material would be lighter, stronger, and durable. Secondly, their orientation and arrangement of asymmetric nanoparticles with ceramic, metal, and polymer leads to the formation of super-thermite materials. For instance, nanosized clays are added to automobile parts and packaging materials to enhance tensile strength, thermal, and flame-retardant properties.

8.2.6 NANOFABRICATION (SYNTHESIS OF NANOMATERIALS)

It is one of the critical steps involved in the design and development of nanomaterials according to the field of application. It is with the advances in synthesis protocol, that nanotechnology has seen such a great leap in a short time span. A great challenge in synthesizing nanomaterials lies in its fidelity, i.e. the process should have complete control over a set of characteristics. There are two common approaches used in the synthesis of nanomaterials, namely bottom-up and top-down.

8.2.6.1 Bottom-Up Approach

In this approach, atoms and/or molecules (in all three states) are arranged and assembled into a single nanostructure unit. They are categorized into chaotic and controlled processes. The chaotic process is that in which the constituent atoms or molecules from a sensitive state are elevated to a chaotic state with an unstable array. In this state of reaction it is difficult to control the mean size and its distribution. The kinetics involved in the formation of end-state products determines the formation of nanoparticles, e.g. laser ablation, pyrolysis, etc. The controlled process contrasts with the chaotic process, where the constituent atoms or molecules are delivered resulting in the formation and growth of nanoparticles in a controlled environment, e.g. self-limiting growth solution, molecular beam epitaxy, etc.

8.2.6.2 Top-Down Approach

A fabrication technique that takes place in a staged manner is one in which bulk material is scaled down to subsequent subsystems until it is reduced to the atomic level, e.g. cutting, milling, and photolithography. Despite a reduction in size, structural defects are likely to appear (Nikolaj et al. 2006) which may or may not have an impact on the surface properties of the reduced nanoparticles.

8.3 INTEGRATION OF NANO WITH BIOLOGICAL SYSTEMS

8.3.1 MAN VS. METALS

Metals, one of the magnificent gifts from nature, have had a great transition period for a huge transformation from stone-age man to modern-day scientific sophistication. Moreover, the development of civilization has relied heavily on the discovery of antique metals such as gold, copper, and silver. This is one of nature's enticements for man's control over his society and also a very important milestone for cultural and ethical refinement. Apart from craftsmanship, the medicinal properties of noble metals such as gold, in the form of "*swarnabhasma*" (gold ash), were explored and practiced in ancient India during the Vedic period (1000 BC – 600 BC) (Brown et al. 2007).

Silver has been equally popular as one of the metals widely used as a disinfectant throughout history. Vessels fabricated with silver lining were used for preserving perishable items and disinfecting water (Richard et al. 2002; Castellano et al. 2007). Macedonians pioneered the use of silver for healing wounds and treating surgical infections. Subsequently, silver formulations were promoted by Hippocrates for early wound healing and the treatment of ulcers. After that, Angelo Sala (1614) administered silver nitrate internally as a laxative and as an anti-inflammatory agent for the treatment of encephalitis and meningitis. Evidence exists for the use of silver in the treatment of epilepsy, as it could halt seizures (Hill et al. 1939).

During the mid-19th century, silver and its products were widely accepted for its anti-spoilage and medical use. A remarkable contribution was made by Mary Sims with her silver wire sutures in reducing puerperal sepsis complications during surgery (Sims 1884). This was followed by Franz Crede, a German obstetrician in the late 18th century, who pioneered the use of silver nitrate at 1% in eye-drop solution for the prevention of ophthalmia neonatorum in newborn infants (Schneider 1984). It is emphasized that even before the microbes were realized to be agents of infections, silver was used empirically over millennia to counteract such medical complications (Alexander 2009).

The implication of nanomaterials and/or nanoparticles in the biomedical field explores the hidden secrets on the physiological aspects of a cell facilitating its impairment and to learn the elegant systems that are naturally created. Moreover, nanoparticles have been used successfully as tools for diagnosing diseases, target-specific delivery of therapeutic agents, and to check bacterial infections. The potent antimicrobial activity with its unique and broad-spectrum mechanism of action, nanoparticles are considered as the greatest contributors of generation-next antibiotics. One such effort to address the most challenging issue of antibiotic resistance is by the use of nanomaterials with improved theranostic potential. Although there are various kinds of nanomaterials, only a few are known to fit healthcare applications. Studies on the use of metallic nanomaterials as an antibacterial material to contain such infectious diseases and disease-causing agents have been the greatest challenge yet to be resolved.

8.3.2 ENGINEERING OF SILVER NANOPARTICLES

In general, silver nanoparticles of different shapes and sizes can be engineered using various techniques based on the field of application. It 1951 silver nanoparticles were first synthesized by a wet chemistry technique using silver nitrate and reducing agent sodium citrate (Turkevich et al. 1951). The addition of solvents and reducing agents could bring about a characteristic change in physical and morphological features of nanoparticles leading to environmental toxicity. Wiley and co-workers (2004) prepared crystalline silver nanoparticles with varied symmetries and controlled the size in the range 20–80 nm using ethylene glycol heated to 148°C in the presence of poly(vinyl pyrrolidone) and sodium chloride. Silver nanospheroids in the size range 20–50 nm using laser ablation at 800 nm with femtosecond laser pulses were prepared. The particles were found less dispersed with low ablation efficiency to those prepared with nanosecond pulse (Tsuji et al. 2003). One of its kind arc discharge methods was adapted to engineer 10 nm-sized silver nanoparticles from silver rod consumption at 100 mg/min (Tien et al. 2008). Although gas condensation and co-condensation techniques were employed to effectively synthesize silver nanoparticles (Baker et al. 2005), chemical reduction remains one of the most common approaches to synthesize nanoparticles using organic and inorganic compounds having reducing properties. Trisodium citrate, N,N-dimethylformamide, ascorbate, and gallic acid have been used to reduce Ag^+ to Ag^0 in aqueous or non-aqueous solutions (Wiley et al. 2005; Evanoff Jr et al. 2004). The tendency to agglomerate demands the use of stabilizers for uniform dispersion. Such stabilizers with distinctive properties retain the surface functionalities of nanoparticles (Merga et al. 2007). Reports on the change in structure, size, and assembly patterns of nanoparticles had been found to be strongly associated with changes in physical property (Kim et al. 2006). The use of silver nanoparticles as an antibacterial material was devised by Kim and co-workers (2007) using sodium borohydride. The inhibitory effect, as determined by electron spin spectroscopy, showed the active participation of free radicals accounted for biocidal activity. Commercial application of antimicrobial metal nanoparticles in paints was investigated by Kumar et al. (2009) using vegetable oil. Surface coating of paint exhibited excellent antibacterial properties against gram-positive and gram-negative bacterial strains. Thenceforth, the synthesis of nanoparticles using biological systems has been the current trend.

8.3.3 Nanobiotechological Aspects (Bio-Nanofactories)

Metals and metalloids present in the environment are effectively used by micro-organisms for their catalytic or structural functions. This is purely based on the interaction of microbes with various kinds of metals which may be beneficial or detrimental. This ability of metal-microbe interaction has formed the basis for bioleaching and bioremediation processes which ensures the bioavailability of various metals in terrestrial and aquatic systems. Though prokaryotes and eukaryotes can bind metal ions and subject them to intracellular processing, only the former could effectively reduce metal ions and conserve energy (Ehrlich 1990).

The uptake of metals by microbes is either used for the activation of metalloenzymes in energy metabolism (Wackett et al. 1989), or in the enzymatic detoxification of pernicious metals such as Hg^{2+} and Ag^+ to Hg^0 and Ag^0 (Summers 1974; Baldi et al. 1989), bacteria such as *Thiobacillus ferrooxidans*, *Pseudomonas arsenitoxidans* (Ilyaletdinov and Abrashitova 1981), *Alcaligenes faecalis*, *Enterobacter clocae* (Wang and Shen 1995), and fungi such as *Aspergillusniger* and *Penicillium* sp. (Alibhai et al. 1991). Shreds of evidence on microbial detoxification and its accumulation on bacterial membrane surfaces have been reported (Beveridge et al. 1981). This important microbial transformation of metals that remained unnoticed has formed the basis for hydrometallurgical applications which the researchers have mimicked to fabricate metals in the laboratory.

Biological synthesis can be divided into two types, namely intra-cellular (Mukharjee et al. 2001) and extra-cellular methods (Ahmad et al. 2003), using microorganisms such as bacteria, fungi, yeasts, and even higher life forms like plants (Mohanpuria et al. 2008). Klaus et al. (1999) have demonstrated synthesized nanoparticles mostly at the periplasmic area of bacteria. The feasibility of recovering the nanoparticles was found to be cumbersome. Initially, the bacterial cell has to be lysed in order to separate the nanoparticles, followed by purification. To avoid such laborious downstream processing, culture supernatants from microbial sources are employed to synthesize metal nanoparticles. The culture supernatant rich in enzymes, metabolites, and metabolic byproducts may aid rapid synthesis of non-toxic, stable silver nanoparticles within optimized environmental conditions.

In contrast to the chemical method, the size and shape of the resulting nanoparticles could also be controlled, provided the parameters were optimized and enzyme concentration maintained depending on the scale of production.

The capability of microorganisms in a stressful environment with limited nutrients seems obvious; the threshold limit at higher metal concentrations induces toxicity. Microorganisms detoxify the effect of metal-induced toxicity by enzymatic coupling using oxidoreductases. Nitrate reductases are one such enzyme involved in assimilatory, respiratory, and periplasmic site-bound biocatalytic agents in reducing NO_3 to NO_2. Soil inhabiting bacteria such as *Thiobacillus*, *Micrococcus*, *Nitrosomonas*, *Bacillus* sp., etc. could effectively reduce nitrates which have been utilized for synthesizing nanoparticles in the laboratory. Studies on the extracellular synthesis of silver nanoparticles using *Staphylococcus aureus*, *Pseudomonas aeruginosa*, *Morganella* sp., *Geobacter sulfurreducens* have been reported. Chakraborty et al. (2012) have investigated *Lactobacillus mindensis*-mediated synthesis of silver oxide nanoparticles, the presence of aldehydes in polysaccharides, and its interaction with silver serves as the reducing and capping agent for polydispersed nature of nanoparticles.

The synthesis of metal nanoparticles by the biological method has received great interest in developing a new technology devoid of adverse effects on the environment (Nanda and Saravanan 2009; Kalishwaralal et al. 2008). This manipulation and fabrication of metals into nanoparticles, for instance, can also be achieved with optimized growth conditions specific for microbes and different metals. The culture supernatant of non-pathogenic bacteria such as *E. coli* has been used for synthesizing silver nanoparticles in a reliable manner within five minutes of its interaction (Shahverdi et al. 2007). Similarly, monodispersed and stable metallic nanoparticles from $[Ag(NH_3)_2]^+$ using *Aeromonas* sp. SH10 and *Corynebacterium* sp. SH09 biomass was reported (Mouxing et al. 2006). Despite pathogenicity, the extracellular metabolites of *K. pneumoniae* could effectively reduce silver nitrate to silver ions in the size range 1–6 nm (Mokhtari et al. 2009). Although the majority of synthesis protocols involved silver nitrate as a metal precursor, nanoparticles synthesized from a different precursor (silver sulfate) using *Salmonella* sp. has also been reported (Ghorbani 2013). It is noteworthy to mention that temperature-dependent nanoparticle synthesis in *Arthrobacter kerguelensis* on par with *P. antartica* has been evidenced (Shivaji et al. 2011), although the physical and chemical factors play a vital role in determining the size, shape, and dispersive nature of nanoparticles formed during bioreduction.

8.3.4 Drug Delivery System

Besides engineering nanostructures intentionally, they are naturally present in almost all organisms ranging from micro to macro level. The prerequisite for any application is determined by thorough knowledge and keen observation of these nanostructures. For instance, the arthropod phyla, comprising insects, show the presence of nanostructures formed via an evolutionary process to help them survive adverse conditions. Similarly, the utilization of nutrients by plants and accumulation of minerals in the biological system occurs in nano-form. In addition, these nanostructures are assembled in small insects and animals protecting them from predatory organisms and they confer a lightweight property to wings via nanowax coatings. The nano-architectural gallery of human organs is primarily made up of nanostructures, viz. bones. Most human secretions such as antibodies, enzymes, and other secretions are found to be in a nanometer size range. It is also evident that the genetic material – DNA/RNA, a major component involved in the growth and reproduction of cells, are typical nanostructures. It is indeed clear that these nanostructures form the basic foundation for all lifeforms. There are some bio-nanostructures which can also be used as drug delivery vectors.

8.3.4.1 Virus as a Vector

Viruses are the largest structurally-characterized intracellular parasites, which can behave as a non-living crystal and a living organism inside host cells causing disease in bacteria, plants, animals, and humans (Duckworth et al. 2002). With the gearing-up of technology, viruses are being tailored for use as catalysts and bio-scaffolds owing to their nanosize, monodispersity, distinct shapes, selective permeability, and ease of handling in genome manipulation, etc. (Flenniken et al. 2006). Nanomaterials of viral origin are prepared by removing their genetic material and making them 'nano-cargoes' for targeted drug delivery (Figure 8.1). Saunders et al. (2009) elaborated on the development of viral nanoparticles (NPs) from cowpea mosaic virus with its RNA removed through a proteolytic process. These nanocages or protein capsids were then used to encapsulate drugs, genes, enzymes, or proteins for targeted delivery with biocompatibility and bioavailability. Research studies are on the way to developing novel nanostructures and cages for compound encapsulation (Wang et al. 2002; Kolliopoulou et al. 2017). In this line, plant viruses are preferred for their non-toxic properties towards human cells at appropriate dosages for effective delivery of the administered drug (Singh et al. 2007).

8.3.4.2 Nanobacteria and Nanobes

In general, bacteria tend to precipitate metals encountered in the environment through their surface-producing metal NPs. The so-called nanobacteria are highly useful in the biosynthesis of low toxicity NPs. For instance, *Pseudomonas stutzeri* A259 was the first bacteria employed to produce silver nanoparticles (AgNPs) (Haefeli et al. 1984). Later, many metals (gold, silver, copper, iron), alloys, nonmagnetic oxide NPs, and metal sulfides quantum dots such as CdS and ZnS were synthesized using different strains of bacteria (Kamaraj SK et al. 2014). More than bacteria, actinomycetes (*Thermomonospora* sp. and *Rhodococcus* sp.) were also employed for nanoparticle production. Although microbial nanofactories were perceived as eco-friendly and biocompatible, they required a longer duration for synthesis and recovery.

Moreover, novel nano-organisms, called nanobes are gaining interest among researchers as they are found during offshore petroleum exploration on Triassic and Jurassic sandstones in Western Australia (Uwins et al. 1998). These nanobes encompass individual cells with 20–150 nm diameter mainly composed of a membrane-bound structure made of carbon, oxygen, nitrogen, DNA, dense cytoplasm, and nuclear area. There were also some mineral compounds present in the cellular structure similar to actinomycetes and fungi. The uniqueness of nanobes is their size, which is well below the range considered to be viable for autonomous life on Earth, and that they were recently found in Martian meteorite ALH84001 (Urwins 2000).

8.3.4.3 Magnetotactic Bacteria

Magnetotactic bacteria are the highly specialized bacteria capable of producing magnetic oxide NPs with unique properties such as super-paramagnetism, high coercive force, and microconfiguration. They are harnessed for biological separation and in biomedicine fields (Bazylinski et al. 1994). These

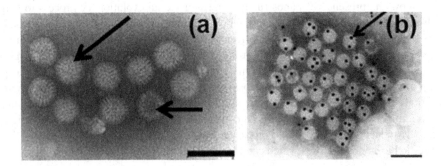

FIGURE 8.1 Negatively stained rotavirus with (a) and without (b) the particles in swine faeces at scale bar 80 nm. Bovine papillomavirus with internationalization of gold at scale bar 60 nm respectively. (Adopted from Catroxo et al. 2015.)

biocompatible magnetites, iron oxide (Fe_3O_4), iron sulfides, and maghemite (Fe_2O_3) are synthesized using magnetotactic bacteria that finds its application in targeted cancer treatment via magnetic hyperthermia, magnetic resonance imaging (MRI), DNA analysis, and gene therapy (Fan et al. 2009). In addition, surface-distributed magnetic iron-sulfide NPs, magnetic octahedral NPs with size range of 12–50 nm, modified iron NPs and super-paramagnetic NPs were also produced by using magnetotactic bacteria. Such bacterial magnetic particles (BMPs) produced are advocated to behave as a bioneedle in a compass and helps those bacteria to migrate under the impact of the Earth's geomagnetic field along with oxygen gradients in aquatic environments as in Figure 8.2 (Chen et al. 2010).

8.3.5 NANOTECHNOLOGY-BASED PRODUCTS

Silver and its compounds have been in use since time immemorial for its potential therapeutic applications in the treatment of burns and open wounds in keeping various bacterial infections at bay. During the 19th century, silver nitrate was used to degranulate the tissues and promote epithelization for quick healing (Castellano et al. 2007). With nanotechnology, silver-based products have been commercialized as nanocrystallized Acticoat™ and Acticoat 7 with silver mesh containing nanocrystals of silver capable of releasing a highly reactive form of silver ions. This controlled and sustained release of silver cations contributes to the effective management in reducing the risk for bacterial colonization and spread of infection (Smith and Nephew 2004). Contreet-H™, yet another product developed, has the tendency to absorb the microbe rather inhibiting its infiltration (Thomas 2003). The fastest killing rate and the broadest antimicrobial activity even against fungi remain the hallmark for its commercialization. The germicidal protection conferred by Acticoat™ has made its extended application along with skin grafts (Holder et al. 2003). *In vitro* studies showed that silver nanocrystal embedded dressings were less toxic to keratinocytes and fibroblasts compared to silver. Similarly, a combined antimicrobial and lower cytotoxic effect has shown significant compatibility on cultured skin grafted to athymic mice (Supp et al. 2005). Although the mechanism by which AgNPs are internalized and taken in remains unexplored, it is speculated that specialized epithelial cells are involved in active absorption (Florence 2005). Indeed, translocation of AgNPs via systemic circulation and their accumulation in vital organs such as the liver, spleen, and kidneys and their potential to cross the blood-brain barrier have been reported (Jimenez-Limana et al. 2014). This homogenous and/or preferential distribution of silver nanoparticles in the kidney or liver limits their prolonged exposure to biological subjects.

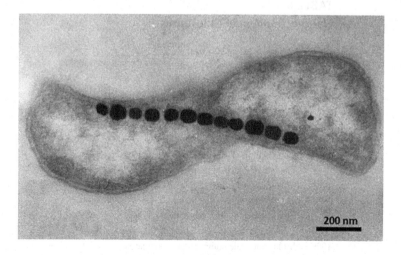

FIGURE 8.2   TEM micrograph of the magnetotactic bacterium at scale bar 300 nm. Magnetite nanoparticles in the form of beads are arranged all along the axis of the cell. (Adopted images with prior permission from Chen et al. 2010.).

8.4 RIGHT TO NANOTECHNOLOGY

Intellectual property rights are a mandate in today's technology-driven age. Building a strategic intellectual property (IP) portfolio is economically important from both an offensive and defensive standpoint (Singh et al. 2016). Applicable areas in nanotechnology to which intellectual property rights can apply, the challenging issues pertaining to the acquisition of IP rights in nanotechnology are as follows:

8.4.1 NANOTECHNOLOGY MILESTONES

Nanotechnology, more descriptively known as molecular manufacturing, involves the design, modeling, fabrication, and manipulation of materials and devices at the atomic scale. It necessitates thorough spatial control of matter at the level of molecules and atoms, with capabilities to process and rearrange them into custom designs. Nanotechnology differs from traditional chemical manufacturing in that the chemical reactions are not left to statistical movements of molecules in solutions, but instead, the molecules are brought into appropriate positions with appropriate speeds and orientations to cause desired reactions.

Nanotechnology also differs fundamentally from micro-manufacturing of silicon chips in that the top-down approach and repeated refinement of bulk materials (e.g. etching silicon) into micro- or even nano-scale designs suffers from defects inherent in the original bulk material. In contrast, nanotechnology's bottom-up approach will build essentially defect-free structures from the atoms up. Table 8.1 highlights some of the milestones in nanotechnology. In 2001 the U.S. government invested a total USD 422 million in nanotechnology, an increase from USD 270 million in 2000, while a total investment of USD 519 million is recorded for 2002. Nanotech research has intensified in universities around the world, and degree programs are increasingly offered at the graduate and undergraduate levels. Venture capital firms are showing a growing interest in nanotechnology, and significant nanotech projects have been initiated by large multinational companies, with some of them allocating up to a third of their research budget to nanotech (www.cmp-cientifica.com).

At present, only the simplest molecular structures can be built in the lab. However, the design and modeling of much larger structures are quite feasible with current computational methods and resources, but there is a demarcation between computational and manufacturing efforts in

TABLE 8.1
Nanotechnological Milestones

Year	Milestones
1959	Feynman delivers 'Plenty of room at the bottom' talk
1974	First molecular electronic device patent filed
1981	Scanning tunneling microscope (STM) invented
1985	Buckyballs discovered
1986	Atomic force microscope (AFM) invented
	'Engines of Creation' published
1987	Quantization of electrical conductance observed
	First single-electron transistor created
1988	First 'designer protein' created
1991	Carbon nanotubes discovered
1993	First nanotechnology lab in the U.S.
1997	DNA-based nanomechanical device created
1999	Molecular-scale computer switch created
2000	U.S. launches National Nanotechnology Initiative
2001	Logic gates made entirely from nanotubes

nanotechnology, with the manufacturing efforts roughly divided into the production of materials and tools. A large portion of nanotechnology's popularity lies in its applications to other already established fields. The various applications to electronics, sensors, aerospace, environmental cleanup and sanitation, and medicine have been tabulated in Table 8.2.

8.4.2 NEED FOR IPR

Emerging nanotechnologies have the potential to impact a wide range of commercially important bioscience and medical applications. For example, researchers are developing nanoscale materials such as diagnostic probes and targeted therapeutic delivery vehicles. By filing patents, companies can protect their intellectual property and leverage their developments over the marketplace in a way that is consistent with antitrust laws. The value and strategic importance of protecting intellectual property cannot be overstated. The costliest example in the U.S. history of R&D is perhaps the case of Eastman Kodak vs. Polaroid which began in the 1970s and was resolved in 1990. Seven patents upheld by Polaroid led to the total destruction of Kodak's instant photography business to the tune of more than USD 3 billion in infringement damages, compensation, and legal fees, research, and manufacturing costs (Rivette and Kline 2000).

More recently in 1990 and 1997, the University of California filed for patent infringement against Genentech for the company's manufacture and sale of the growth hormone product Protropin[R]. A settlement agreement, with a payout of USD 200 million from Genentech was reached in November 1999, with no admission of infringement from Genentech (www.gene.com/pressrelease/1999/11_19_0859AM.html).

After settling patents claims to DEC, IBM, StacElectronics, and Apple at the cost of half a billion dollars, Microsoft has more than a thousand patents to date, 250 times more than the four patents it owned 12 years ago (Kline 2000). Perkin-Elmer owned 160 patents in 1976. Today it has more than 7,000 patents, an increase of more than 45-fold. Clearly, both an offensive and defensive strategy can be used effectively in applying IP rights. Compared with the modest costs in building an effective IP portfolio, a huge drain on resources is involved in resolving patent disputes. With a dispute history reminiscent of David and Goliath, the enormous power of protecting intellectual property should not be overlooked.

8.4.3 HOW DOES NANO IPR VARY FROM BIOTECHNOLOGY IPR

In several respects, nanotechnology appears to be following the model of biotechnology patent policy. The current 'patent land rush' by nanotechnology patent prospectors in many ways mimics the biotechnology experience of the early 1980s (Raj Bawa 2004). As with nanotechnology today,

---

TABLE 8.2

Nanotechnology Efforts for Building IP Strategy in Some Application Areas

Nanotechnology
- Manufacturing
- Materials
- Tools
- Computational

Some Application Areas
- Electronics
- Sensors
- Aeronautics and space travel
- Environment cleanup and sanitation
- Medicine

---

biotechnology involved considerable scientific research funded by the government in several universities and labs. Many biotechnology startups developed out of broad university patents or groups of patents that were licensed to startups following an initial round of venture capital funding. Like nanotechnology now, biotechnology held the promise for a new generation of revolutionary products and treatments in the 1980s. However, 20 years later, the promise of biotechnology potential remains only a promise; the general market perception is that biotechnology still offers more potential than the product (Singh et al. 2019).

Although several significant biotechnology innovations have proven themselves on the market, the pace of the introduction of new biotechnology products and innovations remains far below initial expectations. Arguably, this shortfall in biotechnology innovation is the result of biotechnology anticommons (Michael et al. 2002). Two events in 1980 provided the spark for the biotechnology anticommons: the enactment of the Bayh-Dole Act and the granting of the first U.S. patent on a genetically modified life form (Diamond vs. Chakrabarty 1980). In response to the Bayh-Dole Act, professors quickly patented many aspects of fundamental biotechnology. Lacking significant expertise and prior experience in biotechnology, the understaffed United States Patent and Trademark Office (USPTO) soon began to approve and issue broad and overlapping biotechnology patents to the universities. Professors and researchers subsequently began to leave the universities to find biotechnology startups, and they licensed the biotechnology patents from the universities that held them. Finding that Congress intended to 'include anything under the sun that is made by man' as patentable subject matter, the court in Chakrabarty enabled these early startup firms to attract venture capital financing by providing some measure of certainty that biotechnological inventions could be patented (Chakrabarty 1952).

Seeking to profit from their patents, universities also began using reach-through license agreements to capture returns from future technological developments based on their patented work. A reach-through license agreement basically gives the owner of a patent, used in upstream stages of research, rights in subsequent downstream discoveries. Such rights may take the form of a royalty on sales that result from the use of the upstream research tool, an exclusive or non-exclusive license on future discoveries, or an option to acquire such a license. Because the granting of a singular license was rarely sufficient in conducting the further incremental research necessary to further develop biotechnology applications, other innovators and researchers were required to acquire numerous licenses held by the universities, many of which either were already licensed to certain biotechnology startup firms pursuant to exclusive licenses or were subject to onerous reach-through license agreements. The complexity of the licensing arrangements with the universities, and the concomitant transaction costs, eventually escalated to the point that biotechnology innovation was hampered. Although it is difficult to quantify the effect of anticommons – because delays and outright failures in licensing are not tracked publicly – there is evidence that a biotechnology anticommons exists. Notably, a study conducted by the National Institutes of Health (NIH) Working Group on Research Tools determined that "many scientists and institutions involved in biomedical research are frustrated by growing difficulties and delays in negotiating the terms of access to research tools"; that "case by case negotiations for permission to use research tools and materials create significant administrative burdens that delay research"; that "some users of biomedical research tools have limited resources for paying upfront fees, although their use of the tools could potentially yield valuable future discoveries' and that 'license mechanisms by which tool providers seek to profit from the future discoveries of tool users often involve future royalty obligations or rights to future intellectual property that constrain future opportunities for research funding and technology transfer" (Anon.). Indeed, the chief scientific officer at Bristol-Myers Squibb stated that his company was not able to work on more than 50 proteins that could potentially be involved in cancer because the patent holders either would not allow it or were demanding unreasonable royalties (Michael 2003). Another pharmaceutical executive complained that there is frustration in his company because the basic research performed is made into yet another discovery by introducing a cloned gene. At the end of the day, people are cut off from tools, from making a breakthrough discovery. Outside the pharmaceutical industry, academic scientists report similar problems of access to important biotechnologies in their agricultural research; some owners refuse to grant licenses because they mistrust licensees or wish to retain a field of research for themselves. Given the similarity between nanotechnology and biotechnology, it is likely that continued innovation in nanotechnology will face analogous impediments if numerous and potentially overlapping nanotechnology patent rights are granted and exclusively licensed. Realistically, the traditional concept of a single, strong nanotechnology patent capturing the final value of a product is fairly remote at this early stage of fundamental research. Rather, the early stages of successful nanotechnology innovation are more likely to depend on the cross-pollination of many patents tying together many inventions (Michael et al. 2004). Thus, in order to foster a more innovative environment in nanotechnology, the U.S. patent and licensing system needs to be examined and changed to avoid the innovation-retarding effects of a nanotechnology patent thicket.

This spur in nanotechnology development can be observed by an increase in the number of nano-related patents (Figure 8.3) in recent years (Parisi et al. 2014). Patenting in nanotechnology, in general, presents some important concerns as it is an interdisciplinary field with multiple applications and the risk of overlapping patent claims exist. Further, nano-based patents and traditional patents are not well distinguished. There is a need to understand different regulations and various fields of nanotechnology applications.

Regulation is an important part of a technology to be successful but presently there is no strict regulatory framework for nanotechnology in India. It is imperative to devise a regulatory framework which can look into the pros and cons of a particular invention, study public perspective and identify the associated ethical issues. The nanotechnology research community in India is small and limited to a handful of premier institutes like the laboratories of the Council of Scientific and Industrial Research (CSIR), Defense Research and Development Organization (DRDO), Indian Institute of Science (IISc), and Indian Institutes of Technology (IIT) (Figure 8.4). Therefore, it is essential to escalate institutional capacity in order for nanotechnology to blossom in India.

8.4.4 Pitfalls of Nanotechnology for IPR

Nanotechnology-related inventions are up-surging across all Research & Development and industry sectors. Millions of patents on nanotechnology can be retrieved in the worldwide database. For searching prior filings on nanotechnology, patent offices worldwide have classified nanotechnology under the International Patent Classification (WPN 2015). The USPTO received around 618,330 patent applications in the fiscal year 2014, with a total of 24,090 patents related to nanotechnology granted in 2014, which is a growth of more than 12.68% in the number of nanotechnology patents

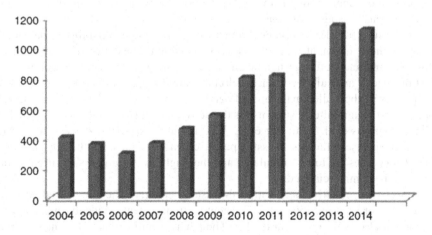

FIGURE 8.3 Increasing trend in patenting of nanotechnology products for the past decade. (Adopted image from http://www.uspto.gov/web/menu/search.html.)

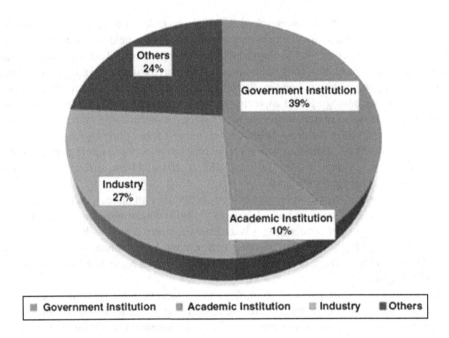

FIGURE 8.4   Percentage of Nanotechnology patents filed by Indian organizations.

in comparison to 2013 (USPTO 2014). Scrutinizing a nano-invention poses serious concerns among the patent grantees to review that the new patent has additional utility as well as a novelty over earlier inventions. Simply submitting a smaller version of a known structure is not patentable.

While nanotechnology opens opportunities for new inventions and discoveries, new challenges are encountered at the patent office as the grantees may lack expertise or adequate domain knowledge to assess nanotechnology patent applications, leading to the disregarding of previous inventions (Paradise 2012). The recent increase of patent applications by universities and the private sector is also another potential challenge for IPR protection for rewarding innovations with patent grants (Heller and Eisenberg 1998). The patent owners can exercise their right to exclude others to undergo research work on its invention, which can seriously restrict future research on nano-inventions. Premature disclosure of trade secrets by pressure from the academic circles to publish or report in government funding projects can reduce the ability to secure patents. Shortcuts in drafting a provisional application also increase the possibility of falling short of minimum disclosure requirements causing limitations in the scope of claims that can be supported by the final specification (Tullis 2012). The recent euphoria for patenting and the inability of patent offices to handle a large number of patent applications has resulted in the rejection of valid claims (Miller et al. 2004). Commercialization of nanotechnology in the agri-food sector requires a thorough assessment of toxicity and safety concerns before hitting the market (Sastry et al. 2011). A large number of patents has been filed in the area of nano-food (e.g. nanoemulsions of vitamins, flavors, and colorants) and in agricultural systems for the delivery of agricultural chemicals and targeted drug delivery (Srinivas et al. 2010; Ehr et al. 2011; Mishra et al. 2017; Fraceto 2018). Measures should be taken to assess those successful patent claims in field trials before granting patent rights. Regulatory rules are to be formulated and thorough investigation is required before these innovations can be commercialized.

Patent protection is effective only within the issuing country. Thus, increased administrative effort is required for securing international patents. Strong patent protection may spur research and invention but also lead to a patent thicket. Though a patent is granted to a nanotechnological invention it may be invalid due to the operation of a previously-granted patent for a similar type of invention, leading to expensive litigation. Innovators lacking the resources to litigate patent validity may be forced to license these patents rather than contest them (Barpujari 2010). A delay in the commercialization of nanotechnology products will give rise to problems such as unintentional infringement of patents, creating business uncertainty and concerns over patents, with some nanotechnology inventions ending up in court rather than by the patent office (Clarkson and DeKorte 2006). The nanotechnology IPR landscape is still somewhat chaotic and limitations can be observed when most of the patents do not result in active commercialization.

8.5 NANOTECHNOLOGY REGULATIONS ACROSS THE GLOBE

For small companies and startups, patents are among the only protections from infringement by large corporations. As companies grow, their ability to keep trade secrets decreases and patents again become a chief method of IP protection. However, it is important to keep in mind that a patent application is published 18 months after filing unless the applicant opts out, in which case a foreign patent may not be pursued for the invention. Hence, unless the applicant opts out and foregoes foreign filing, the description of the invention will end up in the public domain and accessible to competitors whether or not a patent has been issued.

Furthermore, the interdisciplinary nature of nanotechnology presents a special case. The USPTO houses seven different technology centers, including the biotechnology and organic chemistry center and the chemical and materials engineering center, and various units within each center, such as the metallurgy unit and the polymer chemistry unit within the chemical and materials engineering center, but none are dedicated to nanotechnology. This lack of focused expertise combined with the understaffed state of the USPTO is likely to result in: (1) the improper rejection of patents due to a mistaken conclusion that the taught matter is not new; as well as (2) overly broad patents giving the owner excessive control over a particular area, as has happened during the recent flood of information technology patents which overwhelmed the USPTO and resulted in cases such as the 'one-click' Amazon.com patent criticized as too broad and stifling. Challenging an overly broad patent held by a competitor is a costly process, while in the case of an improperly denied patent the company must spend precious time appealing the USPTO's decision, time that could be spent taking advantage of the patent.

The USPTO has reached out to the nanotechnology community for solutions, and recently the Foresight Institute and the USPTO held a patent roundtable addressing these issues. Having a set of nanotechnology specialists within the USPTO and in communication with each other could unify prior art searches and ensure more accurate consideration of nanotechnology patents and increased quality of granted patents (Brown 2002).

8.6 ENVIRONMENTAL AND HEALTH HAZARDS MANAGEMENT

Nanotechnology is an evolving technique that has transformed global industry but certain risks are associated with the environment and human health. Its regulation and mitigation are also not very clear. The guidelines for the application of nanotechnology are introduced at the regional and national level but there is a need to focus on it at the international level (Falkner and Jaspers 2012). The main problem for the decision and policymakers is the vagueness associated with the risks of nanotechnology which limits the development of international synchronization. Nanomaterials can be formed either by natural or synthetic processes.

Nanoparticles are said to have many commercial benefits, but some are toxic in nature. The physico-chemical nature of nanomaterials can have a huge impact on the environment. Insufficient knowledge of nanomaterial toxicity and risks may hinder the pace of industrial application of nano-enabled technologies. Therefore, the potential activities of these materials must be studied comprehensively. In the absence of definitive data, nanomaterial research and regulations could be supported by an efficient characterization of factors leading to toxicity and risks (Linkov et al. 2009a). Studies clustered the nanomaterials into ordered risk categories using Multi-Criteria Decision Analysis by taking into account the variables associated with the toxicity and risk of nanoparticles. However, their research finding does not quantify the environmental risks, but attempts were made for recommendations on precise measurements of nanoparticles (Linkov et al. 2009b).

Although nanotechnology offers significant benefits to mankind, it has detrimental health effects too. Due to their small size, the nanoparticles can disperse into

anatomical barriers to reach the liver, lungs, kidneys, etc. and damage the cells which can cause lesions, granulomas, cancers, Parkinson's disease, Alzheimer's disease, etc. (Chaudhry et al. 2008; Miller and Senjen 2008; Chau et al. 2007; Oberdorster et al. 2005). As a consequence, there is a potential threat to the millions of people working in vast expanses of agricultural lands due to the use of nano-fertilizers and pesticides which can enter into the food chain (Rico et al. 2011). Nanomaterials in foods are detrimental to human consumption because these are chemically more reactive and have greater access to our bodies than larger particles, they can even compromise human immune response, impair DNA replication and transcription, resulting in long term pathological effects (Hoet et al. 2004; Miller and Senjen 2008; Chaudhry et al. 2008; Chaudhry and Castle 2011; Mominet al. 2013).

Detailed understanding of the biological mechanism of nanoparticles is required before application in the field. This will enable the formulation of regulatory rules and help in risk management. Prior to use of products containing nanoparticles in the agricultural sector, a scrupulous analysis is needed for the distribution and number of particles absorbed by the plants in the food chain. The evolution of a participatory, dynamic and responsive nanotechnology policy is required to develop a coordinated risk management strategy in the Indian agriculture and food system for the positive economic impacts of this technology to reach agrarian society (Sastry et al. 2010; Sastry and Rao 2013).

There should be vigorous training of the personnel involved in the nano-agricultural sector. The use of personal protective equipment should be stringent for the workers associated with nanomaterials. To date, researchers have been trained to cater to the needs of the industry but should be skilled enough to carry out new research experiments, analyze and interpret the results and simultaneously be able to incorporate and implement the theory, tools, and techniques of nanotechnology. There should also be strict rules for the nanotechnology industry to adhere to the engineering controls for effective risk management.

Many countries have recommended various precautionary measures to manage the risk associated with nanoparticles. The Health and Environment Alliance, Belgium, recommends a list of measures to manage the risks associated with nanomaterials and stop the commercial sale until appropriate measures are taken. There is a need for a comprehensive policy by a uniform international body to alleviate the risk involved with nanotechnology applications.

8.7 CONCLUSION

There is no doubt that nanotechnology could be a disruptive technology, replacing all other technologies. The need for promoting nanotechnology research and regularizing patent grants is a mandate which can be achieved through stronger public–private partnership. One way to get through this is by a cohesive licensing model where the incentive to research and development-driven companies may be given to promote nanotechnology products. It will help control the price in the future – one of the major hurdles in other fields like the drug industry. Continuous discussions are being conducted on different platforms and policies are being amended but concentrating solely on intellectual property rights will not lead to success stories. Different international and national regulatory agencies on social, health, and economics should come under the same platform to help with policy formulation for promoting innovation on nanotechnology. Countries should increase their gross domestic product investment in nanotechnology research to promote innovation especially in the agriculture sector, where nanotechnology holds potential for a much needed 'Agricultural Revolution'.

REFERENCES

Ahmad A, Mukherjee P, Senapati S et al. (2003) Extracellular biosynthesis of silver nanoparticles using the fungus *Fusarium oxysporum*. *Colloids Surf B Biointerfaces* 28(4): 313–318.

Alexander JW (2009) History of the medical use of silver. *Surginf* 10(3): 289–292.

Alibhai K, Leak DJ, Dudeney AWL et al. (1991) Microbial leaching of nickel from low grade Greek laterites. In: Smith RW, Misra M (eds) *Mineral Bioprocessing*, The Minerals, Metals, and Materials Society, Warrendale, PA, 191–205.

Anon. Nat ' l Inst. of Health, Report of the national institutes of health working group on research tools (June 4, 1998) http://www.nih.gov/news/ researchtools/ # exec.

ASTM Standard E2456 (2006) *Standard Terminology Relating to Nanotechnology*, ASTM International, West Conshohocken, PA.

Baker C, Pradhan A, Pakstis L et al. (2005) Synthesis and antibacterial properties of silver nanoparticles. *J Nanoscinanotechnol* 2(2): 244–249.

Baldi F, Filippelli M, Olson GJ (1989) Biotransformation of mercury by bacteria isolated from a river collecting cinnabar mine waters. *Microb Ecol* 17(3): 263–274.

Barpujari I (2010) The patent regime and nanotechnology: Issues and challenges. *J Intellect Property Rights* 15: 206–213.

Bawa Raj (2004) Nanotechnology patenting in the US, 1. *Nanotechnol L Bus* 36: J 17.

Bazylinski DA, Garratt-Reed AJ, Frankel RB (1994) Electron microscopic studies of magnetosomes in magnetotactic bacteria. *Microsc Res Tech* 27(5): 389–401.

Beveridge TJ, Koval SF (1981) Binding of metals to cell envelopes of Escherichia coli K-12. *Appl Environ Microbiol* 42(2): 325–335.

Brown CL, Bushell G, Whitehouse MW et al. (2007) Characterization of the gold in Swarnabhasma, a microparticulate used in traditional Indian medicine. *Gold Bull* 40(3): 245.

Brown Doug (2002) U.S. patent examiners may not know enough about nanotech. *Small Times*. www.smalltimes.com.Diamond vs. Chakrabarty, 447 U.S. 303 S. Ct. 2204, 1980.

Buzea C, Pacheco II, Robbie K (2007) Nanomaterials and nanoparticles: Sources and toxicity. *Biointerphases* 2(4): MR17–MR71.

Carrier MA (2002) Unraveling the Patent-Antitrust Paradox. 150 U. PA. L. REV. 761, 816–840.

Castellano JJ, Shafii SM, Ko F et al. (2007) Comparative evaluation of silver-containing antimicrobial dressings and drugs. *Int Wound J* 4(2): 114–122.

Catroxo MHB, AMCRPF Martins (2015) *Veterinary Diagnostic Using Transmission Electron Microscopy*. doi:10.5772/61125.

Chakrabarty (1952) 447 U.S. at 309 (citing S. Rep. No. 1979, at 5 (1952); H.R. Rep. No. 1923, 6.

Chau CF, Wu SH, Yen GC (2007) The development of regulations for food nanotechnology. *Trends Food Scitechnol* 18(5): 269–280. doi:10.1016/j.tifs.2007.01.007.

Chaudhry Q, Castle L (2011) Food applications of nanotechnologies: An overview of opportunities and challenges for developing countries. *Trends Food Scitechnol* 22(11): 595–603. doi:10.1016/j.tifs.2011.01.001.

Chaudhry Q, Scotter M, Blackburn J et al. (2008) Applications and implications of nanotechnologies for the food sector. *Food Add Contam A* 25(3): 241–258. doi:10.1080/02652030701744538.

Chen L, Bazylinski DA, Lower BH (2010) Bacteria that synthesize nano-sized compasses to navigate using Earth's geomagnetic field. *Nat Educ Knowl* 3(10): 30.

Clarkson G, DeKorte D (2006) The problem of patent thickets in convergent technologies. *Ann NY Acadsci* 1093: 180–200. doi:10.1196/annals.1382.014.

Duckworth DH, Gulig PA (2002) Bacteriophages: Potential treatment for bacterial infections. *BioDrugs* 16(1): 57–62.

Ehr RJ, Thomas HK, Dale CS et al. (2011) Mesosized capsules useful for the delivery of agricultural chemicals. US Patent Number: US. http://www.ncbi.nlm.nih.gov/pubmed/20110045975A1.

Ehrlich HL, Brierley CL (1990) *Microbial Mineral Recovery*, Mc-Graw-Hill, New York.

Evanoff Jr Chumanov G (2004) Size-controlled synthesis of nanoparticles, measurement of extinction, scattering, and absorption cross sections. *J Phys Chem B* 108(37): 13957–13962.

Falkner R, Jaspers N (2012) Regulating nanotechnologies: Risk, uncertainty and the global governance gap. *Glob Environ Pol* 12(1): 30–55. doi:10.1162/GLEP_a_00096.

Fan TX, Chow SK, Zhang D (2009) Biomorphic mineralization: From biology to materials. *Progr Mater Sci* 54(5): 542–659.

Fenchel T, Blackburn H, King GM, Blackburn TH (2012) Bacterial biogeochemistry. In: *The Ecophysiology of Mineral Cycling*, eds. T Fenchel, H Blackburn, G King, 3rd edn, Academic Press, Boston, MA, 3.

Flemming A (1929) Classics in infectious diseases: On the antibacterial action of cultures of a penicillium, with special reference to their use in the isolation of B. influenza. *Brit J Exp Pathol* 10: 226–236.

Flenniken ML, Willits DA, Harmsen AL et al. (2006) Melanoma and lymphocyte cell-specific targeting incorporated into a heat shock protein cage architecture. *Chem Biol* 13(2): 161–170.

Florence AT (2005) Nanoparticle uptake by the oral route: Fulfilling its potential? *Drug Discov Today Technol* 2(1): 75–81.

Fraceto LF, Maruyama CR, Guilger M et al. (2018) *Trichoderma harzianum* based novel formulations: Potential applications for management of next-gen agricultural challenges. *J Chem Technol Biotechnol* 93(8): 2056–2063.

Ghorbani HR (2013) Biosynthesis of silver nanoparticles using Salmonella typhirium. *J NanostrChem* 3(1): 29.

Haefeli C, Franklin C, Hardy K (1984) Plasmid-determined silver resistance in Pseudomonas stutzeri isolated from a silver mine. *J Bacteriol* 158(1): 389–392.

Heller MA, Eisenberg RS (1998) Can patents deter innovation? The anticommons in biomedical research. *Science* 280(5364): 698–701. doi:10.1126/science.280.5364.698.

Hill WR, Pillsbury DM (1939) *Argyria – The Pharmacology of Silver*, Williams and Wilkins, Baltimore, MD.

Hoet PH, Bruske-Hohlfeld I, Salata OV (2004) Nanoparticles – Known and unknown health risks. *J Nanobiotechnol* 2(1): 12. doi:10.1186/1477-3155-2-12.

Holder I, Durkee P, Supp A, Boyce ST (2003) Assessment of a silver-coated barrier dressing for potential use with skin grafts on excised burns. *Burns* 29(5): 445–448.

Ilyaletdinov AN, Abdrashitova SA (1981) Autotrophic oxidation of arsenic by a culture of *Pseudomonas arsenitoxidans*. *Mikrobiologiya* 50(2): 197–204.

Jiménez-Lamana J, Laborda F, Bolea E et al. (2014) An insight into silver nanoparticles bioavailability in rats. *Metallomics* 6(12): 2242–2249.

Kalishwaralal K, Deepak V, Ramkumarpandian S et al. (2008) Extracellular biosynthesis of silver nanoparticles by the culture supernatant of Bacillus licheniformis. *Mater Lett* 62(29): 4411–4413.

Kamaraj SK, Venkatachalam G, Arumugam P, Berchmans S (2014) Bio-assisted synthesis and characterization of nanostructured bismuth(III) sulphide using Clostridium acetobutylicum. *Mater Chemphy* 143(3): 1325–1330.

Kim D, Jeong S, Moon J (2006) Synthesis of silver nanoparticles using the polyol process and the influence of precursor injection. *Nanotechnology* 17(16): 4019–4024.

Klaus T, Joerger R, Olsson E, Granqvist CG (1999) Silver-based crystalline nanoparticles, microbially fabricated. *Proc Natl Acadsci USA* 96(24): 13611–13614.

Kline D (2000) Net patent fights may YieldSurprises. *Upside*: 175–178.

Kolliopoulou Anna, Clauvis NTT, Smagghe Guy, Swevers L (2017) Viral delivery of dsRNA for control of insect agricultural pests and vectors of human disease: Prospects and challenges. *Front Physiol* 8: 399.

Kelley L, ed. (2003) *Health Impacts of Globalization: Towards Global Governance*, Palgrave Macmillan, Basingstoke, p. 106.

Linkov I, Steevens J, Adlakha-Hutcheon G et al. (2009) Emerging methods and tools for environmental risk assessment, decision-making, and policy for nanomaterials: Summary of NATO Advanced Research Workshop. *J Nanopart Res* 11(3): 513–527. doi:10.1007/s11051-008-9514-9.

Linkov I, Steevens J, Chappel M et al. (2009a) Classifying nanomaterial risks using multi-criteria decision analysis. In: Linkov I, Steevens J (eds) *Nanomaterials: Risks and Benefits. NATO Science for Peace and Security Series C: Environmental Security*, 179–191. doi:10.1007/978-1-4020-9491-0_13.

McNeill WH (1976) *Plagues and Peoples*, Bantam, Doubleday Dell Publishing Group, Inc, New York.

Merga G, Wilson R, Lynn G et al. (2007) Redox catalysis on "naked" silver nanoparticles. *J Phys Chem C* 111(33): 12220–12206.

Michael A (2003) Carrier, resolving the patent-antitrust paradox through tripartite innovation. *Vand L Rev* 56: 1047, 1087.

Michael R, Taylor Jerry Cayford (2004) American patent policy, biotechnology, and African agriculture: The case for Policy Change. *Harv JL & Tech* 17: 321, 350.

Miller G, Senjen R (2008) Out of the laboratory and on to our plates: Nanotechnology in food and agriculture. Friends of the Earth Australia, Europe and USA. http://www.denix.osd.mil/cmrmd/upload/Nanotech_Food_Agric.pdf.

Miller JC, Serrato R, Represas-Cardenas JM et al. (2004) *The Handbook of Nanotechnology: Business, Policy and Intellectual Property Law*, Wiley, Hoboken, NJ.

Mishra S, Keswani C, Abhilash PC et al. (2017) Integrated approach of Agri-nanotechnology: Challenges and future trends. *Front Plant Sci*. doi:10.3389/fpls.2017.00471.

Mohanpuria P, Rana NK, Yadav SK (2008) Biosynthesis of nanoparticles: Technological concepts and future applications. *J Nanopart Res* 10(3): 507–517.

Mokhtari N, Daneshpajouh S, Seyedbagheri S et al. (2009) Biological synthesis of very small silver nanoparticles by culture supernatant of Klebsiella pneumoniae the effects of visible-light irradiation and the liquid mixing process. *Mater Res Bull* 44(6): 1415–1421.

Momin JK, Jayakumar C, Prajapati JB (2013) Potential of nanotechnology in functional foods. *Emir J Food Agric* 25(1): 10–19. doi:10.9755/ejfa.v25i1.9368.

Mouxing FU, Qingbiao LI, Daohua SUN et al. (2006) Rapid preparation process of silver nanoparticles by bioreduction and their characterizations. *Chin J Chemeng* 14(1): 114–117.

Mukherjee P, Ahmad A, Mandal D et al. (2001) Fungus-mediated synthesis of silver nanoparticles and their immobilization in the mycelial matrix: A novel biological approach to nanoparticle synthesis. *Nano Lett* 1(10): 515–519.

Nanda A, Saravanan M (2009) Biosynthesis of silver nanoparticles from Staphylococcus aureus and its antimicrobial activity against MRSA and MRSE. *Nanomedicine* 5(4): 452–456.

Nikolaj L, Kildeby OZ, Andersen RE et al. (2006) *Silver Nanoparticles, P3 Project*, Institute for Physics and Nanotechnology – Aalborg University, Aalborg, Denmark.

Oberdorster G, Oberdorster E, Oberdorster J (2005) Nanotoxicology: An emerging discipline evolving from studies of ultrafine particles. *Environ Health Perspect* 113(7): 823–839. doi:10.1289/ehp.7339.

Paradise J (2012) Claiming nanotechnology: Improving USPTO efforts at classification of emerging nano-enabled pharmaceutical technologies. *Northwest J Technol Intellect Prop* 10(3): 169–208.

Parisi C, Vigani M, Rodríguez-Cerezo E (2014) Proceedings of a workshop on nanotechnology for the agricultural sector: From research to the Fi Eld. JRC scientifi c and policy reports.

Payne DJ, Gwynn MN, Holmes DJ, Pompliano DL (2007) Drugs for bad bugs: Confronting the challenges of antibacterial discovery. *Nat Rev Drug Discov* 6(1): 29–40.

Rakeshkumar T, Dutta T, Gajbhiye V, Jain NK (2009) Exploring dendrimer towards dual drug delivery: PH responsive simultaneous drug-release kinetics. *J Microencapsul* 26(4): 287–296.

Richard JW, Spencer BA, McCoy LF et al. (2002) Acticoat versus silver ion: The truth. *J Burns Surg Wound Care* 1: 11–20.

Rico CM, Majumdar S, Duarte-Gardea M et al. (2011) Interaction of nanoparticles with edible plants and their possible implications in the food chain. *J Agric Food Chem* 59(8): 3485–3498. doi:10.1021/jf104517j.

Rivette KG, Kline D (2000) A hidden weapon for high- Tech Battles. *Upside*: 165–174.

Sastry RK, Rao NH (2013) Emerging technologies for enhancing Indian agriculture-case of nanobiotechnology. *Asian Biotechnol Dev Rev* 15(1): 1–9.

Sastry RK, Rashmi HB, Rao NH (2011) Nanotechnology for enhancing food security in India. *Food Policy* 36(3): 391–400.

Sastry RK, Rashmi HB, Rao NH, Ilyas SM (2010) Integrating nanotechnology into agri-food systems research in India: A conceptual framework. *Technol Forecast Soc Change* 77(4): 639–648. doi:10.1016/j.techfore.2009.11.008.

Saunders K, Sainsbury F, Lomonossoff GP (2009) Efficient generation of cowpea mosaic virus empty virus-like particles by the proteolytic processing of precursors in insect cells and plants. *Virology* 393(2): 329–337. doi:10.1016/j.virol.2009.08.023.

Schneider G (1984) Silver nitrate prophylaxis. *Can Med Assoc J* 131(3): 193–196.

Shahverdi AR, Fakhimi A, Shahverdi HR, Minaian S (2007) Synthesis and effect of silver nanoparticles on the antibacterial activity of different antibiotics against Staphylococcus aureus and Escherichia coli. *Nanomedicine* 3(2): 168–171.

Shivaji S, Madhu S, Singh S (2011) Extracellular synthesis of antibacterial silver nanoparticles using psychrophilic bacteria. *Process Biochem* 46(9): 1800–1807.

Sims MJ (1884) *The Story of My Life*, Marion-Sims H (ed), D. Appleton & Co., New York.

Singh HB, Jha A, Keswani C (Eds.) (2016) *Intellectual Property Issues in Biotechnology*, CABI, Wallingford, UK. 304 pages, ISBN-13: 978-1780646534.

Singh HB, Keswani C, Singh SP (Eds.) (2019) *Intellectual Property Issues in Microbiology*, Springer-Nature, Singapore. 425 pages, ISBN- 978-981-13-7465-4.

Singh P, Prasuhn D, Robert M et al. (2007) Bio-distribution, toxicity and pathology of cowpea mosaic virus nanoparticles in vivo. *J Control Release* 120(1-2): 41–50.

Smith, N (2004) Together at last Dynamic Silver and super-powered absorbency, Acticoat absorbent. Smith and Nephew Pty. Ltd. Product information (https://www.smith-nephew.com/professional/products/advanced-wound-management/acticoat/)

Srinivas PR, Philbert M, Vu TQ et al. (2010) Nanotechnology research: Applications in nutritional sciences. *J Nutr* 140(1): 119–124. doi:10.3945/jn.109.115048.

Summers AP, Sugarman LI (1974) Cell-free mercury (II) reducing activity in a plasmid-bearing strain of *Escherichiacoli. J Bacteriol* 119(1): 242–249.

Supp A, Needy A, Supp D et al. (2005) Evaluation of cytotoxicity and antimicrobial activity of Acticoat burn dressing for management of microbial contamination in cultured skin substitutes grafted to anthymic mice. *J Burn Care Rehabil* 26(3): 238–246.

Tekade RK, Kumar PV, Jain NK (2009) Dendrimers in oncology: An expanding horizon. *Chem Rev* 109(1): 49–87.

Tien DC, Tseng KH, Liao CY et al. (2008) Discovery of ionic silver in silver nanoparticle suspension fabricated by arc discharge method. *J Alloys Compd* 463(1–2): 408– 411.

Tsuji T, Kakita T, Tsuji M (2003) Preparation of nano-size particle of silver with femtosecond laser ablation in water. *Appl Surf Sci* 206(1–4): 314–320.

Tullis TK (2012) Current intellectual property issues in nanotechnology. *Nanotechnol Rev* 1(2): 189–205. doi:10.1515/ntrev-2012-0501.

Turkevich J, Stevenson P, Hillier J (1951) A study of the nucleation and growth processes in the synthesis of colloidal gold. *Discuss Faraday Soc* 11: 55–75.

Urwins P (2000) *Australian Petroleum Production and Exploration Association Conference.* Brisbane, Australia: Australian Petroleum Production & Exploration Association Ltd (APPEA). Novel nano-organisms (nanobes): Living analogues for Martian "nanobacteria"? p. E3.

USPTO (2014) Performance & accountability report. http://www.uspto.gov/about/stratplan/ar/USPTOFY201 4PAR.pdf.

Uwins PJR, Webb RI, Taylor AP et al. (1998) Novel nano-organisms from Australian sandstones. Am Mineral 83: 1541–1550. doi:10.2138/am-1998-11-1242.

Vester B, Douthwaite S (2001) Macrolide resistance conferred by base substitutions in 23S rRNA. *Antimicrob Agents Chemother* 45(1): 1–12.

Wackett LP, Orme-Johnson WH, Walsh CT (1989) Transition metal enzymes in bacterial metabolism. In: Beveridge TJ, Doyle RJ (eds) *Metal Ions and Bacteria*, Wiley, New York, 165–206.

Wang Q, Lin T, Tang L et al. (2002) Icosahedral virus particles as addressable nanoscale building blocks. *Angew ChemInt Ed Engl* 41(3): 459–462.

Wang YT, Shen H (1995) Bacterial reduction of hexavalent chromium. *J Indmicrobiol* 14(2): 159–163.

Wiley B, Herricks T, Sun Y et al. (2004) Polyol synthesis of silver nanoparticles: Use of chloride and oxygen to promote the formation of single-crystal, truncated cubes and tetrahedrons. *Nano Lett* 9: 1733–1739.

WPN (2015) Policy environments and governance for innovation and sustainable growth through nanotechnology DSTI/STP/NANO (2013)13/FINAL. http://www.oecd.org/officialdocuments/publicdisplaydocu mentpdf/?cote=DSTI/STP/NANO(2013)13/FINAL&doclanguage=en.

Yang Z, Zhang Y, Yang Y et al. (2010) Pharmacological and toxicological target organelles and safe use of single-walled carbon nanotubes as drug carriers in treating Alzheimer disease. *Nanomedicine* 6(3): 427–441.

9 Nanocellulose Application in Encapsulation and Controlled Drug Release

Elaine Cristina Lengowski, Eraldo Antonio Bonfatti Júnior, Aline Caldonazo, and Kestur Gundappa Satyanarayana

CONTENTS

9.1 INTRODUCTION

A medicine, or simply drug, is a pharmaceutical product, technically obtained or elaborated, with prophylactic, curative, palliative, or diagnostic purposes [1]. In pharmacology, a drug is a chemical substance, typically of known structure, which, when administered to a living organism, produces a biological effect [2].

The development of new medication is not only focused on the discovery of new drugs but may be linked to the development of new drug encapsulation and drug delivery systems.

A controlled drug delivery system can be seen as a device that allows the introduction of a drug into the human body where the rate, time, and location of its release within the organism is controlled [3]. Thus, this process includes the encapsulation of a therapeutic substance in a matrix, the release of the same therapeutic substance, and in some cases, its transport through the biological membranes to the site of action [3].

Due to the reduction in size, nanomaterials provide unique advantages for application in bio-medical systems [4], such as the high specific surface area, great flexibility, possibility of change in the surface chemistry with multivalent ligands to increase the activity for target molecules, and also can be engineered to interact with specific biological components [5]. One of the nanomaterials that can be used and has been gaining prominence in pharmaceuticals research is nanocellulose.

Nanocellulose can be produced in a number of ways, which would yield different types (par-ticles, short fibers, crystals, etc.), shapes and therefore different properties. There are three types of nanocellulose that have been studied and used in tests for formulating and releasing drugs, bacterial (BNC), nanofibrillated (CNF), and nanocrystalline (CNC). The interest in the use of nanocellulose in the medical area and in the release of drugs is dependent on its non-toxicity, biocompatibility, good mechanical properties, high surface area-to-volume ratio, and potential versatility in terms of chemical modification. There are studies that have used nanocellulose as a means of achieving tar-geted and sustained release of drugs. In practice, the application of nanocellulose can improve the effectiveness of drug delivery systems for different periods of time, as well as improve the ability to increase the amount of water in soluble drugs in an aqueous formulation through the encapsulation.

With this background, this chapter presents an overview of the use of nanocellulose in encap-sulation and controlled drug release, with examples of the use of different types of nanocelluloses using different methods of production in several types of drugs. This chapter also presents some important concepts of nanocelluloses and their biocompatibility with drugs.

9.2 CELLULOSE

Cellulose is the most abundant biopolymer on the planet; however, it is synthesized only by plants, some fungi, bacteria, and algae [6]. It is formed by a non-branched chain of glucose units, with a polymerization degree varying from 13,000 to 14,000 (vegetable cellulose) and from 2,000 to 6,000 (bacterial cellulose) [7].

The cellulose is a biopolymer compound of linearly arranged D-anhydroglucopyranose units linked via β-(1→4) glycosidic bonds [8]. Each replicate of D-anhydroglucopyranose units has three hydroxyl groups being one primary, present in C6 carbon, and two secondary, present in carbons C2 and C3 [8]. These hydroxyl groups are sites that allow the chemical modification of cellulose, espe-cially at the C6 position due to higher reactivity [9]. The cellulose has a reducing end, wherein the hydroxyl of the C1 carbon is in the hemiacetal form and the non-reducing end where the hydroxyl of the C4 carbon is free and forms no bond [10]. Figure 9.1 shows the two cellulose biosynthesis routes on a molecular scale, highlighting the position of the carbons as well as the reducing and non-reducing ends.

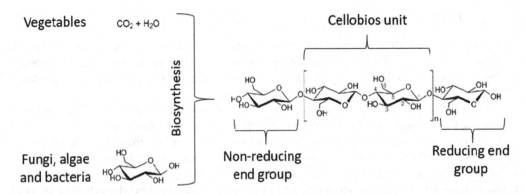

FIGURE 9.1 Biosynthetic routes of cellulose production and demonstration of the cellulose chain on a molecular scale. (Adapted from Lengowski et al. [11] and Börjesson and Westman [12]. With permission.)

It has a complex structure in which various levels of organization exist: the highest level of organization is its morphology that consists of micro and nanofibrils forming a cell wall, and the supramolecular level is comprised of crystalline and amorphous regions [7]. Figure 9.2 shows the levels of organization of cellulose in plants.

In the case of cellulose produced by bacteria, it consists of a complex process. First is the polymerization of glucose residues in the β1-4-glucan chains, which is followed by the extracellular secretion of the chains that form the linear arrangement and are organized and ordered through hydrogen bonds and the van der Waals forces, resulting in the formation of a three-dimensional structure called microfibril [13].

Cellulose has the ability to form intra- and inter-molecular bonds, which provide high mechanical properties to this polymer [8]. The hydrogen bonding formed between the hydroxyl groups of the C2, C6 and C3 carbons and the endocyclic oxygen atom forms the intra-molecular bond, whereas the inter-molecular bond occurs between the C3 and C6 carbons and the successive units of D-anhydroglucopyranose [14].

Because cellulose is a semi-crystalline polymer, it allows the extraction of nanostructures with different morphology (length, diameter, and aspect ratio) and different physical and mechanical properties, depending on the extraction method used [15]. Figure 9.3 shows the amorphous and crystalline structures in the cellulose, as well as the intermolecular bonds between microfibrils.

9.3 NANOCELLULOSE

Nanocelluloses are structures that have at least one of their dimensions smaller than 100 nm [16]. Due to the small dimensions, nanocellulose has attractive properties, such as high strength, excellent stiffness, high surface area, and a great number of hydroxyl groups available for bonding and modification [17].

Cellulose nanofibrils (CNF) are sometimes called cellulose microfibrils (CMF) (II) cellulose nanocrystals (CNC) and (III) bacterial nanocellulose (BNC) [18]. Due to the diameter of these

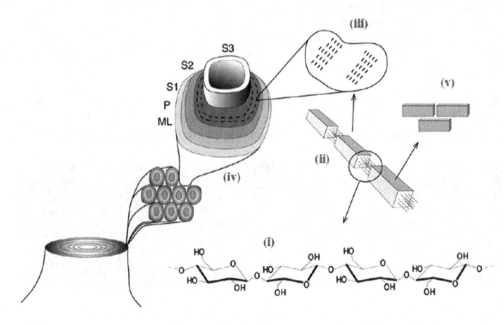

FIGURE 9.2 The cellulose structure: (i) the molecular structure (ii) the polymers ordered into microfibrils with crystalline and amorphous regions, (iii) several microfibrils assembled together to form a macrofibril, and (iv) the different layers in the cell wall. (Reproduced from Börjesson and Westman [12]. With permission.)

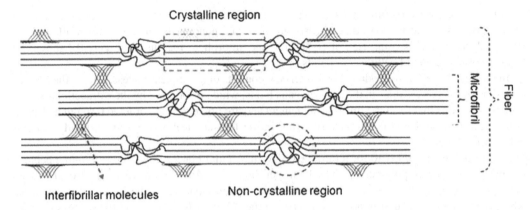

FIGURE 9.3 Part of a cellulose fiber where the crystalline and non-crystalline regions are shown. (Reproduced from Börjesson and Westman [12]. With permission.)

particles being below 100 nm it is challenging to produce nanocellulose due to the high tendency of agglomeration during mixing with other polymers or during drying of aqueous suspensions [15]. However, it should be noted that the chemical, physical, morphological and mechanical properties of these nanostructures are linked to the production process employed. Figure 9.4 illustrates the morphological aspect generally found for these three categories of nanocelluloses.

The methods of production of the nanocelluloses can be mechanical, chemical, or biological [11]. There are several forms of production, with different equipment and also through the combination of production routes, such as enzymatic and mechanical, chemical-mechanical, chemical-enzymatic [21–23].

9.3.1 CELLULOSE NANOFIBERS

CNF is usually extracted by mechanical treatment and may have combinations of chemical or enzymatic agents. The process of obtaining is known as top-down, where the cellulose is moderately degraded, resulting in micro- and nanofibrils with large surface area and diameter ranging from 5 to 100 nm and length of several micrometers. It has a web-like structure with aspect ratio (length/diameter) being very high [11, 15]. The web appearance of cellulose nanofibers is shown in Figure 9.4a.

The CNF has amorphous and crystalline regions, which makes it a more elongated chain in the longitudinal direction. Thus, the long length of the nanofibrillated cellulose chains associated with its surface that contains a wide range of exposed hydroxyl groups ends up in the formation of

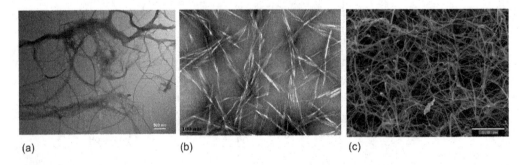

(a)   (b)   (c)

FIGURE 9.4 Transmission electron microscopy of the three categories of nanocellulose. a) Cellulose nanofibrils; b) Cellulose nanocrystals reproduced from Dai et al. [19] with the kind permission of the publishers; c) Bacterial nanocellulose. (Reproduced from Torres et al. [20]. With permission.)

a number of hydrogen bonds [22, 24]. However, longer time or degree of processing of the cellulose causes a significant decrease in the size of the fibers, leading to a decrease in the degree of polymerization and also the crystallinity index of the fibrils, which would result in distinct mechanical properties [25].

CNFs are obtained primarily through mechanical processes or through the combination of some pretreatment followed by mechanical processing. To improve the energy used in the mechanical processes to produce nanocellulose, some pretreatments can be employed to increase the reactivity of the fibrils. The pretreatments act on the fibrils altering the accessibility of the hydroxyl groups, increasing the internal surface area and causing modification in the crystallinity. The pretreatments are different types of chemical hydrolysis (alkali or acid) or enzymatic processes [15].

Among the methods that can be used to produce the CNF is grinding. This method uses a mill having a static and rotating grinding stone (1,400–3,000 rpm). The process involves the passage of the pulp to low consistency between these two stones [15, 26] with a certain number of passes. This leads to fibrillation of the cellulose with the breaking down of hydrogen bonds through shear and compressive forces [27]. The diameters of the NFCs produced by this process ranges from 5 to 157 nm [28].

The high-pressure homogenization (HPH) method also uses low-consistency cellulose, which is cyclically passed through a valve assembly with a very narrow opening at high pressure and low speed. The cellulose is exposed to a pressure drop with the valve opening, resulting in high impact and shear forces, leading to the opening of the fibrils [29]. This method produces fibers with diameters between 20 and 100 nm and lengths of several tens of micrometers [15].

Another method that has been employed is microfluidization, which consists of pumping the cellulose to low consistency under high pressure through chambers of different z-shape sizes [30]. This movement generates high shear and impact forces, producing nanocelluloses of less than 100 nm in diameter [15].

In the cryocrushing method, the cellulose is immersed in liquid nitrogen to freeze and in the sequence follows for processing by high shear forces and impact by mortar and pestle [31]. The ice crystals exert pressure, causing the cell walls to rupture. The CNFs produced by this method have diameters between 30 and 80 nm [15].

In the steam explosion method, the cellulose is exposed to temperatures between 200 and 270°C and pressures between 14 and 16 bar for periods between 20 seconds and 20 minutes. After this period, a sudden decompression is applied, which results in the defibrillation of cellulose. The diameter of the CNF produced by this method is in the range of 10–50 nm [15].

Nanocellulose can also be produced by high intensity ultrasonication, where the hydrodynamic forces of the ultrasound lead to cavitation, causing expansion of the formed gas bubbles and explode, breaking the cellulose fibers [32]. The diameter of the produced nanocellulose is between 5 and 35 nm [23].

As all of these processes are mechanical, they imply expenditure of high energy, the use of pretreatments such as the use of 2,2,6,6-tetramethylpiperidine-1-oxyl oxidation (TEMPO), chemical treatment with sodium hypochlorite or sodium, or enzymatic pretreatment may result in chain openings, facilitating the shredding [33]. The enzymes can also be used to modify or degrade the lignin and the hemicelluloses of biomasses [15] facilitating subsequent hydrolysis or mechanical treatment.

9.3.2 CELLULOSE NANOCRYSTALS

CNC refers to cellulose nanoparticles that underwent hydrolysis under controlled conditions and that lead to the formation of structures in the form of small crystalline cylinders [34]. The nanocrystals are exclusively from the crystalline regions of the cellulose molecule. Depending on the source of extraction, these crystallites will have a diameter of 2–30 nm and length of hundreds of nanometers [15, 35, 36]. The appearance of CNC is shown in Figure 9.4b.

CNC is produced generally by acid hydrolysis. Mechanical methods can be applied prior to acid hydrolysis, facilitating the accessibility of the acid in the inter-fibrillar spaces, a fact that improves solubility and facilitates the hydrolysis of the fibers [36]. Because it has hydroxyl groups available to react, the amorphous regions are easily hydrolyzed by acids when compared to the crystalline regions, which makes it possible to use a variety of acids including sulfuric acid, hydrochloric acid, phosphoric acid, or ammonium persulfate [37]. A typical process of CNC production involves controlled acid hydrolysis, washing, centrifugation, dialysis, and sonication to form a suspension followed by freeze-drying or heat-drying [15, 26]. In addition to depending on the source of origin, the CNCs also depend on the concentration of acid, types of acid, reaction time, and temperature [38].

Depending on the type of acid used, the surface of the nanocrystals will be loaded with different clusters [15]. In the case of methods using sulfuric acid, the temperature is generally used in the range 45–60°C at a concentration of 55–65% for 2 h [36]. Acid hydrolysis with sulfuric acid results in highly soluble crystals having a sulfate group surface, which gives a stable colloidal dispersion when diluted in water [39]; however, surface chemistry is not conducive to covalent functionalization or incorporation in solvents, polymers and hydrophobic resins [37].

Acid hydrolysis using hydrochloric acid or treatment with ammonium persulfate results in CNCs with low aqueous dispersions with hydroxyl or carboxyl surface groups, respectively [36]. However, the surface hydroxyl groups can be modified more easily than the sulfate groups produced by hydrolysis with sulfuric acid [37].

CNCs can also be produced through enzymatic hydrolysis. This type of production does not require harsh chemicals, it has lower energy requirements, and is environmentally sustainable. As in acid hydrolysis, mechanical treatment may be employed to facilitate the access of the enzymes [37]. In the family of cellulases there are two types of enzymes that deserve prominence in the production of CNC, the endoglucanases can be used to digest the amorphous region of cellulose, since the cellobiohydrolases are able to digest the crystalline region [37].

CNCs produced using enzymes having endoglucanase activity have unmodified hydroxyl surface groups similar to CNCs produced using hydrochloric acid digestion.

9.3.3 BACTERIAL NANOCELLULOSE

Unlike the other types of nanocellulose discussed, bacterial nanocellulose (BNC) is produced as a primary extra-cellular metabolite produced by bacteria belonging to the genera *Acetobacter, Agrobacterium, Alcaligenes, Pseudomonas, Rhizobium, Aerobacter, Achromobacter, Azotobacter, Salmonella,* or *Sarcina* [40]. The most efficient bacteria in the production of cellulose are the *Acetobacterxylinum* that have been reclassified and included within the new genus *Gluconacetobcater*, such as *G. xylinus* [41]. Culture of these bacteria is usually carried out at temperatures of 28–30°C, with or without agitation [42].

Bacterial cellulose is secreted in a tape form, with a thickness of approximately 3–4 nm, 70–100 nm wide and 1–9 μm long [42]. The appearance of BNC is shown in Figure 9.4c. These tapes are composed of nanofibrils 2–4 nm in diameter and have excellent properties due to the high degree of polymerization (up to 8,000), high crystallinity (84–89%) and high water retention capacity [15]. Among the advantages of this type of nanocellulose is the purity, since unlike lignocellulosic resources, it does not have hemicelluloses and lignin associated in its composition. Another advantage is that when produced, if they are still wet, bacterial cellulose films can be easily disintegrated into nanofibrils [42].

Bacterial cellulose, as well as plant cellulose, can be hydrolyzed in nanocrystals by acid or enzymatic hydrolysis similar to CNC production [12].

Among the properties that the nanocelluloses possess, it is possible to highlight the large surface area of this material, which may allow high levels of chemical bonds and the possibility of chemical modification. These properties may contribute to a large number of drugs being able to bind to the surface from material and promote the encapsulation of drugs [43]. This use of nanocelluloses is discussed in the following sections.

9.4 CONTROLLED RELEASE OF DRUGS

The controlled release of drugs refers to a method of insertion of drug biomolecules into a living system, having the development of targeted therapy, including the ideal release of the active pharmaceutical ingredient (API) of the formulation as its main focus [44]. This method brings benefits to treatments by increasing bioavailability, decreasing toxicity and reducing the frequency of dosages, as a result of the natural elimination of drugs by human metabolism [45].

Administration of a drug may be affected via the oral, vaginal, rectal, intravenous, subcutaneous, and intramuscular routes. The oral route is the most used method, with the use of tablets, capsules, and syrups to facilitate the release in the gastrointestinal tract [44]. The development of these drugs seeks to favor the bioavailability, following the conventional flow of disintegration rate, dissolution, and permeation by membranes of the gastrointestinal tract, until reaching the systemic circulation [46]. The challenge is to ensure bioavailability and controlled release from formulations that do not affect the properties of the drugs, are non-toxic and viable for the industrial production sector.

The technical resources in the medical and pharmaceutical areas are limited due to the characteristics of poor solubility of drugs in water [46] or high molar mass, which limits large-scale industrial production. The development of formulations using nanoparticles to improve solubility [46], showed efficiency of use as excipients, a base for coatings, or a matrix of tablets [47].

The use of nanoparticles or nanocarriers has attracted considerable attention due to their physical, chemical, and biological properties [48]. There are some criteria that must be met in order for a nanoparticle to be used in a drug delivery system. These include: (a) its long stay within the circulatory system [49]; (b) a higher charge on the surface of the nanoparticle and the drug; and, (c) better binding of the nanoparticle and the drug to control the release mechanism [50], among others.

A new class of sustainable source in the production of a polymeric material arises as a challenge in pharmaceutical development science. The use of natural polymers is studied in pharmacology by promoting the increase of desirable in drugs. Among the possible natural polysaccharides three materials are already widely used as drug delivery systems, viz., cellulose, starch, and chitosan [49, 51].

In the pharmaceutical industry, microcrystalline cellulose (MCC) has a long history of application due to its excellent compaction property and as an excipient to condense drug loaded matrices [52]. However, MCC has some limitations in the release of drugs. For example, when exposed to high humidity, it is subjected to a substantial decrease in the specific surface area in the pores of the powder [53]. Alternatively, MCC can be subjected to a chemical treatment by promoting hydrolysis and forming smaller fragments known as CNC, which increases water solubility [51]. The CNCs provide a higher bioavailability, improving the drug loading capacity, allowed by its structure with open pores and greater surface area [54].

The powder obtained from BNC presents superior bulk density, better flowability, easy particle fragmentation, rearrangement with a lower compression load, and better processability in tablet and capsule formulations compared to MCC [47].

In the development of new excipients, the CNFs have been in great demand in recent years due to their unique colloidal properties, biodegradability, high surface area, rheological properties, barrier, and non-toxicity [55].

A particular feature of the CNF is related to its structure, which has in its monomer three hydroxyl groups on the surface of the fiber, forming hydrogen bonds and showing a hydrophilic character, which in contact with the molecules of the drugs, results in electrostatic interaction [49, 56]. The hydrophilic surfaces and hydrophobics of CNF may be energized for use in the adsorption of hydrophobic drug molecules by preventing them from aggregating or crystallizing [47]. However, it is reported that further research is needed to enable a better understanding of such interactions due to the physico-chemical properties used in the CNF [55]. The electrostatic interactions in the molecules with carboxyl groups present on the surface of CNF are pH-dependent, with these interactions between positively charged drugs seen as the main mechanism of attachment between them [56].

The advantage of using a nanoparticle in a drug delivery system is due to its size, which can be customized since they need to be large enough to prevent rapid penetration into fenestrated blood vessels, and at the same time small enough so as not to affect phagocytosis by macrophages that possess reticuloendothelial structure [49].

9.4.1 OBTAINING PHARMACEUTICAL FORMULAS

The methods most cited in pharmaceutical technology research for obtaining solid dispersions and encapsulation include hot-melt extrusion, spray-drying, and freeze-drying.

9.4.1.1 Hot Melt Extrusion

In the hot melt extrusion process a drug and a polymer are mixed under temperature control and shearing force through an extruder which contains one or more screws [57]. In order to obtain a solid dispersion, the physical mixture (polymer and drug) undergoes a temperature range of 30 to 60°C higher than the glass transition temperature (Tg) of the polymer [58]. Figure 9.5 illustrates this type of extrusion process, wherein the application of high shear force and high temperature in the formulation enables the formation of a solid dispersion besides involving the exchange of thermal energy during the process, followed by instantaneous cooling that affects the thermodynamic and kinetic properties [59].

Hot melt extrusion is an advantageous method for large industrial-scale production because it works as a continuous process that increases poor solubility of water-soluble drugs by producing a solid dispersion without the aid of any solvents [60]. The solid material is obtained through an extruder containing a poor water-soluble drug and a hydrophilic polymer to inhibit the conversion of amorphous to crystalline with the use of some additives such as surfactants to avoid crystallization of the drug during storage[59, 61].

9.4.1.2 Spray-Drying

Spray-drying is an old method for drying materials, highly recommended for thermally sensitive materials such as foods and pharmaceuticals [62]. Due to the continuous mode of operation, the spray-drying method has become very popular in the pharmaceutical and biotechnology industry. This process is advantageous due to its high productivity combined with its high solvent rapid drying capacity [63], which exposes the product to high temperatures in association with subsequent rapid rotary cooling [64]. Spray drying can be used for the production of various powder products, such as solid dispersions, nanoparticles, self-emulsion distribution systems, and eutectic mixtures [63].

Figure 9.6 shows the spray-drying process, which is based on the liquid feed or the suspension, which is pumped into a drying chamber. Parameters such as air inlet temperature and feed rate are determined according to the characteristics of the product. The gas and the dried particles are separated by a cyclone and the powder is collected in a collection vessel [63, 65].

In comparison with hot-melt-extrusion, spray dispersion has some advantages, such as: relatively short time in the development of the formulation; accessibility to thermolabile compounds [64] and polymers; and addition of antioxidants in the incorporation of the solid dispersion [46].

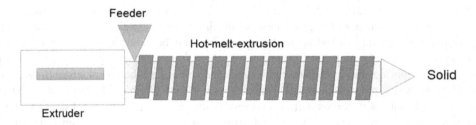

FIGURE 9.5 Hot melt extrusion process.

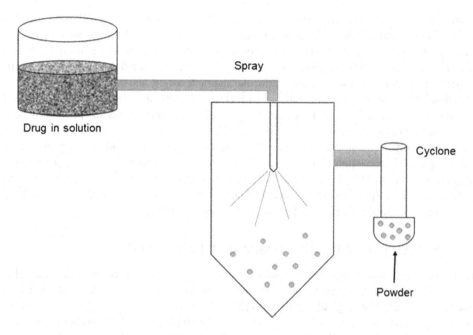

FIGURE 9.6   Spray-drying process to obtain solids.

9.4.1.3   Freeze-Drying

In the freeze-drying process the drug and carrier are dissolved in a solvent and then the solution is frozen in liquid nitrogen to form a lyophilized molecular dispersion [66]. This rapid freezing process provides better molecular distribution and stability, avoiding physical and chemical degradation during manufacturing and storage [67]. The sublimation of the particles occurs under atmospheric pressure, the reduction in the dimensions of the product favoring a lower freezing time caused by the best associated heat and mass transfers [44]. The product is frozen at a temperature below the eutectic temperature or more often at the glass transition temperature (Tg) of the solution in the maximum freezing concentration. During primary drying, the temperature of the product should not exceed Tg to avoid a possible collapse in the amorphous matrix [68].

The freeze-drying process is considered as a 'soft' process in which the solvent is removed without compromising the quality of the product. However, the operating conditions must be carefully selected to obtain an acceptable range of residual moisture and to preserve the characteristics of the formulation during the drying stages [67].

9.4.2   *In Vitro Drug Release Test*

For the release of drugs, even with biodegradable polymers, there is degradation by enzymatic hydrolysis, the combination of diffusion and erosion mechanisms being dependent on the pharmaceutical formulation for the *in vivo* assays [48].

Oral release of drugs still faces obstacles due to the metabolic complex involved in the gastrointestinal system, limitations of bioavailability, and gastric pH [44].

The solubility and permeability of the drug influence the absorption and bioavailability of the oral formulas, which are mostly composed of crystalline solids [69]. The dissolution test is a kinetic study whose purpose is to evaluate the solubility process of the solid formed in a given aqueous solution or dissolution medium.

Some factors affect the rate of dissolution, which include surface of the solid, solubility of the solid in the average dissolution, concentration of solute as a function of time and the rate of dissolution constant. Particle size reduction increases the rate of dissolution [70], the characteristic behavior of solid nanoparticles. The drying processes of the solids increase the specific area of the surface and favor the amorphization of the components of the solid formulation, improving the solubility of the dissolution medium [70]. The rate of dissolution is affected under the solute diffusion coefficient and viscosity of the medium, molecular characteristics, and particle size.

The dissolution test requires the medium with aqueous solutions containing biological components that simulate the gastric and intestinal fluid with diverse compositions and pH [69], which may influence the rate of dissolution of the *in vivo* formulation.

The most widely used equipment for dissolution testing in industrial and research laboratories is shown in Figure 9.7. The equipment has metal blades in a vessel that receives the aqueous solution or dissolution medium. The pharmaceutical formula is positioned in the lower center of the vessel, the metal blades are responsible for stirring at rotational speed which can range from 50 to 150 rpm. The mean vessel volume is 0.9 L and the temperature is 37°C [69, 71].

9.5 NANOCELLULOSE IN THE PHARMACEUTICAL FIELD

With advances in research, the use of nanomaterials, nanoparticles, porous nanomaterials, and nanocellulose has resulted in great interest in the development of new formulas with pharmaceutical, medical, and biotechnological applications [53, 61].

The nanoparticles, viz. CNC, CNF, and BNC may be similar to each other, but there are some particular procedures in obtaining these so that each type of nanocellulose is developed to satisfy the mechanism of release, according to the functions of the drugs [49]. To functionalize the nanocellulose by improving the release profile, modification by chemical or physical incorporation and mixing of coatings, are some of the methods used to increase stability and prolong the release [47].

The nanocellulose can be applied in several other forms or associated, such as surface modification via 2,2,6,6-Tetramethyl-piperidin-1-oxyl (TEMPO), or by the addition of negatively charged carboxylic groups, which allow the immobilization of drugs and macromolecules through the electrostatic interactions. These nanofibers typically have dimensions of 3 to 4 nm in diameter and have been shown to be non-toxic in *in vitro* cellular models [56]. CNF films can be considered as the simplest and most direct material for drug release, which consist of forming a drug-loaded nanocellulose film, which can be prepared by filtration of the drug suspension, followed by drying [55].

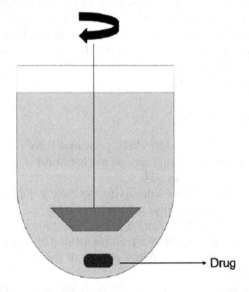

FIGURE 9.7   Equipment for testing of dissolution by blades.

The electrospinning methods, wherein polymers containing drugs are obtained, have gained attention due to their unchanged nanostructure, biocompatibility, biodegradability surface chemistry, and bioactivity of the drug during the spinning process and the biomolecule array.

The nanocomposites are materials that have shown greater compatibility and high efficiency in loading of drugs (98.8%) and superior release (86%), in about 540 hours in composites containing 6% of CNF [51, 54].

9.5.1 PHARMACEUTICAL APPLICATIONS OF NANOCELLULOSE FOR DRUG RELEASE

Nanocellulose can be found in the pharmaceutical field in various forms such as solids, films, membranes, and composites, among others. Table 9.1 lists the applications of solid nanocellulose encapsulation for drug release, together with the tested drugs and drying methods.

The drug release profile containing BNC, CNF, or CNC exhibits rapid early release in the first few hours (0.5 to 10 hours) and a decrease of up to 72 hours [47]. Since the CNC is used more as an auxiliary or support material for other components of the formulations, such as xanthan and chitosan [79] which uses the CNC for the formation of bio-nanocomposite with xanthan and chitosan using the drug 5-fluorouracil. The CNC can potentially be used to prevent aggregation of nanoparticles in aqueous environments, stabilizing crystalline nanoparticles of soluble drugs, preserving their morphology, both in aqueous suspension storage and after lyophilization of the suspensions [55]. The CNC can also be used directly as an excipient or as co-stabilizer to improve the physicochemical and flow properties of the excipient polymer [52], a feature that influences flowability compared to microcrystalline cellulose.

Hot melt extrusion has emerged in recent decades as the most applied industrial process for increasing the solubility and bioavailability of drugs of poor solubility [80]. In obtaining pharmaceutical formulas, hot melt extrusion is most commonly used in drug formulations containing synthetic polymers; however, recent studies have already sought to use nanocellulose in this process of obtaining pharmaceutical formulas as can be seen from the study by Nagalakshmaiah et al. [81], which showed the addition of CNC improved the thermal stability of the polystyrene matrix produced by the hot melt extrusion method.

9.5.2 NANOCELLULOSE IMPROVING DRUG SOLUBILITY

According to the Biopharmaceutics Classification System (BCS), class II drugs, which belong to the largest number of drugs studied, present impaired bioavailability, due to the low solubility in the

TABLE 9.1

Application of Nanocellulose in Solid Form for the Release of Drugs

Nanocellulose	Method of drying	Drugs	References
CNF	Freeze-drying	Beclomethasone dipropionate	[72]
CNF	Freeze-drying	Itraconazole	[73]
CNF	Freeze-drying	Quercetin	[74]
CNF	Spray-drying	Indomethacin, Metoprololtartrate, Verapamilhydrochloride, Nadolo, Ibuprofen, Atenolol	[53]
BNC	Spray-drying	Aspirina	[75]
BNC	Freeze-drying	Albumin	[76]
BNC	Freeze-drying	Famotidine Tizanidine	[77]
CNC	Freeze-drying	Tetracycline hydrochloride	[78]
CNC	Freeze-drying	5-fluorouracil	[79]

variable conditions of the gastrointestinal system [56].To reverse this situation, the CNF and CNC are employed to improve the solubility of drugs by being able to release soluble molecules in water, through ionic interactions [82].

The nanocellulose has advantages in this application due to its barrier properties, rheological, and physicochemical characteristics [47, 83], allowing its large surface area to offer the possibility of positive molecular interactions in the drugs with the lowest levels of solubility [55].

9.5.3 TOXICOLOGICAL TESTS IN NANOCELLULOSE

In general, toxicological tests on nanomaterials are challenging because of the variety of size, shape, coating, and surface reactivity [84]. The toxicity of these materials exhibited very low levels in several models used. However, it is reported that more studies based on *in vitro* and *in vivo*, bio-distribution, cellular interactions, and eco-toxicities have still to be performed [82]. Unlike BNC, CNCs are not pure celluloses because they contain sulfate and ester groups on the surface, which confer important properties such as colloidal stability and enable chemical modifications [47]. Menas et al. [84] observed that the CNC particles were apparently non-toxic, thus implying low or no cytotoxicity in their comparative nanocellulose analysis with different morphologies.

In the case of CNFs, several studies have investigated the biostability [73, 75, 76] and toxicity [82, 84] of this material; however, the observed results indicated they depend on factors such as the surface chemistry and physical form of the solid material, suspension or particle [55]. Thus, use of CNC, BNC, and CNF forms did not report high toxicity cases that compromised the use of these nanoparticles as pharmaceutical materials. However, the need for cytotoxicity assays for the development and obtaining of biomaterials independent of the source is emphasized.

9.6 FINAL CONSIDERATIONS

Nanocellulose has the potential as a biomaterial for use in the pharmaceutical field, especially in the fields of controlled drug release and in improving the solubility of drugs of poor solubility in water. Studies have also shown that the application of nanocellulose, be it CNF, BNC, or CNC in the drugs, increases the bioavailability, consequently bringing benefits to the oral administration of the drug.

The use of nanocellulose needs to be evaluated by *in vitro* dissolution tests and by toxicological tests. The former (dissolution tests) verify the behavior of the matrix with nanocellulose before the pharmaceutical formulations, where the simulations with liquid body indicate the stability and the bioavailability of the drugs. The latter (toxicological tests) are necessary because the nanocellulose is obtained by chemical and physical treatments and it is important to prove the absence of substances that can harm the human organs.

ACKNOWLEDGMENTS

The authors place on record and appreciate the kind permission given by some of the publishers, viz. IN TECH d.o.o., MDPI, and Springer, who have given the authors permission, free of charge, to use some of the figures from their publications. One of the authors (KGS) would like to thank the PPISR, Bangalore, India, with whom he is presently associated, for their encouragement and interest in this collaboration.

REFERENCES

1. Workman ML, LaCharity LA, Kruchko SL (2013) *Understanding Pharmacology: Essentials for Medication Safety.* Elsevier, Amsterdam.
2. Ritter JM, Flower RJ, Henderson G, et al. (2019) *Rang & Dale's Pharmacology*, 9th ed. Elsevier, Amsterdam.

3. Jain KK (2008) Drug delivery systems – An overview. In: Jain KK (ed) *Drug Delivery Systems*. Springer, New York, pp 1–50.

4. Pal U, Pramanik SK (2018) Advances in the application of nanomaterials and Nanosacled materials in physiology or medicine: Now and the future. In: Ganguly BN (ed) *Nanomaterials in Bio-Medical Applications*. Materials Research Forum, Millersville, pp 147–178.

5. Zhang X-Q, Xu X, Bertrand N, et al. (2012) Interactions of nanomaterials and biological systems: Implications to personalized nanomedicine. *Advanced Drug Delivery Reviews* 64(13):1363–1384.

6. Abdul Khalil HPS, Davoudpour Y, Islam MN, et al. (2014) Production and modification of nanofibrillated cellulose using various mechanical processes: A review. *Carbohydrate Polymers* 99:649–665.

7. Siqueira G, Bras J, Dufresne A, et al. (2010) Cellulosic bionanocomposites: A review of preparation, properties and applications. *Polymers* 2(4):728–765.

8. Trivedi P, Fardim P (2019) Recent advances in cellulose chemistry and potential applications. In: Fang Z, Smith RL, Tian X (eds) *Production of Materials from Sustainable Biomass Resources*. Springer, Singapore, pp 99–115.

9. Roy D, Semsarilar M, Guthrie JT, Perrier S (2009) Cellulose modification by polymer grafting: A review. *Chemical Society Reviews* 38(7):2046.

10. Wertz J-L, Mercier JP, Bédué O, Mercier J-P (2010) *Cellulose Science and Technology*. EPFL Press.

11. Lengowski EC, Bonfatti Júnior EA, Kumode MMN, et al. (2019) Nanocellulose in the paper making. In: Inamuddin Thomas S, Mishra RK, Asiri AM (eds) *Sustainable Polymer Composites and Nanocomposites*. Springer International Publishing, Cham, pp 1027–1066.

12. Börjesson M, Westman G (2015) Crystalline nanocellulose — Preparation, modification, and properties. In: Poletto M, Ornaghi Junior HL (eds) *Cellulose – Fundamental Aspects and Current Trends*. InTech, Rijeka, Croatia, pp 159–191.

13. Lustri WR, Barud HGde O, Barud Hda S, et al. (2015) Microbial cellulose – Biosynthesis mechanisms and medical applications. In: Matheus Polletom (ed) *Cellulose – Fundamental Aspects and Current Trends*. InTech, London, pp 133–157.

14. Sixta H (2006) *Handbook of Pulp*. Wiley-VCH Verlag GmbH, Weinheim.

15. Rojas J, Bedoya M, Ciro Y (2015) Current trends in the production of cellulose nanoparticles and nanocomposites for biomedical applications. In: Poletto M, Ornaghi Junior HL (eds) *Cellulose – Fundamental Aspects and Current Trends*. InTech, Rijeka, Croatia, pp 193–228.

16. Qua EH, Hornsby PR (2011) Preparation and characterisation of nanocellulose reinforced polyamide-6. *Plastics, Rubber and Composites* 40(6–7):300–306.

17. Phanthong P, Reubroycharoen P, Hao X, et al. (2018) Nanocellulose: Extraction and application. *Carbon Resources Conversion* 1(1):32–43.

18. Gopakumar DA, Thomas S, Grohens Y (2016) Nanocelluloses as innovative polymers for membrane applications. In: Puglia D, Fortunati E, Kenny JM (eds) *Multifunctional Polymeric Nanocomposites Based on Cellulosic Reinforcements*. William Andrew Publishing, Kidlington, UK, pp 253–275.

19. Dai J, Chae M, Beyene D, et al. (2018) Co-production of cellulose nanocrystals and fermentable sugars assisted by endoglucanase treatment of wood pulp. *Materials* 11(9):1645.

20. Torres FG, Commeaux S, Troncoso OP (2012) Biocompatibility of bacterial cellulose based biomaterials. *Journal of Functional Biomaterials* 3(4):864–878.

21. Saito T, Kimura S, Nishiyama Y, Isogai A (2007) Cellulose nanofibers prepared by TEMPO-mediated oxidation of native cellulose. *Biomacromolecules* 8(8):2485–2491.

22. Pääkkö M, Ankerfors M, Kosonen H, et al. (2007) Enzymatic hydrolysis combined with mechanical shearing and high-pressure homogenization for nanoscale cellulose fibrils and strong gels. *Biomacromolecules* 8(6):1934–1941.

23. He W, Jiang X, Sun F, Xu X (2014) Extraction and characterization of cellulose nanofibers from Phyllostachys Nidularia Munro via a combination of acid treatment and ultrasonication. *BioResources* 9(4):6876–6887.

24. Jonoobi M, Niska KO, Harun J, Misra M (2009) Chemical composition, crystallinity, and thermal degradation of bleached and unbleached kenaf bast (Hibiscus cannabinus) pulp and nanofibers. *BioResources* 4:626–639.

25. Uetani K, Yano H (2011) Nanofibrillation of wood pulp using a high-speed blender. *Biomacromolecules* 12(2):348–353.

26. Lengowski EC, Muniz GIB de, Nisgoski S, Magalhães WLE (2013) Cellulose acquirement evaluation methods with different degrees of crystallinity. *Scientia Forestalis* 41:185–194.

27. Siró I, Plackett D (2010) Microfibrillated cellulose and new nanocomposite materials: A review. *Cellulose* 17(3):459–494.

28. Hassan ML, Mathew AP, Hassan EA, et al. (2012) Nanofibers from bagasse and rice straw: Process optimization and properties. *Wood Science and Technology* 46(1–3):193–205.

29. Chaker A, Mutjé P, Vilar MR, Boufi S (2014) Agriculture crop residues as a source for the production of nanofibrillated cellulose with low energy demand. *Cellulose* 21(6):4247–4259.

30. Ferrer A, Filpponen I, Rodríguez A, et al. (2012) Valorization of residual Empty Palm Fruit Bunch Fibers (EPFBF) by microfluidization: Production of nanofibrillated cellulose and EPFBF nanopaper. *Bioresource Technology* 125:249–255.

31. Hubbe MA, Rojas OJ, Lucia LA, Sain M (2008) Cellulosic nanocomposites, review. *BioResources* 3:929–980.

32. Chen P, Yu H, Liu Y, et al. (2013) Concentration effects on the isolation and dynamic rheological behavior of cellulose nanofibers via ultrasonic processing. *Cellulose* 20(1):149–157.

33. Cheng F, Liu C, Wei X, et al. (2017) Preparation and characterization of 2,2,6,6-tetramethylpiperidine-1-oxyl (TEMPO)-oxidized cellulose nanocrystal/alginate biodegradable composite dressing for hemostasis applications. *ACS Sustainable Chemistry and Engineering* 5(5):3819–3828.

34. Sehaqui H, Allais M, Zhou Q, Berglund LA (2011) Wood cellulose biocomposites with fibrous structures at micro- and nanoscale. *Composites Science and Technology* 71(3):382–387.

35. Elazzouzi-Hafraoui S, Nishiyama Y, Putaux J-L, et al. (2008) The shape and size distribution of crystalline nanoparticles prepared by acid hydrolysis of native cellulose. *Biomacromolecules* 9(1):57–65.

36. Kumar R, Singh S, Singh OV (2008) Bioconversion of lignocellulosic biomass: Biochemical and molecular perspectives. *Journal of Industrial Microbiology and Biotechnology* 35(5):377–391.

37. Anderson SR, Esposito D, Gillette W, et al. (2014) Enzymatic preparation of nanocrystalline and microcrystalline cellulose. *Tappi Journal* 13(5):35–42.

38. Zheng H (2014) *Production of Fibrillated Cellulose Materials – Effects of Pretreatments and Refining Strategy on Pulp Properties*. Aalto University, Espoo, Finland.

39. Wang Q, Zhao X, Zhu JY (2014) Kinetics of strong acid hydrolysis of a bleached kraft pulp for producing cellulose nanocrystals (CNCs). *Industrial and Engineering Chemistry Research* 53(27):11007–11014.

40. Lin S-P, Loira Calvar I, Catchmark JM, et al. (2013) Biosynthesis, production and applications of bacterial cellulose. *Cellulose* 20(5):2191–2219.

41. Yamada Y (2000) Transfer of Acetobacter oboediens and Acetobacter intermedius to the genus Gluconacetobacter as Gluconacetobacter oboediens comb. nov. and Gluconacetobacter intermedius comb. nov. *International Journal of Systematic and Evolutionary Microbiology* 50(6):2225–2227.

42. Charreau H, Foresti ML, Vazquez A (2013) Nanocellulose patents trends: A comprehensive review on patents on cellulose nanocrystals, microfibrillated and bacterial cellulose. *Recent Patents on Nanotechnology* 7(1):56–80.

43. Plackett DV, John KL, Jackson K, Burt HM (2014) A review of nanocellulose as a novel vehicle for drug delivery. *Nordic Pulp and Paper Research Journal* 29(1):105–118.

44. Vishali DA, Monisha J, Sivakamasundari SK, et al. (2019) Spray freeze drying: Emerging applications in drug delivery. *Journal of Controlled Release* 300:93–101.

45. Kolakovic R, Peltonen L, Laukkanen A, et al. (2012) Nanofibrillar cellulose films for controlled drug delivery. *European Journal of Pharmaceutics and Biopharmaceutics* 82(2):308–315.

46. Zhang D, Lee Y, Shabani Z, Frankenfeld Lamm C, Zhu W, Li Y, Templeton A (2018) Processing impact on performance of solid dispersions. *Pharmaceutics* 10(3):142.

47. Klemm D, Cranston ED, Fischer D, et al. (2018) Nanocellulose as a natural source for groundbreaking applications in materials science anocellulose as. *Materials Today* 21(7):720–748.

48. Patel GC, Yadav BK (2018) Polymeric nanofibers for controlled drug delivery applications. In: Grumezescu AM (ed) *Organic Materials as Smart Nanocarriers for Drug Delivery*. William Andrew, Norwich, pp 147–176.

49. Tan TH, Lee HV, Abdulhadi W, et al. (2019) A review of nanocellulose in the drug-delivery system. In: Holban A-M, Grumezescu AM (eds) *Materials for Biomedical Engineering: Nanomaterials-Based Drug Delivery*. William Andrew, Norwich, pp 131–164.

50. Singh I, Sharma A, Park BD (2016) Drug-delivery applications of cellulose nanofibrils. In: Holban A-M, Grumezescu A (eds) *Nanoarchitectonics for Smart Delivery and Drug Targeting*. William Andrew, Norwich, pp 95–117.

51. Gopinath V, Saravanan S, Al-maleki AR, et al. (2018) Biomedicine & Pharmacotherapy A review of natural polysaccharides for drug delivery applications : Special focus on cellulose, starch and glycogen. *Biomedicine and Pharmacotherapy* 107:96–108.

52. Lin N, Dufresne A (2014) Nanocellulose in biomedicine : Current status and future prospect. *European Polymer Journal* 59:302–325.

53. Carlsson DO, Hua K, Forsgren J, Mihranyan A (2014) Aspirin degradation in surface-charged TEMPO-oxidized mesoporous crystalline nanocellulose. *International Journal of Pharmaceutics* 461(1–2):74–81.

54. Du H, Liu W, Zhang M, et al. (2019) Cellulose nanocrystals and cellulose nano fi brils based hydrogels for biomedical applications. *Carbohydrate Polymers* 209:130–144.

55. Löbmann K, Svagan AJ (2017) Cellulose nano fi bers as excipient for the delivery of poorly soluble drugs. *International Journal of Pharmaceutics* 533(1):285–297.

56. Paukkonen H, Ukkonen A, Szilvay G, et al. (2017) European Journal of Pharmaceutical Sciences hydrophobin-nano fibrillated cellulose stabilized emulsions for encapsulation and release of BCS class II drugs. *European Journal of Pharmaceutical Sciences* 100:238–248.

57. Tran P, Pyo Y-C, Kim D-H, et al. (2019) Overview of the manufacturing methods of solid dispersion technology for improving the solubility of poorly water-soluble drugs and application to anticancer drugs. *Pharmaceutics* 11(3):1–26.

58. Gao N, Guo M, Fu Q, He Z (2017) Application of hot melt extrusion to enhance the dissolution and oral bioavailability of oleanolic acid. *Asian Journal of Pharmaceutical Sciences* 12(1):66–72.

59. Fule R, Amin P (2014) Development and evaluation of lafutidine solid dispersion via hot melt extrusion: Investigating drug-polymer miscibility with advanced characterisation. *Asian Journal of Pharmaceutical Sciences* 9(2):92–106.

60. Genina N, Hadi B, Löbmann K (2018) Hot melt extrusion as solvent-free technique for a continuous manufacturing of drug-loaded mesoporous silica. *Journal of Pharmaceutical Sciences* 107(1):149–155.

61. Shuwisitkul D (2016) Hot melt extrusion: An application for enhancing drug solubility. *Asian Journal of Pharmaceutical Sciences* 11(1):45–46.

62. Mantas A, Labbe V, Loryan I, Mihranyan A (2019) Amorphisation of free acid ibuprofen and other profens in mixtures with nanocellulose : Dry powder formulation strategy for enhanced solubility. *Pharmaceutics* 2(2):1–21.

63. Davis M, Walker G (2018) Recent strategies in spray drying for the enhanced bioavailability of poorly water-soluble drugs. *Journal of Controlled Release* 269:110–127.

64. Poozesh S, Bilgili E (2019) Scale-up of pharmaceutical spray drying using scale-up rules: A review. *International Journal of Pharmaceutics* 562:271–292.

65. Marinopoulou A, Karageorgiou V, Papastergiadis E, et al. (2019) Production of spray-dried starch molecular inclusion complexes on an industrial scale. *Food and Bioproducts Processing* 116:186–195.

66. Betageri GV, Makarla KR (1995) Enhancement of dissolution of glyburide by solid dispersion and lyophilization techniques. *International Journal of Pharmaceutics* 126(1–2):155–160.

67. Fissore D, Gallo G, Ruggiero AE, Thompson TN (2019) On the use of a micro freeze-dryer for the investigation of the primary drying stage of a freeze-drying process. *European Journal of Pharmaceutics and Biopharmaceutics* 141:121–129.

68. Meng-Lund H, Holm TP, Poso A, et al. (2019) Exploring the chemical space for freeze-drying excipients. *International Journal of Pharmaceutics* 566:254–263.

69. Radivojev S, Zellnitz S, Paudel A, Fröhlich E (2019) Searching for physiologically relevant in vitro dissolution techniques for orally inhaled drugs. *International Journal of Pharmaceutics* 556:45–56.

70. Alshafiee M, Aljammal MK, Markl D, et al. (2019) Hot-melt extrusion process impact on polymer choice of glyburide solid dispersions: The effect of wettability and dissolution. *International Journal of Pharmaceutics* 559:245–254.

71. Alshehri SM, Shakeel F, Ibrahim MA, et al. (2019) Dissolution and bioavailability improvement of bioactive apigenin using solid dispersions prepared by different techniques. *Saudi Pharmaceutical Journal* 27(2):264–273.

72. Valo H, Arola S, Laaksonen P, et al. (2013) Drug release from nanoparticles embedded in four different nanofibrillar cellulose aerogels. *European Journal of Pharmaceutical Sciences* 50(1):69–77.

73. Valo H, Kovalainen M, Laaksonen P, et al. (2011) Immobilization of protein-coated drug nanoparticles in nanofibrillar cellulose matrices-Enhanced stability and release. *Journal of Controlled Release* 156(3):390–397.

74. Li X, Liu Y, Yu Y, et al. (2019) Nanoformulations of quercetin and cellulose nanofibers as healthcare supplements with sustained antioxidant activity. *Carbohydrate Polymers* 207:160–168.

75. Kolakovic R, Laaksonen T, Peltonen L, et al. (2012) Spray-dried nanofibrillar cellulose microparticles for sustained drug release. *International Journal of Pharmaceutics* 430(1–2):47–55.

76. Müller A, Ni Z, Hessler N, et al. (2013) The biopolymer bacterial nanocellulose as drug delivery system: Investigation of drug loading and release using the model protein albumin. *Journal of Pharmaceutical Sciences* 102(2):579–592.

77. Badshah M, Ullah H, Khan AR, et al. (2018) Surface modification and evaluation of bacterial cellulose for drug delivery. *International Journal of Biological Macromolecules* 113:526–533.

78. Wijaya CJ, Saputra SN, Soetaredjo FE, et al. (2017) Cellulose nanocrystals from passion fruit peels waste as antibiotic drug carrier. *Carbohydrate Polymers* 175:370–376.

79. Madhusudana Rao K, Kumar A, Han SS (2017) Polysaccharide based bionanocomposite hydrogels reinforced with cellulose nanocrystals: Drug release and biocompatibility analyses. *International Journal of Biological Macromolecules* 101:165–171.

80. Repka MA, Bandari S, Kallakunta VR, et al. (2018) Melt extrusion with poorly soluble drugs – An integrated review. *International Journal of Pharmaceutics* 535(1–2):68–85.

81. Nagalakshmaiah M, Nechyporchuk O, El Kissi N, Dufresne A (2017) Melt extrusion of polystyrene reinforced with cellulose nanocrystals modified using poly[(styrene)-co-(2-ethylhexyl acrylate)] latex particles. *European Polymer Journal* 91:297–306.

82. Seabra AB, Bernardes JS, Fávaro WJ, Paula AJ (2018) Cellulose nanocrystals as carriers in medicine and their toxicities : A review. *Carbohydrate Polymers* 181:514–527.

83. Sheikhi A, Hayashi J, Eichenbaum J, et al. (2019) Recent advances in nanoengineering cellulose for cargo delivery. *Journal of Controlled Release* 294:53–76.

84. Menas AL, Yanamala N, Farcas MT, et al. (2017) Chemosphere Fibrillar vs crystalline nanocellulose pulmonary epithelial cell responses: Cytotoxicity or inflammation? *Chemosphere* 171:671–680.

10 Intellectual Property Challenges and Opportunities for India's Nano-Bio-Yatra

Susan K. Finston

CONTENTS

Disclaimer: *The views and opinions expressed in this article are those of the authors and do not necessarily reflect the official policy or position of any agency of the Indian government.*

10.1 INTRODUCTION

The promises of nanotechnology are large, both in terms of the domestic market and opportunities to Indian firms for export-led growth. Given what is at stake and the high expectations put on the burgeoning industry, it is imperative for the current patent doctrines to effectively respond to new technology. Nanotechnology investors face uncertainty about the extent of their patent rights.[1]

This chapter highlights the importance of India's nanomedical research and development, and outlines needed intellectual property (IP) reforms to support India's global nanomedicine ambitions. "Nanotechnology is the future of medicine, offering major opportunities for better, safer diagnostic tools and therapies".[2]

Indian scientists have long worked at the nanoscale,[3] in leading biopharmaceutical companies and at apex public research institutions where the Department of Biotechnology's Biotechnology Industry Research Assistance Council (BIRAC) and the Department of Science the Technology (DST), among others, have made a sustained commitment to nanomedicine research. While the Government of India has demonstrated an enduring commitment to basic research and international partnerships, there can be little return on this investment in nanomedicines without meaningful intellectual property reforms.

India has approached the crossroads of IP reform for innovative life sciences before, at the time of the 1 January 2005 deadline for reintroduction of product patent protection for medicine which coincided with rosy predictions for a new golden age of biopharmaceutical R&D for India. At that time Indian policymakers were ill prepared to design an enabling intellectual property ecosystem and fell far short of the mark. Now, as in the past, India's nanomedicine aspirations cannot come to fruition in the absence of effective protection for intellectual property, including patents and protection of confidential research data. Now, as before, Indian scientists and nanomedicine entrepreneurs look to India's political leadership for meaningful IP reforms. Given ongoing efforts by the Modi government to improve IP administration – and expanding access to IP protection for Start-Up India – there is room for cautious optimism. Time will tell whether the Government of India has learned from history or is destined to repeat it.

10.2   NANOMEDICINE: UNLIMITED POTENTIAL IN A SMALL PACKAGE

While the average medical consumer may remain blissfully ignorant of the important role of nanomaterials in biopharmaceuticals, the nanomedicines market is projected to reach or exceed USD 350 billion by 2025.[4] Nanotechnology tools already have improved imaging and diagnostic tools for earlier diagnosis, individualized treatment options, and better therapeutic success rates for cancer patients. A 2015 World Intellectual Property Organization (WIPO) Working Paper highlights the importance of nanomedicines, noting the advantages provided by the use of nanomaterials to improve drug delivery, potentially converting parenteral or intravenously delivered drugs to oral formulation:

> Drugs with very low solubility possess various biopharmaceutical delivery issues including limited bio accessibility after intake through mouth, less diffusion capacity into the outer membrane, require more quantity for intravenous intake and unwanted after-effects preceding traditional formulated vaccination process. However all these limitations could be overcome by the application of nanotechnology approaches in the drug delivery mechanism.[5]

In addition to new drug delivery systems (NDDS) to improve existing formulations, nanomaterials have also been employed effectively as biosensors for diagnostic purposes (e.g. nanoimaging and lab-on-the-chip) and tissue engineering. Currently study is underway for use of nanoparticles for diagnosis and treatment of atherosclerosis (hardening of the arteries). Future application of particular interest includes gene therapy, oncology (gold nanoparticles as probes and potential cancer treatments), vaccine development and delivery systems, and regenerative medicine.[6] Use of nanomaterials in the area of regenerative medicine holds incredible promise:

> Research in the use of nanotechnology for regenerative medicine spans several application areas, including bone and neural tissue engineering. For instance, novel materials can be engineered to mimic the crystal mineral structure of human bone or used as a restorative resin for dental applications. Researchers are looking for ways to grow complex tissues with the goal of one day growing human organs for transplant.[7]

10.3   NANOMEDICINE DEVELOPMENTS IN INDIA

In 2018, Drs Pooja Bhatia, Suhas Vasaikar, and Anil Wali (Bhatia et al.) published an exhaustive overview of nanomedicine research and development in India, which started in 2007, when the

> Department of Science and Technology (DST) established a Nanomission program to foster basic research, establish research infrastructure, nurture human capital, strike international collaborations, and strengthen the capacity for creating nanoenabled technologies. Other government organizations, such as the Council of Scientific and Industrial Research (CSIR), Defence Research and Development Organization (DRDO), Department of Biotechnology (DBT), and Indian Council of Medical Research (ICMR), also followed suit in funding nanomedicine projects.[8]

To date, nanomedicine research in India is primarily government funded.[9]

Bhatia et al. have also provided helpful diagrams and tables highlighting the large number of public research entities working in the nanomedicines domain, products already brought to market, and other notable milestones/developments.[10]

While the Government of India continues to outpace private investment in nanomedicines at present, the authors identified a significant number of Indian biopharmaceutical companies working to bring nanomedicines to market, where it is notable that these are not regional R&D centers of multinational biopharmaceutical companies but Indian companies recognizing the importance of nanomedicines research.[11]

Bhatia et al.'s research confirms that at the time of publication in 2018, Indian companies had already launched into the market 17 products involving nanomedicine and a further 19 products including nanomaterials are in the process of clinical development (e.g. undergoing clinical trials).[12] And while government funding predominates at this stage of the market,

> angel investors and venture capital (VC) firms such as the Karnataka State Industrial Investment and Development Corporation Limited, Accel India Partners, Navam Capital, Aarin Capital, Nadathur Group, Priaas Investments, ICICI Ventures, and India Science Venture Fund have actively invested in nanomedicine-based ventures in India.[13]

Given the traditional risk-averse nature of venture funds in India, the inclusion of VCs in the nanomedicine sphere is very good news.

Nanomaterials are the future of medicine and the full import of Bhatia et al.'s primary research cannot be overstated: Indian researchers are fully engaged in the important work of medicine at the nanoscale across the board from apex public research institutes to private companies. There is no question that public sector spending has outpaced the Government of India's investment in nanomedicines in large part due to uncertainties relating to whether or not India will provide an enabling environment for commercialization of nanomedicines. Given the lagging state of IP reforms needed to ensure return on investment (ROI) to private sector actors, this is both predictable and unsurprising.

However, the Indian biopharmaceutical industry as well as Indian VCs are cognizant of the potential benefits of commercialization of nanomedicines. The fact that their investments to date have been relatively modest speaks to the need for substantial IP reforms relating to both patents and protection for confidential regulatory data. The relevant question is how policy makers can best support the science to ensure return on investment for the public through effective IP reforms. In this context, the Modi government's continuing support for IP reforms needed to support Start-Up India[14] is encouraging.

10.4 EFFECTIVE INTELLECTUAL PROPERTY (IP) PROTECTION FOR NANOMEDICINES – GIVING SCIENTISTS AND NANOMEDICINE ENTREPRENEURS THE TOOLS FOR SUCCESS

It may be helpful to revisit the past to better understand the complex and politically charged history of patents for medicine in India. This section of the chapter provides a brief overview of the evolution of IP policy in India and the steps needed going forward to give Indian scientists and nanomedicine entrepreneurs the tools they need for success.

Long predating statehood, India had a long and deep history of effective patent protection:

> From the time of the mid-19th century, India offered what was at that time state-of-the-market patent protection. India replaced and amended its patent legislation several times between 1856 and 1965 and remained in the mainstream of patent policy. Taken in parallel with steps to nationalize broad swathes of Indian industry, the Indian government proved receptive to domestic industry lobbying against patent protection for medicines in 1970.[15] The Patents Act 1970 (legislation and rules

implemented in 1972), represented a realignment of national patent priorities, substantially weak-ening patent protection for innovators to meet domestic political and economic interests that were paramount at that time.[16]

From 1970 through the 1990s – even beyond the introduction of India's now famous unilateral trade liberalization in 1991[17] – the previously arcane issue of patents became a political 'third rail' issue for Congress and BJP alike.[18] While positioned as a populist platform to promote access to medi-cines, much has been written (including by me) to document absence of any meaningful commit-ment at that time by either the Government of India or domestic pharmaceutical manufacturers to meet India's daunting public health needs.[19] Beyond rhetoric, India's anti-patent policy did precious little to help the vast majority of Indians lacking access to western medicines.

10.5 REINTRODUCING PRODUCT PATENT PROTECTION FOR MEDICINE: THE WTO TRIPS AGREEMENT

In fact, it took nearly 15 additional years following the 1991 trade liberalization to reinstate product patent protection for medicines at the tail end of 2004 in advance of the 1 January 2005 World Trade Organization (WTO) implementation deadline under the Agreement on Trade Related Aspects of Intellectual Property Rights.[20]

Negotiation of the World Trade Organization (WTO) Agreement on Trade Related Aspects of Intellectual Property Rights (TRIPS) represented an important ideological battle for India and allied non-aligned states, and all aspects of TRIPS were hotly debated. As part of the integrated package of WTO Agreements, WTO negotiators ultimately reached consensus on minimum inter-national standards for intellectual property protection at the national level in all WTO member states. The TRIPS Agreement established the international obligation for WTO members to provide product patent protection without prejudice to the field of technology, i.e. including for pharmaceuti-cal products:

WTO TRIPS Agreement Article 27 – Patentable Subject Matter

1. Subject to the provisions of paragraphs 2 and 3, patents shall be available for any inventions, whether products or processes, in all fields of technology, provided that they are new, involve an inventive step and are capable of industrial application (5). Subject to paragraph 4 of Article 65, paragraph 8 of Article 70 and paragraph 3 of this Article, patents shall be available and patent rights enjoyable without discrimination as to the place of invention, the field of technology and whether products are imported or locally produced.[21]

As a developing country, India availed itself of extended implementation deadlines for WTO Agreements to which it was entitled, including the projected 1 January 2005 deadline for imple-mentation of product patent protection for pharmaceutical products. Rather than use the transitional period to focus on WTO TRIPS implementation however, India and allied WTO members sought to reopen the TRIPS Agreement during the intervening years, as at the WTO Seattle Ministerial in 1999. Meeting in Seattle to set the future direction for WTO negotiations, Organisation for Economic Co-operation and Development (OECD)-level WTO members remained focused on the implementation of TRIPS obligations, while India and other

developing Members of the WTO took a decidedly different view, urging that the Seattle Ministerial initiate the negotiation of rules that take into account their specific interests in large measure to work a rebalancing of the TRIPS Agreement.[22]

The failure of WTO members to reach consensus in Seattle on a forward looking WTO agenda, together with the general breakdown of the talks and the unprecedented public protests and (at times) riot conditions in the host city of Seattle,[23] only encouraged negotiators seeking to unravel

WTO TRIPS obligations ahead of their implementation. Increasing focus on the growing global HIV/AIDS pandemic also created new stresses on the new WTO TRIPS Agreement,[24] leading to delineation of so-called 'TRIPS Flexibilities' including compulsory licensing for export to countries lacking pharmaceutical manufacturing capacity.[25] While restating the principles of the Marrakesh Agreement establishing the WTO, the Doha Ministerial Declaration adopted in November 2001 affirmed that the TRIPS Agreement can be interpreted in a manner supportive of WTO members' public health goals and to address specific concerns relating to global pandemics. However, although the Doha Declaration has been perceived as diluting the standards for intellectual property protection imposed by the TRIPS Agreement,[26] WTO members nonetheless retained the bedrock TRIPS obligation to provide product patent protection for pharmaceuticals.

Accordingly, following a series of incremental amendments to the Patents Act of 1970 as required under the TRIPS Agreement, India had little choice but to proceed with the reintroduction of product patent protection for medicines effective 1 January 2005. And in fact, something unexpected happened in the meantime – led by the Council for Science and Industrial Research (CSIR) Director General Dr R. A. Mashelkar, India's vast science class awakened to the benefits of patent protection for India.[27]

Although there may have an emerging consensus on long term benefit of IP reform for India, the devil was in the detail, and the end result of the Third Amendments to the Patents Act of 1970 left much to be desired. Unfortunately, the grafting of product patent protection on the notably anti-exclusivity slant of the Patents Act (1970) was accompanied by the introduction of additional anti-innovator provisions that did not encourage immediate investment in innovative biopharmaceutical R&D.[28]

10.6   DATA PROTECTION/EXCLUSIVITY FOR COMMERCIALLY VALUABLE CLINICAL DATA

Moreover, apart from the near-universal recognition that the WTO TRIPS Agreement would require India to reintroduce product patent protection for medicines by January 1, 2005, India has yet to recognize and implement intellectual property protections for commercially valuable regulatory data, also known as Data Protection/Exclusivity or Regulatory Data Protection (RDP), and protected in the WTO TRIPS Agreement Article 39.3:

10.6.1   TRIPS Section 7: Article 39

1. In the course of ensuring effective protection against unfair competition as provided in Article 10*bis* of the Paris Convention (1967), Members shall protect undisclosed information in accordance with paragraph 2 and data submitted to governments or governmental agencies in accordance with paragraph 3.
2. Natural and legal persons shall have the possibility of preventing information lawfully within their control from being disclosed to, acquired by, or used by others without their consent in a manner contrary to honest commercial practices so long as such information:
   (a) is secret in the sense that it is not, as a body or in the precise configuration and assembly of its components, generally known among or readily accessible to persons within the circles that normally deal with the kind of information in question;
   (b) has commercial value because it is secret; and
   (c) has been subject to reasonable steps under the circumstances, by the person lawfully in control of the information, to keep it secret.
3. Members, when requiring, as a condition of approving the marketing of pharmaceutical or of agricultural chemical products which utilize new chemical entities, the submission of undisclosed test or other data, the origination of which involves a considerable effort, shall protect such data against unfair commercial use. In addition, Members shall protect such data against disclosure, except where necessary to protect the public, or unless steps are taken to ensure that the data are protected against unfair commercial use.[29]

While India has been slow to recognize the obligation, over time RDP has become entrenched as an international WTO TRIPS obligation independent of patent protection:

> At the international level, regulatory data protection is governed by the Agreement on Trade-Related Aspects of Intellectual Property Rights (TRIPS) of the World Trade Organization (WTO). Article 39.3 of TRIPS requires WTO members to protect test data submitted to regulatory authorities against unfair commercial use and disclosure, except when the public interest so requires or when the data is otherwise protected against unfair commercial use.[30]

India to date has not recognized its TRIPS Article 39.3 obligations and has failed to introduce any formal system of data protection as described above in TRIPS Article 39.3.

Why is this so important for nanomedicine commercialization? While companies in less highly regulated areas may protect data generated in the course of experimentation and product development as trade secrets, this is not permissible in the case of nanomedicines, where regulatory agencies have a compelling public interest in the examination and independent evaluation of regulatory data. Accordingly, scientists and entrepreneurs seeking to bring nanomedicines to market are required to disclose confidential and commercially valuable data to certify the safety and efficacy of new products.[31] Particularly for next generation medicines like nanomedicines, RDP represent a critical form of IP protection, independent of patent, where RDP provides an additional incentive for commercialization and launch via an exclusivity period following the launch of a new product requiring any substantial effort. (RDP runs concurrently to the patent term, eliminating any concerns about evergreening of patent rights through data exclusivity period.)

This much debated issue in India stalled following completion of a much anticipated, lengthy yet unpersuasive report, known as the Satwant Reddy Data Protection Committee Report[32] issued May 2007. Unfortunately the long-awaited Satwant Reddy Report failed to apprehend the importance and value of data protection for biotechnology growth in India, lacking conviction on either side of the issue, and instead providing three separate standards for data exclusivity.[33] With no actionable conclusions, the Satwant Reddy Report fell on its own weight. Over the intervening years, the lack of RDP in India remains as an unresolved irritant that some blame for underinvestment in clinical research and underperformance in biosimilars.[34]

10.7   TAKING SOUNDINGS OF MODI GOVERNMENT'S IP REFORMS

Since taking office in mid-2014, Prime Minister Narendra Modi has sent positive signals with regard to IP policy and administration, including the 2016 issuance of the first ever comprehensive National IP Policy and establishing a dedicated office for national implementation (the Cell for IPR Promotion and Management).[35] The primary focus of these activities is education and awareness building – a very important function that has long been neglected. Without effective public outreach on the importance of effective IP protection for innovative and R&D intensive enterprises, there can be no expectation of meaningful IP reforms. So, these early Modi government IP initiatives are necessary if not sufficient.

In addition the Modi government has made focused efforts to improve IP administration, with measurable gains in terms of reducing patent and trademark application backlogs and improving overall productivity:

> In 2016, the Indian Patent Office (IPO) hired 458 examiners to address the issue of patent and trademark backlog. In 2017, the Patent Rules and the Trademark Rules were revised, to include strict timelines to dispose of cases and streamline examination. Special discounts for filing and an expedited examination for start-ups was introduced. With the hiring of these examiners and these initiatives, the wait-times at the Indian Patent Office were reduced.[36],[37]

The Start-Up India initiative also includes IP support, and has improved access to IP protection at an early stage for eligible companies.[38]

Nonetheless, despite taking a pro-innovation IP policy stance domestically, the Modi government has not mitigated India's ongoing engagement at the multilateral level in the WTO and other fora in ways that would harm the global innovation ecosystem. The Foreign Ministry appears to operate wholly independently and in a time warp, without cognizance of the administration's National IP Policy, Start-Up India or any other innovation priorities or goals.

There have also been other worrisome developments, including the effective weakening of patent rights by the Ministry of Health and Family Welfare (MoHFW) which in April 2017 summarily eliminated the requirement to report the patent status of a health care product, when filing for a manufacturing permit.[39] This effectively undermines the stated intentions of the National IP Policy by reducing coordination and the de-linkage of manufacturing approval and patent status creates the untenable situation where one government department undermines the actions of another, i.e. by the MoHFW approving the manufacture of pharmaceutical products during the period of patent exclusivity granted by the Ministry of Commerce.[40]

Until now the Modi Government has made no meaningful progress towards reform of the Patents Act of 1970, as amended, and also has left unresolved the important issue of Regulatory Data Protection (RDP) as outlined above. There may be cause for some optimism, however, in the recent establishment of a Patent Prosecution Highway (PPH) Agreement between India and Japan that may streamline the patent prosecution process through harmonization of standards. More specifically, through earlier examination by the Japanese Patent Office, a patent applicant meeting patentability requirements in Japan may also receive approval of the application in India.[41] Given that the PPH has only recently been announced by the governments of India and Japan, it may take some time to see its full impact, and/or even if it can withstand inevitable domestic challenges.

10.8   CONCLUSION

While the Modi government, in office from May 2014, has done more in the last five years to promote an enabling environment for next-generation nanomedicines than any previous administration, these efforts have focused in the main on improving public awareness, expanding access to IP services for R&D-intensive startups, and improving patent and trademark administration. It remains to be seen how the GOI will take up the critical challenges of meaningful patent reform, provision of Regulatory Data Protection (RDP) and (reinstatement of) Patent Linkage.

Former CSIR DG Dr R. A. Mashelkar, an early and persistent advocate for India's adoption of effective IP protection, has compared India's past efforts at IP reforms to a series of 'missed buses', yet he also expressed confidence that India would ultimately take the right steps to adopt effective IP policies needed for preeminence in the global knowledge economy.[42] Indian nanomedicine scientists and entrepreneurs deserve a fully modern patent law, together with complementary RDP that provides full rewards and incentives to bring new and useful products to patients in India and around the world.

NOTES

1. Kirthi Jayakumar,"Patenting nanotechnology – the challenges posed to the Indian patent regime," 3 *India Law Journal* 2 (2007), https://www.indialawjournal.org/archives/volume3/issue_2/article_by_ki rhti.html (accessed 10 December 2019).
2. WIPO Economic Research Working Paper No. 29, "Economic growth and breakthrough innovations: A case study of nanotechnology," Lisa Larrimore Ouellette (Nov 2015). (The nanotechnology umbrella also covers many developments in biotechnology and medicine. The biomolecular world operates on the nanoscale: DNA has a diameter of about two nanometers, and many proteins are around ten nanometers in size. Scientists have engineered these biomolecules and other nanomaterials for biological diagnostics and therapeutics, such as for targeted drug delivery for cancer treatment. As of 2013, a few hundred nano-related medical therapies had been approved or had entered clinical trials in the United States.")
3. "Nanoscale" relates to a size measured in microns or nanometers, i.e. a billionth of a meter.

4. "Global nanomedicines outlook and forward-looking opportunities for healthcare," https://www.grandviewresearch.com/press-release/global-nanomedicine-market (accessed 10 December 2019).

5. WIPO Economic Research Working Paper No. 29, "Economic growth and breakthrough innovations: A case study of nanotechnology," Lisa Larrimore Ouellette (Nov 2015).

6. https://www.nano.gov/you/nanotechnology-benefits (See generally; accessed 10 December 2019).

7. Ibid.

8. P. Bhatia et al. "Nanomedicine innovation ecosystem in India," https://doi.org/10.1515/ntrev-2017-0196. (Providing a comprehensive review of nanomedicine R&D in India as relates to drugs, delivery systems, medical devices (including implants), diagnostics (including biosensing and imaging), biomaterials, cosmaceutical, and/or any combination thereof.)

9. Ibid.

10. Ibid.

11. Ibid.

12. Ibid.

13. Ibid.

14. Background and latest Start-Up India initiatives available online at https://www.startupindia.gov.in/ (accessed 10 December 2019).

15. Donald G. McNeil, Jr., "Selling cheap 'generic' drugs, India's copycats irk industry, *N.Y. Times*, 1 Dec 2000, at 1, available at http://www.nytimes.com/library/national/science/health/pharmaceuticals-health.html (accessed Dec 9 2019). ("In fact, India recognizes Western-style intellectual property rights on most products, including computer software, in which it has a thriving industry. But it does not recognize them on chemicals for medicine or agriculture, a position that dates back to its Patents Act of 1970, for which Mr Hamied [managing director of CIPLA, one of India's leading pharmaceutical producers] heavily lobbied Prime Minister Indira Gandhi.")

16. S.K. Finston, "Technology Transfer: Impact and Importance for Indian Biotechnology Growth," *Advances in Biopharmaceutical Technology in India* (Society for Industrial Microbiology, 3929 Old Lee Highway Suite 92A Fairfax, VA 22030-2421, USA, January 2008), p. 558.

17. As a junior U.S. foreign service officer assigned to the Department of State's agricultural trade office, I recall being taken by surprise at India's unexpected unilateral liberalization, not understanding the context of India's balance-of-payments crisis (June 1991), and then Finance Minister Monmohan Singh's strategic trade liberalization strategy.

18. From October 1999 to June 2005 I led the international industry coalition for pharmaceutical patent protection in India in the course of my work for the Pharmaceutical Research & Manufacturers of America (PhRMA). The first time that I visited India in May 2000 as the representative of the U.S. Research-based pharmaceutical industry, I joked that I needed a 10-foot ladder and a shovel – a ladder to be able to see above the hole that I found myself in, and a shovel to try to fill in the deep hole. That was the nature of the relationship between Indian interlocutors and the American research-based pharmaceutical industry at that time.

19. S. Finston, "India: A Cautionary Tale on the Critical Importance of IP Protection," 12 *Fordham Intell. Prop. Media & Ent. L.J.* 887 (2001–2002).

20. On 26 December 2004 while the massive Tsunami wave swept across South Asia, the Indian Cabinet reintroduced product patent protection for pharmaceuticals via Ordinance, e.g. The Patents (Amendment) Ordinance, 2004 with effect from 1 January 2005. This Ordinance was later replaced by the Patents (Amendment) Act 2005 (Act 15 of 2005).

21. WTO TRIPS Agreement, Article 27 – Patentable Subject Matter, https://www.wto.org/english/docs_e/legal_e/27-trips_04c_e.htm (accessed 10 December 2019).

22. F.M. Abbott, "TRIPS in Seattle: The not-so-surprising failure and the future of the TRIPS agenda," 18 *Berkeley Journal of International Law* (2000), p. 167.

23. Lynsi Burton, "WTO riots in Seattle: 15 years ago," *Seattle PI*, 2014, https://www.seattlepi.com/seattlenews/article/WTO-riots-in-Seattle-15-years-ago-5915088.php (accessed 9 December 2019).

24. J. Watal (2000). *Access to Essential Medicines in Developing Countries: Does the WTO TRIPS Agreement Hinder It?* Science, Technology and Innovation Discussion Paper No. 8, Center for International Development, Harvard University, Cambridge, MA. Available online at: https://www.wto.org › res_e › booksp_e › trips_agree_e › chapter_16_e (accessed 10 December 2019). ("The case of HIV/AIDS in developing countries has focused attention on patents and prices. An estimated 95% of those suffering from this disease live in developing countries where the disease is showing no signs of abatement. With over half of such persons belonging to the most productive age of below 25 years, this disease is causing serious social and economic consequences. Being a relatively new disease, many of the medicines are still under 'live' patent protection, with expiry dates of a decade or more.")

25. Declaration on the TRIPS agreement and public health, DOHA WTO MINISTERIAL 2001: TRIPS WT/MIN(01)/DEC/2 (20 November 2001) https://www.wto.org/english/thewto_e/minist_e/min01_e/m indecl_trips_e.htm (accessed 9 December 2019).

26. India, together with Brazil and South Africa, spearheaded the effort at the Doha ministerial meetings to recognize the affordability and availability of medicines as a universal right. See Arvind Panagariya, "India at Doha: Retrospect and prospect." *Economic and Political Weekly*, vol. 37, no. 4, 2002, pp. 279–284. JSTOR, www.jstor.org/stable/4411648. (accessed December 10, 2019). On the heels of its achievement at Doha, India passed the Second Patent Amendments (May 2002 amendments to the Patents Act 1970.2) which, with respect to the definition of invention, patent term, and scope of patentability reiterate the provisions of the TRIPS Agreement. However, taking advantage of the flexibility provided in the Doha Declaration, the 2002 amendments also provide wide compulsory licensing provisions which arguably undermine the existence of exclusive patent rights altogether. Importantly, the amendments have raised the concern among global companies that they will be entitled to patent rights only if they manufacture drugs and medicines in India if importation is found not to amount to "working" of the patent invention.

27. See, e.g., Navi Radjou, "One man's crusade to overhaul India's insular R&D culture," *Harvard Business Review*, 15 July 2008. https://hbr.org/2008/07/one-mans-crusade-to-overhaul-i (accessed December 10, 2019). (Firstly, he recognized the importance of transforming CSIR's amorphous intellectual knowledge into patented intellectual property (IP), which could serve as currency for negotiation with corporate partners. So, Dr Mashelkar changed CSIR researchers' cultural ethos from "publish or perish" into "patent, publish, and prosper." As a result of this shift, the number of patents CSIR received skyrocketed from merely 8 in 1995 to 196 in 2003. Secondly, even though national labs like CSIR account for a massive 80% of total R&D spend in India, Dr Mashelkar wanted taxpayers' money put to hard work. Eager to turn CSIR into a self-funding research institution, he multiplied collaborative projects with the industry, and opened up CSIR labs to international cooperation – two bold acts deemed sacrilegious by the insular Indian scientific community. Yet, the moves paid off. During Dr Mashekar's tenure, the amount CSIR earns from conducting contract research for global corporations like General Electric doubled, reaching $1.26 billion in 2004. Jack Welch was so impressed by the high caliber of CSIR scientists that he eventually opened GE's own R&D lab in Bangalore, the largest ever set up by a multinational in India.) (accessed 10 December 2019).

28. S.K. Finston, "India Brief 2:Technology Transfer: Impact and Importance for Indian Biotechnology Growth," *Advances in Biopharmaceutical Technology in India* (Society for Industrial Microbiology, 3929 Old Lee Highway Suite 92A Fairfax, VA 22030-2421, USA, January 2008) p 558 (Advances in Biopharmaceutical Technology in India (January 2008) ("Although both the Congress-led Union Government and the previous BJP-led Administration expressed bipartisan support for bringing in 2005 TRIPS legislation, the political process limited reforms, and in fact added or left intact TRIPS inconsistent provisions. In the former category I would include the preservation of pre-grant opposition provisions that are out of step with patent norms, expansive compulsory licensing provisions,26 as well as provisions added in the Second Patent Amendments in 2002 which provide for mandatory disclosure of source and origin of genetic resources as a condition for patentability.27 The latter is important to eliminate as it provides a disincentive for investments in clinical research relating to Ayurvedic and other traditional medicines. In addition, the law added or left intact other anomalies including limitations on the Mailbox, local working requirements and compulsory licensing provisions that go beyond inter- national norms.

Discussion of special patent requirements for biotechnology inventions relating to genetic resources and traditional knowledge is included here because of its long-term importance to the biotechnology sector in India. While India has become a top-five global destination for investment in all phases of pharmaceutical development and growth of lower-level biotechnology (e.g. vaccines, monoclonal antibodies, etc.) progresses very well, India lacks an enabling environment and incentives for investment in traditional medicines, other GR/TK inventions and related advanced agricultural and medical technologies. This is due in large part to lack of clarity for patents in these areas. Under current law, there are special additional patent conditions that natural products inventions must meet as relates to mandatory disclosure of source and origin for genetic resources and related traditional knowledge, and there are also additional revocation provisions that weaken clarity of rights after grant.28

Even while the Government of India is trying to establish a special pathway to ensure clinical development of traditional medicines,29 the lack of assured patent rights for these products has scared away investment from this promising field, particularly foreign direct investment. The policies that informed the drafting of these provisions pre-date India's biotechnology prowess and would benefit from review and recalibration to eliminate dilution of patent rights that hinder growth in India's traditional medicines sector.")

29. World Trade Organization (WTO) Uruguay Round Agreements, Agreement on Trade Related Aspects of Intellectual Property Rights (TRIPS), Section 7 Protection of Undisclosed Data Article 39, https://www.wto.org/english/docs_e/legal_e/27-trips_04d_e.htm (accessed 10 December 2019).

30. J. Ellis, "Supporting Innovation in Next-Generation Medicines," *WIPO Magazine*, June 2017, https://www.wipo.int/wipo_magazine/en/2017/03/article_0007.html (accessed 10 December 10 2019).

31. Ibid. ("Regulatory authorities require data from preclinical and clinical trials to be able to approve and certify that a pharmaceutical technology is safe and effective for consumer use before market entry. Clinical trials are painstaking and expensive and add significantly to the cost of developing a new medicine, estimates for which range from USD 1.2 billion (Office of Health Economics, United Kingdom) to USD 2.6 billion (Tufts University, United States).")

32. Satwant Reddy, Report on Steps to be taken by Government of India in the context of Data Protection Provisions of Article 39.3 of TRIPS Agreement, Department of Chemicals & Petrochemicals, Ministry of Chemicals & Fertilizers Government of India (31 May 2007), https://chemicals.nic.in › sites › default › files › DPBooklet (accessed 10 December 10 2019).

33. In the case of Western medicine, there is not only a lesser standard of non-disclosure, but an additional transition period and a possible recommendation (unclear) of some form of non-reliance at some undefined future time.

34. "Poor reputation and lack of regulatory framework hindering Indian biosimilars potential," *The PharmaLetter*, Putney, London, 18 July 2019, https://www.thepharmaletter.com/article/poor-reputation-and-lack-of-regulatory-framework-hindering-indian-biosimilars-potential (accessed 10 December 10 2019).

35. "India: Challenges Faced in the Protection and Enforcement of Patent Rights," 18 September 2019 https://www.managingip.com/Article/3894816/India-Challenges-faced-in-the-protection-and-enforcement-of-patent-rights.html (accessed December 10, 2019).

36. Progress cited in 2017–2018 in reducing patent and trademark backlogs is attributed to the hiring of 458 patent and trademark examiners in 2016. India – Protecting Intellectual Property, Export.gov https://www.export.gov/article?id=India-Protecting-Intellectual-Property (accessed 10 December 2018).

37. Also, during the 2019 election, the Modi Campaign reportedly advertised ongoing improvements in IP administration, increases in electronic filings and related achievements; history – most likely the first time in India that improved IP administration rose to the level of a positive election issue. "Assessing Modi's impact on India's IP landscape," *Managing IP*, 24 July 2019. https://www.managingip.com/Article/3885329/India-Assessing-Modis-impact-on-Indias-IP-landscape.html (accessed 10 December 2019)

38. See generally https://www.startupindia.gov.in/ (accessed 10 December 2019).

39. Ibid.

40. The MoHFW appeared to have acted on its own initiative without coordination with other GOI Departments and without meaningful consultation or prior notice to biopharmaceutical stakeholders.

41. See IPO JPO PPH PROCEDURE GUIDELINES FOR PATENT PROSECUTION HIGHWAY (PPH). Available for download at http://www.ipindia.nic.in/newsdetail.htm?593, 4 December 2019 (accessed 10 December 2019).

42. "Dr Mashelkar has often been called a 'dangerous optimist'; so it was no surprise when he suggested that instead of being rueful of the 'missed bus' we should work on the opportunities offered by the waiting buses of the knowledge economy." *Information Pasteboard*, #IP 428/20–26 November 2000. www.nal.res.in/oldhome/pages/ipnov00.htm.

11 Minimizing Negative Long-Term Occurrences in Nanotechnology Development

Walt Trybula, Deb Newberry, Dominick Fazarro, and Craig Hanks

CONTENTS

11.1 INTRODUCTION

While the focus of this book is to address various intellectual property issues of nanotechnology, the unaddressed question in many approaches to nanotechnology development is the responsibilities to both people and the environment in developing and applying novel materials and products. Responsible development and use of nanomaterials will not only need to comply with regulatory requirements, but because the materials are at the cutting-edge of human knowledge and capacity, will need to go beyond existing regulations and laws to anticipate possible harms and negative consequences (ethical, environmental, health, and so on).[1] Beyond the necessity of considering the impacts of nanotechnology, developers and users of nanomaterials have an obligation to protect the future of nanotechnology. Irresponsible use of nanomaterials could lead to a backlash. But it is not only irresponsible use that must be avoided.

Today, we live in a litigious environment. It is necessary to ensure that every possible action is taken to mitigate the possibility of punitive judgments. Even with all appropriate precautions, there is no guarantee that legal action might not be taken for some unknown supposed infraction. There are two cases that stand out. The first is the suit against Johnson & Johnson (J&J) regarding its talcum powder (talc), which contains nanomaterials.[2] There are areas when talc can be obtained that has some inclusion of asbestos. Other locations provide asbestos-free talc. J&J has certified that it has been using the pure talc for its products since before the 1960s. There was a judgment in the initial case granted against J&J for the potential of its product causing cancer in the claimant. This award was made acknowledging that J&J had not used the talc possibly containing asbestos and the more surprising fact, that the claimant did not know if she had used the J&J product or someone else's product. (Challenging matters for J&J is a fall 2019 news release that J&J is recalling a baby product that might contain tainted talc. Full details have not been released as this chapter goes into print.)

The second case involves the Bayer AG's glyphosate-based Roundup weed killer. The case in question[3] resulted in a USD 2 billion award for the claimed cause of cancer in the two claimants. This jury award was made without the defendants being able to present the fact that the United States Environmental Protection Agency had concluded that there were no risks to public health from the current registered uses of glyphosate. Other studies determined that glyphosate was not carcinogenic to humans. These lawsuits are driven by the World Health Organization's cancer operation that labeled glyphosate as probably carcinogenic to humans.

Complicating the issue of "scientific failings" is the fact that research results have come under serious scrutiny. It has risen to the point that numerous articles have highlighted the fact "scientific mischief" occurs, and a weblog called "Retraction Watch" chronicles research and publication misconduct. Major schools have repaid USD 100s of millions for unverifiable research. In one study, 82% of research results published in peer review journals could not be reproduced by experts including, in some cases, the original researchers. These cases contribute to questions about the process and outcomes of scientific and technical research and development. The Office of Research Integrity at the U.S. Department of Health & Human Services[4] defines three different types of research misconduct. "Fabrication" is making up data or results and recording or reporting them. "Falsification" is manipulating research materials, equipment, or processes, or changing or omitting data or results such that the research is not accurately represented in the research record. "Plagiarism" is the appropriation of another person's ideas, processes, results, or words without giving appropriate credit. When scientists and engineers commit research misconduct, they harm not only themselves, but the scientific and technical research community.

In the early 2000s, there was significant activity in nanotoxicity with various companies providing services for evaluating the toxicity of selected materials. The issue that arises is that the evaluation of each material needs to have a full, lengthy, and expensive test. In an evaluation of possible nanomaterials and combinations that might need to be tested, the surprising result is that between 10^{200} and 10^{800} separate tests would be required.[5] If there had been one material quantified each second since the beginning of the universe, the total materials identified would be less than 10^{18}.

Consequently, there exists a situation today where nanomaterials are being developed that have properties that are not fully understood. Over the lifetime of the material usage, the potential impact on people and their health is not known with certainty, and the long-term impact on the environment is not truly understood. Add to this equation that among non-scientists and some public leaders there is a reluctance to believe a scientific result if someone starts a social media trend questioning the accuracy of the results. A striking, and troubling example, is the continuing unjustified skepticism about the safety and efficacy of vaccines.

This raises the question of what steps need to be taken to provide the best possible assurance and that every effort is taken to minimize the risks. Obviously, not all these issues can be addressed to everyone's satisfaction. What is required is to take steps that provide a trail of all activities taken to minimize the risk to people and the environment. This is where the worldwide effort in nano-safety is directed. The first step is to develop an appreciation of why nanomaterials are "different", and thus require special care, processes, and attention.

The advances in technology and the resultant tools have provided the ability to observe, manipulate, and control material at dimensions that a few years ago were unimaginable. The word "*nano*", which is normally bandied about in the references to novel material properties, is expressed as 10^{-9} meters. It is very difficult to envision the scale of a number of orders of magnitude. Another way to envision the size shrinkage is to consider that one nanometer is one-billionth of a meter. One could compare one U.S. dollar to one billion U.S. dollars. The difference in buying power is obvious. However, the thickness of a single bill is 0.0043 inches. One billion U.S. dollar bills stacked on top of each other would reach almost 68 miles. This difference between a nanometer and a meter is not as obvious without some means of comparing it to something that can be visualized.

Working at this small scale we are learning many things about the properties of materials that we previously did not have capabilities to examine. As civilization developed, people learned about the characteristics of materials based on the bulk properties and applied this knowledge to create new devices, tools, etc. Now, the capability exists to observe and manipulate materials down to the sizes that consist of a small number of atoms. Individual atoms are of a size that is below one nanometer, but small clusters of these atoms are in the single digit nanometer range. One of the consequences of the smaller dimensions of materials is that the percentage of atoms that are exposed to the surface of the material becomes a significant fraction of the total number of atoms. There have been ongoing efforts to characterize the properties of the various materials. The explosive difference between 80 nm aluminum particles and 25 nm aluminum particles is but one example.

11.1.1 SAFETY

Paramount in the workplace and the environment, safety is especially important for developments in nanotechnology, in which case material properties may be completely unknown. Nanotechnology Safety (NANO-SAFETY) requires many things: knowledge of effects, understanding of particle behavior, toxic effects depending on the application, residual impact on the environment, etc. The areas of understanding can be characterized in a number of categories, which include: 1) Material Properties; 2) Impact on People and the Environment; 3) Handling of Nanomaterials; and 4) Business Focus.[6]

1) Material Properties: Obviously, one aspect is the development of an understanding of the properties of all the materials, a situation complicated by the lack of availability of metrology tools. A more fundamental question is what properties should be investigated. If the starting point is to ensure an understanding of the impact of the nanomaterials on people and the environment, then investigations can be focused on material properties within the expected operational parameters, like room temperature, atmospheric pressures, etc. Understanding the properties is necessary before it is possible to understand their impact.

2) Impact on People and the Environment: Another aspect is a greater understanding of the impact of nanomaterials on the human body. For instance, significant advances are being made in the treatment of cancer by employing customized molecules that incorporate nanoparticles and deliver them to cancerous sites. These specialized molecules either deliver specific chemicals or other materials like gold or carbon nanotubes to the site requiring treatment. The chemicals will react with the cancer and begin destroying it. The other materials can be heated by many different methods and destroy the cancerous cells through the elevated temperatures. These approaches promise significant advances in the treatment of diseases; however, the long-term impact on the body is under investigation and no definitive answers exist.

3) Handling of Nanomaterials: The question of handling and storing nanomaterials is important from both the implications for the people involved and the impact on the environment.

Yet, without any knowledge of the basic properties, the extent of the precautions required is unknown. One always wants to err on the side of safety, but potentially onerous procedures, based on worst-case scenarios, will diminish the progress being made in applying nanotechnology to everyday problems. Procedures are required based on fundamental evaluations and historical efforts.

4) Business Focus: The business aspect is critical. "Given the fact that businesses need to protect their workers as well as their corporate liability, they need to operate according to established guidelines. These guidelines do not exist! Consequently, there is the potential for significant corporate liability."[7]

11.1.1.1 Material Characterization

Nanomaterials are not new. They have existed in nature all along. What is new is our ability to control the manufacture of these materials and to create novel combinations of atoms in materials that have specialized properties. Historically, material properties are those physical characteristics of a material that are considered invariant. Some of these characteristics include, in alphabetical order: boiling point, chemical potential, color, density, ductility, elasticity, electrical (conductivity and resistivity), hardness, melting point, specific heat capacity, and viscosity. The entire list is a function of the state of the material, whether gas, liquid, or solid.

What we know today is that when the size of the material in the lower double-digit nanometer range, there are other properties that dominate the characteristics of the nanomaterials. In this region, the number of atoms that are available to the surface becomes a significant fraction of the total number of atoms and the reactivity increases. Quantum effects can impact both electrical and optical characteristics. Crystal orientation can have an impact on the resultant properties.

If these were not enough additional considerations, there are new types of materials that can be produced in volume. Carbon nanotubes, which are a form of the carbon fullerene structure family and were first produced in quantity at Rice University, occur only rarely in nature. This nanotube structure is a carbon allotrope that is a cylindrical molecule with electrical properties that are based on the orientation of the atomic structure (chirality). While graphite is the most common form of carbon, it has very different characteristics at atomic dimensions. A single atomic layer of this material is called graphene and has extremely interesting electrical properties. A double atomic layer of graphene has different properties from the single layer. The list of novel properties and "interesting" material is constantly expanding.

11.1.1.2 Nanomaterial Properties

Current regulatory and legislative efforts try to provide specific guidelines for which materials must be considered "nano" and which do not. In most cases only "size" counts and that size is usually set as 100 nm. The potential issue from these approaches to regulation is that both over-regulation of unimportant elements and under-regulation of critical items will occur.

It is important to understand that the changes in the various properties of nanomaterials are significantly different from the bulk material properties. The following is a partial list of material properties that need to be considered:

- **Size**: As discussed earlier, aluminum nanoparticles are highly reactive in the range of 30 nm. These particles are employed in certain types of rocket fuel. So, size makes a difference.
- **Shape:** A recent report has shown that carbon nanotubes do not have a negative impact on cells. However, a similar carbon nanotube particle, with the exception of its being unrolled, has been shown to have a detrimental effort on the cells.[8]
- **Thickness:** Graphene, sometimes referred to as a two-dimensional material, has significant electrical properties and is being considered for future transistor work. The properties change with the addition of each layer through five layers where the bulk properties start to dominate. Add hydrogen to the graphene matrix and the material becomes an insulator.

- **Temperature:** The melting point of gold is 1,064.18°C and this has been known for centuries. Work on gold nanoparticles[9] has shown that the melting point of gold changes significantly when particle size approaches 15 nm and continues to decrease as the size decreases.
- **Electrical Properties:** The semiconductor industry has found that the conductivity of nanoscale copper is dependent on the crystal orientation and the grain boundaries as the size of this conductor approaches 40 nm. Gold changes from a conductor to a semiconductor below 1 nm.[10]
- **Color:** Gold is known as a yellow material, but if it is put in suspension, the color will change dependent on the size of the particles in solution. The red color in stained glass windows is dependent on the particular size of gold nanoparticle introduced into the glass during manufacturing. This characteristic was employed during the Middle Ages.
- **Duration:** A recent study found that time can alter the properties of materials. Testing the same material at different locations, which involved shipping time, resulted in different results after removing the equipment variables from the equation.[11]
- **Equipment:** Equipment may deteriorate over time and there needs to be a common means of measuring. But this is not sufficient, because the previously referenced study showed that different designs of sonicators produced different results.[12] Another report from Georgia Tech found that graphene oxide actually changes over the first month after manufacture due to interaction with hydrogen.[13]

One frequently overlooked characteristic is that of "aggregation". Aggregation has been classified as mutual attraction, caused by van der Waals forces, between particles. This attraction between the particles creates significant challenges to accurately evaluate the material properties of the particles being investigated. Some of the manufacturing processes may promote the aggregation of the particles. Subsequent evaluations will have the bias of evaluating the aggregation and not the material per se. Adding to the challenges is the fact that equipment for testing and evaluation of the nanomaterials, while leading-edge, is still not fully capable of providing the precision required for adequate control.

Note: For adequate control of dimensions, the equipment must be able to measure at least an order of magnitude smaller than the minimum particle dimension. Ideally, measurements should be made using equipment capable of resolving two orders of magnitude smaller than what is being measured.

11.1.1.3   Potential Hazards

When people discuss the issues regarding nanotechnology safety, they will probably be focusing on the issue of nanotoxicity. Unfortunately, in a number of cases, the time required to determine toxicity effects is of at least a moderate duration. In many cases, searches for effects of various nanomaterials result in no information. Lack of information does not negate responsibility.

Another aspect of the risk assessment concern is that people are quick to "blame" new technologies for the cause of incidents. There were newspaper reports in August 2009 indicating that two Chinese women had "the dubious honor of being the first humans to be killed by nanotechnology".[14] There was considerable media coverage about the fact these two women died and five others were permanently disabled due to "nanoparticles". A more detailed report indicated that the women were working in a "70 square meter workshop where they put about six kilos of polyacrylic paste into a pressure sprayer and sprayed it on polystyrene boards or glass". The workshop had a door but no windows, and for five months before the women became ill, the sprayer's vent was inoperative, and the door kept closed to conserve heat.[15] Autopsies performed found nanoparticles encased in epoxy in the victims' lungs. Since the acrylate used is a low toxicity adhesive, the cause had to be the nanoparticles. A follow up to the study's results indicated that some toxicologists have stated that if normal safety precautions had been followed, there would have been no health issues.[16] The point made is that while working with unknowns, it is important to take precautions.

Having established that there are dangers in working with materials that may contain nanoparticles, it is advantageous to examine a number of potentially dangerous conditions involving nanoparticles. The list of issues and dangers are numerous and would definitely include toxicity issues, chemical dangers, fire, explosion, dust, electrostatic, size, shape, volume, density, and concentration.

The issue that appears most often is that of toxicity. Is the nanomaterial toxic or not? In most cases, the answer will be "we do not yet know". A report on the impact of nanoparticles on health evaluated three particle types, titanium dioxide, carbon black, and diesel particles, using a rat model due to the sensitivity of the species. Negative effects were found that could eventually provide tumor growth.[17] However, carbon black is known as a problem in larger than nano sizes. It is a known danger and these types of materials need to be handled with great care.

Chemical dangers are numerous, and a significant amount of safety precautions are available for existing "macro" sizes of the materials. Therefore, in many cases, it is not only the "nano" size aspect that can result in toxicity issues. The additional issue that arises with nanomaterials is when the size of the particle permits a large percentage of the total number of atoms to be near the surface so as to precipitate a reaction. Aluminum is an example of this condition. As mentioned earlier, 80 nm aluminum particles are relatively inert, reaction-wise, when exposed to air/oxygen. However, at the smaller sizes, the particles become very reactive and will cause explosions when exposed to air (oxygen is needed for the reaction).

Fire is always an issue. A part of the problem is that fire burns material and causes the burned particles to float in the air. The hazard can be double because these airborne particles can be inhaled and cause problems and may be able to create a chemical reaction.

Explosions are a concern if the material is very reactive at small dimensions. The other situation is that when material that is known to be capable of causing explosions is used in increasingly small devices, the barriers for the prevention of unwanted occurrences as devices become smaller/thinner, and failures can more readily occur.

Dust is a problem if the entity is light enough to float in the air. Floating particles can travel where air currents move them. This means the particles can enter the air supply that workers are breathing. Often overlooked, the size of the nanoparticles makes them very capable of electrostatically clinging to garments, equipment, cleaning items, etc. Careful control of the material, the equipment employed, and the work protection clothing is needed to prevent accidental contamination of people or other work areas.

Nanomaterial size enters into consideration due to the fact that material properties may change with size. There are concerns by medical researchers that below a certain size, specific nanoparticles are dangerous because they may readily enter the body. The shape of the particles causes concerns. Carbon nanotubes (CNT) are shaped like needles and typically are less than 5 micrometers long. Asbestos has primarily the same shape and is known to be dangerous to human health if ingested. Even though the asbestos is much longer, the similarity of shape has caused some concerns about the application of CNT in products. The volume of the material and the ability to control it is also a concern. Silver nanoparticles have been shown to promote healing and reduce infection if applied to wound dressings. The issue is that if the same silver nanoparticles are released into the environment, they can be potentially highly toxic to bacteria and very small life forms. The concentration of the material is important because there are points above which materials become dangerous.

Another consideration is the area when materials can be absorbed into the food chain. The recent issue of arsenic in the food supply raised a lot of questions.[18] The rice plant acquires minute amounts of various elements from its growing environment. The primary element it needs is silicon to provide the smooth structure. The danger is that arsenic is close enough chemically to silicon that the rice plants also take up arsenic. The absorption of the arsenic by rice is in its more toxic inorganic form rather than the organic (carbon-based) form. The inorganic form bonds into living cells and destroys them. This dependency on the form of the nano entity and chemical structure applies not only to arsenic and carbon. For example, quantum dots, some of which contain cadmium in combination with one or more other elements, are relatively benign. However, there has been some preliminary research that shows after a period of time in the environment, the materials may separate. This has the potential of leaving cadmium in its elemental form in the environment.

The potential for some significant issues to arise is always a possibility. Careful control of the material is required to ensure that risks to people and the environment are minimized.

11.1.2 Nanotechnology Safety (Nano-Safety)

In the early 2000s, there was significant activity in the field of nanotoxicity with various companies providing services for evaluating the toxicity of selected materials. The issue that arises is that the evaluation of each material needs to have a full, lengthy, and expensive test. It was obvious that it would be impossible to test for every possible condition and size of material. An example: is 30 nm aluminum the same as 40 nm aluminum? The answer, based on reactivity, is "NO". Is a distribution of 35 nm aluminum particles with a half-width of 1 nm the same as a distribution with a half-width of 10 nm? Again, the answer is "NO". This concern led to the 2007 release of a Nano-Safety white paper addressing the initial steps that needed to be taken. This effort was followed a couple of years later by an Occupational Health and Safety Administration (OHSA) funded nanotechnology eight-hour seminar.[19]

The white paper raises the question of what steps need to be taken to provide the best possible assurance that every effort is taken to minimize the risks. This would not be the first time that technology has been developed that contains unknown elements of possible danger. The development of nuclear energy in the U.S. Manhattan Project faced such challenges. Barbara Foster covered that topic in the first chapter of the Nano-Safety textbook.[20] She points out that she is living proof that her father, who managed the safety effort, did a good job of preventing radiation damage to humans.

The key element is to protect the organization along with the general population and the environment. This requires having a qualified workforce. Safety does not happen by accident. Lack of safety can provide many opportunities for legal challenges for any organization. Proper understanding of the requirements for safety in the nanotechnology realm is critical for the successful deployment of any items incorporating nanomaterials. This must be done while recognizing the fact that the vast majority of materials that are or will be employed will have unknown impacts. Consequently, some of the efforts will have to take an approach employed by firefighters. That is, to classify the type of situation, to identify the potential hazardous situation, and to employ the appropriate mitigation methods.

The process of developing mitigation techniques is a multi-step process. The first step is the identification of potential risks. This involves developing an overall assessment of the proposed application of the nanoparticles, the known issues with the materials at the bulk state, and potentially related findings of related (similar composition) materials. Once specific concerns are identified, an evaluation of the potential severity must be made. At this point, the evaluation of risk/rewards may begin.

The difficult part of the evaluation occurs when a prototype is being developed. What process works for a lab experiment will probably not be cost-effective at high volume manufacturing. A variety of both cost and risk containment options must be allocated for future volume development. Associated issues involve the potential regulations that may be in the country of origin or in the country where the product will be employed.

There are a number of terms that refer to various aspects regarding the evaluation of risk. "*Risk Assessment*" as employed in this document refers to the evaluation of potential conditions that could involve the development of a risk (potential for harm, injury, or worse to people or the environment) based on the application, usage, disposal, or even storage of a nanomaterial. "*Risk Management*" is considered the methodology that is employed to handle the potential situations that could arise from the presence of nanomaterials being improperly contained. "*Risk Mitigation*" refers to the methodology that is employed to resolve issues that have a potential for occurring or have occurred.

11.2   ISSUES REQUIRING NANO-SAFETY

As with any investigation addressing potential hazards and safety, there are fundamental items that must be addressed. This can be considered the who, what, and why of nano-safety. The "who" is not only the people investigating, producing, handling, or developing the nanomaterials, but also the end user, whether the materials are employed in a product or part of a medical treatment. The concern for the environment not only includes the environmental health and safety concerns, but also additional uncontrolled or inadvertent physical and chemical events/processes during the life cycle of the material.

The "what" is a more difficult question to answer. The obvious selection of the "materials" is much too limiting. The difference in handling and controlling 25 nm and 75 nm aluminum nanoparticles is only one example of the need to have a much broader approach. A vial of aluminum nanoparticles needs to be handled carefully due to the fact the fine dust can be inhaled or absorbed through the skin. However, the 25 nm aluminum nanoparticles will create an energetic reaction when exposed to the oxygen in the air. This happens when the vial is opened if the environment is not controlled. The "surprising" properties are also interesting. Both platinum and silver, which are non-magnetic, have a magnetic moment in a 13-atom configuration. Transition metal can exist in five different states: as hydrated atoms; metal complexed in a small protein; metal adsorbed to the surface of a 1 nm mineral particle; metal adsorbed to the surface of a 20 nm particle; and the same except to a 200 nm particle.[21] Metal nanomaterial crystal orientation preferences can be a function of size. CeO_2 < 10 nm habit of truncated octahedron with {100} and {111} faces. CeO_2 > 10 nm shifts toward {111} octahedron.[22] Nanomaterials may be shape-shifters. There is a lot more that we do not know about the nanomaterial properties as a function of size. Consequently, any developed plans must accommodate the unknowns.

The "why" has many more questions and answers. The first goal is the protection of both people and the environment. The focus of understanding the "why" is the development of processes and procedures that will minimize potential legal actions for improper handling or application of nano-materials. There are too many variations in the properties of the nanomaterials as their size changes to be able to evaluate the potential short- and long-term effects. This last statement requires that there be an ethical overview of programs that are developed to create new applications and processes whether for medical cures or for new technological applications.

11.2.1   How Are People Impacted?

In order to consider the means of protecting people, one must understand what people need to be protected from. This includes the possible routes for nanomaterials to enter the person. The three primary means are inhalation, ingestion, and skin absorption.

Airborne nanoparticles can be inhaled and will deposit in the respiratory tract. This can be similar to the issues caused by asbestos. Carbon nanotubes, which are thin cylinders, have been compared to five of the six types of asbestos. (The sixth type of asbestos is curved, and it is more difficult for that type to be inhaled.) The inhaled particles may also enter the bloodstream and move to other organs. The potential hazards from inhalation are that certain nanomaterials can induce cancers including mesothelioma. These particles could be deposited in the respiratory tract and can cause inflammation and damage to lung cells and tissues. Nanofibers may create a situation which can cause pulmonary inflammation and fibrosis. There is also concern that these particles can migrate in the body to cause issues in both the heart and the brain.

Ingestion is not only caused by poor hygiene practices. Hand-to-mouth transfer is definitely a cause of ingestion. As mentioned, it is possible for ingested nanoparticles to migrate to other organs. In the late 2000s, carbon nanotubes inserted into the gut in an experiment for treating Alzheimer's disease were found in the heart, liver, and brain. The hazards of ingestion include nanosilver damage to the liver and causing a slight, but permanent bluish discoloration of the skin. Titanium dioxide and silicon dioxide have caused immune responses in intestinal dendritic cells. Zinc oxide has been shown to damage the DNA of intestinal cells.[23]

Skin absorption or dermal exposure has conflicting data as to the ability of nanoparticles to penetrate the skin. But, this route of contamination has not been ruled out. Studies have shown little penetration of nano-oxides beyond surface layers. However, metal nanoparticles have been shown to penetrate damaged or diseased skin. Quantum dots may penetrate skin at levels that might be prevalent in occupational situations.[24]

There is a long list of nanomaterials including carbon nanotubes, titanium oxide (TiO_2), iron oxide, silver, and quantum dots that have been observed as possible sources of damaging cells. The National Institute for Occupational Safety and Health (NIOSH) has stated that nanoscale TiO_2 particles have a higher mass-based potency than larger particles. Consequently, nanoscale TiO_2 particles are classified as a potential occupational carcinogen.

A fourth type of danger can be created by the nanoparticles. Some nanomaterials can act as chemical catalysts and produce undesired reactions, which can create a risk of explosion and fire. Nanomaterials may float in the air and provide a combustible situation that requires slight amounts of energy to combust. These include both wood and flour dust.

Working with newly created materials may open up areas of considerable unknowns for the long-term effects on both people and the environment.

11.2.2 How to Address the Challenges? – Ethics

The obvious starting point is "first do no harm". In order to move forward developing materials with unknown properties, caution must be predominant in all phases of activities. Decisions made without being able to know the potential impact on people and the environment requires direction. Products are developed and marketed quickly. As mentioned previously, full testing and evaluations could take seven to ten years or more. Having the concept of an ethical approach to the situation will permit some guidance in decision making.

In general, people want to do quality work and make sound decisions. What does it mean to make sound decisions? Are technical knowledge, skill in testing and evaluating results, and a basic understanding of the underlying principles sufficient? How does one maintain a current level of information and still have time to do quality work? This is especially true when working with nano-materials that have applications that cross materials science, biology, engineering, and chemistry. The multidisciplinary expert in all these fields in one person is rare.

The fact that engineering works in new areas does not mean that the people involved understand the full considerations for the application of that technology. Without guiding constraints there will be incidents like the Tuskegee experiments or the recent Boeing 737/MAX8 aircraft. Even the Challenger disaster was an example of usage beyond the anticipated design specifications. Routine applications can be influenced by schedules or budget constraints. Without a directive to include ethical considerations in the ongoing effort, it is possible that people will make choices without fully considering the potential results.

The current state of technology requires introducing novel concepts as quickly as possible with further refinements to be added after the initial release. The application of technology innovation in a rapid manner leads to potential disconnects with ethical considerations. Normal engineering efforts may not put ethical considerations at a very high level. The applications and the possible impact are left to "upper management". The technologist possesses special knowledge. There are historical cases where professionals have purposely not stated all the facts. Physicians in the 1950s and 1960s might not tell patients about their cancer. Part of the reason was the fact that during that time the prognosis for curing the cancer was relatively small. The same line of reasoning could possibly be used by a technologist to not discuss the potentials of inhaling carbon nanotubes because he or she thought the risk of the inhalation was small. This is along the lines of logic where the technologist believes there is a potential benefit from pursuing the work and it justifies the risk while being concerned that others might not continue to work due to the potential risk. In any of these cases not talking about the risks, whether explaining or identifying them, is not an ethical procedure.

There is a significant concern about today's artificial intelligence (AI) programming. The unanswered portion of this work is who is making the decisions on how the AI evaluates the situations. The program is basically a series of choices, which takes the decisions down a particular branch within the program. The expert systems of the 1980s reinforced the decision-making process based on evaluations of the results that were fed back into the program. Depending on where the program was being used would provide localized decisions on which branches to take. An example of this is the selection of a driving route. If the decision is being made in upstate New York and it is snowing, the selection of a route would try to avoid steep hills. The same decision made in a location that treated the roads to mitigate the impact of snow and consequential slippery roads would not avoid that route. As usage would build up over time each location would have its own set of rules. AI is developing a generalized set of rules in an attempt to apply decisions throughout all situations. A programmer sitting in an office somewhere is trying to develop the AI guidance to cover various conditions. It is quite possible that the programmer misidentifies some situations which could cause significant issues.

Engineering and technology also have ethical and communicative dimensions that go beyond following law and regulation and the skillful application of up-to-date knowledge. A technologist who makes decisions for another is unethical when the person involved is capable of making the decisions that impact their life. This fact is a mainstay in both clinical and research bioethics, which requires obtaining, informed consent. It should also exist in all the work that is being done with nanotechnology.[22]

11.3   WHAT NEEDS TO BE DONE?

The immediate consideration is protecting the workers who are the front line of developing the nanomaterials. It is imperative that the process to protect people and to measure and monitor their exposure to nanomaterials is the first step.

Each employee, who will work with the nanomaterial, should have a baseline health record prior to beginning any activities with the nanomaterials. There should be a folder in the Human Resources operation for each individual involved with nanomaterials that not only includes the baseline, but also all training activities for both handling the materials and operating any equipment that processes the nanomaterials. The employer should make available medical screening and surveillance for workers exposed to nanomaterials, if appropriate. The organization should have current Occupational Safety and Health Administration (OSHA) practices enforced, and periodically review medical surveillance requirements under OSHA standards (e.g. cadmium, respiratory protection).

Personal protection equipment needs to be provided for each person involved with the nanomaterials. The type of equipment needs to be appropriate for the type of material they will be handling/using. In order to provide the necessary protection equipment, employers need to evaluate the working conditions available based on the type of material that is anticipated to be employed.

Employers need to assess potential worker exposure to nanomaterials to identify the control measures needed and determine if the controls used are effective in reducing exposures by:

- Identifying and describing processes and job tasks where workers may be exposed to nanomaterials;
- Determining the physical state of the nanomaterials such as dust, powder, spray, or droplets;
- Determining routes of exposure (e.g., inhalation, skin contact, or ingestion) of particulates, slurries, suspensions, or solutions of nanomaterials;
- Identifying the most appropriate sampling method to determine the quantities, airborne concentrations, durations, and frequencies of worker exposures to nanomaterials;

- Determining what additional controls may be needed based on the exposure assessment results and evaluating the effectiveness of controls already in place. Employers should adopt the most effective controls available to limit worker exposure;
- Ensure that all employees involved with nanomaterials are trained in handling and storage of the materials. Regular training and refresher courses should be held periodically;
- All employees need to have training on equipment being employed for nanomaterials development and/or manufacture.

Because the research and use of nanomaterials continues to expand, and information about potential health effects and exposure limits for these nanomaterials is still being developed, employers should use a combination of the following measures and best practices to control potential exposures:

- Work with nanomaterials in ventilated enclosures (e.g. glove box, laboratory hood, process chamber) equipped with high-efficiency particulate air (HEPA) filters;
- Where operations cannot be enclosed, provide local exhaust ventilation (e.g., capture hood, enclosing hood) equipped with HEPA filters and designed to capture the contaminant at the point of generation or release;
- Provide hand washing facilities and information that encourages the use of good hygiene practices;
- Establish procedures to address the cleanup of nanomaterial spills and decontamination of surfaces to minimize worker exposure. For example, prohibit dry sweeping or use of compressed air for cleanup of dusts containing nanomaterials, use wet wiping and vacuum cleaners equipped with HEPA filters;
- Have annual health evaluations of each of the employees involved with nanomaterials.

11.3.1 Material and Equipment

The single most important item in handling the nanomaterials is maintaining the material history. Records need to be maintained on every step the material is handled or modified from the time the material initially arrives until it ships. The records then need to be archived. Information included is the original source of the material, the date of the material arrival, the date and time of material movement to any processing or experimental operations, the time and date the actions were completed and the material moved to another operation of back to storage, and the final date or material shipment or disposition. Lot numbers and specific equipment employed for manufacture or testing need to be identified.

The nanomaterials should be handled as a potentially dangerous item with a very high value. This implies that the material containers are not left out in the open. The containers need to be clearly marked with contents, and if appropriate, the processing steps already finished. The materials should be kept in a limited access area with specific personnel identified and trained to disperse the materials to the trained employees. Procedures for handling mishaps with the nanomaterials need to be posted, and the employees trained to handle potential situations.

Material control means more than putting something in a storage cabinet. As mentioned in a previous paragraph, maintaining records of both the material movement and the employees involved at each step is important. This not only tracks the material by lot number, but also has the employees identified with the various steps in the process. This latter tracking could become very important if an annual health evaluation determines there might have been a contamination that created a potential health issue.

There are three different categories of equipment that need to be maintained. The first is the safety equipment for the employees. Their training, or refresher course, for the proper usage needs to have a traceable record. This equipment needs to be inspected and verified for proper operation on a regularly scheduled basis. The testing and/or manufacturing equipment needs to have regular maintenance with replacement of parts and consumables as indicated by the supplier. This category includes both hardware and software as appropriate. The last category is the monitoring of the environment. The author is aware of one case where the particulate monitoring device had a high level of contamination in the laboratory air. The alarms sounded and everyone went into the nanomaterial contamination prevention procedure. Fortunately, the company also had monitors at other locations including outside the entrance to the facility. After the immediate check for contamination from the process yielded no source for the contamination, the other monitors were checked. The situation was caused by a very high level of springtime pollen and not an accidental release of nanomaterials. All of the monitoring records are stored.

Records need to be stored off-site as well as maintained at the facility. Many organizations regularly store important materials, which include computer backups, at a safe facility some distance from the working facility.

11.4   IMPLEMENTATION AND TRAINING OF NANO-SAFETY

While there are some training courses on Nano-Safety, the required focus may not be available in any particular area. Educational organizations tend to focus their specific emphasis on the needs of their locality. This is especially true for the two-year advanced education institutions. Consequently, it may not be possible for an organization to obtain the type of skill-level required for the implementation of Nano-Safety in their organization.

The remainder of this chapter provides an overview of the initial and successful Nano-Safety programs in the hope that individual local implementation can draw from experience in creating these programs. The initial effort developed a seminar through OHSA funding. The two courses, in modular format, were funded by the National Science Foundation (NSF). The two-year technical course was developed with a number of funding sources including NSF. Details of the courses are provided in the following section. Pulling elements from the content may provide some guidance in attempts to develop an internal program to address the risks and rewards of nanotechnology.

11.4.1   DEVELOPED EDUCATIONAL PROGRAMS

Nanotechnology has evolved from narrowly focused, equipment oriented, research topic to a multidisciplinary, tabletop equipment operation, research and application topic in the brief period of time since Richard Feynman introduced the concept of the nanoscale and challenged a group of physicists to write the Library of Congress on the head of a pin.

This evolution has impacted multiple businesses; large and small, every market segment, regulatory agencies worldwide, and education. More recently, since 2012, the topic and vocabulary of sustainability has entered both the corporate and educational arenas. Sustainability applies not only to the longevity of a company, product, or educational program but also to the facets of a technology that impact economy, environment, and society. Educational content that involves sustainability facets must address specifically health and safety impacts (positive and negative) on both humans and the environment. The societal aspects of sustainability include local, national, and international citizen responsibility. It is necessary for companies, their employees, educators, and students to have a sense of responsibility and legal/regulatory understanding and adherence to fully encompass the breadth of "sustainability". Educators, when including sustainability aspects in the curriculum, must also be aware of the philosophy of their "customer" companies (those that will hire their graduates), with regard to sustainability and how it may apply, in a broad sense, to a particular company. This is clearly a challenge for educators, since the philosophies for different companies with regard to sustainability can vary greatly.

Considerations for sustainability will also include corporate level influences that often are not considered in an educational curriculum. For example, consider a company that manufactures plasticware, forks, knives, and spoons; perhaps part of a company's philosophy is to become more "ecofriendly" and the desire is to make a portion or all of their products biodegradable. The process, at a high level that the company follows, includes starting material procurement, material mixing, molding/manufacturing, and packaging. After investigation of requirements, materials, and the physical constraints of the manufacturing process, a biodegradable option is determined. However, it is also determined that the manufacturing process needs to be modified with the potential requirement for new equipment. It is also found that the starting material is more expensive than the non-biodegradable material. Hence, for this example, the economic considerations have a significant impact on the proposed introduction of a biodegradable product which fulfills the company philosophy. The company may then decide to assess packaging options and determines that a biodegradable packaging material is a viable approach.

The above is an example of the multiple paths that addressing sustainability can take in a corporate world, and company's desire or expectation of their employees to be able to think broadly and consider multiple facets and influences. In a nanotechnology educational program at Dakota County Technical College, the Department Director, Deb Newberry, addressed this requirement by introducing a research course. The research topic was: "The impact of TiO_2 nanoparticles on freshwater bi-valve mollusks". Although the topic, from a research standpoint, was extremely viable and had the potential for published research papers for the students, the main intent of the class was to allow the students to develop the project plans, timelines, Gantt charts, interdependencies, troubleshoot issues, define data collection requirements, and approach, etc. In addition to these "soft skills" to be developed, the students were also charged with looking at the sustainability aspects such as environmental impact, regulatory aspects, and societal influences. These students then graduated with a portfolio which included not only technical skills but also non-technical and broad-reaching sustainability knowledge and skills.

While educators are being required, and educational institutions evaluated, on the number of graduates produced within a two- or four-year period, they face the challenge of introducing and teaching the multiple facets of nanotechnology within these constraints. An additional challenge for the educational community, coupled with the complexity imposed because of the multi-disciplinary nature of nanotechnology, is the fact that required or desired employee knowledge and capabilities are dependent upon the size of the company.

The first question that must be addressed when considering the creation of a nanotechnology educational program is: "who/which companies will be hiring the program graduates?" Making the assumption that the majority of program graduates will desire to remain in the region, a first step is to develop strong relationships with the local companies and obtain their participation in defining requirements for knowledge and skills that they desire in employees. These requirements are then translated into student outcomes and competencies that result from the educational content. Awareness of local and regional industries will also support the design and direction for the curriculum content. For some regions the focus may be on nano-electronics or photonics, while for others the focus may be nano-biotechnology (device or drug development), and for other regions environmental or energy generation involving nanotechnology may be the focus. Independent of the specific "nano" area of focus, it is critical to know the regional companies, their area of focus and employee requirements. After this information is obtained, evaluated and agreed upon by the educators and company representatives, curriculum creation can begin. This step addresses the multi-disciplinary aspect of nanotechnology and allows the educational institution to serve the customer base.

A second aspect for consideration is understanding the laboratory and equipment expertise required by the company customer community. If, for example, the majority of companies that may hire program graduates are large, often international, companies then the educational institution may choose to focus on a specific set of equipment or laboratory practices and capabilities. This is especially true when students may find a career at a nano-biotechnology company developing pharmaceuticals or working in another rather "narrow" area such as tissue engineering. In these cases, the educational content may be very focused and of a relatively short duration with fewer pieces of equipment and procedures to be taught. However, if these large companies have multiple areas of focus and/or research and development, then the educational program may have to be broader, cover more pieces of equipment and include more in-depth coverage of traditional sciences such as physics and chemistry. Material science topics may also need to be covered. Finally, if the local region is populated by many smaller, startup companies that are looking for nano savvy employees the educational content to satisfy their requirement may be extremely broad. Often, all employees at smaller companies are required to be a "jack of all trades", performing chemistry experiments creating new materials one day, then using multiple pieces of equipment to assess the properties of those materials the next.

Finally, independent of industry or company size, the educational curriculum must include aspects of teamwork, effective written and oral communication, work planning, problem solving, and critical thinking. These skills, once called "soft skills", are becoming more important in all industries due to multiple factors.

In any event, the changing landscape of nanotechnology must be taken into account with a strong and concrete integration and relationships with regional and local industries established. Educational institutions and their educators must learn how to work with industry and industry must learn how to define and delineate their requirements and expectations for their employees. The following section covers programs that address Nano-Safety as: a) a short seminar; b) a full semester college course; c) modules that can be inserted into existing college courses; d) a two-year program; and, e) certification. The material is taken from the initial institutions that incorporated Nano-Safety into their efforts.

11.4.1.1   Nanotechnology Safety Seminar

In 2010, the Occupational Safety and Health Administration, U.S. Department of Labor, awarded a grant (SH-21008-10-60-F-48) to Rice University[25] to develop both a four- and an eight-hour course that addressed the requirements for ensuring that workers understand the need for Nano-Safety. The resulting effort was evaluated after sampling technical personnel at appropriate technical conferences. The eight-hour offering was selected as the better of the two options. The program consisted of seven modules:

- Module 1 focuses on introducing nanotechnology and nanomaterials. Just because the term "nano" has been around for a number of years does not mean everyone understands what happens at the nanoscale.
- Module 2 provides an insight into worker needs regarding both nanomaterial toxicology and environmental impacts.
- Module 3 covers evaluating exposure to nanomaterials in the workplace.
- Module 4 addresses controlling exposure to nanomaterials.
- Module 5 discusses risk management.
- Module 6 covers existing rules, regulations, and standards.
- Module 7 wraps up the seminar and provides guidance on how to explore for new regulations and other developments.

11.4.1.2   College/University Courses

The majority of National Science Foundation-Nanotechnology Undergraduate Education (NSF-NUE)-funded projects have focused on teaching students about the development of products, devices, systems, and/or nanomaterials. Few projects focus on the societal dimensions of nanotechnology.[26]

This Nano-Safety project developed, implemented, and assessed two modular courses dealing with "nanotechnology safety" that include societal, ethical, environmental, health, and safety issues related to nanotechnology for undergraduates in engineering and engineering technology. The courses were developed with the guidance of a Nanotechnology Advisory Council (NAC) comprised of nanotechnology leaders from academia and industry in order to improve the focus of the course material. The effort had strong support from the NSF Nano-Link Center at Dakota County Technical College (DCTC) and the Center for the Environmental Implications of NanoTechnology (CEINT) at Duke University. The effort was a collaborative project between Texas State University-San Marcos (Texas State), a Hispanic Serving Institution (HSI) and the University of Texas at Tyler (U.T. at Tyler), whose student population is 60% female.

The effort brought together an interdisciplinary team of faculty with project expertise from mechanical and manufacturing engineering, civil engineering, electrical engineering, industrial education and technology, physics, biology, and philosophy, and ethics.

The first course, "Introduction to Nanotechnology Safety," is an introductory level, building foundational knowledge and stimulating interest in the subject. The second course, "Principles of Risk Management for Nanoscale Materials" is an advanced course that covers issues related to assessing and managing health and environmental risks, as well as ethical aspects of nanotechnology, at a deeper level. These full courses can be included in the curriculum in their entirety, or specific modules can be inserted separately into existing courses.

"Introduction to Nanotechnology Safety" introduces students to nanotechnology, nanomaterials and manufacturing, national security implications, and societal and ethical issues of nanotechnology. This course is a freshman/sophomore level course. After completing this course, students will be able to: (a) understand the ethical and societal impact of nanotechnology; (b) understand fundamental concepts in sustainable nanotechnology; and (c) understand the nature and development of nanotechnology.

The nine modules in the "Introductory Course" cover the following topics:

- Module 1A: "What is Nanotechnology & Ethics in NanoTechnology?" This introductory approach provides both a historical overview to establish a common understanding of all the key developments in nanotechnology as well as an introduction to the standard terminology used in nanotechnology. An overview of the worldwide societal implications and dimensions of nanotechnology are covered along with examples of potential dangers that have been identified.
- Module 2A: "Ethics of Science and Technology" This module builds on the impacts of Scientific and Technological Change and develops the Concepts of Moral Agency and Moral Consequences within Engineering and Business goals.
- Module 3A: "Social Impacts" The challenge addressed in this module includes understanding the Ethical Framework as applied to engineering, the consequences of actions and learn how ethical reasoning depends on what we know and what we can know. This is an area of uncertainty that must be addressed to make appropriate decisions.
- Module 4A: "Ethical Methods and Processes" covers approaches to Ethical Evaluation including Framing Ethical Questions (add pragmatist model) and assessing possible courses of action, language, terminology, and rhetoric of ethics are the key outcome of this module.
- Module 5A: "Nanomaterials and Manufacturing" covers reviews of existing manufacturing processes along with the environmental concerns. The specialized handling requirements for med-bio nanomaterials and the general handling, storage, and tracking of nanomaterials in general.
- Module 6A: "Environmental Sustainability" addresses the question of what sustainability needs to include: i) environmental, ii) economic, iii) political, and iv) technological impacts. This includes determining what are the sustainability issues (risks and benefits) with nanotechnology including: i) research, ii) development, and iii) deployment – Making Decisions with Incomplete Knowledge of Outcomes.
- Module 7A: "Nanotechnology in Health and Medicine" is a growing field. There is a need to cover the potential issues, including pharmaceuticals and therapeutics, diagnostics and imaging, nanoscale surgery, implants and tissue engineering, multifunctional nanodevices

and nanomaterials, personalized medicine, and the broader health care system, Bioethics and Nanotechnology in Human Research Trials.

- Module 8A: "Military and National Security Implications" provides a high-level view of nanotechnology applied to homeland security, discussions of the potential for developing defensive and offensive weapons to protect oneself and one's country.
- Module 9A: "Nanotechnology Issues in the Distant Future". The last module reviews challenges and pitfalls of exponential manufacturing along with the current interest in nanotechnology and life extension, Coverage of the development of new materials, scaling process, and the required controls for volume manufacturing provide a basis for evaluating future developments.

The "Advanced Course", "Principles of Risk Management for Nanoscale Materials", addresses the health and environmental risks of nanotechnology. This course is a junior/senior level course. After successfully completing this course, students are capable of: (a) understanding the health and environmental risks of nanotechnology; (b) understanding how to work in a group and conduct systematic research to write a group-based term paper on case studies and/or research topic; and (c) understanding approaches to assessing life-cycle risk assessment of nanotechnology. This course meets Accreditation Board for Engineering and Technology (ABET) student-learning outcomes: (a) an ability to apply knowledge of mathematics, science, and engineering; (b) an ability to function on multidisciplinary teams; (c) a recognition of the need for, and an ability to engage in, life-long learning.

The eight modules in the "Advanced Course" cover the following topics:

- Module 1B: "Overview of Occupational Health & Safety" delves into i) methods and practices, ii) theories of accident causation, iii) accident investigation & reporting, and iv), hazards control & communication.
- Module 2B: "Applications of Nanotechnology" reviews nanomaterials for groundwater remediation and nanoparticle use in pollution control. Detailed applications are presented in the i) health field, ii) in energy applications, iii) in the information and communication sector, iv) the current and coming applications in heavy industry, and, v) the myriad of consumer applications.
- Module 3B: "Assessing Nanotechnology Health Risks" requires dose-response assessment, dose-response evaluation, and risk characterization, and human health and toxicology: Also included are overviews of nanotechnology safety programs in the workplace and training and incentives.
- Module 4B: "Sustainable Nanotechnology Development" presents how to develop environmental regulations pertaining to nanotechnology including analyses of nanoparticles in the environment and life cycle risk assessment (LCRA) for sustainable nanotechnology applications.
- Module 5B: "Environmental Risks Assessment" considers i) nanoparticle transport, aggregation, and deposition, ii) treatment of nanoparticles in wastewater, iii) potential ecological hazard of nanomaterials iv) environmental toxicology and risk assessment, and, v) balancing risks and rewards.
- Module 6B: "Ethical and Legal Aspects of Nanotechnology" provides case scenarios in private industry and government that address legal duties and regulations, managers' responsibility and workers' compensation, and role of the Occupational Safety and Health Administration (OSHA), National Institute for Occupational Safety and Health (NIOSH), and Environmental Protection Agency (EPA)
- Module 7B: "Developing a Risk Management Program" includes American Society for Testing and Materials (ASTM) and OSHA guidelines for working with nanomaterials, prevention and control strategies for engineering, and administrative and personnel protection.

The goal is developing nanotechnology risk management in Total Safety Management (TSM) and Quality Management (QM) frameworks

- Module 8B: "Case studies" requires the students to demonstrate their learning based on actual academic visits to industrial sites.

11.4.1.2.1 Full Course Offerings[27]

UT at Tyler offered two full-semester courses, as part of the development of the Nano Management Concentration in the College of Business and Technology. TECH2303: "Introduction to Nanotechnology Safety" was the introductory course described in the previous section with all aspects of the course being covered during the semester. TECH 3303: "Principles of Risk Management for Nanoscale Materials" is the advanced course and was also covered in detail during the semester.

UT at Tyler offered these courses initially online, administering a satisfaction survey at the end of the semester to assess students' perceptions, which were critical to the continuous improvement and success of this project. UT at Tyler has also made these courses available online to industry professionals and community leaders.

11.4.1.2.2 Modular Insertion into Courses[28]

Texas State picked appropriate modules for incorporation into existing courses. These modules were taught face-to-face. Administrators at both institutions (Texas State and UT Tyler) are committed to educational effort in Nano-Safety. This required that the department chairs work with curriculum committees of individual programs for effective and smooth injection of courses and/or teaching modules. For a number of courses, this module insertion presents an opportunity to have experts provide the lectures on the various Nano-Safety topics.

11.4.1.3 Two-Year Program[29]

Dakota Country Technical College (DCTC) in Rosemount Minnesota created a two-year NanoScience Technologist Program, which became a model for other two-year programs around the country. The curriculum has been used either in its entirety or by portions in over 50 universities, colleges and high schools worldwide. Experiments and activities created within this program have been used by over 700 educators and has reached over 92,000 students. The program graduates (100+) were immediately hired and approximately 20% have pursued BS, MS, and Ph.D. degrees.

The need for the program developed due to the requirement of a workforce with knowledge and skills in the nanotechnology arena that could be utilized at local companies. These companies included 3M, Honeywell, Medtronic, General Mills, Entegris, Hysitron, Smart Skin Technologies, Lloyds Foods, HB Fuller, Valspar Paints, and Surmodics. The input and cooperation of these companies led to the creation of a four-semester program that included electronics, materials, biotechnology, manufacturing (6 Sigma included), soft skill development, and sustainability considerations. Nano-Safety was added into the courses as the material became available. This effort was integral to provide an evaluation of a diverse workforce.

The challenge for the creation of the DCTC NanoScience Technologist program was to provide a significantly diverse program that would also enable the students to have competent skills in using the various types of analytical equipment. The employers wanted these technicians not to be "same process every day" technicians. They wanted the technicians to understand the nanoscale concepts as they worked with the materials. It was also required that the students have a hands-on approach to working with various materials and equipment. Probably, the most challenging was the requirement that the students develop the skills that would enable them to be able to perform critical analyses of various situations and planned operations.[30]

DCTC applied for and received a National Science Foundation grant in 2004. This grant provided for the finalization of the curriculum content and initial program offering. The result was a 72 credit Associate Degree (AAS) multi-disciplinary program. The nanotechnology content is offered in each semester, allowing students to gradually increase their depth of understanding and application of the concepts of nanoscience. Also, half of the program credits are for nano-specific content. Finally, students are required to do a 320-hour minimum, participatory internship. For most students the internship is much longer, and they are often hired by the company for which they interned.

Under the leadership of Deb Newberry, the program enjoyed strong industry support through internships and jobs for program graduates. In 2008, DCTC was awarded a National Science Foundation grant for the Nano-Link Center. Nano-Link is a Regional Center for Nanotechnology Education. The program has been successful in creating a technical technician workforce for the local businesses. In 2011, Texas State University partnered with Newberry at DCTC to become an evaluation facility for the Nano-Safety educational material that was being developed.

The cultural backgrounds of central Texas with Hispanic heritage and the Minnesota Mid-Eastern immigrants were considered a perfect test of the NSF sponsored Nano-Safety course development. The concept demonstrated the ability of the developed course modules to successfully reach the students in both locations and in programs with a different focus.

11.4.1.4 Certification

Working with the ATMAE[31] a Nano-Safety Certificate program has been created [Ref.26]. The purpose of this effort is to provide an assurance of competence in the working with and handling of nanomaterials. The material covered includes the following:

Introduction to Nanotechnology and Nanomaterials

- Definition of "Nanomaterial"
- Contrast objects at the nanoscale with larger and smaller forms of matter
- Define key terminology associated with nanotechnology
- Explain ways that nanomaterial properties differ from molecules and micro-scale particles
- Describe major classes of nanomaterials produced, their properties and potential benefits

Identification of Occupational and Environmental Health Issues

- Potential for nanomaterials to cause occupational illness
- Setting occupational exposure limits (OELs)
- Nanomaterial categorization and handling practices systems
- Control banding in the nanotechnology environment
- Ventilation systems and nanoparticle control
- Administrative controls and procedures
- Hierarchy of controls and how they apply to nanoparticles

Hazard Communication and Workforce Orientation

- Hazcom Program Elements
- Safety Data Sheet Development
- Discussion of the limitations associated with nanoparticle sampling and analysis
- NIOSH's Personal Protective Equipment (PPE) recommendations for nanotechnology workers
- Medical Surveillance

Managing Nanomaterial Exposures and Occupational Health Programs

- Integration of environmental health and safety (EHS)elements into nanomaterial development activities
- Selection of personal protective equipment

- Respiratory protection for nanotechnology workers
- Differentiation between qualitative and quantitative fit testing
- Donning and doffing elastomeric half-face and N-95 filtering facepiece respirators

Spill Control and Emergency Planning

- Incident investigation and risk communication
- Sampling and analytical methods
- Framework for managing nanomaterial risks in the workplace
- Limitations of current hazard communication efforts around engineered nanoparticles
- Emergency response and contingency planning

Nanotechnology Regulations and Standards

- Regulatory requirements (in the United States, Canada, and abroad)
- State your rights under the Occupational Safety and Health (OSH) Act
- Articulate which OSHA standards apply to nanomaterial workplaces
- Articulate other regulations and standards that are applicable to nanomaterial workplaces

Tools and Resources for Further Study
- Freely available authoritative resources for keeping up to date with nano safety and EHS

11.5 SUMMARY

This chapter does not provide answers, because in many cases the problems, much less the answers, are not yet understood. Consequently, the professional working with nanotechnology needs to be competent in how to work with nanomaterials and to recognize the situations where there might be a danger. What has been presented is an overview of the work that has been done to establish the fundamentals to address any unknown issue with nanomaterials. The course development efforts have provided a guideline for selecting the key elements that need to be understood for each individual issue. Every situation is different, but trained professionals are able to address the unknown situations in a manner to mitigate risk and protect both people and the environment. Always start with the people.

How is this accomplished? The immediate consideration is protecting the workers who are the front line of developing the nanomaterials. It is imperative that the process to protect people and to measure and monitor their exposure to nanomaterials is the first step.

Each employee, who will work with the nanomaterial, should have a baseline health record prior to beginning any activities with the nanomaterials. There should be a folder in the Human Resources operation for each individual involved with a nanomaterial that not only includes the baseline, but also all training activities for both handling the materials and operating any equipment that processes the nanomaterials.

Personal protection equipment needs to be provided for each person involved with the nanomaterials. The type of equipment needs to be appropriate for the type of material they will be handling/using. Most importantly, do not reinvent the wheel. There are professionals who have worked through the challenges to develop what exists today. Learn from their work. If possible, attend meetings and talk with people. We need to ensure the safety of all efforts to minimize problems in the future.

ACKNOWLEDGMENTS

As this book is published, there have been thirteen years of work on developing the basis for Nano-Safety. There are many people who have been involved. It is important to recognize their contribution. The official effort started with a grant from OHSA to Rice University. Kristen Kulinowski was the Principal Investigator (PI) on that contract that created the world's first training on nanotechnology safety. Walt Trybula was the PI at Texas State. The NSF award for the development of two college courses was made to Texas State University where Jitendra Tate was the PI. Dominick Fazarro was the PI at the University of Texas at Tyler. Texas State implemented the courses developed in a modular fashion, while Dominick Fazarro taught the material as complete courses. Jitendra Tate included Craig Hanks, who is at Texas State, to bring the dimension of ethics to the Nano-Safety effort. Deb Newberry, Newberry Technology Associates, Inc., did yeoman's work in developing the 72-credit nanotechnology technician at DCTC. As the Nano-Safety education effort evolved, she incorporated modular elements into the existing technician program. There are many more people who have contributed to the development of this activity, creating a safe place for people to work and creating safe conditions that will not have negative effects on the environment. The need to maintain a critical understanding of the requirements for having a safety mentality is critical for future safety and may help mitigate circumstances where legal issues arise.

NOTES

1. Craig Hanks, J. and Kay Hanks, Emily. (2017). Ethics and communication: The essence of human behavior: What we need to know to protect workers. In: Dominick E. Fazarro, Walt Trybula, Jitendra Tate and Craig Hanks (eds), *Nano-Safety: What We Need to Know to Protect Workers*. Walter de Gruyter, Boston, MA, p. 196.
2. http://www.thedailybeast.com/articles/2016/02/24/can-baby-powder-really-cause-cancer.html (accessed 25 November 2019).
3. https://www.reuters.com/article/us-bayer-glyphosate-lawsuit/california-jury-hits-bayer-with-2-billion-award-in-roundup-cancer-trial-idUSKCN1SJ29F (accessed 25 November 2019).
4. https://ori.hhs.gov/definition-misconduct (accessed 25 November 2019).
5. http://news.rice.edu/2010/09/29/osha-bolsters-rice-based-safety-program-on-eve-of-buckyball-discovery-conference/NSF award (accessed 25 November 2019).
6. http://www.tryb.org/a_white_paper_on_nano-safety.pdf (accessed 25 November 2019).
7. Ibid.
8. Carbon nanotubes and graphene – properties, applications and market. https://www.graphene-info.com/carbon-nanotubes (accessed 25 November 2019).
9. Klabunde, Kenneth J. (2001). Introduction to Nanotechnology. In: K. J. Klabunde (ed.), *Nanoscale Materials in Chemistry*, Wiley Inter Science, New York, USA, pp. 1–13.
10. Ibid.
11. Advanced Energy Consortium private communications.
12. Ibid.
13. Ibid.
14. http://www.reuters.com/article/idUSN19810(accessed 25 November 2019).
15. From Meridian Institute Nanotechnology Portal on Tuesday, 18 May 2010, http://www.merid.org/NDN/.
16. http://chem.pitt.edu/documents/201002111222210.Deaths%20Linked%20to%20Nanoparticles.htm (accessed 25 November 2019).
17. Hurt, Evelyn and Trybula, Walt. (2017). What is considered reliable information. In: Dominick E. Fazarro, Walt Trybula, Jitendra Tate and Craig Hanks (eds), *Nano-Safety: What We Need to Know to Protect Workers*. Walter de Gruyter, Boston, MA, p. 196.
18. Bhattacharya, S., Gupta, K., Debnath, S., Ghosh, U.C., Chattopadhyay, D. and Mukhopadhyay, A., (2012). Arsenic bioaccumulation in rice and edible plants and subsequent transmission through food chain in Bengal basin: A review of the perspectives for environmental health. *Toxicological & Environmental Chemistry* 94(3), 429–441.
19. http://news.rice.edu/2010/09/29/osha-bolsters-rice-based-safety-program-on-eve-of-buckyball-discovery-conference/NSF award (accessed 25 November 2019).
20. Foster, Barbara. (2017). The world of nanotechnology. In: Dominick E. Fazarro, Walt Trybula, Jitendra Tate and Craig Hanks (eds), *Nano-Safety: What We Need to Know to Protect Workers*. Walter de Gruyter, Boston, MA, p. 196.
21. Waychunas, G.A., Zhang, H. (2008). Structure, chemistry, and properties of mineral nanoparticles. *Elements* 4:381–387.

22. Hochella, M.F. Jr. (2008). Nanogeoscience: From origins to cutting edge applications. *Elements* 4:373–379.

23. Sunscreen ingredient may increase skin cancer risk – https://www.sciencedaily.com/releases/2012/05/120507131951.htm (accessed 25 November 2019).

24. Tang, L., Zhang, C., Song, G., Jin, X. and Xu, Z. (2013). In vivo skin penetration and metabolic path of quantum dots. *Science China Life Sciences* 56(2), 181–188.

25. National Science Foundation Award Abstract. http://nsf.gov/awardsearch/showAward?AWD_ID=1242087 (accessed 25 November 2019).

26. NSF-NUE (Nanotechnology Undergraduate 446 Education) award #1242087

27. Dominick Fazarro was responsible for the full course offerings at the University of Texas at Tyler.

28. Jitendra Tate was the primary PI and responsible for the modular implementation at Texas State University.

29. Deb Newberry was responsible for the complete development and implementation of the program at Dakota County Technical College.

30. This section is adapted from previous work by Deb Newberry. A more detailed explanation of the entire process is available in Feather, J.L. and Aznar, M.F. (2011). *Nanoscience Education, Workforce Training, and K-12 Resources*. CRC Press, Boca Raton, Florida, USA, pp. 181–186. ISBN 978-1-4200-5394-4.

31. ATMAE (Association of Technology, Management, and Applied Engineering) is working with Fazarro to complete the offering of a Nano-Safety Certification program.

12 Impact of Nanomaterials on Human Health through/via Food Chain

Sudhir Shende, Vishnu Rajput, Aniket Gade,
Tatiana Minkina, Mahendra Rai, and Shishu Pal Singh

CONTENTS

12.1 INTRODUCTION

The nano-era began the late 1990s and the production of nanoparticles (NPs) could reach 1.663 million tons by 2020 (Rajput et al. 2020). The global market for NPs was valued at USD 10.92 billion in 2016 and is projected to reach USD 25.26 billion by 2022 (Rajput et al. 2020). There are more than 1,600 nanotechnology-based consumer products, and an online database lists 3,036 products in various categories that contain NPs, and that number is continuously increasing (www.nanodb.dk). Due to specific characteristics such as physical and chemical properties and high surface–volume ratio, application of NPs is fast increasing, for instance, in agriculture, chemicals, cosmetics, coatings, electronics and optics energy, environmental remediation, fuel additives, food packing, paints, plastics, textile, and next-generation medicine, etc. Nanotechnology is becoming an interesting and active area of research and how it affects the environment is still a question; as the applications are increasing, a number of issues are raised, such as ecological, ethical, health and safety, policy and regulatory, and related to intentionally releases into an environment each year with a majority of them end up in the soil, and have become a priority research area in recent years (Keller and Lazareva, 2013; Garner and Keller, 2014; Keller et al. 2017; Mishra et al. 2017; Servin et al. 2017; Rajput et al. 2018).

Once NPs end up into the soil, they may perhaps have effects on soil chemical and physical properties, and could form new kinds of toxic compounds, interact with other pollutants, and be capable of existing for an extended time in the soil, which disturbs the soil microbial community, plant growth, and consequently human health via the food chain. However, to date, no reliable methods to detect, identify, and quantify NPs in the presence of natural NPs have been developed. Major gaps in knowledge regarding the behavior and potential toxicological risks of NPs in soils are emerging. These engineered NPs can be metal- or carbon-based. Metal-based NPs can be grouped as metals, metal oxides, and quantum dots (Verma et al. 2019). Reviews on the basis of data and modeling analysis indicate that the production, use, and disposal of various NPs leads to release of thousands of tons of the most common NPs (e.g. Ag, Al, Ce, Cu, Fe, Si, Ti, Zn, etc.) into the environment each year with the majority of them ending up in the soil, directly or through landfills from sludge and other waste (Gottschalk et al. 2009; Keller and Lazareva, 2013; Keller et al. 2017; McGillicuddy et al. 2017; Bundschuh et al. 2018). Air and water also get a considerably higher quantity of distribution of NPs (Keller et al. 2013). A number of research studies, as well as reviews, recommended that the metal NPs are extremely toxic to a broad variety of organisms, in particular for plants and human beings. The high accumulation of metal-based NPs was observed in the edible part of plant tissues which is a concerning issue for human health (Arora et al. 2008, Kim et al. 2008, Jang et al. 2010, Luo et al. 2015, Yan and Chen, 2019).

Until now, the potential negative impacts of nanomaterials (NMs) on human health and the environment have been rather speculative and uncorroborated (Gogotsi, 2003). In the review, Wiesner and co-workers (2006) reported the risks of manufactured NMs and their potentially hazardous effects on human and animal health. In their study, they found that fullerene-based NPs showed cytotoxicity to human cells because of their antibacterial effects, which induces DNA damage, accumulates in livers of rats, and generates reactive oxygen species (ROS). Oberdorster (2004) demonstrated the effects of fullerene and C_{60} based NMs on the oxidative stress generation in the brain cells of rats and other animals (Oberdorster, 2004). In addition to this, the inorganic NPs like silicon dioxide (SiO_2), titanium dioxide (TiO_2), and zinc oxide (ZnO) induced pulmonary inflammation in humans and animals (Gogotsi, 2003; Oberdorster, 2004). This ground-breaking information generated a great deal of interest in the scientific community and government agencies like the Royal Society and the Royal Academy of Engineering are publishing the risks of NPs on the environment in its annual reports to generate public awareness (The Royal Society London, 2004). Even though the Federal Drug Administration (FDA) in the United States has permitted some non-toxic NPs for their exploitation in paints and sunscreen lotions, there are toxic NPs and chemicals, such as asbestos, ultrafine particles, diesel particulate matter, and lead, which are known to accumulate in the food chain. It is well known that it is very complicated to extrapolate the experience with bulk to NMs due to their reactivity, chemical, and physical property differences, etc. For example, antibacterial silver NPs dissolve in acids that would not dissolve bulk silver, which indicates their increased reactivity (The Royal Society London, 2004). This chapter will contribute to understanding the possible effects of NPs on human health.

12.2 ACCUMULATION OF NANOPARTICLES IN CROP PLANTS

In the ecosystem, plants are the producers and represent the key factor in the food chain as a primary trophic level. Concerning the food security issue, the majority of the harvested edible tissues or organs of vegetables or cereals are consumed by livestock and humans. Since NPs can be taken up and accumulated in plants, they are capable of creating a further menace to human health by invading the food chain and finally transferring to the human body. In fact, it was demonstrated that several NPs possibly will cycle in the ecosystem throughout various trophic levels in a terrestrial or aquatic food chain (Monica and Cremonini, 2009; Rico et al. 2011; Kalman et al. 2015; Tangaa et al. 2016; Tripathi et al. 2017; Yan and Chen, 2019).

In an aquatic ecosystem, planktonic algae function as primary producers, which are positioned at the foundation of the aquatic food chain; hence, in some studies, algae were particularly selected as the basic trophic level in the investigation of the trophic transfer of NPs. McTeer and co-workers (2014) investigated the bioavailability, toxicity, and trophic transfer of silver nanoparticles (AgNPs) between the grazing crustacean *Daphnia magna* and alga *Chlamydomonas reinhardtii*, which are the members of two dissimilar trophic levels. Nanosilver resulting from AgNPs was accumulated into microalgae. *Daphnia magna* was feeding on Ag-containing algae that accumulated nano-derived Ag in the organism and it confirmed the trophic transfer of AgNPs between *Daphnia magna* and algae (McTeer et al. 2014). Likewise, Kalman and co-workers (2015) demonstrated bioaccumulation along with the trophic transfer of NPs in a simplified freshwater food chain encircling *Daphnia magna* and the green alga *Chlorella vulgaris*. AgNPs were accumulated in algae; then these AgNP-contaminated algae were fed to *Daphnia magna*. A few days later Ag uptake was observed in *Daphnia magna*, directing more investigation into the diet as a prevailing pathway for the uptake of Ag in *Daphnia magna* (Kalman et al. 2015). Park et al. (2018), used paddy microcosm systems to calculate approximately the trophic transfer of AgNP- Polyvinylpyrrolidone (PVP) and AgNP-citrate among diverse trophic level organisms for example aquatic plants, biofilms, river snails, and Chinese muddy loaches. After exposure, AgNPs were quickly coagulated, in addition to precipitated, on the sediment. The examination of the stable isotope indicated a very close relationship among the Ag content in the prey, and in their subsequent predators, representing the impact of AgNPs on ecological receptors and food chains as well (Park et al. 2018).

There were very limited studies on the probable trophic transfer of AgNPs in terrestrial food chains; however, the toxicological effect on plants was closely studied. Besides this, the terrestrial trophic transfer of other metal NPs, like gold nanoparticles (AuNPs), (Judy et al. 2012), cerium oxide nanoparticles (CeO_2-NPs) (Hawthorne et al. 2014), and lanthanum oxide nanoparticles (La_2O_3-NPs) was also investigated (De la Torre Roche et al. 2015). In a test terrestrial food chain, the caterpillars of tobacco hornworm (*Manduca sexta*) were fed on tomato leaves, which were surface contaminated with the treatment of AuNPs on the leaves. Subsequently, the transfer of AuNPs from the tomato leaves to the tobacco hornworm was observed (Judy et al. 2012). Therefore, these studies present the possibility that NPs will also be transferred in terrestrial food chains. *In vitro* as well as *in vivo* both the studies established the AgNPs toxicity on mammalian cells. For instance, AgNP exposure reduced lung function with the production of inflammatory lesions in the lungs of rats (Sung et al. 2008), the consequences of this resulted in AgNP accumulation in the olfactory bulbs together with the rat's brain (Kim et al. 2008). As AgNPs are can accumulate and transfer in the food chain, it may become hazardous to humans. The exposure of AgNPs to human cells stimulated inflammatory as well as immunological reactions, which ultimately caused oxidative stress and led to cellular damage (Arora et al. 2008, Jang et al. 2010, Luo et al. 2015). The accumulation of NPs in edible plant tissues is another concerning issue, as these accumulations in crops could reduce the food and feed quality, and transfer to the food chain, which could be a serious threat to human health, as well as cause economic losses. Several studies showed that the NPs can be taken up and accumulated in plant cells (Da Costa and Sharma, 2015; Peng et al. 2015; Rajput et al. 2018). These observations indicate accumulation in both roots and shoot tissues of crop plants. The root tissues could be the main storage of metals especially for naturally grown plants in highly polluted areas (Fitzgerald et al. 2003; Ghazaryan et al. 2018).

The accumulation of NPs can alter plants' physiological processes and affect the integrity of cellular and sub-cellular organelle organization. A number of studies reported that the NPs reduced germination, decreased root and shoot growth, altered photosynthesis, induced oxidative stress, and caused an imbalance in the nutritional content of crop and quality yields (Rajput et al. 2017; Gao et al. 2018; Ochoa et al. 2018). Plants can accumulate NPs in their original form or as metal ions but could vary with the varying physicochemical features (Rico et al. 2011; Peng et al. 2015; Cota-Ruiz et al. 2018). The high accumulation of NPs in-plant could adversely modify proteins, lipids, and nucleic acid content by generating hydroxyl radicals (Halliwell and Gutteridge, 1985). Thus, the bioaccumulation and trophic transfer of NPs in the food chain is very critical for assessing and mitigating its potential harm to human health. Therefore, there is an urgent need to increase knowledge and understanding related to this particular issue. The accumulation of NPs greatly depends on the growth medium, types of plant species, type of NPs, and applied concentrations (Table 12.1).

NP adsorption is considered to be the initial step for NP bioaccumulation in plants (Nair et al. 2010). For example, the strong adsorption of copper oxide nanoparticles (CuO-NPs) can take place on the plant root surface partially through mechanical adhesion; the adsorbed CuO-NPs cannot be desorbed by competing for ions (Zhou et al. 2011). The plants grown in a different medium

TABLE 12.1

Accumulation of Various NPs in Different Crop Plants

Type of Nanoparticle (NPs)	Crops	Concentration (mg/L* mg/kg)	Medium	Accumulation of NPs (mg/kg)		References
				Root	Above-Ground	
CuO	Lettuce	20	Nutrient solution	9941.3	28.8	Hong et al. (2015)
CuO	Alfalfa	100		3977.7	147.1	Peng et al. (2015)
CuO	Rice		Nutrient solution	9000	------	
Ag	Rice	------		------	------	Thuesombat et al. (2014)
Ag	Triticum aestivum	0.5, 1.5, 2.5, 3.5, 5		------	------	Dimkpa et al. (2013)
Ag	Oryza sativa	0, 0.2, 0.5, 1		------	------	Nair and Chung (2014)
Ag	Phaseolus radiates; Sorghum bicolor	0, 5, 10, 20, 40		------	------	Lee et al. (2012)
Ag	Vicia faba	800		------	------	Abd-Alla et al. (2016)
Ag	Solanum tuberosum	0, 2, 10, 20		------	------	Bagherzadeh et al. (2016)
Ag	Triticum aestivum; Vigna sinensis; Brassica juncea	50, 75		------	------	Mehta et al. (2016)
Ag	Raphanus sativus	0, 125, 250, 500		------	------	Zuverza-Mena et al. (2016)
Ag	Cowpea (Vigna unguiculata L. Walp.); Wheat (Triticum aestivum L.)	0–1.6		------	------	Wang et al. (2015)
Ag	Oryza sativa L.	0.1, 1, 10, 100, 1000		------	------	Thuesombat et al. (2014)
Ag	Zea mays; Brassica oleracea	0.05, 0.1, 1, 18.3, 36.7, 73.4		------	------	
Ag	Lycopersicon esculentum	0, 100, 1000		------	------	Song et al. (2013)
Ag	Cucurbita pepo	250, 750		------	------	Hawthorne et al. (2012)
Ag	Allium cepa	0–80		------	------	Panda et al. (2011)
Ag	Allium cepa	100		------	------	Kumari et al. (2009)

(Continued)

TABLE 12.1 (CONTINUED)
Accumulation of Various NPs in Different Crop Plants

Type of Nanoparticle (NPs)	Crops	Concentration (mg/L* mg/kg)	Medium	Accumulation of NPs (mg/kg)		References
				Root	Above-Ground	
Ag and Ag$_2$S-	Lettuce		-------	-------	-------	Doolette, et al. (2015)
Ag$_2$S	*Vigna unguiculata; Triticum aestivum*	0–20	-------	-------	-------	Wang et al. (2015)
Au	Radish and Ryegrass	-------	-------	14–900 ng mg−1	-------	Zhu et.al (2012)
Au	Rice and Pumpkin	-------	-------	7–59 ng mg−1	-------	Zhu et.al (2012)
CeO$_2$	Cucumber	-------	-------	-------	-------	Zhang, et al. (2011)
CeO$_2$	Corn	-------	-------	-------	-------	Zhao et al. (2012)
CeO	Soybean	-------	Soil	-------	-------	Hernandez-Viezcas et al. (2013)
CuO	Kidney bean	100	Soil	800	-------	Apodaca et al. (2017)
CuO	Lettuce and Cabbage	250	Foliar spray	3773 4448	-------	Xiong et al. (2017)
Cu	Mungbean and Wheat	-------	-------	-------	-------	Lee et al. (2008)
CuO	Maize	-------	-------	-------	-------	Wang et al. (2012)
CuO	Mustard	500	MS medium	>500	-------	Nair and Chung (2015)
CuO	Wheat	500	Sand	-------	375	Dimkpa et al. (2013)
CuO	Rice	1000	Hydroponic	1544.1	17.27	Da Costa and Sharma (2015)
CuO	Cucumber	1000	Soil	500	-------	Kim et al. (2013)
CuO	Ipt-cotton	1000	Nutrient solution	7000	-------	Le Van et al. (2016)
CuO	Mustard	1500	Hydroponic	190.4	-------	Rao and Shekhawat (2016)
Fe$_3$O$_4$	Pumpkin	-------	Soil	-------	-------	Zhu et al. (2008)
TiO$_2$	Wheat	-------	-------	-------	-------	Larue et al. (2012)
TiO$_2$	*Lycopersicon esculentum*	-------	-------	-------	-------	Song et al. (2013)
TiO$_2$	Cucumber	-------	-------	-------	-------	Servin et al. (2012)
ZnO	Soybean	-------	Soil	-------	-------	Hernandez-Viezcas et al. (2013)
ZnO	Maize and Rice	2000	-------	-------	-------	Yang et al. (2015)

(Continued)

TABLE 12.1 (CONTINUED)

Accumulation of Various NPs in Different Crop Plants

Type of Nanoparticle (NPs)	Crops	Concentration (mg/L* mg/kg)	Medium	Accumulation of NPs (mg/kg)		References
				Root	Above-Ground	
ZnO	Rice	--------	-------	------	------	Chen et al. (2015)
ZnO	Tomato	10–100	-------	------	------	Li et al. (2016)
ZnO	*Pisum sativum*	25, 250, 500	Soil	------	------	Mukherjee et al. (2014)
ZnO	*Pisum sativum*	250 and 1000	Soil	------	------	Mukherjee et al. (2016)
ZnO	Maize	400 and 800	Soil	------	------	Wang et al. (2016b)
ZnO	*Medicago sativa*	500 and 750	--------	------	------	Bandyopadhyay et al. (2015)
ZnO-	*Brassica nigra*	500–1500	-------	------	------	Zafar et al. (2016)

showed moderate to high accumulation of Cu in a dose-dependent manner, for example, in rice roots 1,544–9,000 mg/kg (Da Costa and Sharma, 2015; Peng et al. 2015), in wheat roots: 132.27 mg/kg (Zhang et al. 2018) and shoots: 375 mg/kg (Dimkpa et al. 2013); in mustard roots >500 mg/kg (Nair and Chung, 2015) in whole plants 190.4 mg/g (Rao and Shekhawat, 2016), in alfalfa roots 3,977.7 mg/kg (Hong et al. 2015), in whole plant tissue of lettuce 3,773 mg/kg (Xiong et al. 2017) and 7,000 mg/kg in cotton (Le Van et al. 2016). The uptake of nano-CuO in tomato, alfalfa, cucumber, and radish seedlings was also noticed in the range of 4–1,748 mg/kg dry biomass when grown on semi-solid agar media (Ahmed et al. 2019). Cu-based nanopesticide applied to lettuce showed 1,353–2,501 mg/kg Cu accumulation after 30 days of exposure which was 82–140 times higher in vascular and 115–184 times higher in mesophyll cells relative to the control (Zhao et al. 2016). The accumulation of Cu content was 20% lower, and 33% lower in roots and shoots of tomato than in hydroponic conditions due to the formation of homo/hetero aggregates of nano-CuO in soil (Ahmed et al. 2018; Rajput et al. 2020). Larue et al. (2012) reported titanium oxide nanoparticles (TiO_2-NPs) are able to accumulate in wheat roots and then distribute within the whole plant tissues without dissolution or crystal phase modification. Zhu and co-workers (2012) conducted the comparative study of AuNP accumulation in the roots of four plant species indicated radish and ryegrass roots generally accumulated higher amounts of the AuNPs (14–900 ng/mg) than rice and pumpkin roots (7–59 ng/mg) (Zhu et al. 2012). AgNPs have also been observed to be capable of easy adsorption by rice roots (Thuesombat et al. 2014). NP accumulation rate by plant roots can also be influenced by the properties of NPs and the environmental conditions (Chen, 2018). Doolette et al. (2015) deliberated uptake of Ag from biosolid-amended soil containing silver sulfide nanoparticles (Ag_2S-NPs) by the plants, in this, authors reported that ammonium thiosulfate and potassium chloride fertilization significantly increased the AgNP concentrations in the roots and shoots of the plant. Zhang et al. (2011) reported that smaller NPs have high accumulation rates in cucumber roots. However, soil organic matter may affect NP accumulation by plants (Zhao et al. 2012). Characteristics of both NPs and plants play an important role in NP translocation and accumulation. For instance, a study conducted by Zhu et al. (2012) indicated AuNPs can accumulate in rice shoots, whereas they cannot be accumulated in the shoots of radish and pumpkin. Besides, it has also been shown that the positively charged AuNPs are most easily taken up by plant roots, though the negatively charged AuNPs are most competently translocated into plant shoots (including stems and leaves) from the roots (Zhu et al. 2012).

12.3 HUMAN EXPOSURE TO NANOMATERIALS

The impacts of NPs on human health are associated with the effects of materials and devices fabricated using nanotechnological techniques, and can be classified into two major groups: (1) the potential for nanotechnological innovations that have medical applications in curing diseases, and (2) the potential hazardous effect to health posed by exposure to NMs (Thomas et al. 2013). NP interaction inside the body or how the NPs behave inside the human body is one of the crucial issues that needs to be determined. A most important concern is the accumulation of non-degradable or slowly degradable NPs in the organs. Phagocytes, cells which ingest and destroy foreign matter, can be overloaded by NPs thereby triggering stress reactions, which lead to inflammation and weakness in the body's defense against the other pathogenic organisms. Moreover, upon exposure to tissue and fluids NPs will absorb some of the macromolecules onto their surface that may affect the regulatory mechanisms of enzymes and other proteins as well. Other than these, chemical compositions, surface structure, surface charge, aggregation and solubility, and the presence and absence of functional groups are the other properties of NMs that influence the toxicity (Nel et al. 2006; Herzog et al. 2007).

Toxicity is influenced by a large number of variables, which means that it is very complicated to oversimplify the health risks coupled with exposure to NPs, and each novel NP must be assessed individually along with all the properties taken into consideration. The human body has developed a tolerance towards most of the naturally occurring elements and molecules with which contact occurs, as the absence of natural immunity towards new substances, and hence NPs, is more likely found to be toxic. NPs can enter the body through the skin, lungs, and digestive system, which might help in the creation of 'free radicals' that can cause cell damage. There is also concern that once NPs are in the bloodstream, they will be able to cross the blood-brain barrier (BBB) (Thomas et al. 2013). Kirchner et al. (2005) noted the three main causes of NP toxicity following contact with living cells are, (a) chemical toxicity of material from which they have been made, for instance, Cd^{2+} is released from cadmium selenide NPs; (b) small size: NPs may stick to cellular membranes and enter the cells. Attachment of NPs to the membranes and storage of NPs inside the cells can impair cellular function even in the case of chemically inert NPs, which neither decompose nor react with other matrix components; (c) shape, for instance, carbon nanotubes could easily pierce the cell membrane. Recently, Gupta and Xie (2018) reviewed NP application in daily life, with special reference to their toxicity and the regulation issue. The authors discussed a variety of gateway routes for NMs in the human body, its application in everyday life, as well as the mechanisms by which the NPs showed toxicity. Moreover, they further explore the passage of NPs into the air, soil, and aquatic ecosystems, resulting in diverse environmental impacts (Gupta and Xie, 2018).

It is not just an assumption that the contact of NPs is dangerous to human beings and the environment. Several kinds of research have shown that NPs can cause lung damage in rats, e.g. Titanium oxide NPs (20 nm) and iron NPs (Gonzalez et al. 2008; Fraser et al. 2010). Titanium dioxide and fullerene have shown to lead to brain damage in fish and dogs (Shaw and Handy, 2011). An overview of some exposure cases for humans and the environment is shown in Table 12.2.

It is evident from Table 12.2, that the release of NPs may perhaps come from point sources as well as non-point sources, for example, wet deposition from the atmosphere, factories, landfills, stormwater runoff, and abrasion from products which have NMs. Biochemical cycling of NPs engages photochemical reactions in the atmosphere, accumulation, aggregation or uptake, transformation, and degradation in organisms. Respirators and air filters have to be used to eradicate NMs from the air, which occurs because of the variable means of atmospheric transport. Human exposure to NMs is most probably during NM manufacturing, however, inhalation of NMs released into the atmosphere and ingestion of drinking water or food (e.g. fish, seafood, etc.) that have accumulated NPs could also be possible. Likewise, dermal exposure from sunscreens and cosmetics is also possible. For effective delivery of iron NPs to the cells, unique exterior coatings will be mandatory to make

TABLE 12.2

Paths of NPs into the Environment

Product	Examples of NMs	Potential Release and Exposure
Cosmetics	TiO_2 or ZnO	Directly applied to skin, washing off to the environment and the disposal of containers
Fuel additives	Cerium oxide	Exhaust emission
Paints and coatings	Silver NPs coatings and hydrophobic nanocoatings	Wear and washing
Clothing	Silver NPs coatings and hydrophobic nanocoatings	Absorption by the skin
Electronics	Carbon nanotubes	E-Waste will lead to emission of CNT
Toys and utensils	Carbon nanotubes	E-Waste will lead to emission of CNT
Combustion processes	Ultrafine particles	Emission with the exhaust
Soil regeneration	Calcium- and phosphorus-based NPs	High local emission and exposure
Production of NMs	Byproducts in the nano range	Byproducts could be emitted locally

Source: The Royal Society London, 2004.

the particles portable in the sub-surface (The Royal Society London, 2004). These alterations possibly will, in turn, enhance their prospects for unnecessary exposure to humans and other organisms. A lot of other types of NMs will also necessitate surface coatings for their projected application, for example, metal oxide NPs, and quantum dots used as a magnetic resonance imaging contrast agents (The Royal Society London, 2004).

Exposure of NMs to workers, consumers, and the environment looks inevitable with the increasing number of commercially available nanotechnology-based products and their production (Wijnhoven et al. 2009). Exposure is a key element in risk assessment of NMs as it is a prerequisite for the probable toxicological and eco-toxicological effects to be acquired. There are multiple routes through which NP exposure can take place. The exposure of NMs to humans can be categorized as, (a) dermal (e.g. through the use of cosmetics with NPs); (b) inhalation (of NPs, for instance, in the workplace); (c) ingestion (such as ingestion of food products having NPs); and (d) injection (of, for instance, nanotechnology-based medicine). Although there are many different kinds of NMs, concerns have mainly been raised about free NPs (The Royal Society London, 2004; Marquis et al. 2009). Figure 12.1 shows the probable routes of uptake and translocation of NPs in the human body (Oberdorster, 2004). The possible routes for the entry of NPs into the body are from water, air, food, clothes or drug delivery systems. The majority of these NPs pass through the bloodstream to reach organs, for example the kidneys, brain, and liver (Oberdorster et al. 2005).

Recently, Rajput and co-workers (2019) reviewed the toxicological effect of zinc oxide (ZnO) and CuO-NPs on human health via the plant system through the soil. In their study, the authors concluded that the application of these NPs was established to pass through variable chemical and biochemical reaction pathways, which might affect biological nitrogen fixation, as well as damage to the plant cells, and may perhaps cause a serious threat to human health as well. To overcome these situations the authors suggested safety evaluation and toxicological risk assessment studies, including study of exposure routes and safe exposure doses of the ZnO- and CuO-NPs.

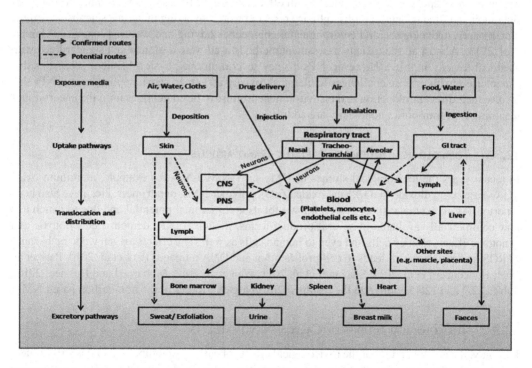

FIGURE 12.1 Different pathways of nanoparticles (NPs): central nervous system (CNS), gastrointestinal tract (GI), and in the peripheral nervous system (PNS). Although many uptake and translocation routes have been demonstrated in literature, others still are hypothetical and need to be investigated.

It has been carefully examined that the diseases related to NPs may perhaps be Alzheimer's disease, asthma, bronchitis, Parkinson's disease, lung cancer, Crohn's disease, heart diseases, colon cancer, rheumatoid arthritis, urticaria, scleroderma, and diseases related to spleen, liver, etc. (Buzea et al. 2007). Free NPs will possibly either disperse into an environment through the direct outlet or during the degradation of NMs (e.g. surface-bound NPs or nanosized coatings). There are multiple routes of exposure of NPs to the environment. The studies have shown that the drugs, which are often coated, and these coatings could be degraded through either metabolism inside the human body or by the transformation in the environment because of UV-radiation (Chow et al. 2008). In aquatic environments, the transport and fate of NPs has attracted comparatively little attention. This only emphasizes the necessity to study the various probable routes, which will alter the properties of NPs once they are released in nature. Supplementary ways of environmental exposure are spills or leakage from production, transport, and disposal of NPs or products. The Royal Society and The Royal Academy of Engineering recommended that the exploitation of free NPs in environmental uses like remediation must be proscribed until it has been revealed that the benefits are more important than the risks (Wijnhoven et al. 2009). It is unknown to date how much exposure to 'nanolitter' may affect living organisms, as well as the exact mechanisms of NM toxicity. Dhawan and Sharma (2010) reported the toxicological studies of biological systems employing metal and metal oxide nanoparticles (MO-NPs), fullerenes and carbon nanotubes (Dhawan and Sharma, 2010). They have concluded that further studies to conclusively establish the safety/toxicity of NMs are still needed, as there are many experimental challenges and issues encountered when assessing the toxicity of NMs.

In another review, Thomas et al. (2013) explored the impact of NMs on health and the environment, they reviewed all the aspects of the risk of exposure of NMs through several routes such as inhalation, ingestion, dermal, and drug delivery by engineered NMs. Again, they suggested the toxicological effects of NMs need to be examined during a product's lifecycle including manufacture, use, and disposal. The hazardous health effect and safety issues of NMs require further consideration and fundamental interdisciplinary research work, involving materials scientists, medical practitioners, toxicologists, and environmental engineers studying and working together (Thomas et al. 2013). Almost all the toxicity assessment methods used were designed and standardized with chemical toxicity in mind. However, NPs display several unique physicochemical properties that can interfere with or pose challenges to classical toxicity assays. In short, applications of NPs that involve their direct introduction to the environment promise to be contentious until the uncertainties regarding fate, transport, and toxicity are addressed.

12.3.1 Toxicological Studies of Metallic Nanoparticles

In recent years the toxicological studies of NMs such as MO-NPs for example, aluminum oxide (Al_2O_3), titanium dioxide (TiO_2), zinc oxide (ZnO), polymer NMs, dendrimers, etc., have also been reported. As an example, silver is well known for its antimicrobial potential and due to which it is now commercially available in several product forms, though AgNPs demonstrate cytotoxic and genotoxic effects to mice cells and even to human cells as well. The effects are shown to be because of ROS generation, which leads to cell proliferation and DNA damage (Park et al. 2010; Park et al. 2011; Hackenberg et al. 2011; Shavandi et al. 2011; Wan et al. 2012; Mohamed and Hussien, 2016). Tables 12.3 and 12.4 summarize the results of toxicological studies of the non-carbon based NMs.

12.3.2 Toxicological Studies of Carbon Nanomaterials

Presently, the understanding of the toxicological effects of carbon nanotubes (CNTs) is still limited, even though a large number of ongoing studies. As the carbon NMs present in different CNT species i.e. (multiwall carbon nanotube (MWCNT) and single-wall carbon nanotube (SWCNT)) with differences in length, diameter, surface area, size, oxidation and/or functionalization, and model

TABLE 12.3

Summary of Toxicological Studies with Non-Carbon Based NMs

Type of Carbon Nanomaterial	Model Used for the Study	Methods	Results Obtained	References
Nanosize TiO_2 (Degussa Aeroxide P25)	BV2 microglia were grown in 225-cm² cell culture flasks in 10% Fetal Calf Serum (FCS) and 1% Penicillin streptomycin	Fluorescent and chemiluminescent probes, measures of intracellular ATP using the CellTiter-Glo assay	P25 promoted the instantaneous "oxidative burst" response in microglia and also hampered the mitochondrial energy production. Small groups of NPs and micron-sized aggregates were overwhelmed by the microglia. Mouse microglia reacted to Degussa P25 with cellular as well as morphological expressions of free radical formation	Long et al. (2006)
The TiO_2 Engineered NPs from DeGussa AG	Fish cells	SEM, TEM, single-cell electrophoresis or comet assay, cytokinesis-blocked micronucleus assay, neutral red retention (NRR) assay	Elevated levels of cytotoxicity established	Vevers and Jha (2008)
Superparamagnetic iron oxide NPs	Rat and mouse macrophages in vitro	Cytokine assay, Iron determination, Statistical analysis	Shifted macrophages towards an anti-inflammatory, less responsive phenotype	Siglienti et al. (2006)
Ferumoxtran-10, that is, dextran-coated, very tiny superparamagnetic iron oxide particles	Human monocyte-macrophages (HMMs) in vitro	Ferrozine assay Neutral red and (3-(4,5- imethylthiazol-2-yl)-2,5-diphenyl tetrazolium bromide) (MTT) assay. Cytokine assay, Nitro blue tetrazolium (NBT) assay for oxidative burst	Even high concentrations of Ferumoxtran-10 are not acutely toxic to HMMs in vitro, do not activate them to produce pro-inflammatory cytokines or superoxide anions, are not chemotactic or interfere with Fc-receptor-mediated phagocytosis	Müller et al. (2011)
Oleic acid-luronic block copolymer coated iron oxide MNPs	Male rats (240-260 g)	Magnetic susceptibility, oxidative stress analysis, histology	Injected iron easily metabolized, did not cause long term changes in liver enzyme levels or oxidative stress	Jain et al. (2008)
Superparamagnetic iron oxide	The murine macrophage cell line	Phagocytosis assay, migration assay, viability and proliferation test	No changes in viability and cell proliferation	Hsiao et al. (2008)

(Continued)

TABLE 12.3 (CONTINUED)

Summary of Toxicological Studies with Non-Carbon Based NMs

Type of Carbon Nanomaterial	Model Used for the Study	Methods	Results Obtained	References
Cationic PAMAM dendrimers surface-functionalized with lauroyl chains or polyethylene glycol (PEG) 2000	Human epithelial cells	Cell viability studies, cytotoxicity	Marked decrease in cytoxicity due to surface modification	Jevprasesphant et al. (2003)
PAMAM dendrimers, DAB dendrimers and DAE dendrimers, and CSi-PEO dendrimer	Rats cell culture media	Hemolysis assay, in vitro cytotoxicity, ^{125}I-Radiolabelling	Cationic dendrimers are hemolytic and cytotoxic. Anionic ones are not	Malik et al. (2000)
TiO_2, 32 nm and 45 m² g⁻¹; ZnO, 40–100 nm and 10–25 m² g⁻¹; CuO, 23–37 nm and 25–40 m² g⁻¹; and Fe_2O_3, 20–40 nm, 30–60 m² g⁻¹	Three *X. laevis* mating pairs to ensure adequate supply of viable embryos FETAX solution	FETAX assay, mortality, malformations, stage, snout vent length (SVL), and total body length (TBL)	Exposures up to 1,000 mgL⁻¹ for TiO_2, Fe_2O_3, CuO, and ZnO do not increase mortality, but did induce developmental abnormalities. Gastrointestinal, spina, abnormalities were observed in CuO and ZnO NMs exposures at concentrations as low as 3.16 mgL⁻¹ (ZnO). The minimum concentration to inhibit growth of tadpoles exposed to CuO or ZnO NMs was 10 mgL⁻¹, indicating select NMs can negatively affect amphibians during development	Nations et al. (2011)
TiO_2 25–70 nm, ZnO 50–70 nm, CuO 30 nm, Bulk TiO_2, ZnO and CuO	*S. cerevisiae* S288C (fungal genome)	Inhibition of growth rate (*I*) and reduction in viable cell count (CFU/mL) at the 8th and 24h hour of cultivation	Nano and bulk TiO_2 were not shown toxicity even at 20,000 mgL⁻¹. Nano and bulk both ZnO were of comparable toxicity	Kasemets et al. (2009)
ZnO (13 nm) and CeO_2 (8 nm) NPs synthesized by flame spray pyrolysis	Macrophage and epithelial cells or rats, mice and rabbit	Cell viability and microscopy	Induction of ROS, apoptosis and inflammation by ZnO but inhibition of ROS by CeO_2	Xia et al. (2008)
Gold (2, 40 nm) three morphologies, spheres, rods, and urchins, coated with poly (ethylene glycol) (PEG) or cetyl trimethylammonium bromide (CTAB)	Microglia and neurons. Microglia are the resident immune cells of the brain	Dark-field microscopy, two-photon-induced luminescence (TPL)	Internalization of gold NP by both microglial cells and primary neurons	Hutter et al. (2010)

(Continued)

TABLE 12.3 (CONTINUED)
Summary of Toxicological Studies with Non-Carbon Based NMs

Type of Carbon Nanomaterial	Model Used for the Study	Methods	Results Obtained	References
Iron oxide NPs (5.3 ± 3.6 nm)	Male mice of six weeks age, intratracheal instillation	Cell cycle analysis, histopathology, gene expression in tissue	Iron oxide NPs might induce chronic lung inflammatory responses by a single intratracheal instillation in mice	Park et al. (2010)
CdTe quantum dots with mercaptopropionic acid, SiO_2 or PEG coatings	Female mice, penetration across placental barrier, intravenous injection in vivo	Cytotoxicity analyses Toxic behavior	NPs were accumulated in the fetus. When the size of NP decreased, more accumulation took place. However, coating decreased the transfer through the placental barrier	Chu et al. (2010)
CdTe quantum dots	Human umbilical vein endothelial cells (HUVECs), in vitro	Cytotoxicity, genotoxicity studies	These quantum dots have cytotoxic and genotoxic effects independent of dose. Surface treatment of QD reduces the effects	Wang et al. (2010)
TiO_2 NPs (35 nm) Silica (70–1,000 nm)	Female mice, penetration across placental barrier, intravenous injection in vivo	Cytotoxicity studies	Both the nanoparticles were found in the placenta, fetal liver and fetal brain. These damaging effects are concurrent to structural as well as functional deformities in the placenta. When the surfaces of the silica NPs are modified with carboxyl and amine groups the effects were reduced	Yamashita et al. (2011)
ZnO-NPs (40 ~ 100 nm) TiO_2-NPs (~ 32 nm)	Earthworms (Eisenia fetida), Artificial soil	Acute toxicity tests, reproductive toxicity tests	ZnO on filter paper was acutely toxic, while TiO_2 was non-toxic. Both were non-toxic in sand. Both caused reproductive effects in artificial soil ZnO-NPs exhibited greater toxicity than TiO_2 NPs in Eisenia fetida	Casnas et al. (2011)
MgO-NP (100 nm)	Human umbilical vein endothelial cells (HUVECs), in vitro	Cytotoxicity studies (MTT Assay, NO release) microscopy	Does dependent toxicity	Ge et al. (2011)

(Continued)

TABLE 12.3 (CONTINUED)
Summary of Toxicological Studies with Non-Carbon Based NMs

Type of Carbon Nanomaterial	Model Used for the Study	Methods	Results Obtained	References
ZnO-RT (4.6±0.6 nm width, 18.0±4.2 nm length, aspect ratio of 3.9) rod shaped ZnO-60°C (13.1±3.9 nm width, 80.5±6.8 nm length, aspect ratio of 6.1) rod shaped	Normal human lung fibroblast cells, inhalation of the NMs	Cell viability studies, MTT assay	Toxicity depends on the ROS generation, size independent, morphology and physicochemical properties of the NMs influences toxicity	Park et al. (2011)
AgNPs (20 and 200 nm), TiO_2NPs (21 nm)	Male mice (8 –10 weeks), human testicular embryonal carcinoma cell line	Cell viability studies, MTT assay, cell death analysis, comet assay, cytokine assay	Ag NPs are more cytotoxic than TiO_2NP. Both have effects on reproduction human health and environment	Asare et al. (2012)
AgNPs (45 nm)	Adult male rats (300–350 g), smooth muscles of trachea	Nitric oxide production and analysis, western blot analysis	AgNPs modify the contractile action through NO production and possibly induce hyper-reactivity of tracheal smooth muscle	Gonzalez et al. (2011)
TiO_2 NPs (5, 10, and 32 nm)	African clawed frogs (*Xenopus laevis*)	Cytotoxicity, cell viability, cell growth, interaction with UV-radiation	Whether alone or with UV light, the TiO_2NPs concentrations significantly affected growth of tadpoles as determined by total body length, snout-vent length, and developmental stage	Zhang et al. (2012)

Source: Thomas et al. 2013.

TABLE 12.4

Various Experimental Models Used to Demonstrate NPs' Toxicity

Nanoparticles (Size/Dose/Exposure Time)	Experimental Model/Route	Results	References
AgNPs 730 nm; 0.2% to 2%	Human skin; HaCaT keratinocytes	• Cytotoxicity was not observed against keratinocytes, also there was no cell death induced by UVB which confirmed the potential of AgNPs to be utilized as a preservative in cosmetics • Though, AgNPs may perhaps penetrate the disrupted human skin at concentrations about 0.002–0.02 ppm	Kokura et al. (2010)
AgNPs 7–10 nm; stabilized with polyethylenimine	Human hepatoma cell line, HepG2	• AgNPs and ionic form Ag^+ contributed to the toxic effects • Stronger chromosome damages were caused by AgNPs than by polystyrene NPs as well as ionic Ag^+ • Abnormal changes in cell morphology and cellular shrinkage at higher doses (> 1.0 mg/L) was observed in cytotoxicity analysis	Kawata et al. (2009)
AgNPs Less than 50 nm; 0.1, 10 μg/ml; 1, 3, and 24 h	Human mesenchymal stem cells (hMSCs)	• In hMSCs, cytotoxicity and genotoxicity were demonstrated at notably higher concentrations, when compared with an antimicrobial effective level • At concentrations of 10 μg/mL, cytotoxic effects were observed for all test exposure stages • DNA damage was observed in comet assay and the chromosomal aberration test after 1, 3, and 24 h at 0.1 μg/mL exposure	Hackenberg et al. (2011)
AgNPs 35 nm; 0.1, 0.5, 1.0, 2.5 and 5 μg/mL	Human microvascular endothelial cells (HMEC); endothelial colony forming cells (ECFC)	• Statistically significant cytotoxicity was observed at a concentration 0.1 μg/mL (P < 0.01) as well as dose dependent viability reduction of HMEC after 72 h at the concentration 1.86 ± 0.13 μg/mL was observed as inhibitory concentration 50 (IC50) • In comet assay, after 24 h treatment of ECFC by AgNPs (2.5 and 5 μg/mL) induced a dose dependent but reversible genotoxic stress	Castiglioni et al. (2014)
AgNPs 5 nm; 0.1–100 μg/mL; stabilized with ammonia (SNA) or PVP (SNP)	Mouse fibroblast L929; Wistar albino rats (implant of polyethylene tubes filled with a fibrin sponge embedded in 100 mL SNA or SNP)	• No cytotoxicity to cell line L929 at 25 μg/mL or lower concentrations of treatment of SNA and SNP • Cytotoxic effect to L929 was observed at higher concentrations, SNA was significantly more toxic than SNP • A sensible inflammatory response consisting mostly of neutrophils was observed in rats, after 7 days, implantation of polyethylene tubes and this inflammatory response was abridged with period	Takamiya et al. (2016)

(Continued)

TABLE 12.4 (CONTINUED)

Various Experimental Models Used to Demonstrate NPs' Toxicity

Nanoparticles (Size/Dose/Exposure Time)	Experimental Model/Route	Results	References
AuNPs 2 nm; quaternary ammonium functionalized AuNP and carboxylatesubstituted AuNP	Red blood cells, Cos-1 cells, and bacteria (*E. coli*).	• Anionic (carboxylate substituted) AuNPs were nontoxic, whereas Cationic (quaternary ammonium functionalized) AuNPs were moderately toxic • Cell-lysis, dependent on concentration was observed, which was mediated by initial electrostatic binding	Goodman et al. (2004)
AuNPs 5, 10, 30, and 60 nm; 4000 (μg/kg; 200 μL; Polyethylene glycol (PEG)-coated	Mice (intraperitoneal injection)	• In different organs, depending on size the accumulation of PEG-coated AuNPs was takes place • In vivo toxicity was size independent. The toxicity of 10 nm and 60 nm particles was found to be superior than that of 5 nm and 30 nm particles • Significant enhancement in the levels of alanine aminotransferase (ALT) and aspartate aminotransferase (AST) caused by 10 nm and 60 nm AuNPs, signify the liver damage in mice	Zhang et al. (2011)
AuNPs 20 nm; (200 μM Au); citrate (cit) and 11-mercaptoundecanoic acid (11-MUA) coatings	Human liver HepG2 cells	• Both AuNPs were internalized in a concentration dependent manner but without a noteworthy variation in their degree of internalization • No significant cytotoxicity was induced by two differently coated AuNPs • cit-AuNP exhibited DNA damage at lower concentrations besides absence of cytotoxicity cit-AuNP exhibited DNA damage at lower concentrations besides absence o cytotoxicity	Fraga et al. (2013)
AuNPs 10 nm; (13 mM Au)	Dendritic cells (extracted from bone marrow of C57BL/6 mice)	• AuNPs were not cytotoxic for dendritic cells, however, these AuNPs reduced the secretion of interleukin-1β (IL-β) when the cells were concurrently stimulated by addition of lipopolysaccharides (LPS)	Villiers et al. (2010)
SiO₂NPs 70, 300, and 1000 nm; 10, 30, 100 mg/kg doses	BALB/c male mice (intra-venous injection)	• 70 nm NPs induced liver injury at 30 mg/kg dose, whereas 300/1000 nm NPs did not show any effect, even at 100 mg/kg dose • 70 nm NPs augmented serum markers of liver injury, serum aminotransferase along with inflammatory cytokines in dose-dependent manner • Repetitive administration of 70 nm NPs twice a week for four weeks, even at 10 mg/kg, caused hepatic fibrosis	Nishimori et al. (2009)

(Continued)

TABLE 12.4 (CONTINUED)
Various Experimental Models Used to Demonstrate NPs' Toxicity

Nanoparticles (Size/Dose/Exposure Time)	Experimental Model/Route	Results	References
SiO_2-NPs Long rods Aspect Ratios (ARs) of ~5, length of 720 nm (65 nm) and short rods (ARs) of ~1.5; length of 185 nm (22 nm)	Mice (intravenous injection)	• Long-rods SiO_2-NPs distributed in spleen, whereas short-rods were trapped in liver • Short-rods NPs showed a more rapid clearance rate than that of long-rods in both urine and feces • MSNs with both ARs were present in lungs in higher concentration after PEG modification • Serum biochemistry, hematology, and histopathology results did not showed significant toxicity in vivo	Huang et al. (2011)
SiO_2-NPs 200–460-nm; four different SiO_2-NP food additives	Gastrointestinal cell lines GES-1 and human colorectal adenocarcinoma cell Caco-2; (in vitro gastro-intestinal toxicology)	• At lower concentration i.e.100 µg-mL^{-1} and less, adverse effects were not observed in gastrointestinal cell lines, even after 72 h exposure making them safe to be utilized as additives for foods • At higher concentrations i.e. 200 µg/ml or more, a significant change in cell functions was observed after 48 h exposure • SiO_2-NP was not change the cell morphology or induce apoptosis/necrosis, whilst, they entered the cells and inhibited the cell growth by cell cycle arrest	Yang et al. (2014)
SiO_2-NPs 20–200 nm; 10–500 µg/mL	Three different cell types A549 and HepG2 epithelial cells and NIH/3T3 fibroblasts	• A549 cells: size and dose-independent toxicity for doses ≥ 50 µg/mL for all but the 60 nm SiO_2-NP after 72 h of exposure. For the 60 nm particles, high doses (200 and 500 µg/mL) resulted in drastically reduced viability in comparison to all other SiO_2-NP doses and sizes tested (p > 0.05) • NIH/3T3 cells: size and dose dependent reductions in viability at both 24 and 72 h of exposure to SiO_2-NP at doses ≥ 50 (µg/ml (p > 0.05). At the highest dose (500 µg/ml). 60 nm SiO_2-NP caused an extreme reduction in NIH/3T3 cell viability to 4.55 ± 1.75% and 1.30 ± 0.26% of control after 24 and 72 h of exposure, respectively • HepG2 cells: no size dependent changes in viability in response to SiO_2-NP doses greater then 50 µg/mL, except for the highest dose of 60 nm SiO_2-NP which caused a greater reduction in viability than any other SiO_2-NP treatment (p > 0.05) significantly	Kim et al. (2015)

(Continued)

TABLE 12.4 (CONTINUED)

Various Experimental Models Used to Demonstrate NPs' Toxicity

Nanoparticles (Size/Dose/Exposure Time)	Experimental Model/Route	Results	References
SiO₂ NPs Spherical SiO₂ NP AR of 1,diameter of 83 nm; short-rods AR of 1.75, length of 146 nm (83 nm); long-rods AR of 5, length of 483 nm (96 nm); 40 mg/kg 100 μL.	Male mice (oral)	• Unlike intravenous administration, there were no intact • SiO₂-NPs (AR = 1, 1.75, and 5) in urine at 24 h post oral administration • Most of the orally administrated SiO₂-NPs were rapidly excreted from faeces, but some intact SiO₂-NPs or their degradation products were absorbed through intestinal mucosa, entered into systematic circulation, and finally excreted via renal excretion • With the decrease of aspect ratio, the systematic absorption by small intestine and other organs increased and the urinary excretion decreased • Renal tubular necrosis, hemorrhage and vascular congestion in the renal interstitium were observed for all three SiO₂-NPs irrespective of the geometrical features, suggesting possible kidney damage • All three SiO₂-NPs showed no obvious hematological toxicity	Li et al. (2015)
TiO₂ NPs 10–30 nm; 10 and 100 mg/kg	Goldfish (*Carassius auratus*)	• Accumulation of TiO₂ NP increased from 42.71 to 110.68 ppb in the intestine and from 4.10 to 9.86 ppb in the gills with increasing exposure dose from 10to 100 mg/L TiO₂-NP. • MDA, a biomarker of lipid oxidation was detected in the liver of goldfish. • TiO₂-NP exposure also inhibited growth of goldfish	Ates et al. (2013)
TiO₂-NPs 21 nm; 6.0–10 mg/m³	Female C57BL/6 mice (intratracheal instillation)	• ELISA examination shown activation of harmonize cascade and inflammatory progressions in heart along with specific activation of complement factor 3 in blood	Husain et al. (2015)
TiO₂ NPs 1–200 nm; uncoated (anatase and rutile); polyacrylate-coated; 10 and 100 mg/L for 24 h	Chinese hamster lung fibroblast (V79) cells	• Coated and uncoated both the TiO₂-NPs decreased the cell viability in a mass- as well as size-dependent manner. Polyacrylate coated particles were only cytotoxic at higher concentration (100 mg/L) • DNA damage was exhibited only by uncoated TiO₂-NP, in addition a more genotoxicity effect was shown in nano-sized anatase compared to rutile	Hamzeh and Sunahara (2013)

(Continued)

TABLE 12.4 (CONTINUED)

Various Experimental Models Used to Demonstrate NPs' Toxicity

Nanoparticles (Size/Dose/Exposure Time)	Experimental Model/Route	Results	References
TiO$_2$ NPs Less than 50 nm; 500 mg/kg; for 5 consecutive days and sacrificed after 24 h, 7 days, or 14 days	Mice (oral)	• Animals showed mild to moderate changes in the cytoarchitecture of brain tissue in a time dependent manner. • Comet assay revealed apoptotic DNA fragmentation. • Point mutation of presenilin 1, gene linked to Alzheimer's disease (PCR-SSCP and direct sequencing)	Mohamed and Hussien (2016)
TiO$_2$-NPs 154 nm;1.0 g/L; 24 h, 48 h, 7 days	Human HaCaT keratinocytes	• TiO$_2$ NPs did not infuse intact/damaged skin and exerted a low cytotoxicity effect only at a higher dose and extended exposure	Crosera et al. (2015)
TiO$_2$-NPs 20500 nm; 5%; uncoated and coated particles	Skin of Yucatan mini pigs	• TiO$_2$-NPs did not penetrate through the intact epidermis from sunscreen formulations	Sadrieh et al. (2010)
TiO$_2$ NPs 4 nm, 60 nm; 5%; 30, 60 days	Pig ear, hairless mice (topical)	• 30 days after topical application on pig ear, TiO$_2$.NP penetrated into deep layer of epidermis. • In vivo experiments with hairless mice showed penetration of particles in various tissues such as brain (10–15 µg/g), lung (12–18 µg/g), and spleen (22–30 µg/g) after 60 days of exposure.	Wu et al. (2009)
TiO$_2$-NPs 100 nm; (0, 2, 4, 6, 8, and 10 mM) (0, 0.25, 0.50, 0.75, 1, 1.25,1.50, 1.75, 2 mM)	*Allium cepa*; Human lymphocytes	• MDA (Malondialdehyde) concentration increased at 4 mM treatment dose in *Allium cepa* indicated lipid peroxidation as possible mechanism for DNA damage • In human lymphocytes, genotoxicity was shown at a low dose of 0.25 mM followed by a decrease in extent of DNA damage at elevated concentrations	Ghosh et al. (2008), Ghosh et al. (2010)
- 50 nm and 100 nm	HaCaT keratinocytes, 3D-epidermis	• Abridged cell viability of HaCaT keratinocytes in dose- and time-dependent manner • No irritation in 3D-epidermis model at 1 mg/mL after 24 h of exposure	Vinardell et al. (2017)

(Continued)

TABLE 12.4 (CONTINUED)

Various Experimental Models Used to Demonstrate NPs' Toxicity

Nanoparticles (Size/Dose/Exposure Time)	Experimental Model/Route	Results	References
ZnO-NPs 70 nm; 0–10 mg/mL	*Drosophila* larvae (in vivo) Human lung fibroblast MRC5 cells (in vitro)	• A dramatic increase in the ROS levels in gut cells of *Drosophila* larvae was observed, which leading to a decrease in viability in vivo • MRC5 cells missing their membrane integrity in addition released the Lactate dehydrogenase in a dose-dependent manner • Cell viability was significant diminished as early as 24 h after treatment, and ZnO-NPs at a concentration of 50 µg/mL (617 µM) caused overall cell death • Noteworthy decrease in entire DNA content of treated cells at both S and G2/M phases of cell cycle • Accumulation of 8-OHdG i.e. an oxidized DNA nucleoside indicating oxidative DNA damage due to production of ROS.	Ng et al. (2017)
ZnO-NPs 72 nm; 61, 123 µM	Madin-Darby canine kidney (MDCK) cells	• Genotoxic effect dependent on size was established using Methyl methane sulfonate solution (0.45 µM) as a positive control • A statistically considerable increase in double- and single-strand DNA breaks in cells at sub-cytotoxic concentrations accompanied by a reduction of Catalase and Glutathione S- transferase activity	Kononenko et al. (2017)
ZnO-NPs Below 100 nm; 5, 50, and 300 mg/kg for 14 days	Rats (oral administration)	• Significant increase in oxidative stress at 5 mg/kg ZnO-NPs dose induced enhancement in MDA content and reduced in SOD and Glutathione peroxidase enzymes activity in the liver, and elevation in plasma AST, ALT, and Alkaline phosphatise (ALP) levels • 5 mg/kg ZnO-NPs dose caused hepatocytes swelling, congestion of RBC, accumulation of inflammatory cells and extensively increased apoptotic index • 300 mg/kg dose of ZnO-NPs had poor hepatotoxicity effect	Mansouri et al. (2015)
ZnO-NPs 50 nm; 5, 25, 50, and 100 µg/mL for 24 h	Human lung epithelial cells (L-132)	• Glutathione (GSH) level depletion and an enhance in ROS levels recommended incidence of oxidative stress • DNA fragmentation i.e. apoptotic cell death • Metallothionein gene expression touted as a biomarker for metal induced toxicity	Sahu et al. (2013)

Source: Gupta and Xie, 2018.

systems are used, comparisons of these different studies are not easy. Moreover, CNTs contain contaminants, which might be active too. *In vivo* studies on CNTs are limited (Chiaretti et al. 2008), however, these studies found a reduction of cell proliferation and cytotoxic effects, which are associated with cell death, apoptosis, and/or generation of ROS (Cui et al. 2005; Monteiro-Riviere et al. 2005; Manna et al. 2005; Bottini et al. 2006; Wick et al. 2007; Kang et al. 2009; Wadhwa et al. 2011), whereas, MWCNTs showed the effect of treatment as an alteration of the paracellular permeability of the impairment of human airway epithelial cells, which was described by Rotoli and co-workers (2008). Conversely, there are other reports available, which demonstrate low or the absence of cytotoxic effects due to the exposition of CNTs (Pulskamp et al. 2007; Yacobi et al. 2007; Zeni et al. 2008).

Higher concentrations of both SWCNTs and MWCNTs weaken the viability and this effect is due to the strong tendency of CNTs to agglomerate (Davoren et al. 2007). There are very few reports available on the genotoxic potential, mutagenic, or carcinogenic effects of different types of CNTs. The direct interactions of NPs with the cells may result in genotoxic events, which are defined as primary genotoxicity, or through the capacity of NPs to stimulate inflammatory reaction and to create an excess of ROS that may be known as secondary genotoxicity (Schins, 2002). CNTs have been initiated to cross biological membranes and penetrate the epithelial cells. Due to the small size of CNTs, the specific surface area enlarges, leading to escalating interaction as well as uptake of NMs across the living cells (Jia et al. 2005). Sotto and co-workers (2009) evaluated a kind of MWCNT, differentiated by an extremely small surface/volume ratio and low level of metal contamination (<0.1% Fe) (Sotto et al. 2009). In conclusion, the purity (<0.1% Fe) of these NMs could have reduced the threat of oxidative damage as well as the toxic effects attributable to the presence of metal contaminants. Therefore, the MWCNTs selected come into view to be devoid of mutagenic effects in addition to the lack of mutagenicity, which might be related to their structure and/or purity. Some of the toxicological studies of carbon nanotubes are summarized in Table 12.5.

From the above tables, it can be concluded that the toxicity of NMs depends on the conditions employed i.e. physico-chemical properties and charges, which can be changed by any alteration in these conditions by modifying the NMs such as functionalization or coating. The toxicity of the NMs employed also depends on the size as well as shape. Usually as the size decreases, the toxicity increases. *In vivo* NMs that pass through barrier membranes like blood-brain or blood-ocular are capable of causing several special metabolic activities and thus generating toxicity in the organisms. To minimize cell toxicity, the cell proliferation mechanism needs to be clearly understood (Thomas et al. 2013).

12.4 CONCLUSION AND FUTURE PERSPECTIVES

Nanomaterials have different levels of influence on plants and animals, as well as human beings; due to this reason the specificity has to be studied in great detail. The research methodologies used to evaluate the toxicity are based on conventional materials that are larger and have better results; these methodologies have to be customized so that the mechanism of action can be completely implicit. However, until now, reliable methods to detect, identify, and quantify NPs in the presence of natural NPs, their behavior and potential associated toxicological risks are not well developed. The dangerous effect of NPs on humans is still unknown. Therefore, it is essential to explore in depth and systematically the mechanism of toxicity, accumulation level of NPs in plant tissues, interactive forms and their bio/geo-transformation, and effects on human health via the food chain or other mode of exposure. Complicating matters is the fact that the fate of NPs may vary with the varying physiology/anatomy/organism species, types of medium, and mode of application. It is also crucial to develop a unified methodology for testing NP toxicity in natural environments, and joint efforts should be conducted to determine the toxicity of NPs under different climatic conditions, and in various components of the biosphere, especially in soil, to develop the permissible levels for applications, and to determine the threshold levels of their contents in the respective ecological

TABLE 12.5

Summary of the Toxicological Studies of Carbon-Based NMs

Type of Carbon Nanomaterial	Model Used for the Study	Methods	Results Obtained	References
MWCNT (Nanostructured and amorphous materials) ϕ; 10-20, 30-50 nm, SWCNT (Nanostructured and Amorphous Materials), ϕ; 1-2 nm, Carbon black (Printex 90; Evonik Degussa) ϕ; 14 nm	Rat alveolar macrophage cells human lung epithelial cells	Cell viability assay, ROS analysis, Cytokine assay, nitric oxide assay	No sign of acute toxicity	Pulskamp et al. (2007)
SWCNT (electric arc discharge method) ϕ; 1.4 nm, L; 1 μm, MWCNT (CVD, Shenzhen Nanotech Port) ϕ; 50 nm, L; 10 μm Fullerenes (C_{60})	Adult pathogen-free healthy guinea pigs (250–300 g) (inhalation exposure)	Bronchoalveolar lavage (BAL) analysis, cytotoxicity studies, microscopy studies	The tested carbon NMs (SWNTs, MWNT10, and C_{60}) exhibit quite different Cyto-toxicity	Jia et al. (2005)
MWCNT (Baytubes; Bayer material science) acid functionalization, ϕ; 10–16 nm	Male rats (inhalation xposure)	Particle-induced pulmonary toxicity, BAL analysis on post-exposure days of 7, 28 and 90, Gene expression analysis	The pulmonary inflammogenicity following exposure to MWCNT was concentration-dependent with evidence of regression over time	Ellinger-Ziegelbauer et al. (2009)
MWCNT ϕ; 20–50 nm, L; 5.9, 0.7 μm (grinded, un-grinded, heated to 600°C, 2,400 °C; 2,400°C then grinded)	Female rats, injected directly into the lungs by intratracheal (i.t.) instillation	BAL analysis, histopathology, micronucleus analysis	The genotoxicity and the acute pulmonary toxicity of CNTs were abridged upon heating however, restored upon grinding, indicating that the CNTs intrinsic toxicity is chiefly mediated by the presence of defective sites in their carbon framework	Muller et al. (2008)
MWCNT (Nanocyl NC 7000; Nanocyl S.A.) ϕ; 5-15 nm, L; 0.1-10 μm	Rats (inhalation exposure)	90-day inhalation toxicity study BAL analysis, histopathology analysis, biomedical test	No systemic toxicity but multifocal granulomatous inflammation, diffuse histiocytic and neutrophilic inflammation, and intra-alveolar lipoproteinosis detected	Ma-Hock et al. (2009)

(Continued)

TABLE 12.5 (CONTINUED)
Summary of the Toxicological Studies of Carbon-Based NMs

Type of Carbon Nanomaterial	Model Used for the Study	Methods	Results Obtained	References
MWCNT φ; 11.3 nm, L; 0.7 μm MWCNT (MWNT-7; Mitsui)	Rats, (i.t. instillation)	Histopathology analysis	No mesothelioma, no sustained inflammatory reaction, carcinogenic effect of MWCNT identified	Sakamoto et al. (2009)
MWCNT, φ; 11.3 nm, L; 0.7 μm	Mice, 30-day and 60-day inhalation exposure	Biochemical indices in BAL fluid analysis and pathological examination	Did not induce obvious pulmonary toxicity in 30-day exposure group, but induced severe pulmonary toxicity in 60-day exposure group	Li et al. (2009)
Ultrafine carbon black (ufCB) 14 nm in diameter and fine carbon black (CB) 260 nm in diameter	Rats, (i.t. instillation)	BAL analysis, histopathology analysis	ufCB has greater ability than CB to produce lung inflammation and oxidant stress	Li et al. (1999)
MWCNTs, produced by CVD, φ: 20 nm and aspect ratio: 80–90; carbon nanofibers, (CNFs, obtained from Pyrograf Products, Inc.) φ: 150 nm and aspect ratio: 30–40; and Flake like-shaped carbon NPs (carbon black, obtained after grinding graphite) aspect ratio: about 1	Three different human lung-tumor cell lines	Cell proliferation studies, cytotoxicity	Carbon nanotubes are less toxic than carbon fibers and NPs	Magrez et al. (2006)
SWCNT (HiPco; Carbon Nanotechnologies) φ; 0.8–1.2 nm, L; 0.1–1 μm	The human alveolar carcinoma epithelial cell line medium depletion	Cytotoxicity studies (Alamar Blue (AB) and the clonogenic assay	SWCNT can induce an indirect cytotoxicity by alteration of cell culture medium (in which they have previously been dispersed)	Casey et al. (2008)
SWCNT (HiPco; carbon nanotechnologies), φ; 0.8–1.2 nm, L; 0.1–1 μm SWCNT which was cut chemically, φ; 0.8–1.2 nm, L; 20–80 nm	Cells from a human lung adenocarcinoma with the alveolar type II phenotype, normal human primary bronchial epithelial cells (NHBE)	Cell viability assay, luciferase reporter gene assay, real-time RT-PCR, biomedical test and enzyme linked immunosorbent assays (ELISA)	HiPco SWCNT samples suppress inflammatory responses of A549 and NHBE cells. The employ of di-palmitoyl phosphatidyl choline (DPPC) enhanced the degree of SWCNT dispersion and in order, leads to better toxicity responses of particles	Herzog et al. (2009)

(Continued)

TABLE 12.5 (CONTINUED)
Summary of the Toxicological Studies of Carbon-Based NMs

Type of Carbon Nanomaterial	Model Used for the Study	Methods	Results Obtained	References
Carbon nanotubes (product no. 636797; >50 % SW, ~40 % other nanotubes; 1.1 nm × 0.5–100 μm) and graphite nanofibers (product no. 636398; outer ϕ: 80–200 nm, inner ϕ: 30–50 nm, L 5-20 μm) purchased from Sigma-Aldrich (Steinheim, Germany)	Transformed human bronchial epithelial cell line exhibiting an epithelial phenotype	Cell viability assay, comet assay (single cell gel electrophoresis), micronucleus assay	Both CNTs and GNFs are genotoxic in human bronchial epithelial BEAS 2B cells in vitro may be due to the fibrous nature of these NMs with a possible contribution by catalyst metals present in the materials	Lindberg et al. (2009)
Coarse carbon black (CB), Huber 990, ϕ:260 nm; nanoparticulate CB (NPCB), Printex 90, ϕ: 14 nm	Lung cells derived from Human adenocarcinoma	Cell viability assay, DNA strand breakage analysis, protein assay	Exposure of cultures of cells to NP resulted in transient arrest of cells passing through the cell cycle	Mroz et al. 2007
MWCNT (XNRI WMVT-7) ϕ: 67 nm	A human bronchial epithelial cell line and a Chinese hamster ovary cell line	Cytotoxicity primer array assay for cytokine and cytokine-elated mRNA levels Western blot Reporter gene assay	MWCNT were highly cytotoxic in human bronchial epithelial cells. The new method to quantitatively measure cellular uptake of MWCNT indicated that cells started to associate with MWCNT after approximately 1-2 h lag	Hirano et al. (2010)
Pure C_{60} (99.9%) ϕ: 0.7 nm. The EliCarb1 SWCNT ϕ: 0.9–1.7 nm, L: <1 l m. The CB (Printex 90), ϕ: 14 nm	Epithelial cell from the mouse	Cytotoxicity, cell cycle analysis, DNA damage, mutagenicity analysis, reactive oxygen species (ROS) production	SWCNT and C_{60} are less genotoxic in vitro than carbon black and diesel exhaust particles	Jacobsen et al. (2008)
SWCNT (National Institute of Standards and Technology) ϕ: 1.4 nm, L: 2-5 μm	Normal and malignant human mesothelial cells	Cell viability assay, ROS analysis, comet assay	SWCNTs be capable to cause potentially unfavorable cellular responses in mesothelial cells, through activation of molecular signaling linked with oxidative stress	Pacurari et al. (2008)

(Continued)

TABLE 12.5 (CONTINUED)

Summary of the Toxicological Studies of Carbon-Based NMs

Type of Carbon Nanomaterial	Model Used for the Study	Methods	Results Obtained	References
MWCNT Mitsui & Company (MWNT-7, Lot #0507001K28). φ; 0.8–1.2 nm, L; 0.1–1 μm	Male mice (7 weeks old) MWCNT pharyngeal aspiration exposure, BAL analysis, Histopathology, Immunofluorescence	Exposure of mice to MWCNT caused dose- and time-dependent pulmonary inflammation and damage	MWCNT exposure also caused rapid development of pulmonary fibrosis. Furthermore, MWCNT can reach the pleura after pulmonary exposure	Porter et al. (2010)
SWCNT (HiPco; carbon nanotechnologies)	Specific pathogen-free adult female mice (8–10 week), inhalation and pharyngeal aspiration exposures of SWCNT	Cell viability assay, cytokine assay	Outcomes of inhalation exposure to respirable SWCNT were very similar to those seen after pharyngeal exposure route leading to pulmonary toxicity. The chain of pathological events was realized through synergized communications of inflammatory response as well as oxidative stress terminating in the development of multifocal granulomatous pneumonia, mutagenesis, and interstitial fibrosis	Shvedova et al. (2008)
MWCNT, length around 10–20 μm with an aspect ratio of more than three	Mice, intraperitoneally administration	LDH assay, cell cycle analysis, ROS analysis, comet assay, mutagenicity analysis	Carbon-made fibrous or rod-shaped micrometer particles may share the carcinogenic mechanisms postulated for asbestos	Takagi et al. (2008)
MWCNT prepared by decomposition of ethylene on a cobalt and iron catalyst supported on an alumina bed and ground during 6 h. and by thermal treatment (2,400°C under argon)	Specific pathogen-free male rats (9–12 weeks old)	Histopathological analysis	The incidence of tumors other than mesothelioma was not significantly increased across the different MWCNT groups. The length of the MWCNT tested (<1 μm on average), the absence of a sustained inflammatory reaction to MWCNT, and the capacity of these MWCNT to quench free radicals might have caused different levels of carcinogenic activities	Muller et al. (2009)

(Continued)

TABLE 12.5 (CONTINUED)
Summary of the Toxicological Studies of Carbon-Based NMs

Type of Carbon Nanomaterial	Model Used for the Study	Methods	Results Obtained	References
MWCNT (Graphistrength C100, ARKEMA, France) were produced by CVD on a supported catalyst in a fluidized bed	Human alveolar epithelial cells and mesothelial cells	Cell viability assay, oxidative stress, cell proliferation analysis	MWCNT produced by CVD for industrial purposes as spherical sets of several hundreds of micrometers exert adverse biological effects without being internalized by human epithelial and mesothelial pulmonary cells	Tabet et al. 2009
Raw CNTs carbon nanotechnologies, Inc. Φ; 1.0 nm, L; 1–2 μm Purified CNTs prepared from the raw SWNTs by a non-acidic treatment, Φ; 1.0 nm, L; 1–2 μm US tubes prepared from the raw SWNTs by fluorination and pyrolysis, Φ; 1.0 nm, L; 20–80 nm	Pathogen-free male mice (22 ± 2 g) Intraperitoneal administration	Oral toxicity, biochemical tests and blood counts, microscopy (optical, TEM and SEM)	After intraperitoneal administration, SWNTs, irrespective of their length or dose combine inside the body to form fiber-like structures. Smaller aggregates did not induce granuloma formation, but they persisted inside cells for up to 5 months after administration. Short (<300 nm) well-individualized SWNTs can escape the reticuloendothelial system to be excreted through the kidneys and bile ducts. It is suggested that if the potential of SWNTs for medical applications is to be realized, they should be engineered into discrete, individual "molecule-like" species	Kolosnjaj-Tabi et al. (2010)
Aerosolized single-wall (SW) CNT and multiwall (MW) CNT (5 μg/g of mice) for 7 consecutive days in a nose-only exposure system	Mice	Microscopic studies, BAL analysis, Polymorphonuclear leukocytes	Inhaled CNTs induce inflammation, fibrosis, alteration of oxidant and antioxidant levels, and induction of apoptosis-related proteins in the lung tissues to trigger cell death	Ravichandran et al. (2011)
MWCNT, prepared by CVD, Φ; 30 nm, L; <1 μm	Human umbilical vein endothelial cells (HUVECs), in vitro	Cell viability studies, ROS assay	Induce cytotoxicity and genotoxicity via oxidative damage pathways	Guo et al. (2011)

(Continued)

TABLE 12.5 (CONTINUED)
Summary of the Toxicological Studies of Carbon-Based NMs

Type of Carbon Nanomaterial	Model Used for the Study	Methods	Results Obtained	References
N-SWCNT (ϕ: 1.8 nm, surface area: 878 m^2/g) SG-SWCNT (ϕ: 3.0 nm, surface area: 1,064 m^2/g) N-MWCNT (ϕ: 44 nm, surface area: 69 m^2/g) MWNT-7 (ϕ: 60 nm, surface area: 23 m^2/g)	Male rabbits and male guinea pigs	Dermal irritation test, eye irritation experiment, skin sensitization studies	MWCNT is weak acute irritant to the skin and eyes, however, no skin sensitization for all types of CNTs studied	Ema et al. (2011)
SWCNTs (catalytic chemical vapor deposition), ϕ: 0.7–1.3 nm, L: 450–2,300 nm, COOH-SWCNTs (electric arc discharge), ϕ: 1.4 $\pm$ 0.1 nm	Human endothelial cells (HUVEC)	Cytotoxicity assays	Both types of CNTs induced toxic effects in HUVEC cells in a concentration- and time-dependent way. In addition, the carboxylic acid functionalization results in a higher toxicity compared to the SWCNTs	Gutiérrez-Praena et al. (2011)
Graphene oxide and reduced graphene oxide, three dispersants: polyethylene glycol (PEG), polyethylene glycol-polypropylene glycol-polyethylene glycol (Pluronic P123), and sodium deoxycholate (DOC)	Mice fibroblast cells	Cytocompatibility studies	The materials exhibit toxicity, which depends on the concentration and the type of dispersant. Both materials show relatively good cytocompatibility when the concentration is between 3.125 µg/mL and 12.5 µg/mL. The lowest toxicity is shown by graphene oxide suspended in PEG	Wojtoniszak et al. (2012)

Source: Thomas et al. 2013.

region. Therefore, it is recommended that the series of safety assessments and toxicological threat standards should be formulated for NP applications that include the exposure route, safe exposure doses, kinetics of dissolution and migration to groundwater.

CONFLICT OF INTEREST

The authors have no conflicts of interest to declare.

ACKNOWLEDGMENTS

The authors are grateful to the Russian Science Foundation (No. 20-14-00317) for supporting the research.

REFERENCES

Abd-Alla MH, Nafady NA, Khalaf DM (2016) Assessment of silver nanoparticles contamination on faba bean *Rhizobium leguminosarum* bv. viciae-Glomus aggregatum symbiosis: Implications for induction of autophagy process in root nodule. *Agric Ecosyst Environ* 218:163–177.

Ahmed B, Khan MS, Musarrat J (2018) Toxicity assessment of metal oxide nano-pollutants on tomato (*Solanum lycopersicon*): A study on growth dynamics and plant cell death. *Environ Pollut* 240:802–816.

Ahmed B, Rizvi A, Zaidi A, Khan MS, Musarrat J (2019) Understanding the phyto-interaction of heavy metal oxide bulk and nanoparticles: Evaluation of seed germination, growth, bioaccumulation, and metallo-thionein production. *RSC Adv* 9(8):4210–4225.

Apodaca SA, Tan W, Dominguez OE, Hernandez-Viezcas JA, Peralta-Videa JR, Gardea-Torresdey JL (2017) Physiological and biochemical effects of nanoparticulate copper, bulk copper, copper chloride, and kinetin in kidney bean (*Phaseolus vulgaris*) plants. *Sci Total Environ* 599–600:2085–2094.

Arora S, Jain J, Rajwade JM, Paknikar KM (2008) Cellular responses induced by silver nanoparticles: In vitro studies. *Toxicol Lett* 179(2):93–100.

Asare N, Instanes C, Sandberg WJ, Refsnes M, Schwarze P, Kruszewski M, Brunborg G (2012) Cytotoxic and genotoxic effects of silver nanoparticles in testicular cells. *Toxicologist* 291(1–3):65–72.

Ates M, Demir V, Adiguzel R, Arslan Z (2013) Bioaccumulation, sub-acute toxicity, and tissue distribution of engineered titanium dioxide (TiO_2) nanoparticles in goldfish (*Carassius auratus*). *J Nanomater* 2013:460518.

Bottini M, Bruckner S, Nika K et al. (2006) Multi-walled carbon nanotubes induce T lymphocyte apoptosis. *Toxicol Lett* 160(2):121–126.

Bundschuh M, Filser J, Luderwald S et al. (2018) Nanoparticles in the environment: Where do we come from, where do we go to? *Environ Sci Eur* 30(1):6.

Buzea C, Pacheco I, Robbie K (2007) Nanomaterials and nanoparticles: Sources and toxicity. *Biointerphases* 2(4):MR17–MR72.

Casey A, Herzog E, Lyng FM, Byrne HJ, Chambers G, Davoren M (2008) Single walled carbon nanotubes induce indirect cytotoxicity by medium depletion in A549 lung cells. *Toxicol Lett* 179(2):78–84.

Casnas JE, Qi B, Li S, Maul JD, Cox SB, Das S, Green MJ (2011) Acute and reproductive toxicity of nano-sized metal oxides (ZnO and TiO_2) to earthworms (*Eisenia fetida*). *J Environ Monit* 13(12):3351–3357.

Castiglioni S, Caspani C, Cazzaniga A, Maier JA (2014) Short-and long-term effects of silver nanoparticles on human microvascular endothelial cells. *World J Biol Chem* 5(4):457–464.

Chen H (2018) Metal based nanoparticles in agricultural system: Behavior, transport, and interaction with plants. *Chem Spec Bioavailab* 30(1):123–134.

Chen J, Liu X, Wang C et al. (2015) Nitric oxide ameliorates zinc oxide nanoparticles induced phytotoxicity in rice seedlings. *J Hazard Mater* 297:173–182.

Chiaretti M, Mazzanti G, Bosco S et al. (2008) Carbon nanotubes toxicology and effects on metabolism and immunological modification in vitro and in vivo. *J Phys Condens Matter* 20(47):474203–474212.

Chow EK-H, Pierstorff E, Cheng G, Ho D (2008) Copolymeric nanofilm platform for controlled and localized therapeutic delivery. *ACS Nano* 2(1):33–40.

Chu M, Wu Q, Yang H et al. (2010) Transfer of quantum dots from pregnant mice to pups across the placental barrier. *Small* 6(5):670–678.

Cota-Ruiz K, Delgado-Rios M, Martínez-Martínez A, Núñez-Gastelum JA, Peralta-Videa JR, Gardea-Torresdey JL (2018) Current findings on terrestrial plants-Engineered nanomaterial interactions: Are plants capable of phytoremediating nanomaterials from soil? *Curr Opin Environ Sci Health* 6:9–15.

Crosera M, Prodi A, Mauro M et al. (2015) Titanium dioxide nanoparticle penetration into the skin and effects on HaCaT cells. *Int J Environ Res Public Health* 12(8):9282–9297.

Cui D, Tian F, Ozkan CS, Wang M, Gao H (2005) Effect of single wall carbon nanotubes on human HEK293 cells. *Toxicol Lett* 155(1):73–85.

Da Costa MVJ, Sharma PK (2015) Effect of copper oxide nanoparticles on growth, morphology, photosynthesis, and antioxidant response in *Oryza sativa*. *Photosynthetica* 54(1):110–119.

Davoren M, Herzog E, Casey A, Cottineau B, Chambers G, Byrne HJ, Lyng FM (2007) *In vitro* toxicity evaluation of single walled carbon nanotubes on human A549 lung cells. *Toxicol in Vitro* 21(3):438–448.

De la Torre Roche R, Servin A, Hawthorne J et al. (2015) Terrestrial trophic transfer of bulk and nanoparticle La_2O_3 does not depend on particle size. *Environ Sci Technol* 49(19):11866–11874.

Dhawan A, Sharma V (2010) Toxicity assessment of nanomaterials: Methods and challenges. *Anal Bioanal Chem* 398(2):589–605.

Dimkpa CO, McLean JE, Martineau N, Britt DW, Haverkamp R, Anderson AJ (2013) Silver nanoparticles disrupt wheat (*Triticum aestivum* L.) growth in a sand matrix. *Environ Sci Technol* 47(2):1082–1090.

Doolette CL, McLaughlin MJ, Kirby JK, Navarro DA (2015) Bioavailability of silver and silver sulfide nanoparticles to lettuce (*Lactuca sativa*): Effect of agricultural amendments on plant uptake. *J Hazard Mater* 300(Supplement C):788–795.

Ellinger-Ziegelbauer H, Pauluhn J (2009) Pulmonary toxicity of multi-walled carbon nanotubes (Baytubes(R)) relative to [alpha]-quartz following a single 6 h inhalation exposure of rats and a 3 months post-exposure period. *Toxicologist* 266:16–29.

Ema M, Matsuda A, Kobayashi N, Naya M, Nakanishi J (2011) Evaluation of dermal and eye irritation and skin sensitization due to carbon nanotubes. *Regul Toxicol Pharmacol* 61(3):276–281.

Fitzgerald EJ, Caffrey JM, Nesaratnam ST, McLoughlin P (2003) Copper and lead concentrations in salt marsh plants on the Suir Estuary, Ireland. *Environ Pollut* 123(1):67–74.

Fraga S, Faria H, Soares ME et al. (2013) Influence of the surface coating on the cytotoxicity, genotoxicity and uptake of gold nanoparticles in human HepG2 cells. *J Appl Toxicol* 33(10):1111–1119.

Fraser TWK, Reinardy HC, Shaw BJ, Henry TB, Handy RD (2010) Dietary toxicity of single-walled carbon nanotubes and fullerenes (C60) in rainbow trou (*Oncorhynchus mykiss*). *Nanotoxicol* 5:98–108.

Gao X, Avellan A, Laughton S, Vaidya R, Rodrigues SM, Casman EA, Lowry GV (2018) CuO nanoparticle dissolution and toxicity to wheat (*Triticum aestivum*) in rhizosphere soil. *Environ Sci Technol* 52(5):2888–2897.

Garner KL, Keller AA (2014) Emerging patterns for engineered nanomaterials in the environment: A review of fate and toxicity studies. *J Nanopart Res* 16(8):2503.

Ge S, Wang G, Shen Y et al. (2011) Cytotoxic effects of MgO nanoparticles on human umbilical vein endothelial cells *in vitro*. *IET Nanobiotechnol* 5(2):36–40.

Ghazaryan KA, Movsesyan HS, Khachatryan HE et al. (2018) Copper phytoextraction and phytostabilization potential of wild plant species growing in the mine polluted areas of Armenia. *Geochem Explor Environ Anal* 19(2):155.

Ghosh M, Bandyopadhyay M, Mukherjee A (2010) Genotoxicity of titanium dioxide (TiO_2) nanoparticles at two trophic levels: Plant and human lymphocytes. *Chemosphere* 81(10):1253–1262.

Ghosh PS, Kim CK, Han G, Forbes NS, Rotello VM (2008) Efficient gene delivery vectors by tuning the surface charge density of amino acid-functionalized gold nanoparticles. *ACS Nano* 2(11):2213–2218.

Gogotsi Y (2003) How safe are nanotubes and other nanofilaments? *Mat Res Innovat* 7(4):192–194.

Gonzalez C, Salazar-Garcia S, Palestino G et al. (2011) Effect of 45 nm silver nanoparticles (AgNPs) upon the smooth muscle of rat trachea: Role of nitric oxide. *Toxicol Lett* 207(3):306–313.

Gonzalez L, Lison D, Kirsch-Volders M (2008) Genotoxicity of engineered nanomaterials: A critical review. *Nanotoxicology* 2(4):252–273.

Goodman CM, McCusker CD, Yilmaz T, Rotello VM (2004) Toxicity of gold nanoparticles functionalized with cationic and anionic side chains. *Bioconjug Chem* 15(4):897–900.

Gottschalk F, Sonderer T, Scholz RW, Nowack B (2009) Modeled Environmental concentrations of engineered nanomaterials (TiO_2, ZnO, Ag, CNT, fullerenes) for different regions. *Environ Sci Technol* 43(24):9216–9222.

Guo Y-Y, Zhang J, Zheng Y-F, Yang J, Zhu X-Q (2011) Cytotoxic and genotoxic effects of multi-wall carbon nanotubes on human umbilical vein endothelial cells in vitro. *Mutat Res* 721(2):184–191.

Gupta R, Xie H (2018) Nanoparticles in daily life: Applications, toxicity and regulations. *J Environ Pathol Toxicol Oncol* 37(3):209–230.

Gutiérrez-Praena D, Pichardo S, Sanchez E, Grilo A, Camean AM, Jos A (2011) Influence of carboxylic acid functionalization on the cytotoxic effects induced by single wall carbon nanotubes on human endothelial cells (HUVEC). *Toxicol Vitro* 25(8):1883–1888.

Hackenberg S, Scherzed A, Kessler M et al. (2011) Silver nanoparticles: Evaluation of DNA damage, toxicity and functional impairment in human mesenchymal stem cells. *Toxicol Lett* 201(1):27–33.

Halliwell B, Gutteridge JMC (1985) Free radicals in biology and medicine. *J Free Radic Biol Med* 1(4):331–332.

Hamzeh M, Sunahara GI (2013) In vitro cytotoxicity and genotoxicity studies of titanium dioxide (TiO_2) nanoparticles in Chinese hamster lung fibroblast cells. *Toxicol Vitro* 27(2):864–873.

Hawthorne J, De la Torre Roche R, Xing B et al. (2014) Particle-size dependent accumulation and trophic transfer of cerium oxide through a terrestrial food chain. *Environ Sci Technol* 48(22):13102–13109.

Hawthorne J, Musante C, Sinha SK, White JC (2012) Accumulation and phytotoxicity of engineered nanoparticles to *Cucurbita pepo*. *Int J Phytoremediat* 14(4):429–442.

Hernandez-Viezcas JA, Castillo-Michel H, Andrews JC et al. (2013) In situ synchrotron X-ray fluorescence mapping and speciation of CeO(2) and ZnO nanoparticles in soil cultivated soybean (*Glycine max*). *ACS Nano* 7(2):1415–1423.

Herzog E, Byrne HJ, Casey A et al. (2009) SWCNT suppress inflammatory mediator responses in human lung epithelium in vitro. *Toxicol Appl Pharmacol* 234(3):378–390.

Herzog E, Casey A, Lyng FM, Chambers G, Byrne HJ, Davoren M (2007) A new approach to the toxicity testing of carbon based nanomaterials- the clonogenic assay. *Toxicol Lett* 174(1–3):49–60.

Hirano S, Fujitani Y, Furuyama A, Kanno S (2010) Uptake and cytotoxic effects of multi-walled carbon nanotubes in human bronchial epithelial cells. *Toxicol Appl Pharmacol* 249(1):8–15.

Hong J, Rico CM, Zhao L, Adeleye AS, Keller AA, Peralta-Videa JR, Gardea-Torresdey JL (2015) Toxic effects of copper-based nanoparticles or compounds to lettuce (*Lactuca sativa*) and alfalfa (*Medicago sativa*). *Environ Sci Proc Impact* 17(1):177–185.

Hsiao J-K, Chu H-H, Wang Y-H et al. (2008) Macrophage physiological function after superparamagnetic iron oxide labeling. *NMR Biomed* 21(8):820–829.

https://www.marketsandmarkets.com/Market-Reports/metal-nanoparticle-market-38262033.html (accessed on 26 April 2019).

https://www.prnewswire.com/news-releases/the-global-market-for-metal-oxide-nanoparticles-to-2020-210 803631.html (accessed on 26 April 2019).

Huang X, Li L, Liu T, Hao N, Liu H, Chen D, Tang F (2011) The shape effect of mesoporous silica nanoparticles on biodistribution, clearance, and biocompatibility in vivo. *ACS Nano* 5(7):5390–5399.

Husain M, Wu D, Saber AT et al. (2015) Intratracheally instilled titanium dioxide nanoparticles translocate to heart and liver and activate complement cascade in the heart of C57BL/6 mice. *Nanotoxicology* 9(8):1013–1022.

Hutter E, Boridy S, Labrecque S, Lalancette-Hébert M, Kriz J, Winnik FM, Maysinger D (2010) Microglial response to gold nanoparticles. *ACS Nano* 4(5):2595–2606.

Jacobsen NR, Pojana G, White P et al. (2008) Genotoxicity, cytotoxicity, and reactive oxygen species induced by single-walled carbon nanotubes and C(60) fullerenes in the FE1-Mutatrade mark Mouse lung epithelial cells. *Environ Mol Mutagen* 49(6):476–487.

Jain TK, Reddy MK, Morales MA, Leslie-Pelecky DL, Labhasetwar V (2008) Biodistribution, clearance, and biocompatibility of iron oxide magnetic nanoparticles in rats. *Mol Pharm* 5(2):316–327.

Jang J, Lim D-H, Choi I-H (2010) The impact of nanomaterials in immune system. *Immune Netw* 10(3):85–91.

Jevprasesphant R, Penny J, Jalal R, Attwood D, McKeown NB, D'Emanuele A (2003) The influence of surface modification on the cytotoxicity of PAMAM dendrimers. *Int J Pharm* 252(1–2):263–266.

Jia G, Wang H, Yan L et al. (2005) Cytotoxicity of carbon nanomaterials: Single-wall nanotube, multi-wall nanotube, and fullerene. *Environ Sci Technol* 39(5):1378–1383.

Judy JD, Unrine JM, Rao W, Bertsch PM (2012) Bioaccumulation of gold nanomaterials by *Manduca sexta* through dietary uptake of surface contaminated plant tissue. *Environ Sci Technol* 46(22):12672–12678.

Kalman J, Paul KB, Khan FR, Stone V, Fernandes TF (2015) Characterisation of bioaccumulation dynamics of three differently coated silver nanoparticles and aqueous silver in a simple freshwater food chain. *Environ Chem* 12(6):662–672.

Kang S, Mauter MS, Elimelech M (2009) Microbial cytotoxicity of carbon-based nanomaterials: Implications for river water and wastewater effluent. *Environ Sci Technol* 43(7):2648–2653.

Kasemets K, Ivask A, Dubourguier H-C, Kahru A (2009) Toxicity of nanoparticles of ZnO, CuO and TiO_2 to yeast *Saccharomyces cerevisiae*. *Toxicol Vitro* 23(6):1116–1122.

Kawata K, Osawa M, Okabe S (2009) In vitro toxicity of silver nanoparticles at noncytotoxic doses to HepG2 human hepatoma cells. *Environ Sci Technol* 43(15):6046–6051.

Keller AA, Adeleye AS, Conway et al. (2017) Comparative environmental fate and toxicity of copper nanomaterials. *NanoImpact* 7:28–40.

Keller AA, Lazareva A (2013) Predicted releases of engineered nanomaterials: from global to regional to local. *Environ Sci Technol Lett* 1(1):65–70.

Keller AA, McFerran S, Lazareva A, Suh S (2013) Global life cycle releases of engineered nanomaterials. *J Nanopart Res* 15(6), 1692. Doi: 10.1007/s11051-013-1692-4.

Kim IY, Joachim E, Choi H, Kim K (2015) Toxicity of silica nanoparticles depends on size, dose, and cell type. *Nanomedicine* 11(6):1407–1416.

Kim YS, Kim JS, Cho HS et al. (2008) Twenty-eight-day oral toxicity, genotoxicity, and gender-related tissue distribution of silver nanoparticles in Sprague-Dawley rats. *Inhal Toxicol* 20(6):575–583.

Kirchner C, Liedl T, Kudera S et al. (2005) Cytotoxicity of colloidal CdSe and CdSe/ZnS nanoparticles. *Nano Lett* 5(2):331–338.

Kokura S, Handa O, Takagi T, Ishikawa T, Naito Y, Yoshikawa T (2010) Silver nanoparticles as a safe preservative for use in cosmetics. *Nanomedicine* 6(4):570–574.

Kolosnjaj-Tabi J, Hartman KB, Boudjemaa S et al. (2010) In vivo behaviour of large doses of ultra short and full-length single-walled Swiss mice. *ACS Nano* 4(3):1481–1492.

Kononenko V, Repar N, Marušič N, Drašler B, Romih T, Hočevar S, Drobne D (2017) Comparative in vitro genotoxicity study of ZnO nanoparticles, ZnO macroparticles and $ZnCl_2$ to MDCK kidney cells: Size matters. *Toxicol Vitro* 40:256–263.

Kumari M, Mukherjee A, Chandrasekaran N (2009) Genotoxicity of silver nanoparticles in *Allium cepa*. *Sci Total Environ* 407(19):5243–5246.

Larue C, Laurette J, Herlin-Boime N et al. (2012) Accumulation, translocation and impact of TiO_2 nanoparticles in wheat (*Triticum aestivum* spp.): Influence of diameter and crystal phase. *Sci Total Environ* 431:197–208.

Le Van N, Rui Y, Cao W, Shang J, Liu S, Nguyen Quang T, Liu L (2016) Toxicity and bio-effects of CuO nanoparticles on transgenic Ipt-cotton. *J Plant Interact* 11(1):108–116.

Lee WM, An YJ, Yoon H, Kweon HS (2008) Toxicity and bioavailability of copper nanoparticles to the terrestrial plants mung bean (*Phaseolus radiatus*) and wheat (*Triticum aestivum*): Plant agar test for water-insoluble nanoparticles. *Environ Toxicol Chem* 27(9):1915–1921.

Lee W-M, Kwak JI, An Y-J (2012) Effect of silver nanoparticles in crop plants *Phaseolus radiatus* and *Sorghum bicolor*: Media effect on phytotoxicity. *Chemosphere* 86(5):491–499.

Li JG, Li QN, Xu JY et al. (2009) The pulmonary toxicity of multi-wall carbon nanotubes in mice 30 and 60 days after inhalation exposure. *J Nanosci Nanotechnol* 9(2):1384–1387.

Li L, Liu T, Fu C, Tan L, Meng X, Liu H (2015) Biodistribution, excretion, and toxicity of mesoporous silica nanoparticles after oral administration depend on their shape. *Nanomedicine* 11(8):1915–1924.

Li M, Ahmed GJ, Li C et al. (2016) Brassinosteroid ameliorates zinc oxide nanoparticles-induced oxidative stress by improving antioxidant potential and redox homeostasis in tomato seedling. *Front Plant Sci* 7:615.

Li XY, Brown D, Smith S, MacNee W, Donaldson K (1999) Short-term inflammatory responses following intratracheal instillation of fine and ultrafine carbon black in rats. *Inhal Toxicol* 11(8):709–731.

Lindberg HK, Falck GC-M, Suhonen S et al. (2009) Genotoxicity of nanomaterials: DNA damage and micronuclei induced by carbon nanotubes and graphite nanofibres in human bronchial epithelial cells in vitro. *Toxicol Lett* 186(3):166–173.

Long TC, Saleh N, Tilton RD, Lowry GV, Veronesi B (2006) Titanium dioxide (P25) produces reactive oxygen species in immortalized brainmicroglia (BV2) implications for nanoparticle neurotoxicity. *Environ Sci Technol* 40(14):4346–4352.

Luo Y-H, Chang LW, Lin P (2015) Metal-based nanoparticles and the immune system: Activation, inflammation, and potential applications. *BioMed Res Int* 2015:143720.

Magrez A, Kasa S, Salicio V et al. (2006) Cellular toxicity of carbon based nanomaterials. *Nano Lett* 6(6):1121–1125.

Ma-Hock L, Treumann S, Strauss V et al. (2009) Inhalation toxicity of multiwall carbon nanotubes in rats exposed for 3 months. *Toxicol Sci* 112(2):468–481.

Malik N, Wiwattanapatapee R, Klopsch R et al. (2000) Dendrimers: Relationship between structure and biocompatibility in vitro, and preliminary studies on the biodistribution of ^{125}I labelled polyamidoamine dendrimers in vivo. *J Control Release* 65(1–2):133–148.

Manna SK, Sarkar S, Barr J et al. (2005) Single-walled carbon nanotubes induces oxidative stress and activates nuclear transcription factor-kB in human keratinocytes. *Nano Lett* 5(9):1676–1684.

Mansouri E, Khorsandi L, Orazizadeh M, Jozi Z (2015) Dose-dependent hepatotoxicity effects of zinc oxide nanoparticles. *Nanomed J* 2(4):273–282.

Marquis BJ, Love SA, Braun KL, Haynes CL (2009) Analytical methods to assess nanoparticle toxicity. *Analyst* 134(3):425–439.

McGillicuddy E, Murray I, Kavanagh S et al. (2017) Silver nanoparticles in the environment: Sources, detection and ecotoxicology. *Sci Total Environ* 575:231–246.

McTeer J, Dean AP, White KN, Pittman JK (2014) Bioaccumulation of silver nanoparticles into *Daphnia magna* from a freshwater algal diet and the impact of phosphate availability. *Nanotoxicology* 8(3):305–316.

Mehta CM, Srivastava R, Arora S, Sharma AK (2016) Impact assessment of silver nanoparticles on plant growth and soil bacterial diversity. *3 Biotech* 6(2):254.

Mishra S, Keswani C, Abhilash PC, Fraceto LF, Singh HB (2017) Integrated approach of Agri-nanotechnology: Challenges and future trends. *Front Plant Sci.* doi:10.3389/fpls.2017.00471.

Mohamed HR, Hussien NA (2016) Genotoxicity studies of titanium dioxide nanoparticles (TiO_2NPs) in the brain of mice. *Scientifica (Cairo)* 2016:6710840.

Monica RC, Cremonini R (2009) Nanoparticles and higher plants. *Caryologia* 62(2):161–165.

Monteiro-Riviere NA, Nemanich RJ, Inman AO, Wang YY, Riviere JE (2005) Multi-walled carbon nanotube interactions with human epidermal keratinocytes. *Toxicol Lett* 155(3):377–384.

Mroz RM, Schins RP, Li H, Drost EM, Macnee W, Donaldson K (2007) Nanoparticle carbon black driven DNA damage induces growth arrest and AP-1 and NFkappaB DNA binding in lung epithelial A549 cell line. *J Physiol Pharmacol* 58(Supplement 5):461–470.

Mukherjee A, Peralta-Videa JR, Bandyopadhyay S, Rico CM, Zhao L, Gardea-Torresdey JL (2014) Physiological effects of nanoparticulate ZnO in green peas (*Pisum sativum* L.) cultivated in soil. *Metallomics* 6(1):132–138.

Mukherjee A, Sun Y, Morelius E et al. (2016) Differential toxicity of bare and hybrid ZnO nanoparticles in green pea (*Pisum sativum* L.): A life cycle study. *Front Plant Sci* 12(6):1242.

Muller J, Delos M, Panin N, Rabolli V, Huaux F, Lison D (2009) Absence of carcinogenic response to multiwall carbon nanotubes in a 2-year bioassay in the peritoneal cavity of the rat. *Toxicol Sci* 110(2):442–448.

Muller J, Huaux F, Fonseca A et al. (2008) Structural defects play a major role in the acute lung toxicity of multiwall carbon nanotubes: Toxicological aspects. *Chem Res Toxicol* 21(9):1698–1705.

Nair PM, Chung IM (2014) Physiological and molecular level effects of silver nanoparticles exposure in rice (*Oryza sativa* L.) seedlings. *Chemosphere* 112:105–113.

Nair PM, Chung IM (2015) Study on the correlation between copper oxide nanoparticles induced growth suppression and enhanced lignification in Indian mustard (*Brassica juncea* L.). *Ecotoxicol Environ Saf* 113:302–313.

Nair R, Varghese SH, Nair BG, Maekawa T, Yoshida Y, Kumar DS (2010) Nanoparticulate material delivery to plants. *Plant Sci* 179(3):154–163.

Nations S, Wages M, Caas JE, Maul J, Theodorakis C, Cobb GP (2011) Acute effects of Fe_2O_3, TiO_2, ZnO and CuO nanomaterials on *Xenopus laevis*. *Chemosphere* 83(8):1053–1061.

Nel A, Xia T, Mädler L, Li N (2006) Toxic potential of materials at the nanolevel. *Science* 311(5761):622–627.

Ng CT, Yong LQ, Hande MP, Ong CN, Yu LE, Bay BH, Baeg GH (2017) Zinc oxide nanoparticles exhibit cytotoxicity and genotoxicity through oxidative stress responses in human lung fibroblasts and *Drosophila melanogaster*. *Int J Nanomed* 12:1621–1637.

Nishimori H, Kondoh M, Isoda K, Tsunoda S, Tsutsumi Y, Yagi K (2009) Silica nanoparticles as hepatotoxicants. *Eur J Pharm Biopharm* 72(3):496–501.

Oberdorster E (2004) Manufactured nanomaterials (fullerenes, C-60) induce oxidative stress in the brain of juvenile largemouth bass. *Environ Health Perspec* 112(10):1058–1062.

Oberdorster G, Oberdorste E, Oberdorster J (2005) Nanotoxicology: An emerging discipline evolving from studies of ultrafine particles. *Environ Health Perspect* 113(7):823–839.

Ochoa L, Zuverza-Mena N, Medina-Velo IA, Flores-Margez JP, Peralta-Videa JR, Gardea-Torresdey JL (2018) Copper oxide nanoparticles and bulk copper oxide, combined with indole-3-acetic acid, alter aluminum, boron, and iron in *Pisum sativum* seeds. *Sci Total Environ* 634:1238–1245.

Pacurari M, Yin XJ, Zhao J et al. (2008) Raw single-wall carbon nanotubes induce oxidative stress and activate MAPKs, AP-1, NF-kappaB, and Akt in normal and malignant human mesothelial cells. *Environ Health Perspect* 116(9):1211–1217.

Panda KK, Achary VMM, Krishnaveni R, Padhi BK, Sarangi SN, Sahu SN, Panda BB (2011) In vitro biosynthesis and genotoxicity bioassay of silver nanoparticles using plants. *Toxicol Vitro* 25(5):1097–1105.

Park H-G, Kim JI, Chang K-H et al. (2018) Trophic transfer of citrate, PVP coated silver nanomaterials, and silver ions in a paddy microcosm. *Environ Pollut* 235:435–445.

Park E-J, Kim H, Kim Y, Yi J, Choi K, Park K (2010) Inflammatory responses may be induced by a single intratracheal instillation of iron nanoparticles in mice. *Toxicologist* 275(1–3):65–71.

Park MVDZ, Neigh AM, Vermeulen JP et al. (2011) The effect of particle size on the cytotoxicity, inflammation, developmental toxicity and genotoxicity of silver nanoparticles. *Biomaterials* 32(36):9810–9817.

Peng C, Duan D, Xu C et al. (2015) Translocation and biotransformation of CuO nanoparticles in rice (*Oryza sativa* L.) plants. *Environ Pollut* 197:99–107.

Pulskamp K, Diabate S, Krug HF (2007) Carbon nanotubes show no sign of acute toxicity but induce intracellular reactive oxygen species in dependence on contaminants. *Toxicol Lett* 168(1):58–74.

Rajput V, Minkina T, Sushkova S et al. (2019) ZnO and CuO nanoparticles: A threat to soil organisms, plants, and human health. *Environ Geochem Health*. doi:10.1007/s10653-019-00317-3.

Rajput V, Minkina T, Suskova S et al. (2017) Effects of copper nanoparticles (CuONPs) on crop plants: A mini review. *BioNanoSci* 8:36–42.

Rajput V, Minkina T, Ahmed B et al. (2020) Interaction of copper-based nanoparticles to soil, terrestrial, and aquatic systems: Critical review of the state of the science and future perspectives. *Rev Environ Contaminat Toxicol* 252:51–96.

Rajput V, Minkina T, Fedorenko A et al. (2018) Toxicity of copper oxide nanoparticles on spring barley (*Hordeum sativum* distichum). *Sci Total Environ* 645:1103–1113.

Rao S, Shekhawat GS (2016) Phytotoxicity and oxidative stress perspective of two selected nanoparticles in *Brassica juncea*. *3 Biotech* 6(2):244.

Rico CM, Majumdar S, Duarte-Gardea M, Peralta-Videa JR, Gardea-Torresdey JL (2011) Interaction of nanoparticles with edible plants and their possible implications in the food chain. *J Agric Food Chem* 59(8):3485–3498.

Rotoli BM, Bussolati O, Bianchi MG, Barilli A, Balasubramanian C, Bellucci S, Bergamaschi E (2008) Non-functionalized multi-walled carbon nanotubes alter the paracellular permeability of human airway epithelial cells. *Toxicol Lett* 178(2):95–102.

Sadrieh N, Wokovich AM, Gopee NV et al. (2010) Lack of significant dermal penetration of titanium dioxide from sunscreen formulations containing nano- and submicron-size TiO_2 particles. *Toxicol Sci* 115(1):156–166.

Sahu D, Kannan GM, Vijayaraghavan R, Anand T, Khanum F (2013) Nanosized zinc oxide induces toxicity in human lung cells. *ISRN Toxicol* 2013:316075.

Sakamoto Y, Nakae D, Fukumori N et al. (2009) Induction of mesothelioma by a single intrascrotal administration of multi-wall carbon nanotube in intact male Fischer 344 rats. *J Toxicol Sci* 34(1):65–76.

Schins RP (2002) Mechanisms of genotoxicity of particles and fibers. *Inhal Toxicol* 14(1):57–78.

Servin AD, Castillo-Michel H, Hernandez-Viezcas JA, Diaz BC, Peralta-Videa JR, Gardea-Torresdey JL (2012) Synchrotron micro-XRF and micro-XANES confirmation of the uptake and translocation of TiO(2) nanoparticles in cucumber (*Cucumis sativus*) plants. *Environ Sci Technol* 46(14):7637–7643.

Servin AD, Pagano L, Castillo-Michel H et al. (2017) Weathering in soil increases nanoparticle CuO bioaccumulation within a terrestrial food chain. *Nanotoxicology* 11(1):98–111.

Shavandi Z, Ghazanfari T, Moghaddam KN (2011) In vitro toxicity of silver nanoparticles on murine peritoneal macrophages. *Immunopharmacol Immunotoxicol* 33(1):135–140.

Shaw BJ, Handy RD (2011) Physiological effects of nanoparticles on fish: A comparison of nanometals versus metal ions. *Environ Int* 37(6):1083–1097.

Shvedova AA, Kisin E, Murray AR et al. (2008) Inhalation vs. aspiration of single-walled carbon nanotubes in C57BL/6 mice: Inflammation, fibrosis, oxidative stress, and mutagenesis. *Am J Physiol Lung Cell Mol Physiol* 295(4):L552–L565.

Siglienti I, Bendszus M, Kleinschnitz C, Stoll G (2006) Cytokine profile of iron-laden macrophages: Implications for cellular magnetic resonance imaging. *J Neuroimmunol* 173(1–2):166–173.

Song U, Jun H, Waldman B, Roh J, Kim Y, Yi J, Lee EJ (2013) Functional analyses of nanoparticle toxicity: A comparative study of the effects of TiO_2 and Ag on tomatoes (*Lycopersicon esculentum*). *Ecotoxicol Environ Saf* 93:60–67.

Sotto AD, Chiaretti M, Carru GA, Bellucci S, Mazzanti G (2009) Multi-walled carbon nanotubes: Lack of mutagenic activity in the bacterial reverse mutation assay. *Toxicol Lett* 184(3):192–197.

Sung JH, Ji JH, Yoon JU et al. (2008) Lung function changes in Sprague-Dawley rats after prolonged inhalation exposure to silver nanoparticles. *Inhal Toxicol* 20(6):567–574.

Tabet L, Bussy C, Amara N et al. (2009) Adverse effects of industrial multi-walled carbon nanotubes on human pulmonary cells. *J Toxicol Environ Health A* 72(2):60–73.

Takagi A, Hirose A, Nishimura T et al. (2008) Induction of mesothelioma in p53 ± mouse by intraperitoneal application of multi-wall carbon nanotube. *J Toxicol Sci* 33(1):105–116.

Takamiya AS, Monteiro DR, Bernabe DG et al. (2016) In vitro and in vivo toxicity evaluation of colloidal silver nanoparticles used in endodontic treatments. *J Endod* 42(6):953–960.

Tangaa SR, Selck H, Winther-Nielsen M, Khan FR (2016) Trophic transfer of metal-based nanoparticles in aquatic environments: A review and recommendations for future research focus. *Environ Sci Nano* 3(5):966–981.

The Royal Society and the Royal Academy of Engineering (2004) *Nanoscience and Nanotechnologies: Opportunities and Uncertainties.* The Royal Society, London.

Thomas SP, Al-Mutairi EM, De SK (2013) Impact of nanomaterials on health and environment. *Arab J Sci Eng* 38(3):457–477.

Thuesombat P, Hannongbua S, Akasit S, Chadchawan S (2014) Effect of silver nanoparticles on rice (*Oryza sativa* L. cv. KDML 105) seed germination and seedling growth. *Ecotoxicol Environ Safe* 104(Supplement C):302–309.

Tripathi DK, Singh S, Singh S et al. (2017) An overview on manufactured nanoparticles in plants: Uptake, translocation, accumulation and phytotoxicity. *Plant Physiol Biochem* 110:2–12.

Verma SK, Das AK, Gantait S, Kumar V, Gurel E (2019) Applications of carbon nanomaterials in the plant system: A perspective view on the pros and cons. *Sci Total Environ* 667:485–499.

Vevers WF, Jha AN (2008) Genotoxic and cytotoxic potential of titanium dioxide (TiO_2) nanoparticles on fish cells in vitro. *Ecotoxicology* 17(5):410–420.

Vinardell MP, Llanas H, Marics L, Mitjans M (2017) In vitro comparative skin irritation induced by nano and non-nano zinc oxide. *Nanomater (Basel)* 7(3):56.

Wadhwa S, Rea C, O'Hare P et al. (2011) Comparative in vitro cytotoxicity study of carbon nanotubes and titania nanostructures on human lung epithelial cells. *J Hazard Mater* 191(1–3):56–61.

Wan R, Mo Y, Feng L, Chien S, Tollerud DJ, Zhang Q (2012) DNA damage caused by metal nanoparticles: Involvement of oxidative stress and activation of ATM. *Chem Res Toxicol* 25(7):1402–1411.

Wang L, Zhang J, Zheng Y, Yang J, Zhang Q, Zhu X (2010) Bioeffects of CdTe quantum dots on human umbilical vein endothelial cells. *J Nanosci Nanotechnol* 10(12):8591–8596.

Wick P, Manser P, Limbach LK et al. (2007) The degree and kind of agglomeration affect carbon nanotube cytotoxicity. *Toxicol Lett* 168(2):121–131.

Wang P, Menzies NW, Lombi E et al. (2015) Silver sulfide nanoparticles (Ag_2S-NPs) are taken up by plants and are phytotoxic. *Nanotoxicology* 9(8):1041–1049.

Wang Z, Xie X, Zhao J, Liu X, Feng W, White JC, Xing B (2012) Xylem- and phloem- based transport of CuO nanoparticles in maize (*Zea mays* L.). *Environ Sci Technol* 46(8):4434–4441.

Wiesner MR, Lowry GV, Alvarez P, Dionysiou D, Biswas P (2006) Assessing the risks of manufactured nanomaterials. *Environ Sci Technol* 40(14):4336–4345.

Wijnhoven SWP, Peijnenburg WJGM, Herberts CA et al. (2009) Nano-silver a review of available data and knowledge gaps in human and environmental risk assessment. *Nanotoxicology* 3(2):109–138.

Wojtoniszak M, Chen X, Kalenczuk RJ et al. (2012) Synthesis, dispersion, and cytocompatibility of graphene oxide and reduced graphene oxide. *Colloids Surf B Biointerf* 89:79–85.

Wu J, Liu W, Xue C et al. (2009) Toxicity and penetration of TiO_2 nanoparticles in hairless mice and porcine skin after subchronic dermal exposure. *Toxicol Lett* 191(1):1–8.

www.nanodb.dk (accessed on 25 April 2019).

Xia T, Kovochich M, Liong M et al. (2008) Comparison of the mechanism of toxicity of zinc oxide and cerium oxide nanoparticles based on dissolution and oxidative stress properties. *ACS Nano* 2(10):2121–2134.

Xiong T, Dumat C, Dappe V et al. (2017) Copper oxide nanoparticle foliar uptake, phytotoxicity, and consequences for sustainable urban agriculture. *Environ Sci Technol* 51(9):5242–5251.

Yacobi NR, Phuleria HC, Demaio L et al. (2007) Nanoparticle effects on rat alveolar epithelial cell monolayer barrier properties. *Toxicol Vitro* 21(8):1373–1381.

Yamashita K, Yoshioka Y, Higashisaka K et al. (2011) Silica and titanium dioxide nanoparticles cause pregnancy complications in mice. *Nat Nanotechnol* 6(5):321–328.

Yan A, Chen Z (2019) Impacts of silver nanoparticles on plants: A focus on the phytotoxicity and underlying mechanism. *Int J Mol Sci* 20(5):1003.

Yang YX, Song ZM, Cheng B et al. (2014) Evaluation of the toxicity of food additive silica nanoparticles on gastrointestinal cells. *J Appl Toxicol* 34(4):424–435.

Yang Z, Chen J, Dou R, Gao X, Mao C, Wang L (2015) Assessment of the phytotoxicity of metal oxide nanoparticles on two crop plants, maize (*Zea mays* L.) and rice (*Oryza sativa* L.). *Int J Environ Res Public Health* 12(12):15100–15109.

Zafar H, Ali A, Ali JS, Haq IU, Zia M (2016) Effect of ZnO nanoparticles on *Brassica nigra* seedlings and stem explants: Growth dynamics and antioxidative response. *Front Plant Sci* 7:535.

Zeni O, Palombo R, Bernini R, Zeni L, Sarti M, Scarf MR (2008) Cytotoxicity investigation on cultured human blood cells treated with single-wall carbon nanotubes. *Sensor* 8(1):488–499.

Zhang J, Wages M, Cox SB et al. (2012) Effect of titanium dioxide nanomaterials and ultraviolet light coexposure on African clawed frogs (*Xenopus laevis*). *Environ Toxicol Chem* 31(1):176–183.

Zhang XD, Wu D, Shen X et al. (2011) Size dependent in vivo toxicity of PEG-coated gold nanoparticles. *Int J Nanomed* 6:2071–2081.

Zhang Z, Ke M, Qu Q et al. (2018) Impact of copper nanoparticles and ionic copper exposure on wheat (*Triticum aestivum* L.) root morphology and antioxidant response. *Environ Pollut* 239:689–697.

Zhao L, Ortiz C, Adeleye AS, Hu Q, Zhou H, Huang Y, Keller AA (2016) Metabolomics to detect response of lettuce (*Lactuca sativa*) to $Cu(OH)_2$ nanopesticides: Oxidative stress response and detoxification mechanisms. *Environ Sci Technol* 50(17):9697–9707.

Zhao L, Peralta-Videa JR, Varela-Ramirez A et al. (2012) Effect of surface coating and organic matter on the uptake of CeO_2 NPs by corn plants grown in soil: Insight into the uptake mechanism. *J Hazard Mater* 225–226:131–138.

Zhou D, Jin S, Li L, Wang Y, Weng N (2011) Quantifying the adsorption and uptake of CuO nanoparticles by wheat root based on chemical extractions. *J Environ Sci* 23(11):1852–1857.

Zhu H, Han J, Xiao JQ, Jin Y (2008) Uptake, translocation, and accumulation of manufactured iron oxide nanoparticles by pumpkin plants. *J Environ Monit* 10(6):713–717.

Zhu ZJ, Wang H, Yan B et al. (2012) Effect of surface charge on the uptake and distribution of gold nanoparticles in four plant species. *Environ Sci Technol* 46(22):12391–12398.

Zuverza-Mena N, Armendariz R, Peralta-Videa JR, Gardea-Torresdey JL (2016) Effects of silver nanoparticles on radish sprouts: Root growth reduction and modifications in the nutritional value. *Front Plant Sci* 7:90.

13 Patenting Issues in the Development of Nanodrugs

Theivasanthi Thirugnanasambandan

CONTENTS

ABBREVIATIONS

FDA	Food and Drug Administration
USPTO	United States Patent and Trademark Office
MIRIBEL	Minimum Information Reporting in Bio-Nano Experimental Literature
CDER	Center for Drug Evaluation and Research
OTR	Office of Testing and Research
MAPP	Manual of Policies and Procedures
IPR	intellectual property rights
TRIPS	Trade Related Aspects of Intellectual Property Rights
PCT	Patent Cooperation Treaty
WIPO	World Intellectual Property Organization
WTO	World Trade Organization
UN	United Nations
EPC	European Patent Convention
ECHA	European Chemical Agency
ISO	International Standard Organization
API	active pharmaceutical ingredient
NPs	nanoparticles
SLN	solid lipid nanoparticles
P-GQDs	polyethyleneglycol-graphene quantum dots
GQDs	graphene quantum dots
PEI	polyethyleneimine
PAMAM	polyamidoamine
PEG	polyethyleneglycol
PLGA	polylacticglycolicacid
ADME	absorption, distribution, metabolism, and excretion
PBPK	physiologically based pharmacokinetic
EPR effect	enhanced permeability and retention effect
EC$_{50}$	half maximal effective concentration
ROS	reactive oxygen species
MRI	magnetic resonance imaging
TEM	transmission electron microscope
SEM	scanning electron microscope
XRD	X-ray powder diffraction
DLS	dynamic light scattering
BET	Brunauer–Emmett–Teller
TGA	thermogravimetric analysis
ICP-MS	inductively coupled plasma mass spectrometry

13.1 INTRODUCTION

Nanotechnology is an emerging field that is highly interdisciplinary, and application-oriented in nature. Nanotechnologists synthesize novel nanomaterials in different methods with different properties such as size and morphology. The quality and properties of the synthesized nanomaterials determine the applications in various fields. The researchers can protect their intellectual property by obtaining patents for the synthesized nanomaterials, including nano-drugs.

Nanotechnology research is in its infancy. Hence, the researchers are not under any obligation and there is no need for them to confine their research in a particular direction. They can attempt new ways to synthesize, characterize and study the properties of nanomaterials. Bawa et al. have reported that at present, more money is invested in the research and development of new drugs, however, approval for the number of new drugs is very low [1].

A nanopharmaceutical is a product that contains materials in the size range of 1 to 100 nm in at least one dimension, and the nanopharmaceutical itself is defined as having particles in the range of 100 and 1,000 nm. However, it has different or altered pharmaceutical characteristics due to application of nanotechnology when compared to the active pharmaceutical ingredient (API) [2–4]. Presently, some of the nanomedicines related to cancer treatment are in clinical research and another proportion of them are under clinical investigation [5]. Figure 13.1 illustrates the involvement of the various types of nanomaterials (such as inorganic, drug-conjugated, lipid-based, and polymer-based) in the establishment of nanomedicines.

Nanotechnology is the novel technology that has many ideas and approaches to solve the problems arising in the pharmaceutical industry. It is utilized in the bio-pharmaceutical industry for many applications and to produce nanopharmaceutical products, as well as technologies such as nanomedicine, diagnostic tools, biosensors, biomarkers, implant technology, nanorobots, and vector or carrier molecules for theranostics. Nanomedicines, nanoparticles, nanobiotechnology and, nanopharmacology are some of the general terms utilized in nanotechnology-based pharmaceutical applications [6].

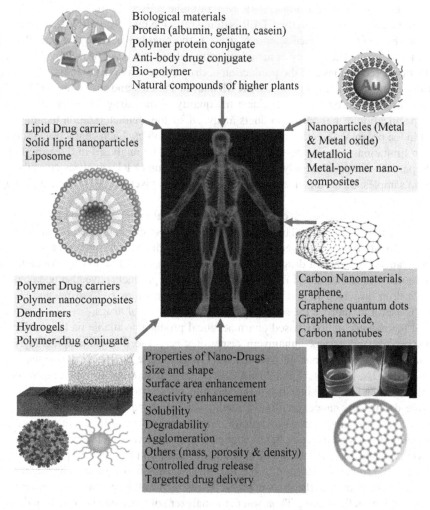

Biological materials
Protein (albumin, gelatin, casein)
Polymer protein conjugate
Anti-body drug conjugate
Bio-polymer
Natural compounds of higher plants

Lipid Drug carriers
Solid lipid nanoparticles
Liposome

Nanoparticles (Metal
& Metal oxide)
Metalloid
Metal-poymer nano-
composites

Polymer Drug carriers
Polymer nanocomposites
Dendrimers
Hydrogels
Polymer-drug conjugate

Carbon Nanomaterials
graphene,
Graphene quantum dots
Graphene oxide,
Carbon nanotubes

Properties of Nano-Drugs
Size and shape
Surface area enhancement
Reactivity enhancement
Solubility
Degradability
Agglomeration
Others (mass, porosity & density)
Controlled drug release
Targetted drug delivery

FIGURE 13.1 Schematic diagram shows the overview of the establishment of nanomedicines for various treatments. Presently, many nanomedicines are under clinical investigation. Lipids/polymers based, inorganic and drug-conjugated nanoparticles have been established in clinical research.

Since 1996, approximately 3,000 patents related to nanopharmaceuticals were issued in the United States. In these patents, diagnosis and treatment of cancer has been the main focus. Nanomaterials have advantages such as increased drug absorption, controlled release of drugs, and reduced side effects [7]. The Center for Drug Evaluation and Research (CDER) of the Food and Drug Administration (FDA) monitors the utilization of nanomaterials in drug products. Liposomal formulations are the nanomaterials that have more attention. They are utilized in the treatment of cancer. The average particle size of these products is around 300 nm [8].

The National Institute of Health (NIH) has invested approximately USD 4.4 billion for nanotechnology research since 2001. The NIH is investigating the clinical translation of nanopharmaceutical products [9]. Discussions among industry, academia, and regulatory agencies are needed to obtain approval for the nanodrug products from government agencies [10].

The FDA received more than 80 applications in nanocrystal technology to improve the solubility and bioavailability of water-insoluble drugs [11]. The CDER has identified the three different analyses of nanodrug products, i.e. characterization of nanodrug products, in vitro tests, and toxicity studies, and verifies these analyses when giving approval to nanodrug products [12].

Nanodrug products have advantages like target specificity. Such a feature is not available in small molecule drugs. Target specificity can be achieved with a polymer coating over nanodrug products. This coating prevents the production of toxicity in immune cells and avoids side effects. The Office of Testing and Research (OTR) in the CDER is doing research on the quality issues associated with nanodrug products. The controlled release of drugs is thoroughly studied to evaluate the safety and efficacy of the product. The quality of nanodrug formulations can be improved by the advancement of manufacturing techniques and the production of drug carriers like liposomes and polymers [13].

The FDA guidance document and CDER Manual of Policies and Procedures (MAPP) are the two documents that can be used to evaluate the quality of nanodrug formulations and products. The size specification of nanodrug products analyzed by the characterization techniques is more important in the pharmacokinetic and pharmacodynamic profiles. Each characterization technique has its own limitations. For example, colloidal samples can be analyzed in dynamic light scattering (DLS) particle size analysis only. X-ray powder diffraction (XRD) analysis gives the crystallite size of solid samples. Scanning electron microscopy (SEM) gives the 3-D image and shape of the nanoparticles only [14].

The number of patents filed decides the commercial value of any technology. This has also become true for nanotechnology. The majority of the patents have been filed on nanocrystal-based drug carriers. In addition, nanosuspensions prepared using the bottom-up approach have been patented [15]. Nanocrystals based drug carriers are the first marketed nanodrug product. They have also been more commercialized [16]. Some solid dispersion products have also been patented [17]; a solid-state drug dispersed in a water-soluble carrier is known as solid dispersion. This type of dispersion is useful to improve the solubility and bioavailability of drugs.

Some of the nanotechnology-based pharmaceutical products available on the market are listed here: liposomal as nanocarriers (annamycin, cisplatin, vincristine, doxorubicin, fentanyl); pegylated proteins and polypeptides as nanopharmaceuticals (pegademase bovine, certolizumab pegol, PEG Interferon ALFA-2A, PEG Interferon ALFA-2B); polymer-based nanoformulations as pharmaceuticals (PEG anti TNF-α antibody fragment, calcium phosphate nanoparticles vaccine adjuvant, poly-l-lysine dendrimer, nanocrystalline 2 methoxyestardiol, paclitaxel nanoparticles in porous, hydrophilic matrix); nanocrystals (megasterol, fenofibrate); protein-drug nanoconjugates (abraxane, denileukin diftitox); surfactant-based nanoproducts (propofol); metal-based nanoformulations (ferumoxytol, ferumoxide); and virosomes (gendicine, rexin-G) [6].

Overlapping of patents occur due to the interdisciplinary nature of the nanotechnology field. Hence, inventors face many difficulties in patent searches as well as patent information retrieval. A technology mining method using Thomson data analyzer software has been reported earlier. The results of that technology show that countries like Korea, Japan, and China are doing well in this regard [18].

13.2 PATENTS

13.2.1 INNOVATION VS INVENTION

An innovation is something that has novelty. It has better properties than exist already. It has been adopted to create positive value. The main difference between an innovation and an invention is the creation of positive value. An invention may not create positive value, but it exhibits further development in a particular area. The theory related to the adopting of an innovation is called *diffusion of innovations* [19].

Scientific skills are essential for an invention to create new ideas but in the case of innovation, skills such as technical and marketing strategy are required to commercialize the idea. Innovation and invention both play major roles in the patents or the intellectual property rights of the inventor. Patenting strategy protects the inventions and innovations from revealing the secrets related to the inventions. Figure 13.2 enumerates the different aspects related to invention and innovation.

13.2.2 INTELLECTUAL PROPERTY RIGHTS

A patent is intellectual property with vested rights. It protects the legal rights of its owner by excluding others from unauthorized making, using, selling, and importing an invention for a limited period (generally for 20 years). It is an exclusive right granted by the government to the inventor in exchange for publishing or public disclosure of the invention. Many countries deal with patent rights under civil law. Patents are essential in some industries and they are not essential, as well as irrelevant, in others [20].

Patent is a negative right; it is not a positive right or privilege. It means that the patent does not give a right to make, use, or sell the invention. This right empowers the patentee to prevent or stop

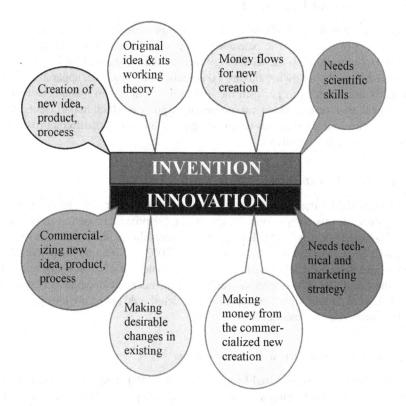

FIGURE 13.2 Schematic diagram reveals the various aspects of invention and innovation.

the utilization (such as manufacture, sale, use, and importation or offer for sale of the patented product, process, or composition) of that invention by others without obtaining permission from the patentee [21].

13.2.3 PATENTS IN VARIOUS COUNTRIES

The patent procedures, requirements, and exclusive rights of each country differ widely according to their laws and international agreements. A patent has one or more claims that define the invention and the specific property right. These claims should also fulfill the patentability requirements like novelty, usefulness, and non-obviousness [22–23]. As per the TRIPS (Trade Related Aspects of Intellectual Property Rights) agreement of World Trade Organization (WTO), patents of any invention (whether product or process, in all fields of technology) should be available in WTO member states. The invention should be new, involve an inventive step and be useful for industrial application [24].

13.2.4 LIMITATION OF PATENT RIGHTS

A patent right has a territorial boundary. It is applicable only within the country which granted the respective patent. Enforcement of a patent and taking legal action against the infringement or violation of the patent rights can be done only within the territory, i.e. the country where the patent was granted. The Patent Cooperation Treaty (PCT) enables the filing of an international patent through a single patent application. However, the discretion of the individual patent office is final when granting a patent against the PCT application.

13.2.5 CLASSIFICATION OF PATENTABILITY

If an invention has novelty, usefulness, and non-obviousness, limited legal monopoly will be granted in the form of a patent. The patent will disclose all the details related to the invention. Hence, other people can develop the technology further and can bring new inventions. Patentable items are classified into the following classes: (a) invention of a machine, apparatus, or device or its interrelated parts are designed to perform the intended functions; (b) manufactured or fabricated items; (c) chemical, mechanical, electrical, or any other process which makes changes in chemical or physical properties; and (d) materials having properties or characters different from their original composition or constituent ingredients.

13.2.6 CRITERIA OF PATENTABILITY

Patents will be granted for the inventions that fulfill the criteria of patentability. As per the Indian Patent Act, a patentable invention means, "a new product or process involving an inventive step and capable of industrial application" [25]. The criteria of patentability are newness, inventive step, and industrial applicability. The Patents (Amendment) Act, 2002, reports that the patentable inventions must not have been known earlier. It will be considered as new, if it has not been published or used anywhere in the world [26]. The invention should have technical advancements better than existing ones or should have economic significance, or both, which should not be obvious [27]. It should have a property like industrial applicability [28].

13.2.7 TYPES OF PHARMACEUTICAL PATENTS IN INDIA

Patents are a major form of Intellectual Property Rights (IPRs). They are utilized by the pharmaceutical industry mainly for trading and Research & Development (R&D) activities. The criteria of patentability, and different types of pharmaceutical patents, provide the fundamental, essential, and relevant knowledge about the pharmaceutical patenting to the researchers, and therefore issues arising during the development and patenting of nanodrugs.

The pharmaceutical industry is a most intense and knowledge-driven field. R&D activities of this industry are very costly and unpredictable. The results of such activities may be a new, inventive, and useful product or process. It can be protected from unauthorized use by proper patent rights. It is an essential right for pharmaceutical companies to compete in the highly competitive market. Pharmaceutical patents in India are classified as: drug compound patents; formulation/composition patents (to prepare a formulation and/or quantity of its key ingredients); synergistic combination patents (enhancement of actions by two or more drugs); technology patents (techniques used to solve problems like stabilization, taste-masking, and solubility increasing); polymorph patents (to reduce impurities or increase stability); biotechnology patents (diagnostic, therapeutic and immunological products); and process patents, i.e. new and inventive process [29].

13.2.8 Transfer of the Patent Rights

A patent right is one kind of property. Hence, the patentee can transfer the vested rights to others by making a patent assignment or by granting a patent license. The Indian Patent Act instructs that an assignment or license of a patent must be in writing with clear terms and conditions, as well as the rights and obligations of the parties [30].

13.3 NANOTECHNOLOGY IN DRUG DESIGN AND DEVELOPMENT

13.3.1 Emerging Technology in Medicine

Nanotechnology can support faster drug discovery and drug development and can reduce the cost. The solubility enhancement and target specificity are the two important things that are possible with nanotechnology. There is no need to adhere to the size definition of nanoparticles, i.e. the size range of the particles should be between 1 and 100 nm. Bioavailability improvement, toxicity reduction, and low dose are the other parameters to be considered in developing a nanodrug. These parameters can also be achieved with a drug of a size greater than 100 nm.

Nanotechnology can provide a solution to cancer therapy. Solubility and bioavailability are some of the challenges and problems related to cancer drugs that are faced by the pharma industry. They can be solved by nanotechnology. The cost and time consumption during drug development can be reduced by using nanotechnology. These solutions bring some benefits to the drug manufacturer, and nanotechnology also offers innovative therapy, with high quality products. Mir et al. have reported on the various roles of nanotechnology in the medical field: from imaging systems to emerging controlled drug delivery (nanomedicine). Figure 13.3 exhibits the properties of NPs as well as the effects on lungs [31].

During the process of clinical translation of nanodrugs, procedures like biological challenges, large scale manufacturing, intellectual property rights, and government regulations should also be considered [32]. The clinical success of iron oxide NPs has not achieved yet. Iron oxide NPs will be the ideal material for drug delivery systems in the future by overcoming the challenges with the help of new technologies [33].

In addition to the encapsulation of the drug internally, the polymer NPs have other advantages like promoting the drug permeation along the gastro-intestinal mucus barrier and increasing the residence time. The solubility of the drug can be increased by reducing the particle size, which in turn increases the surface area of the drug. To reduce the particle size, top-down approaches like high pressure homogenization and bead and pearl milling technologies are applied in the pharmaceutical industry. As an example, nanocrystals produced by a wet milling process are useful to improve the solubility of drugs. Liposomes and iron carbohydrate complexes are usually used for drug loading. Dexferrum, INFeD, Ferrlecit, Venofer, Ferinject, and Ferheme are some of the products with iron

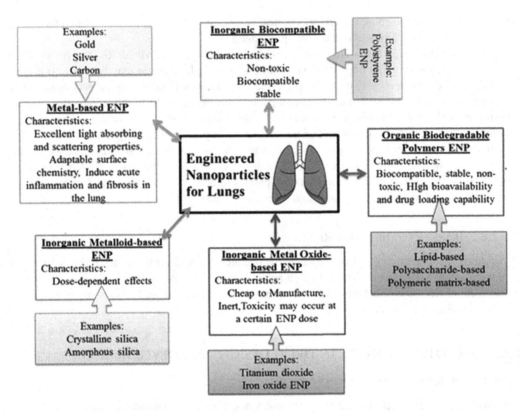

FIGURE 13.3 Schematic diagram shows the properties of various engineered nanoparticles (ENPs). The efficacious effects produced by the ENPs on lungs are illustrated. (From Mir Maria et al. 2017.)

carbohydrate complex available in the market. These iron carbohydrate drugs are characterized to properly understand the absorption, distribution, metabolism, and excretion (ADME) processes of the drug in the body [10].

Liposomes act as a drug carrier. Many drugs are loaded inside the liposomes. When comparing them, doxorubicin maintains the top position in loading inside liposomes. The liposomes are usually administered through an intravenous route. Theoretical studies like physiologically based pharmacokinetic (PBPK) modeling and simulation methods are used to analyze the ADME mechanism of nanodrug products [34].

In developing countries there is a higher dependence on herbal medicines. Improving the standard of these medicines will be useful for such countries. The efficacy of herbal medicines is lower because their bioactive compounds (like flavonoids, tannins, and terpenoids) are not able to enter the lipid membranes of the cell. This challenge will be avoided by nanotechnology. If this technology is combined with herbal medicines, it will support the enhancement of the actions of the related herbal medicines [35].

13.3.2   SYNTHESIS OF NANOPARTICLES

A method for large scale preparation of drug loading NPs has recently been reported. The name of the method is the flash nanoprecipitation method. It supports production of up to 1 kg of NPs per day. Antimalarial drug lumefantrine-loaded NPs have been prepared by this method with the size of around 200 nm. Figure 13.4 shows the mixing chambers using for nanoparticle synthesis [36]. Large scale preparation of NPs for pharmaceutical use is a major challenge. Flash nanoprecipitation solves that challenge. It is also more technically advanced compared to existing methods. Technical advancement is an issue faced by inventions when patenting.

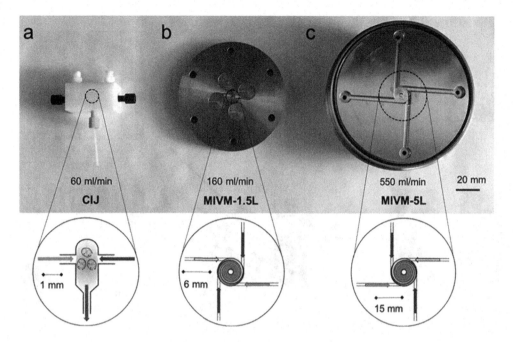

FIGURE 13.4 Images of the mixers (used for nanoparticle synthesis). (a) Confined impinging jet mixer (CIJ). (b) Multi-inlet vortex mixer (MIVM)-1.5 L and (c) MIVM-5 L. Schematic diagram of the mixing chambers of CIJ, MIVM-1.5 L, and MIVM-5 L are shown below. (From Feng Jie et al. 2019.)

Biosynthesis of NPs utilizing biological materials (plant extracts, fungi, bacteria, and algae) has many benefits like ecofriendliness and compatibility for pharmaceuticals and other biomedical applications because of the synthesis process. In this process, the synthesis has been done without the inclusion of any toxic chemicals [37]. Metal NPs obtained from the plant-mediated synthesis method are utilized as potential pharmaceutical agents and nanopharmaceutical products for the treatment of malaria, HIV, cancer, hepatitis, and other diseases. Some of the nanopharmaceutical products have already been commercialized and are available on the market. Figure 13.5 exhibits the plant-mediated synthesis (involvement of bioreduction mechanisms) of the metal NPs [6].

13.3.3 Nanotoxicity

The toxic effect of nanodrugs can cause oxidative stress, inflammation, and genotoxicity. The mechanism behind the toxicity of NPs should be studied clearly to confirm the safety measures of nanodrugs. polyethyleneglycol, polyethyleneoxide, polaxamer, polaxamine, and polysorbate 80 are some of the polymers in which the drugs are loaded. They are administered through different routes such as injection, oral, and pulmonary. The coated polymers can also cause cell proliferation, differentiation, necrosis, and apoptosis. Inhaled NPs are deposited in the lung periphery which causes inflammation [38]. Nanotoxicology studies the toxicity produced by the nanomaterials at various levels, i.e. from synthesis to applications. Figure 13.6 exhibits the biomedical applications of nanoparticles with interrelationships among nanotechnology, nanobiotechnology, and nanotoxicology.

The drugs loaded inside the polymer NPs attain a particle size of around 300 nm. This kind of innovative formulation reduces the toxicity of the drug by accumulation in target cells. It also offers a prolonged effect. For example, the chemotherapy drug temozolomide is encapsulated inside the polymer NPs (utilized for the treatment of brain cancer, gliomas, and melanomas). In vitro cytotoxic studies show that temozolomide loaded polymer NPs not only inhibit tumor growth but also increase the life span of the animal [39].

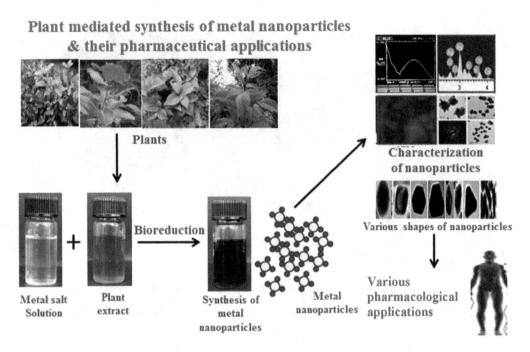

FIGURE 13.5 Schematic diagram shows the biogenesis of metal nanoparticles. It illustrates the various steps related to the metal nanoparticles from plant-mediated synthesis (bioreduction mechanism) to pharmacological application level. (From Khandel Pramila et al. 2018.)

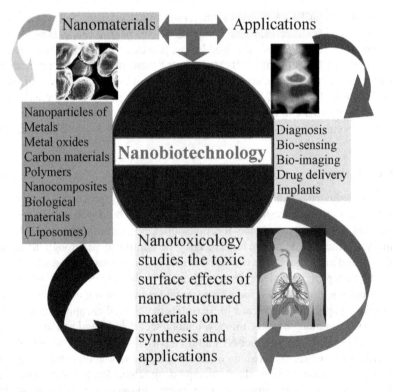

FIGURE 13.6 Schematic diagram shows the biomedical applications of nanoparticles. Also, it illustrates the interrelations among nanotechnology, nanobiotechnology, and nanotoxicology.

13.4 NANOPARTICLES IN DRUG DESIGN AND DEVELOPMENT

13.4.1 NANOPARTICLES

Nanoparticles are used to improve the pharmacokinetic and pharmacodynamic properties of the drugs that are utilized in various treatments. Slowing et al. have reported that mesoporous silica NPs are used for drug delivery because of their efficient cell internalization in animal and plant cells. Their morphology can be controlled easily. Silica NPs also have more surface area and pore volume [40].

Active targeting of the NPs can be done by attaching ligands (like proteins, antibodies, or small molecules) to them. These attached ligands are also called biomarkers. They can reach the receptors on specific cells. This kind of active targeting can enhance the intracellular drug accumulation and cellular uptake. The NPs can circulate for a long time by giving more exposure of the drug to tumor sites or affected parts of the body. Tumors have increased permeability of blood and poor lymphatic drainage which is known as the enhanced permeability and retention (EPR) effect. Passive targeting is done to target solid tumors due to the EPR effect on tumors. The clinical validation of active targeting is less because of the toxicity associated with the biomarkers. The increased accumulation of the drug at the tumor rather than normal cells reduces the amount of the drug used. It leads to a reduction in the side effects. Less than 0.01% of the drug is sufficient in this case when compared to 1 to 5% of a conventional method of drug delivery. Doxil (doxorubicin hydrochloride) was approved by the FDA during 1995. It is the first FDA approved drug for the treatment of Kaposi's sarcoma in HIV patients. Doxil is still used for the treatment of ovarian and metastatic breast cancer. Many chemotherapy drugs for cancer are hydrophobic. They are dissolved in toxic agents like polyethoxylated castor oil to avoid the hydrophobic effect. Hence, dose reduction is important to reduce toxic effects and it is possible with nanotechnology [41].

The EPR effect is not present in all tumors. Before reaching the target, the drug crosses several barriers like immunological clearance, renal clearance, enzymatic and mechanical degradation, and extracellular matrix. Some harmful solvents are used to dissolve the water-insoluble drugs. Dimethyl sulfoxide (DMSO) is one of the harmful solvents. Its toxicity has been confirmed from various analyses and studies. Polyethoxylated castor oil (Cremophore EL) and dehydrated ethanol mixture and polysorbate 80 (Tween 80) are the solvents used to dissolve paclitaxel and docetaxel. To avoid the side effects of these solvents, patients are treated with antihistamines and corticosteroids. Therefore, NPs have become a suitable solution for these difficulties. For example, Abraxane is a product in which the drug paclitaxel is bound inside albumin NPs. This product can be used at higher doses due to the biocompatible nature of the albumin NPs. However, the safety issues of NPs are also to be considered. Some of the NPs used for drug loading are toxic. Carbon nanotubes induce mesothelioma similar to asbestos. The toxicity of NPs may vary based on their size and shape. So, a complete characterization of the NPs and their toxicity studies are essential for their approval in patenting, commercialization and clinical translation. The toxic properties of NPs can be used to destroy cancer cells instead of using drugs. Surface modification of NPs can reduce their toxic effects. For example, gadolinium fullerene particles decorated with hydroxyl groups prevent ROS production in cells [42].

13.4.2 PARTICLES THERAPY

Worldwide, radiation therapy is used in the treatment of cancer. X-rays are involved in this therapy to produce radiation. However, ions have more ballistic properties and are biologically more effective when compared to X-rays. Nowadays, X-rays are replaced with ion beams. Protons and carbon ions are involved to produce the ion beams for medical uses. This technology is found to be more effective and economically viable. However, these ions have some limitations. They create negligible radiation effects on the healthy cells at the entrance of the tumor. These limitations (without affecting the healthy cells) can be avoided by using NPs [43].

The utilization of charged particle beams for cancer radiation therapy is known as particle therapy. It is a recent cancer therapy that has more advantages than conventional radiotherapy. It reduces the radiation dose and supports ion radiation effects at the target only (without damaging the healthy tissues before reaching the tumor). Figure 13.7 explains the particle therapy and nanomedicines that are applied to brain tumors as well as their effects [43]. The ion beams can be tuned by modulating the energy. Hence, during this radiation therapy, with the support of NPs, the radiation effect can be delivered at the target only. It also avoids the undesirable effects on the healthy tissues that are present near and around the tumor.

13.4.3 SOLID LIPID NANOPARTICLES (SLNs)

Solid lipid nanoparticles (SLN) are lipid bilayers; they can encapsulate the drug. This bilayer can be attached to the bilayer of cell membranes to deliver the drug. Mishra et al. have reported that SLNs are emerging as one of the colloidal nanodrug delivery systems Figure 13.8 (a) explains the synthesis procedure of SLNs. Figure 13.8 (b) illustrates the various applications of SLNs in drug delivery systems [44].

SLNs consist of glycerides, fatty acids, and hydrophilic drugs along with an emulsifier [45]. Even though more SLN products are patented, marketed products are still few. Regulatory restrictions and economic issues are the major reasons for the low number of marketed products. Muller and Lucks have filed a patent application for an SLN product (nanoemulsion) during 1996. It is the first patented SLN product that describes nanoemulsion preparation using a high-pressure homogenization process [46].

13.4.4 POLYMER NANOPARTICLES

Proteins are used as drugs which are easily digested in the stomach before they reach the target. Hence, proteins are coated with polymers to release in a controlled way. This controlled release technique is known as polymer therapeutics. Dendrimers are branched polymers and are used as drug carriers. Cancer medicine like methotrexate is attached to dendrimers for drug delivery.

Polymer therapeutics can improve the pharmacokinetic properties of the drug [47]. Anticancer drugs, therapeutic DNAs, and Small interfering RNA (Ribonucleic Acid) (siRNAs) are attached on the surface of dendrimers due to the presence of more functional groups on the surface [48]. Many dendrimers like polyamidoamine (PAMAM) have been patented for various applications related to drug delivery, imaging, and diagnosis [49]. Polymer NPs like polyhydroxyalkanoate are

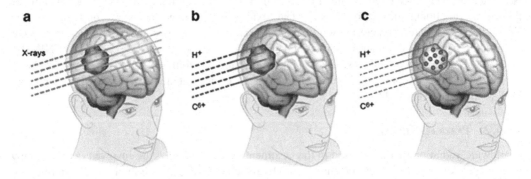

FIGURE 13.7 Schematic diagram shows the particle therapy and nanomedicine. (a) Highly penetrating X-ray radiation damaging the healthy tissues. (b) Ballistic effects of ions: negligible radiation effects around the tumor or healthy cells and considerable effects at the entrance of the tumor. (c) Improvement of ion radiation effects by nanoparticles without affecting the healthy cells (nanoparticles are present inside the tumor or target). (From Lacombe Sandrine et al. 2017.)

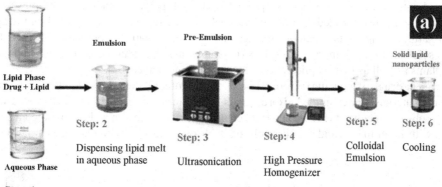

FIGURE 13.8 Schematic illustration of solid lipid nanoparticles (SLNs). (a) Step by step preparation of SLNs in the hot homogenization method. (b) Various applications of SLNs in medicine. (From Mishra Vijay 2018.)

used as drug carriers. They are also utilized in protein purification and immobilization of matrices. These polymer NPs have biodegradable and good mechanical flexibility properties [50]. Polymeric micelles are one kind of drug delivery carrier formed by block co-polymers with an inner hydrophobic core and outer hydrophilic nature. The stability of polymeric micelles is affected by polymer composition, drug encapsulation, and environmental conditions [51].

Polymers with more water content are known as hydrogels. These hydrogel NPs are used as drug carriers. Biopolymers like chitosan and alginate are utilized to prepare hydrogel NPs for the loading of drugs. Some synthetic polymers like polyvinyl alcohol, polyethyleneimine, and polyN-isopropylacrylamide are also utilized to prepare hydrogel NPs. Patra et al. have reported on the applications of natural compounds in nanomedicines and sources of natural compounds. Figure 13.9 (a) exhibits sources of biopolymers. Figure 13.9 (b) enumerates the natural compounds obtained from higher plants as well as their applications related to nanomedicines [52].

Anticancer drugs like paclitaxel, doxorubicin, 5- fluorouracil, and dexamethasone are prepared using nanomaterials. For example, doxorubicin is loaded inside liposomes and then coated with the polymer polyethyleneglycol (PEG). The PEG coating process is known as PEGylation. Paclitaxel is attached with albumin NPs and it is currently available on the market with the trade name Abraxane.

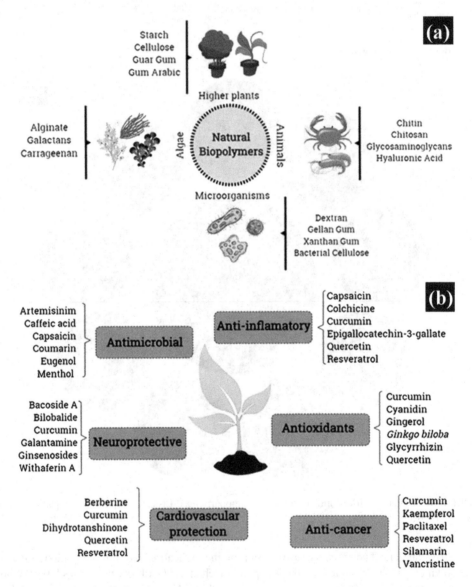

FIGURE 13.9 Schematic diagram shows the natural materials utilized in nanomedicine. (a) Various sources of natural biopolymers, i.e. higher plants, animals, microorganisms and algae (b) Some of the natural compounds obtained from higher plants for nanomedicine applications. Among these natural compounds, some are already available on the market, others are in clinical trials and others are undergoing research. (From Patra et al. 2018.)

Pillai et al. reported that the cancer nanomedicines like Abraxane, Doxil, DaunoXome, Oncaspar, and DepoCyt are in the advanced stage of clinical development. The FDA has given approval for these nanomedicines due to their advanced stage in clinical development. [53].

Paclitaxel and doxorubicin are the well-known cancer drugs formulated with liposomes and albumin nanoparticles. Dexamethasone and 5-fluorouracil are some other cancer drugs designed with nanoparticles. Quantum dots, chitosan nanoparticles, and polylacticglycolic acid (PLGA) NPs are some of the drug carriers used to encapsulate iRNA for in vitro delivery [39].

Polymers are also used to mask the taste of drugs. As an example, Eudragit E100 is employed for the coating of potassium chloride (KCl). This drug is utilized for the treatment of hypokalemia. Eudragit E100 is a water-insoluble polymer that effectively (2 times) masks the unpleasant taste and improves the palatability of the KCl. After administration of this drug, i.e. within 30 minutes, it releases the entire (100%) KCl. The graph (Figure 13.10) compares the in vitro analysis result related to the KCl releasing time of marketed KCl syrup formulation and Eudragit E100 coated KCl syrup formulation [54].

13.4.5 GOLD NANOPARTICLES

The combined activity of anticancer drugs (bleomycin and doxorubicin) and drug carriers (gold NPs) on HeLa cancer cell lines is analyzed. This combination decreases drug resistance and improves the outcomes of chemotherapy. It also improves the therapeutic efficacy and reduces the drug concentration. This drug carrier has excellent stability, high drug loading, an excellent drug release behavior, and half maximal effective concentration (high EC 50) values. Figure 13.11 shows the synthesis and characterizations of anticancer drugs loaded with gold nanoparticles (AuNPs) [55].

AuNPs are promising for miRNA delivery by decorating the NPs with polyethyleneimide and liposomes. These polymers with ultra-small gold NPs can offer high transfection efficiency, more cellular uptake and less toxicity [56]. AuNPs have been synthesized for curcumin drug delivery. In this process, initially, gold precursor has been chemically reduced to prepare AuNPs. In the next step, gelatin (biopolymer) has been coated on the prepared AuNPs. Finally, the resulted nanocomposite (curcumin drug carrier) has been loaded with curcumin (for curcumin delivery applications) [57]. It is observed that the gelatin coating on AuNPs is the major issue in the drug delivery system.

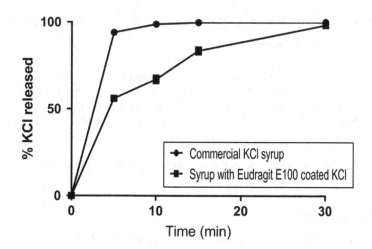

FIGURE 13.10 Graph shows the comparison of in vitro dissolution of marketed KCl syrup formulation and Eudragit E100 (a water-insoluble polymer) coated KCl syrup formulation. Eudragit E100 is coated on KCl to mask the unpleasant taste effectively (2 folds) that improves the palatability of KCl. It releases the KCl completely (100%) in 30 minutes but the conventional KCl syrup releases in 15 minutes. (From Kulkarni Madhur et al. 2019.)

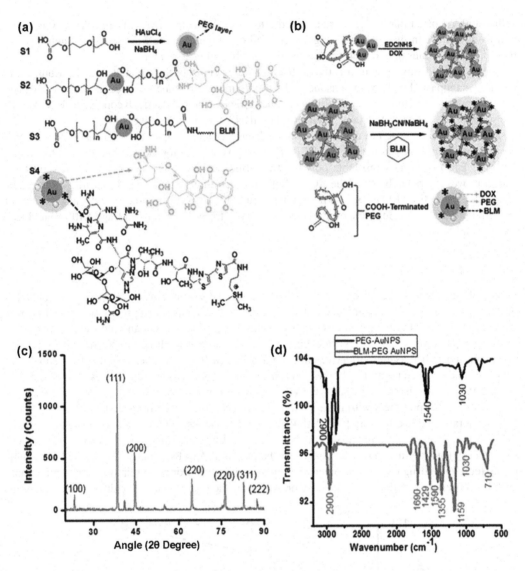

FIGURE 13.11 Synthesis of gold nanoparticles (AuNPs) loaded with anticancer drugs bleomycin (BLM) and doxorubicin (DOX). (a) Synthesis and conjugation steps: PEG-capped AuNPs (S1); PEG-AuNPs linked to DOX (S2); PEG-AuNPs linked to BLM (S3); PEG-AuNPs linked to both DOX and BLM (S4). (b) Formation of S2, S3 and S4 NPs. (c) XRD spectrum of S1 (d) FTIR spectra of S1 and S3. (From Farooq Muhammad et al. 2018.)

The size of the AuNPs plays a key role in the release of curcumin. Figure 13.12 exhibits the loading of curcumin on gelatin-gold nanocomposite.

13.4.6 Iron Oxide Nanoparticles

Iron oxide NPs have attractive magnetic and biological properties. Higher-level drug loading is possible with them. [58]. According to the literature, iron oxide NPs can be utilized as nanodrugs safely. There is no histopathological damage or any other damage to the organs while administering the iron oxide NPs. Also, the cytotoxic effect is at a very much lower level.

Smart NPs release the drugs depending on the pH level. Mohanta et al. designed the pH-responsive smart NPs using iron oxide NPs with PEGylation. These smart NPs will deliver the drug as and

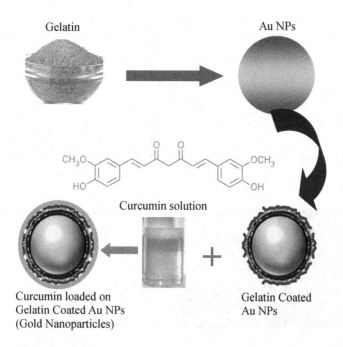

FIGURE 13.12 Schematic diagram shows the steps involved in the curcumin loading on the gelatin-coated gold nanoparticles.

when needed. To prepare the smart drug, initially spherical iron oxide NPs with a size of 8–20 nm are synthesized by precipitation method. After that, the synthesized NPs are coated with PEG-400. Finally, the PEG-coated iron oxide NPs are loaded with anticancer drug daunorubicin. Optimization of the PEG/iron oxide NPs ratio can enhance the drug loading. A high amount of drug release at low level pH confirms the smart nature of the NPs [59].

Rat studies are performed for the targeted delivery of PEGylated iron oxide NPs for lung cancer (H460). Targeting can be achieved by decorating the NP with the anti-epidermal growth factor receptor monoclonal antibodies [60]. The quantity of iron oxide NPs accumulated at the tumor site in rats can be analyzed in MRI by the magnetic targeting method. A magnetic field of 0.4 Tesla is applied for 30 minutes and the NPs of 12 mg/kg are injected intravenously. Applying the magnetic targeting for four hours increases the NP exposure to glioma up to five-fold [33].

Flavonoids from natural materials have low bioavailability. Flavonoid quercetin is loaded inside iron oxide NPs in order to enhance its bioavailability property. Using computational methods, quercetin flavonoid is targeted towards proteins involved in the learning and memory functions of rats. Better learning and memory functions are identified in quercetin loaded iron oxide NPs. Figure 13.13 shows the characterization studies of the quercetin loaded superparamagnetic iron oxide nanoparticles using different tools [61].

Biodistribution and toxicity studies are more important for clinical translation of newly designed drugs. Polyethyleneimine (PEI) coated iron oxide NPs are used for low interference RNA delivery. This NP drug conjugate is administered via intravenous injection. Accumulation studies are performed by Prussian blue straining in the heart, liver, spleen, lung, and kidney. The results show that the cytotoxic effect is very low. The NP drug is deposited in large amounts in the liver and spleen. However, there is no evidence for any histopathological damage or any other damage to the organs. It is concluded that PEI-coated iron oxide NPs are the best in vivo carrier for siRNA [62].

Iron oxide NPs are utilized in biomedical applications. The polymer coated on the NPs, and size of the NPs, decides the uptake of the drug. PEG-coated iron oxide NPs exhibit less uptake in cells when compared to polyethyleneimine coated iron oxide NPs. PEG-coated iron oxide NPs induce

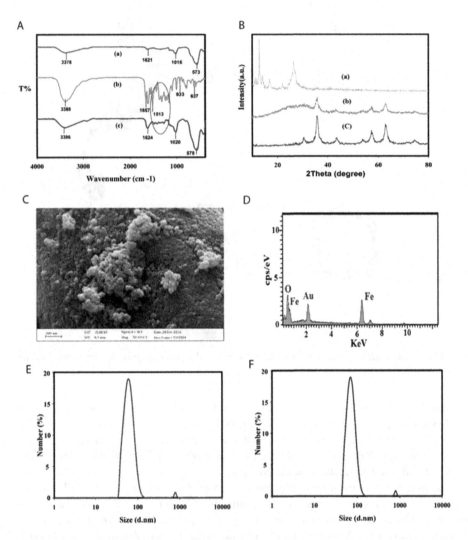

FIGURE 13.13 Characterization of quercetin conjugated with superparamagnetic iron oxide nanoparticles (QT-SPION). (A) FT-IR spectra (a) dextran-coated, (b) pure QT, (c) QT- SPIONs. (B) XRD pattern (a) QT, (b) dextran-coated SPIONs (c) QT- SPIONs. (C and D) SEM and EDX analyses of QT- SPIONs respectively. (E and F) Dynamic light-scattering spectra of dextran-coated SPIONs and QT-SPIONs. (From Amanzadeh Elnaz et al. 2019.)

autophagy in cells which prevents cytotoxicity of iron oxide NPs. In the case of comparing two different sizes of PEG-coated iron oxide NPs, tumors take up more NPs of 10 nm size than of 30 nm. No toxicity is observed with PEG-coated iron oxide NPs. However, some dose-dependent lethal toxicity is observed with PEI-coated iron oxide NPs. Figure 13.14 shows the TEM and DLS images of polymer-coated iron oxide NPs. It also shows the distribution curves related to the distribution of iron oxide NPs in tumors and in the internal organs obtained from the ICP-MS analysis [63].

13.4.7 QUANTUM DOTS

Quantum dots are used for both diagnosis and therapeutic applications of cancer cells. Nowadays graphene quantum dots are used for the treatment of breast cancer cells using herceptin (antibody) and cyclodextrin (polymer) as labels. Doxorubicin is loaded as the cancer drug inside the drug

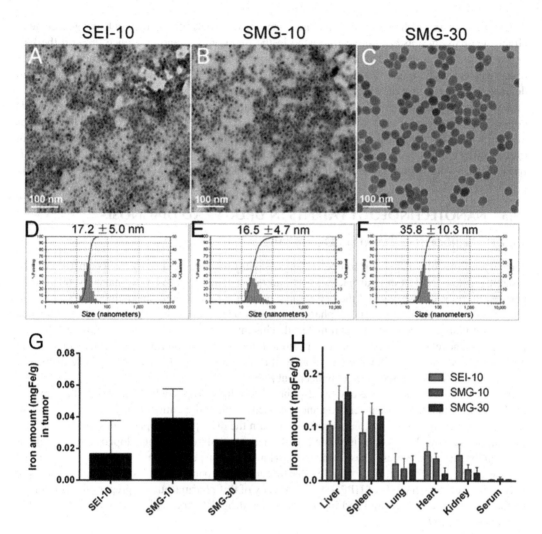

FIGURE 13.14 TEM images and dynamic light scanning (DLS) size distribution graphs of SEI-10, SMG-10, and SMG-30. TEM images of SEI-10, SMG-10, and SMG-30 are shown in (A), (B) and (C) respectively. The related DLS size distributions are shown in (D), (E) and (F). In vivo bio-distribution of different IONPs (1.5 mg Fe/kg) in SKOV-3 tumor of mice is shown as graphs (G) and (H). The distribution of IONPs (G) in tumors (H) in the internal organs. The distribution has been analyzed by ICP-MS after 24 hours of the intravenous injection administration of IONPs. Magnetic iron oxide nanoparticles (IONPs); SKOV-3 ovarian cancer cells; 10 nm size IONPs with PEI coating (SEI-10); 10 nm and 30 nm size IONPs with PEG coating (SMG-10 and SMG-30). (From Feng Qiyi et al. 2018.)

carrier graphene quantum dots. Cancer cells are more acidic in nature. This allows the quantum dots drug carrier system to control the release of the drug. Cell viability studies and confocal laser scanning microscopy analysis show that this system can offer potential anticancer activity against breast cancer cells [64].

Carboxylated graphene quantum dots are analyzed in vivo for their biodistribution and toxicity studies. Graphene quantum dots are highly water-soluble in nature. They were administered through intravenous injection to mice. It was found that graphene quantum dots accumulated in the liver, spleen, lung, kidney, and tumor sites. They were administered at 5 mg/kg or 10 mg/kg dosage level for 21 days. The results showed that there was no toxicity, and no organ damage was found in mice [65].

Cytotoxicity is much lower in graphene quantum dots because of the high oxygen content. Graphene quantum dots are not accumulated in the main organs, and clearance through the kidneys is fast. There is no in vivo and in vitro toxicity while using graphene quantum dots [66]. It is observed from the literature that graphene quantum dots can be used in cancer treatment with a high degree of safety because they have no side effects or have very few side effects.

It is observed that cancer cells are more acidic in nature. However, acidity or alkalinity of foods does not alter the cancer risk. The American Institute for Cancer Research argues that acidity or alkalinity of foods neither decreases nor increases the pH level in the human body. Modifications in the cell surroundings in the body, i.e. making them less-acidic or alkaline (high pH) and less cancer-friendly, are virtually impossible. Slight changes in the body's pH level lead to a threat to life [67].

13.5 NANOTECHNOLOGY PATENTS IN DRUGS AND DIAGNOSIS

13.5.1 NANOCRYSTALS PATENTS IN DRUG DELIVERY

Nanocrystalline solid dispersion compositions possessing discrete particles have been patented. The discrete particles consist of pharmaceutically and nutraceutically active veterinary crystals. These particles have been dispersed in the crystallization inducer and/or are coexisting with crystals of the crystallization inducer matrix. A pharmaceutically acceptable excipient is added (as and when required) to this matrix as an optional material. This invention generates nanocrystalline solid dispersions in a novel one-step process. It is useful to improve the dissolution of pharmaceutically and nutraceutically active veterinary material as well as to improve the bioavailability. The decreased crystallite size leads to dissolution enhancement [68].

Pyropheophorbides such as 2-devinyl-2-(1-hexyloxyethyl)pyropheophorbide (HPPH) is a hydrophobic photosensitizing anticancer drug and it is used for photodynamic therapy. Paras Prasad et al. have patented their invention of nanocrystals (contain the drug pyropheophorbides) useful for photodynamic therapy. This patent is related to the nanocrystals or polymer doped nanocrystals of hydrophobic drug molecules. These nanocrystals have been dispersed in an aqueous system without any stabilizer or surfactant. For example, pharmaceutical preparation consisting of nanocrystals or polymer-doped nanocrystals of HPPH. Drug efficacy of this pharmaceutical preparation is found to be the same with a drug formulated in a conventional drug delivery method in both in vitro and in vivo conditions [69].

13.5.2 IRON OXIDE NANAOPARTICLE PATENTS IN DRUG DELIVERY

Inventors Jonathan Leor et al. have patented iron oxide NPs. In accordance with that patent, iron oxide NPs can be utilized in the treatment of non-infectious inflammatory disorders. They also disclose a treatment method for those disorders utilizing iron oxide NPs, pharmaceutical compositions, and other kits made with such particles [70].

Milk protein casein-coated iron oxide NPs have been patented recently for drug delivery applications. Drug-loaded iron oxide NPs are coated with an inner polymeric layer and an outer casein layer, with a layer-by-layer deposition method. Oral administration is possible with casein coating. The casein layer is degraded by intestinal protease, leading to a controlled or slow release of the drug [71].

13.5.3 GRAPHENE QUANTUM DOTS PATENTS

This patent explains the simple preparation of biocompatible polyethyleneglycol-graphene quantum dots (P-GQDs) using minimally hazardous chemicals. P-GQDs material is electrochemically synthesized (at room temperature) and embedded inside the polyethyleneglycol (PEG) matrix. In this method, a PEG coating over GQDs is avoided; instead GQDs are embedded inside a PEG matrix.

This patent has mainly focused on the two aspects of GQDs, i.e. size and ROS. Small size damages cell organelles; ROS hamper the cells and activities of cells in many ways. On the other side, GQDs material is used for bioimaging and drug delivery in HeLa cells. Hence, it is essential to maintain the fluorescence properties of the GQDs. Intracellular ROS assay suggests that ROS are involved in cytotoxicity activities. After considering these factors, the embedding of GQDs inside the PEG matrix is done to prepare the P-GQDs. The prepared P-GQDs reduce the cytotoxicity by controlling ROS production without affecting the fluorescence properties. A high concentration of P-GQDs (5.5 mg/mL) shows 60% cell viability and reduces ROS production [72].

13.5.4 GQDs Patents in Imaging

Graphene quantum dots are also used as fluorophores in imaging applications. Other semiconductor quantum dots have high brightness and photo-stability. However, they are toxic, and their large molecules contain heavy metal ions. Hence, biomolecules conjugated graphene quantum dots are now emerging as fluorophores for imaging applications. They have advantages such as good photoluminescent properties, chemical inertness, and low cost [73].

13.6 ISSUES IN THE NANODRUG PATENTS

13.6.1 Monitoring of Nanodrugs

Nanotechnology develops innovative therapies and high-quality products at low cost. The medical and pharmaceutical fields should support these developments by adopting them. Pharmaceutical companies should make such innovative products available to the public instead of looking for shareholder profits. The U.S.-based FDA and USPTO are facing challenges to grant approval for new drugs and medical-related products designed with the principles of nanotechnology. Recently, the Indian government has formulated regulations (Guidelines for Evaluation of Nanopharmaceuticals in India) to monitor nanopharmaceuticals.

The FDA of CDER is monitoring drug products with nanomaterials. The impact of nanomaterials on safety and efficacy is assessed to form the regulations. To ensure safety when using nanomaterials, the characterizations of these nanomaterials should be conducted properly and in a scientifically reliable manner. The characterization techniques have some limitations, which can be identified with the help of the existing regulations [74, 75]. Necessary solutions can also be devised for such limitations in accordance with the regulations. The characterization techniques are discussed in the subsequent sections.

Fornaguera and Maria have reported that nanomedicine is a revolution at the nanoscale. They have explained about the world market status of the nanomedicine (in 2017) by comparing nanomedicine availability with corresponding nanomedicine literature. They have also studied the involvement of the stakeholders and the most common challenges in the commercialization of nanomedicines. Figure 13.15 (a) exhibits a comparison study between nanomedicines and literature. Figure 13.15 (b) exhibits the stakeholders' involvement and the challenges in the lab-to-market translation of nanomedicines [76].

As per the guidelines of the Indian government, nanopharmaceutical means a pharmaceutical preparation that contains nanomaterials intended for internal or external application on the body for the purpose of therapeutics, diagnostics, and any health benefit. A nanopharmaceutical should contain particles in the 1 to 1000 nm size range. Less than 1% of the particles with a size above 1000 nm cross the exemption limit. However, during the stable period, less than 10% variation in the particle size range is permitted, i.e. the nanopharmaceutical should not have varied or altered particle size range more than 10% due to instability. These guidelines apply to the finished nanopharmaceutical formulation and active pharmaceutical ingredient of a new molecule or an already approved molecule with altered dimensions, properties, or phenomenon (due to the application of

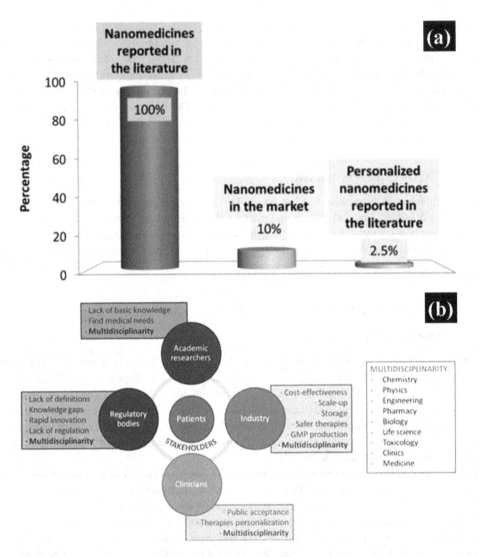

FIGURE 13.15 Schematic diagram related to nanomedicine. (a) Plot of nanomedicine situations in 2017 – nanomedicine available on the world market vs percentages corresponding to nanomedicine literature. (b) Stakeholders' involvement in moving nanomedicines from lab to market and the most common challenges faced by them in this translation. (From Fornaguera, Cristina et al. 2017.)

nanotechnology) intended to apply in the diagnosis, treatment, mitigation, or prevention of diseases in humans [4].

13.6.2 Challenges Related to Nanodrugs

Solubility, target specificity, and toxicity problems are some of the identified challenges that are associated or related to the new nanotechnology products or nanodrugs. To obtain the possible solutions for such problems in a scientific manner, proper understanding of the applications of nanotechnology in food, pharmaceutical, and medicinal fields is essential. The general definition of nanoparticles, i.e. particles in the size range of 1 to 100 nm are suitable only for products having industrial applications. This size range can be relaxed in the case of pharmaceutical products (during formulation, delivery, and efficacy) to avoid undesirable, unknown effects and toxicity.

During the approval process of nanotechnology-based food and medicinal products, it is necessary to consider the accumulation and elimination of nanoparticles in the organs of humans and animals as safety and precautionary measures. For example, the functionalization of iron oxide NPs with folic acid helps with the complete elimination of NPs from the body.

13.6.3 Issues in the Interactions of Nanodrugs and Biology

The interactions between nanoengineered materials and biological systems develop the biological applications of nanotechnology. They are also essential to study as well as to understand the bio–nano interface. Variability of published literature is an important barrier in this multidisciplinary area. The development of Minimum Information Reporting in Bio–Nano Experimental Literature (MIRIBEL) is based on four principles, i.e. reusability, quantification, practicality, and quality. If the development of MIRIBEL is combined with journal and community adoption, it will improve the various outcomes in the bio-nano research field. The improvement in the outcomes may be in the form of data exchange, communication, reproducibility, deeper analysis of published data, and systematic comparison between approaches and materials [77]. MIRIBEL can be applied to acquire necessary information to solve various issues in the development of nanodrugs or any other nanotechnology products relating to pharmaceutical, medical, biotechnology, and allied fields.

Tumor biology and tumor interaction with NPs are difficult to understand. Large scale manufacturing of drugs encapsulated with nanomaterials is a very big challenge in clinical translation and commercialization [78]. Nanomedicines have a long history in cancer drug approval from 1995 to 2015, i.e. from liposomal doxorubicin (Doxil) to liposomal irinotecan (Onivyde). The difficulties in clinical translation of these medicines are the identification of appropriate biomarkers for specific targeting, large scale manufacturing, characterization of these products, and reproducibility. Academia, industry, and regulatory agencies should collaborate to overcome these difficulties and to eradicate cancer [5].

13.6.4 Issues Related to IPR

Intellectual property rights (IPR) are the rights given to an invention to protect it for a specific period of time. In the case of infringement of intellectual property, the rights holder can sue in order to enforce the rights. Trade secrets, trademarks, copyrights, industrial design, geographical, and patents are the types of IPR. The World Intellectual Property Organization (WIPO) protects IPR all over the world. WIPO is an agency of the United Nations (UN) and it has 186 members.

The inventors can harvest the commercial benefits of their invention through IPR. An invention can be patented, if it has global novelty, non-obviousness, and industrial applications [79]. Inventors can patent a new nanomaterial, or an existing nanomaterial based on its synthetic method, properties, applications, and the cost of preparation. Even a small difference in any of these qualities or parameters can generate a patent.

Nanotechnology faces major challenges like overlapping of patents when obtaining patents for the new materials or products. A single nanomaterial has applications in various fields. This leads to the overlapping of patents. For example, gold nanoparticles are applied in drug delivery and biosensors. The USPTO is now granting more nanomaterial patents. Hence, it is essential to implement necessary regulations and guidelines for the patents related to the nanotechnology products.

Like the patents of every invention, patents of nanodrug inventions have territorial boundary issues. These issues can be managed with the help of the Patent Cooperation Treaty by filing an international patent through it. In the case of patents related to a nanodrug invention that belongs to a European country, as per the European Patent Convention (EPC), a European patent can be filed as a single patent application at the European Patent Office.

The EPC is a multilateral treaty that establishes the European Patent Organisation to grant European patents. Although the term European patent refers to patent granted under the EPC, such a patent is not a unitary right, but a group of essentially independent nationally enforceable, nationally revocable patents [80]. The EPC frames the legal structure for the granting of European patents [81]. A single patent application can be filed at the European patent office in one language [82] or can be filed at the national patent office (if the national law permits) of European nations [83].

13.6.5 ISSUES IN THE STANDARDS OF NANOMATERIALS

The reproducibility is more important for nanomaterials. The European Chemical Agency (ECHA) has listed some properties of nanomaterials. In order to get a patent for a nanomaterial, the listed properties should be analyzed [84]. Properties of the nanomaterials such as particle size, shape, specific surface area, crystal structure, density, volume, hydrophobicity, chemical identity, presence of organic coating/impurities, and bandgap are listed as the major properties in the standardization process of nanomaterials. The issues relating to the standards of the nanomaterials arising during patent filing can be solved by analyzing the nanomaterials. The characterization techniques mentioned below are utilized for such analysis.

For a material to be approved as a nanomaterial, the particle size should be in the range of 1 to 100 nm, but different characterization techniques yield different particle sizes. A transmission electron microscope (TEM) with automated image analysis is now considered as a reliable method. The gas adsorption technique is known as BET analysis; it is a well-established method to measure the specific surface area of the nanomaterials. Personick et al. have reported that the size, shape of the NP, and number of particles in a sample can be analyzed by TEM. SEM provides 3-D information about the shape, whereas TEM gives only 2-D information about the shape of the material [85]. SEM lacks lateral information of the sample [86]. The sample preparation is somewhat difficult in TEM analysis.

The crystal structure of a nanomaterial can be analyzed using the X-ray diffraction (XRD) method and TEM analysis [87]. The limitation of XRD is the difficulty in analyzing amorphous samples [88]. The hydrophobicity of a nanomaterial affects its characters such as dispersibility in water and toxicity [89]. The hydrophobicity of a surface coated with nanomaterials (i.e. the famous lotus effect in nanotechnology) can be studied using contact angle measurements [90, 91]. Formation of a smooth surface layer is a serious limitation in this application. Hydrophobicity plays a major role in the nanotoxicity.

The presence of an organic coating in a nanomaterial is identified by thermogravimetric analysis (TGA). The presence of impurities in nanomaterials is identified by using the ICP-MS method [84]. Confirmation of the chemical identity is obtained by the coupling of TGA with mass spectrometry [92]. The density of the material determines the sedimentation rate of NPs when they form agglomerations [93, 94]. The volume of a nanomaterial (in powder form) is measured by using gas pycnometry [95]. It is essential to measure the density of a nanomaterial when it is applied in nanofluids or nano dispersions. The density can be calculated if the volume is known or from the measured value of the volume.

Bandgap is an important property that has to be measured for a semiconducting nanoparticle. This property is related to the biological properties. If the Fermi level of a biological fluid lies in the energy level of the highest valence band or the energy level of the lowest conduction band, then there will be more ROS generation. Bandgap can be measured using a tauc plot obtained from UV-Vis spectroscopy. The analysis and standardization of nanomaterials using the characterization techniques is essential to obtaining a patent for nanomaterials.

The European Commission suggests, if the environment, health, safety, or competitiveness warrants it, that the proportion of nano-sized particles in nanomaterials can be reduced to lower than 50% [96]. Cells are in the micrometer size range and NPs are in the nanometer size range (size range of 1 –100 nm). Hence, NPs are able to diffuse into the cells easily and exhibit action different from their bulk materials. The different actions lead to untoward effects or toxicity of the nanomaterials [97]. These effects or toxicity will be a major issue in obtaining a patent for the nanodrug.

Nanomaterials with particle sizes of more than 100 nm do not exhibit any strange properties. Utilizations of the materials that have large size particles, i.e. above 100 nm (instead of NPs with sizes between 1 and 100 nm) in nanodrug or nanopharmaceutical products will avoid the untoward effects. Such materials can be utilized like other conventional materials without any hesitation. They have a higher surface area that is close to the nanomaterials, which improves their activity. As per the European Commission suggestion, the proportion of NPs in the nanodrugs can also be reduced to avoid the untoward effects or the toxicity.

The interactions between the nanomaterials and biological organs are yet to be understood and standardized. The International Standard Organization (ISO) has an active technical committee (ISO/TC 229) for this kind of standardization. For the standardization of nanomaterials, knowing about the detailed synthesis procedure of a patented material is the only way to get the composition of the material; for the size and shape analysis of NPs, the size measured from all dimensions should be analyzed; average particle size should not be considered to analyze particle size distribution; zeta potential analysis works well to characterize the NPs present in a colloidal solution [77].

13.7 CONCLUSION

This work has made an attempt to review nanotechnology-based nanopharmaceutical products and their patents. Nanotechnology is an emerging technology. Although it is in the initial stage of development, it has produced many nanopharmaceutical products as well as technologies to solve problems arising in the pharmaceutical industry. The interdisciplinary and application-oriented nature of nanotechnology helps in such activities. Nanotechnology is utilized in the biopharmaceutical industry to produce nanomedicine, diagnostic tools, biosensor, biomarker, implant technology, nanorobots, and vector or carrier molecules for theranostics.

A nanopharmaceutical is defined as a product that contains nanomaterials in the 1 to 100 nm size range (in any one dimension) and has different or altered pharmaceutical characteristics comparing to the API. Also, nanopharmaceuticals have particles of 100 to 1,000 nm in size will also be known as nanopharmaceutical products. Applying nanotechnology in nanopharmaceutical products alters the pharmaceutical characteristics. Solid lipid NPs, polymer NPs, metal NPs, metal oxide NPs, and quantum dots are mainly utilized in nanopharmaceuticals.

The different or altered pharmaceutical characteristics create ideas for new inventions or innovative nanodrug products. The researchers can protect their intellectual properties by getting patents. To overcome the territorial limitations of the patents, it is essential to file patent applications in different countries. In such cases, the PCT helps to file an international patent through a single patent application. Likewise, the EPC facilitates the filing of a European patent as a single patent application at the European Patent Office.

Nanotoxicity is a major challenge that restricts nanopharmaceutical products during patent filing. Production of nanomaterials with no side effects or very few side effects (like graphene quantum dots) avoids such challenges. In the case of cancer treatment, using particle therapy instead of radiation therapy avoids the side effects.

The FDA of CDER monitors the safety and efficacy of the drug products with nanomaterials. The Indian government has created guidelines to monitor and to evaluate nanopharmaceuticals. In accordance with the regulations, necessary solutions can be made for the limitations and challenges related to the nanopharmaceuticals. Although issues and challenges exist, nanotechnology-based pharmaceutical products such as liposomal nanocarriers, pegylated proteins/polypeptides, nanocrystals, protein–drug nanoconjugates, surfactant-based nanodrugs, and nanoformulations based on metals/polymers have been patented and marketed.

ACKNOWLEDGMENTS

The author expresses thanks to her husband, Mr G. Sankar, for his assistance in this work. She also acknowledges the assistance of the International Research Center, Kalasalingam Academy of Research and Education (Deemed University), Krishnankoil – 626 126, (India), for providing necessary support and facilities.

REFERENCES

1. Bawa, Raj, Srikumaran Melethil, William J. Simmons, and Drew Harris. "Nanopharmaceuticals: Patenting issues and FDA regulatory challenges." *The Scitech Lawyer* 5(2) (2008): 10–15.
2. Fatehi, L., S.M. Wolf, J. McCullough, R. Hall, F. Lawrenz, J.P. Kahn, C. Jones et al. "Recommendations for nanomedicine human subjects research oversight: An evolutionary approach for an emerging field." *The Journal of Law, Medicine & Ethics: A Journal of the American Society of Law, Medicine & Ethics* 40(4) (2012): 716–750. doi:10.1111%2Fj.1748-720X.2012.00703.x.
3. Ragelle, H., F. Danhier, V. Préat, R. Langer, and D.G. Anderson. "Nanoparticle-based drug delivery systems: A commercial and regulatory outlook as the field matures." *Expert Opinion on Drug Delivery* 14(7) (2017): 851–864. doi:10.1080/17425247.2016.1244187.
4. "Guidelines for evaluation of nanopharmaceuticals in India." https://www.aiims.edu/images/pdf/notice/Nanopharmaceuticals%20guidelines%20book%20uploaded%20on%20AIIMS%20&%20Pharm.pdf.
5. Tran, Stephanie, Peter-Joseph DeGiovanni, Brandon Piel, and Prakash Rai. "Cancer nanomedicine: A review of recent success in drug delivery." *Clinical & Translational Medicine* 6(1) (2017): 44. doi:10.1186/s40169-017-0175-0.
6. Khandel, Pramila, Ravi Kumar Yadav, Deepak Kumar Soni, Leeladhar Kanwar, and Sushil Kumar Shahi. "Biogenesis of metal nanoparticles and their pharmacological applications: Present status and application prospects." *Journal of Nanostructure in Chemistry* 8(3) (2018): 217–254. doi:10.1007/s40097-018-0267-4.
7. Claire du Toit, Lisa, Viness Pillay, Yahya E. Choonara, Samantha Pillay, and Sheri-lee Harilall. "Patenting of nanopharmaceuticals in drug delivery: No small issue." *Recent Patents on Drug Delivery & Formulation* 1(2) (2007): 131–142.
8. D'Mello, Sheetal R., Celia N. Cruz, Mei-Ling Chen, Mamta Kapoor, Sau L. Lee, and Katherine M. Tyner. "The evolving landscape of drug products containing nanomaterials in the United States." *Nature Nanotechnology* 12(6) (2017): 523.
9. Henderson, Lori A., and Lalitha K. Shankar. "Clinical translation of the National Institutes of Health's investments in nanodrug products and devices." *The AAPS Journal* 19(2) (2017): 343–359.
10. Zheng, Nan, Dajun D. Sun, Peng Zou, and Wenlei Jiang. "Scientific and regulatory considerations for generic complex drug products containing nanomaterials." *The AAPS Journal* 19(3) (2017): 619–631.
11. Chen, Mei-Ling, Mathew John, Sau L. Lee, and Katherine M. Tyner. "Development considerations for nanocrystal drug products." *The AAPS Journal* 19(3) (2017): 642–651.
12. Tyner, Katherine M., Nan Zheng, Stephanie Choi, Xiaoming Xu, Peng Zou, Wenlei Jiang, Changning Guo et al. "How has CDER prepared for the nano revolution? A review of risk assessment, regulatory research, and guidance activities." *The AAPS Journal* 19(4) (2017): 1071–1083.
13. https://www.fda.gov/drugs/news-events-human-drugs/advancing-science-nanotechnology-drug-development (accessed 15 December 2019).
14. Sadrieh, N., (2014). Overview of CDER Nanotechnology-Related Drug Database. Available at: http://pqri.org/wp-content/uploads/2015/08/pdf/Nakissa.Sadrieh.Presentation.pdf.
15. Chin, William Wei Lim, Johannes Parmentier, Michael Widzinski, Hui Tan, and Rajeev Gokhale. "A brief literature and patent review of nanosuspensions to a final drug product." *Journal of Pharmaceutical Sciences* 103(10) (2014): 2980–2999.
16. Ahire, Eknath, Shreya Thakkar, Mahesh Darshanwad, and Manju Misra. "Parenteral nanosuspensions: A brief review from solubility enhancement to more novel and specific applications." *Acta Pharmaceutica Sinica B* 8(5) (2018): 733–755.
17. Zhang, Jinglu, Han Run, Weijie Chen, Weixiang Zhang, Ying Li, Yuanhui Ji, Lijiang Chen et al. "Analysis of the literature and patents on solid dispersions from 1980 to 2015." *Molecules* 23(7) (2018): 1697.
18. Ozcan, Sercan, and Nazrul Islam. "Patent information retrieval: Approaching a method and analysing nanotechnology patent collaborations." *Scientometrics* 111(2) (2017): 941–970.
19. Rogers, E.M. *Diffusion of Innovations.* 4th edition, Simon and Schuster, New York, 2010.
20. "WIPO intellectual property handbook: Policy, law and Use. WIPO 2008." ISBN 9789280512915. https://www.wipo.int/edocs/pubdocs/en/intproperty/489/wipo_pub_489.pdf.
21. Shear, R.H., and T.E. Kelley. "A researcher's guide to patents." *Plant Physiology* 132(3) (2003): 1127–1130. doi:10.1104/pp.103.022301.
22. "Patents: Frequently Asked Questions." World Intellectual Property Organization. from the original on 20 June 2015. Retrieved 24 June 2015.
23. Lemley, Mark A., and Carl Shapiro. "Probabilistic patents." *Journal of Economic Perspectives*, Stanford Law and Economics Olin Working Paper No. 288 19(2) (2005): 75. doi:10.2139/ssrn.567883.

24. WTO ANALYTICAL INDEX-TRIPS Agreement. "Article 27- Patentable Subject Matter." https://www.wto.org/english/res_e/publications_e/ai17_e/trips_art27_jur.pdf.

25. Mukherjee, S. "The new Indian patent law: a challenge for India." *International Journal of Strategic Property Managemen* 1(1/2) (2006): 131–149.

26. Section 2(1)(j) of the Patents (Amendment) Act, 2002, No. 38 of 2002 (June 25, 2002).

27. Section 2(1)(l) of the Patents (Amendment) Act, 2005, No. 15 of 2005 (April 4, 2005).

28. Section 2(1)(ja) of the Patents (Amendment) Act, 2005, No. 15 of 2005 (April 4, 2005).

29. Mathur, Vipin. "Patenting of pharmaceuticals: AnIndian perspective." *International Journal of Drug Development & Research* 4(3) (July–September 2012): 27–34.

30. Harrington, P.J., L.M. Hodges, K. Puentener, and M. Scalone. "Synthesis of 3,6-Dialkyl-5,6-Dihydro-4-Hydroxy-2h-Pyran-2-One." Indian Patent IN, 2007. http://www.ncbi.nlm.nih.gov/pubmed/206678.

31. Mir, Maria, Saba Ishtiaq, Samreen Rabia, Maryam Khatoon, Ahmad Zeb, Gul Majid Khan, Asim ur Rehman et al. "Nanotechnology: from in vivo imaging system to controlled drug delivery." *Nanoscale Research Letters* 12(1) (2017): 500. doi:10.1186/s11671-017-2249-8.

32. Hua, Susan, Maria B.C. De Matos, Josbert M. Metselaar, and Gert Storm. "Current trends and challenges in the clinical translation of nanoparticulate nanomedicines: Pathways for translational development and commercialization." *Frontiers in Pharmacology* 9 (2018): 1–14.

33. Chertok, Beata, Bradford A. Moffat, Allan E. David, Yu Faquan, Christian Bergemann, Brian D. Ross, and Victor C. Yang. "Iron oxide nanoparticles as a drug delivery vehicle for MRI monitored magnetic targeting of brain tumors." *Biomaterials* 29(4) (2008): 487–496.

34. Li, Min, Peng Zou, Katherine Tyner, and Sau Lee. "Physiologically based pharmacokinetic (PBPK) modeling of pharmaceutical nanoparticles." *The AAPS Journal* 19(1) (2017): 26–42.

35. Bonifacio, Bruna Vidal, Patricia Bento da Silva, Matheus Aparecido dos Santos Ramos, Kamila Maria Silveira Negri, Taís Maria Bauab, and Marlus Chorilli. "Nanotechnology-based drug delivery systems and herbal medicines: A review." *International Journal of Nanomedicine* 9 (2014): 1.

36. Feng, Jie, Chester E. Markwalter, Chang Tian, Madeleine Armstrong, and Robert K. Prud'homme. "Translational formulation of nanoparticle therapeutics from laboratory discovery to clinical scale." *Journal of Translational Medicine* 17(1) (2019): 200. doi:10.1186/s12967-019-1945-9.

37. Dhillon, G.S., S.K. Brar, S. Kaur, and M. Verma. Green approach for nanoparticle biosynthesis by fungi: Current trends and applications. *Critical Reviews in Biotechnology* 32(1) (2012): 49–73.

38. Onoue, Satomi, Shizuo Yamada, and Hak-Kim Chan. "Nanodrugs: Pharmacokinetics and safety." *International Journal of Nanomedicine* 9 (2014): 1025.

39. Suri, Sarabjeet Singh, Hicham Fenniri, and Baljit Singh. "Nanotechnology-based drug delivery systems." *Journal of Occupational Medicine & Toxicology* 2(1) (2007): 16.

40. Slowing, Igor I., Brian G. Trewyn, Supratim Giri, and V.S.-Y. Lin. "Mesoporous silica nanoparticles for drug delivery and biosensing applications." *Advanced Functional Materials* 17(8) (2007): 1225–1236.

41. Ventola, C. Lee. "Progress in nanomedicine: approved and investigational nanodrugs." *Pharmacy & Therapeutics* 42(12) (2017): 742.

42. Wolfram, Joy, Motao Zhu, Yong Yang, Jianliang Shen, Emanuela Gentile, Donatella Paolino, Massimo Fresta et al. "Safety of nanoparticles in medicine." *Current Drug Targets* 16(14) (2015): 1671–1681.

43. Lacombe, Sandrine, Erika Porcel, and Emanuele Scifoni. "Particle therapy and nanomedicine: State of art and research perspectives." *Cancer Nanotechnology* 8(1) (2017): 9. doi:10.1186/s12645-017-0029-x.

44. Mishra, Vijay, Kuldeep Bansal, Asit Verma, Nishika Yadav, Sourav Thakur, Kalvatala Sudhakar, and Jessica Rosenholm. "Solid lipid nanoparticles: Emerging colloidal Nano drug delivery systems." *Pharmaceutics* 10(4) (2018): 191. doi:10.3390/pharmaceutics10040191.

45. Kaur, Indu P., and Rouit Bhandari (2012). Solid lipid nanoparticles entrapping hydrophilic/amphiphilic drug and a process for preparing the same. PCT application number: PCT/IN2012/000154 dated, 5(03), 2012.

46. Battaglia, Luigi, and Elena Ugazio. "Lipid nano-and microparticles: An overview of patent-related research." *Journal of Nanomaterials* 2019 (2019): 2834941. Doi: 10.1155/2019/2834941.

47. Wagner, Volker, Anwyn Dullaart, Anne-Katrin Bock, and Axel Zweck. "The emerging nanomedicine landscape." *Nature Biotechnology* 24(10) (2006): 1211.

48. Cai, Xiaopan, Jingjing Hu, Jianru Xiao, and Yiyun Cheng. "Dendrimer and cancer: A patent review (2006–present)." *Expert Opinion on Therapeutic Patents* 23(4) (2013): 515–529.

49. Baker Jr, James R., and Baohua Huang, University of Michigan, 2015. Dendrimer compositions and methods of synthesis. U.S. Patent 8,945,508.

50. Lu, Xiao-Yun, Dao-Cheng Wu, Zheng-Jun Li, and Guo-Qiang Chen. "Polymer nanoparticles." *Progress in Molecular Biology & Translational Science* 104 (2011): 299–323.

51. Owen, Shawn C., Dianna PY Chan, and Molly S. Shoichet. "Polymeric micelle stability." *Nano Today* 7(1) (2012): 53–65.

52. Patra, Jayanta Kumar, Gitishree Das, Leonardo Fernandes Fraceto, Estefania Vangelie Ramos Campos, Maria del Pilar Rodriguez-Torres, Laura Susana Acosta-Torres, Luis Armando Diaz-Torres et al. "Nano based drug delivery systems: Recent developments and future prospects." *Journal of Nanobiotechnology* 16(1) (2018): 71. doi:10.1186/s12951-018-0392-8.

53. Pillai, Gopalakrishna. "Nanomedicines for cancer therapy: An update of FDA approved and those under various stages of development." *SOJ Pharmacy and Pharmaceutical Sciences* 1(2) (2014): 13.

54. Kulkarni, Madhur, Brijesh Vishwakarma, Samik Sen, Sandhya Anupuram, and Abhijit A. Date. "Development and evaluation of taste masked dry syrup formulation of potassium chloride." *AAPS Open* 5(1) (2019): 1. doi:10.1186/s41120-019-0030-z.

55. Farooq, Muhammad U., Valentyn Novosad, Elena A. Rozhkova, Hussain Wali, Asghar Ali, Ahmed A. Fateh, Purnima B. Neogi et al. "Gold nanoparticles-enabled efficient dual delivery of anticancer therapeutics to HeLa cells." *Scientific Reports* 8(1) (2018): 2907. doi:10.1038/s41598-018-21331-y.

56. Yu, Meng, Bo Lei, Chuanbo Gao, Jin Yan, and Peter X. Ma. "Optimizing surface-engineered ultrasmall gold nanoparticles for highly efficient miRNA delivery to enhance osteogenic differentiation of bone mesenchymal stromal cells." *Nano Research* 10(1) (2017): 49–63.

57. Khodashenas, Bahareh, Mehdi Ardjmand, Mazyar Sharifzadeh Baei, Ali Shokuhi Rad, and Azim Akbarzadeh Khiyavi. "Gelatin–gold nanoparticles as an ideal candidate for curcumin drug delivery: Experimental and DFT studies." *Journal of Inorganic & Organometallic Polymers & Materials* (2019). doi:10.1007/s10904-019-01178-0.

58. Vangijzegem, Thomas, Dimitri Stanicki, and Sophie Laurent. "Magnetic iron oxide nanoparticles for drug delivery: Applications and characteristics." *Expert Opinion on Drug Delivery* 16(1) (2019): 69–78.

59. Mohanta, Subas Chandra, Arindam Saha, and P. Sujatha Devi. "Pegylated iron oxide nanoparticles for pH responsive drug delivery application." *Materials Today: Proceedings* 5(3) (2018): 9715–9725.

60. Wang, Zhongling, Ruirui Qiao, Na Tang, Ziwei Lu, Han Wang, Zaixian Zhang, Xiangdong Xue et al. "Active targeting theranostic iron oxide nanoparticles for MRI and magnetic resonance-guided focused ultrasound ablation of lung cancer." *Biomaterials* 127 (2017): 25–35.

61. Amanzadeh, Elnaz, Abolghasem Esmaeili, Rezvan Enteshari Najaf Abadi, Nasrin Kazemipour, Zari Pahlevanneshan, and Siamak Beheshti. "Quercetin conjugated with superparamagnetic iron oxide nanoparticles improves learning and memory better than free quercetin via interacting with proteins involved in LTP." *Scientific Reports* 9(1) (2019): 6876. doi:10.1038/s41598-019-43345-w.

62. Yu, Qin, Xiao-qin Xiong, Lei Zhao, Ting-ting Xu, Hao Bi, Rong Fu, and Qian-hua Wang. "Biodistribution and toxicity assessment of superparamagnetic iron oxide nanoparticles in vitro and in vivo." *Current Medical Science* 38(6) (2018): 1096–1102.

63. Feng, Qiyi, Yanping Liu, Jian Huang, Ke Chen, Jinxing Huang, and Kai Xiao. "Uptake, distribution, clearance, and toxicity of iron oxide nanoparticles with different sizes and coatings." *Scientific Reports* 8(1) (2018): 2082. doi:10.1038/s41598-018-19628-z.

64. Ko, N.R., M. Nafiujjaman, J.S. Lee, H.-N. Lim, Y.-K. Lee, and I.K. Kwon. "Graphene quantum dot-based theranostic agents for active targeting of breast cancer." *RSC Advances* 7(19) (2017): 11420–11427.

65. Nurunnabi, Md., Zehedina Khatun, Kang Moo Huh, Sung Young Park, Dong Yun Lee, Kwang Jae Cho, and Yong-kyu Lee. "In vivo biodistribution and toxicology of carboxylated graphene quantum dots." *ACS Nano* 7(8) (2013): 6858–6867.

66. Chong, Yu, Yufei Ma, He Shen, Xiaolong Tu, Xuan Zhou, Jiaying Xu, Jianwu Dai et al. "The in vitro and in vivo toxicity of graphene quantum dots." *Biomaterials* 35(19) (2014): 5041–5048.

67. Alice, R.D. Another Cancer and Diet Claim: The Alkaline Diet. American Institute of Cancer Research (2010). https://www.aicr.org/patients-survivors/healthy-or-harmful/alkaline-diets.html (accessed 15 December 2019).

68. Bansal, Arvind Kumar, Ajay Kumar Raju Dantuluri, Shete Ganesh Bhaskarao, and Pawar Yogesh Bapurao. "Nanocrystalline solid dispersion compositions and process of preparation thereof." U.S. Patent 9,801,855, issued October 31, 2017.

69. Prasad, Paras, Haridas Pudavar, Koichi Baba, Indrajit Roy, Tymish Ohulchanskyy, Ravindra Pandey, and Allan Oseroff. "Method for delivering hydrophobic drugs via nanocrystal formulations." U.S. Patent Application 11/471,075, filed June 14, 2007.

70. Leor, Jonathan, Tamar Ben-Mordechai, Shimrit Adutler-Lieber, Rimona Margalit, Inbar Elron-Gross, and Yifat Glucksam-Galnoy. "Iron oxide nanoparticles for use in treating non-infectious inflammatory disorders." U.S. Patent 9,492,480, issued November 15, 2016.

71. Mao, Hui, and Jing Huang. "Casein coated drug-loaded iron oxide nanoparticles." U.S. Patent 9,737,492, issued August 22, 2017.
72. Singh, Neetu, and Anil Chandra. "Biocompatible graphene quantum dots for drug delivery and bioimaging applications." U.S. Patent 9,642,815, issued May 9, 2017.
73. Chen, Peng, and Xin Ting Zheng. "Biomolecule-Graphene Quantum Dot Conjugates and Use Thereof." U.S. Patent Application 14/471,928, filed March 5, 2015.
74. Sadrieh, Nakissa, Anna M. Wokovich Neera V. Gopee, et al. "Lack of significant dermal penetration of titanium dioxide from sunscreen formulations containing nano- and submicron-size TiO$_2$ particles." *Toxicological Sciences* 115(1) (2010): 156–166.
75. Cruz, Celia N., Katherine M. Tyner, Lydia Velazquez, Kenneth C. Hyams, Abigail Jacobs, Arthur B. Shaw, Wenlei Jiang et al. "CDER risk assessment exercise to evaluate potential risks from the use of nanomaterials in drug products." *The AAPS Journal* 15(3) (2013): 623–628.
76. Fornaguera, Cristina, and Maria García-Celma. "Personalized nanomedicine: A revolution at the nanoscale." *Journal of Personalized Medicine* 7(4) (2017): 12. doi:10.3390/jpm7040012.
77. Faria, Matthew, Mattias Björnmalm, Kristofer J. Thurecht, Stephen J. Kent, Robert G. Parton, Maria Kavallaris, Angus P.R. Johnston et al. "Minimum information reporting in bio–nano experimental literature." *Nature Nanotechnology* 13(9) (2018): 777.
78. Shi, Jinjun, Philip W. Kantoff, Richard Wooster, and Omid C. Farokhzad. "Cancer nanomedicine: Progress, challenges and opportunities." *Nature Reviews Cancer* 17(1) (2017): 20.
79. Saha, Chandra Nath, and Sanjib Bhattacharya. "Intellectual property rights: An overview and implications in pharmaceutical industry." *Journal of Advanced Pharmaceutical Technology & Research* 2(2) (2011): 88.
80. Scourfield, Tom. "Jurisdiction and Patents: ECJ rules on forum for validity and cross-border patent enforcement." *The CIPA Journal* 35(8) (2006): 535.
81. Article 2(1) of EPC. https://www.epo.org/law-practice/legal-texts/html/epc/2013/e/ar2.html.
82. Article 14 of EPC. https://www.epo.org/law-practice/legal-texts/html/epc/2013/e/ar14.html.
83. Article 75(1)(b) of EPC. https://www.epo.org/law-practice/legal-texts/html/epc/2013/e/ar75.html.
84. Gao, Xiaoyu, and Gregory V. Lowry. "Progress towards standardized and validated characterizations for measuring physicochemical properties of manufactured nanomaterials relevant to nano health and safety risks." *NanoImpact* 9 (2018): 14–30.
85. Personick, M.L., M.R. Langille, J. Zhang, and C.A. Mirkin. "Shape control of gold nanoparticles by silver underpotential deposition." *Nano Letters* 11(8) (2011): 3394–3398.
86. Rades, S., V.-D. Hodoroaba, T. Salge, T. Wirth, M.P. Lobera, R.H. Labrador, K. Natte et al. "High-resolution imaging with SEM/T-SEM, EDX and SAM as a combined methodical approach for morphological and elemental analyses of single engineered nanoparticles." *RSC Advances* 4(91) (2014): 49577–49587.
87. Thamaphat, K., P. Limsuwan, and B. Ngotawornchai. "Phase characterization of TiO2 powder by XRD and TEM." *The Kasetsart Journal* 42 (2008): 357–361.
88. Moreau, L.M., D.-H. Ha, H. Zhang, R. Hovden, D.A. Muller, and R.D. Robinson. "Defining crystalline/amorphous phases of nanoparticles through X-ray absorption spectroscopy and X-ray diffraction: The case of nickel phosphide." *Chemistry of Materials* 25(12) (2013): 2394–2403.
89. Moyano, D.F., M. Goldsmith, D.J. Solfiell, D. Landesman-Milo, O.R. Miranda, D. Peer, and V.M. Rotello. "Nanoparticle hydrophobicity dictates immune response." *Journal of the American Chemical Society* 134(9) (2012): 3965–3967. doi:10.1021/ja2108905.
90. Galet, L., S. Patry, and J. Dodds. "Determination of the wettability of powders by the Washburn capillary rise method with bed preparation by a centrifugal packing technique." *Journal of Colloid & Interface Science* 346(2) (2010): 470–475.
91. Xiao, Y., and M.R. Wiesner. "Characterization of surface hydrophobicity of engineered nanoparticles." *Journal of Hazardous Materials* 215 (2012): 146–151.
92. Kango, S., S. Kalia, A. Celli, J. Njuguna, Y. Habibi, and R. Kumar. "Surface modification of inorganic nanoparticles for development of organic–inorganic nanocomposites—A review." *Progress in Polymer Science* 38(8) (2013): 1232–1261.
93. DeLoid, G., J.M. Cohen, T. Darrah, R. Derk, L. Rojanasakul, G. Pyrgiotakis, W. Wohlleben, and P. Demokritou. "Estimating the effective density of engineered nanomaterials for in vitro dosimetry." *Nature Communications* 5 (2014): 3514.
94. Lee, S.Y., H. Chang, T. Ogi, F. Iskandar, and K. Okuyama. "Measuring the effective density, porosity, and refractive index of carbonaceous particles by tandem aerosol techniques." *Carbon* 49(7) (2011): 2163–2172.

95. Park, J.H., M.A. Lee, B.J. Park, and H.J. Choi. "Preparation and electrophoretic response of poly (methyl methacrylate-co-methacrylic acid) coated TiO 2 nanoparticles for electronic paper application." *Current Applied Physics* 7(4) (2007): 349–351.

96. EU. "EU Commission recommendation on the definition of nanomaterial (2011/696/EU)." *Official Journal of the European Union* L 275 (2011): 38–40. https://ec.europa.eu/research/industrial_technolog ies/pdf/policy/commission-recommendation-on-the-definition-of-nanomater-18102011_en.pdf.

97. Thirugnanasambandan, Theivasanthi. "Advances and trends in nano-biofertilizers." (2018). Available at SSRN: https://ssrn.com/abstract=3306998 or doi:10.2139/ssrn.3306998.

14 A Survey of Recent Nanotechnology-Related Patents for Treatment of Tuberculosis

Tanmayee Nayak, Rakesh Kumar Singh, Lav Kumar Jaiswal, Ankush Gupta, and Juhi Singh

CONTENTS

14.1 INTRODUCTION

Tuberculosis (TB) is one of the deadliest infectious diseases caused by bacteria from the *Mycobacterium tuberculosis* complex. This causative agent of TB is known as one of the most successful pathogens capable of thriving in its host for a long time period without causing the active disease. In 1882, Robert Koch was awarded the Nobel Prize for the discovery of the causative agent of TB. Although the pathogen primarily attacks the pulmonary system, the disease can spread through almost any part of the body. The symptoms of active TB include chronic cough, fever, night sweats, and weight loss, and can be lethal if left untreated. TB spreads through the contaminants from active TB patients through air droplets released from coughs or sneezes. However, people with latent TB do not spread the infection. People more susceptible to active infection are those with HIV/AIDS, malnutrition, and those who smoke (Nanotechnology-Based Approach in Tuberculosis Treatment, Cooper, 2009) Globally, TB is one of the main causes of mortality. According to the World Health Organization (WHO), one-third of the world's population is affected by *M. tuberculosis* (Mtb), creating approximately 9 million new cases and deaths of 2 million yearly while the remainder of the population is unaffected (World Health Organization, 2013a). By mortality rate, TB comes second after AIDS among infectious diseases. Therefore, in 1993, TB was declared a global health emergency by the WHO; there are several challenges in the field of treatment of the disease. The critical problem with chemotherapy is the development of a resistant strain of the pathogen against the drugs (Lienhardt et al., 2012). Apart from this, with drugs are taken intravenously or orally, most of the drug molecules do not reach the target as it is distributed through the whole body via the circulatory system creating adverse side effects (Greenblatt, 1985).

14.2   GLOBAL TB REPORT

Globally, TB is one among the top 10 causes of death and ranked second among infectious disease after HIV/AIDS. Millions of people are infected with TB annually. According to estimation by the WHO in 2017, TB resulted in ~1.3 million deaths among HIV-negative people and there were an extra 300,000 deaths from TB among HIV-positive people. In 2017, it is estimated that around 10.0 million people were infected with the disease (Sairam et al., 2017). Among affected individuals ~5.8 million were men, ~3.2 million were women, and ~1.0 million were children. These cases cover all countries and age groups, overall 90% were adults, 9% were HIV positive people (72% in Africa) and two-thirds were from eight countries, i.e. India (27%), China (9%), Indonesia (8%), the Philippines (6%), Pakistan (5%), Nigeria (4%), Bangladesh (4%), and South Africa (3%). Apart from these, 22 other countries in the WHO list of 30 countries with a high TB rate accounted for 87% of the world's cases. Only 6% of global cases were in the WHO European Region (3%) and WHO Region of the Americas (3%) (World Health Organization, 2013a).

Drug-resistant TB has emerged as a major public health crisis. In 2017 it is estimated that 483,000–639,000 people developed TB that was resistant to rifampicin, the most effective first-line drug, and of these, 82% had multidrug-resistant TB (MDR-TB). Three countries accounted for almost half of the world's cases of MDR-TB: India (24%), China (13%), and the Russian Federation (10%) (Floyd et al., 2018).

14.3   *M. TUBERCULOSIS* (MTB)

M. tuberculosis, the pathogenic organism for TB, is a weak gram-positive, rod-shaped, non-flagellar and non-spore forming bacterium. It is a slow-growing microbe with a generation time of ~24 hours (Sparks et al., 1998). The structure of the *M. tuberculosis* cell wall is quite complex, comprising both gram-positive and gram-negative bacteria. Its cell wall is thick and composed of a thin peptidoglycan layer surrounded by an outer thick mycolic acid layer which is esterified to an arabinogalactan layer above the peptidoglycan (Daffé, 2015).

The *M. tuberculosis* complex (MTbC) consists of the following members which infect numerous animal species: *M. africanum, M. bovis, M. caprae, M. microti, M. mungi, M. orygis, and M. pinnipedii* apart from *M. tuberculosis*. These species are very closely related and fall under the *M. tuberculosis* phylogeny (Comas et al., 2013).

14.4   DRUG RESISTANCE

M. tuberculosis is a clonal microorganism and therefore exchange of DNA via horizontal gene transfer is not prevalent here. This phenomenon has resulted in a relatively low rate of evolution as well as a low rate of resistance as compared to other classes of bacterial pathogens (Eldholm and

Balloux, 2016). However, Multidrug-resistant Tuberculosis (MDR-TB) has become a major health concern, which was caused when Mtb developed resistance against at least two first-line drugs like isoniazid and rifampin. MDR-TB requires a new line of drugs and takes a longer time to cure (Skrahina et al., 2012). Factors which can cause drug resistance are improper medications, premature treatment, etc., which can spread from person to person (Pontali et al., 2016).

14.4.1 MDR (MULTIDRUG-RESISTANT) TB

MDR-TB is caused when the pathogen strain is resistant to at least two drugs, viz. isoniazid and rifampicin, which are also known as first-line drugs (World Health Organization, 2016). Between 2016 and 2017 the number of MDR-TB cases has increased by more than 30% in six of the 30 countries having high risk of MDR-TB (World Health Organization, 2006).

14.4.2 XDR (EXTENSIVELY DRUG-RESISTANT) TB

XDR-TB is caused when the pathogenic strain is, apart from the first line drugs isoniazid and rifampicin, resistant to at least one of the second line drugs like amikacin, kanamycin, or capreomycin (World Health Organization, 2006). MDR and XDR-TB do not show response in starting six months of standard TB treatment with first-line anti-TB drugs. Treatment for these can often take two years or more and treated with drugs which are less effective, more toxic, and also expensive (World Health Organization, 2016).

14.4.3 TDR (TOTALLY DRUG-RESISTANT) TB

The third type of resistant TB is referred to as totally drug-resistant (TDR-TB/XXDR-TB or extremely drug-resistant TB. It is resistant to both the first and second line of drugs and extremely difficult to treat, but not always impossible to cure.

14.5 MECHANISM OF INFECTION

The first stage of tuberculosis starts with inhalation of contaminated droplets containing the pathogen. The majority of the droplets are trapped in the upper respiratory tract and the pathogen is destroyed, but a few manage to penetrate further to reach the alveoli, where the macrophages phagocytize them. The macrophages are the major host cells for Mtb. The survival strategies of Mtb inside macrophages involve the prevention of phagosome-lysosome fusion and suppression of pro-inflammatory responses (Nasiruddin et al., 2017). In the second stage, Mtb multiplies inside the macrophage, and lyses them. This cellular damage attracts the monocytes and inflammatory cells, where monocytes are differentiated into macrophages which in turn ingest the released pathogen and again the lysis of pathogen loaded macrophages starts. The third stage starts after 2–3 weeks of infection, where the T cell response is generated, and lymphocytes migrate to the area of infection. Here antigen presentation takes place and the cytokines released by the macrophages are IL-12, TNF-α, IL-, etc. including proinflammatory cytokines, and cell-mediated immunity is developed by the host cell. At this stage the pathogen is enclosed by macrophages within an acidic and anaerobic tubercle where growth of the pathogen is prevented. This phage is called latent TB which is a major characteristic of the disease. In the last stage, the tubercles are destroyed following many causes such as malnutrition, compromised immunity and/or HIV infection, steroid abuse, etc. (Sherman et al., 1980).

14.6 CURRENTLY AVAILABLE TREATMENT FOR TB

14.6.1 FIRST-LINE DRUG THERAPY

At present the standard treatment of six months for TB consists of the combined oral dosage of different anti-TB drugs which are as follows: rifampicin (RIF), isoniazid (INH), pyrazinamide (PYR), ethambutol (ETB), and streptomycin (STM). These are the first-line anti-TB drugs (World Health Organization, 2013b). As the therapy may pose a complexity, the WHO and the International Union against TB and Pulmonary diseases have proposed the strategies to combine at least two of these first-line drugs in one dosage pattern, called fixed dose combinations (FDC) (World Health Organization, 2017).

14.6.2 Second-Line Drug Therapy

When the TB pathogen strain develops resistance against at least two first-line drugs, isoniazid and rifampin, it proceeds towards a complex form of TB known as MDR-TB. This form of TB has continued to be a major concern. It is estimated, in 2016, around 500,000 people globally, were diagnosed with MDR-TB. Two of the major possible reasons for the development of multidrug-resistant strains are incomplete treatment using anti-TB drugs and poor quality medicines.

Second-line drugs such as amikacin (AMK), kanamycin (KNM) or capreomycin (CPM), rifabutin (RFB), rifapentine, aminosalicylic acid, ethionamide (ETM), protionamide (PTM), cycloserine, terizidone, ciprofloxacin (CPX), ofloxacin, levofloxacin (LVX) and moxifloxacin (MOX). These drugs are given in a particular combination to treat MDR-TB (Mehta et al., 2012).

14.6.3 Third-Line Drugs

The third-line drugs used for the treatment of TB are rifabutin, linezolid, thioridazine, arginine, vitamin D, some macrolides, i.e. clarithromycin and thioacetazone. These third-line drugs are not listed by the WHO. Unlike first- and second-line drugs, these third-line drugs have reduced effect and efficacy. In order to increase the effectiveness of third-line drugs, new technologies are being used, e.g. conjugation with carriers, biodegradable polymers, liposomes, etc. (Hari et al., 2010).

14.6.4 Bacteriophage Enzymes as Anti-TB Agents

Bacteriophages or phages are the viruses that infect and replicate within bacteria in a specific manner, consequently killing the host bacteria. Phages perform this act by means of two proteins, holins which create holes on bacterial cell walls, and endolysin which is an enzyme targeting the bonds in the cell wall. These endolysins have emerged as antibacterials as they can lyse the bacteria when applied externally.

Phages against mycobacteria are known as mycobacteriophages and produce two types of endolysins, i.e. lysin A which targets the peptidoglycan layer and lysin B which targets the mycolyl arabinogalactan layer. Research is in progress to use these enzymes to clear Mtb pathogen in infected individuals (Borysowski et al., 2006).

14.6.5 Nanotechnology Approach

Advancements in the field of healthcare and medicine are the most promising outcomes of nanotechnology, which have enabled potential developments in pharmaceuticals viz., drug synthesis and delivery, diagnosis, and treatment of various diseases (Singh and Nalwa, 2011). Nanobiotechnology based drug delivery applications are accomplished through the use of designed nanomaterials, as well as creating delivery systems from nanoscale molecules (Suri et al., 2007). As the second- and third-line drugs are less effective and impart more toxicity as compared to first-line drugs, new strategies are required to be developed to increase the effectiveness and proper delivery of these drugs. In this scenario, nanotechnology has paved the way for the development of more effective drug delivery systems as well as the development of next-generation vaccines for the treatment of the disease.

14.6.5.1 Nanoliposomes (NLPs)

Nanoliposomes, submicron lipid bilayer vesicles of sizes ranging from 20 nm to several micrometers, are a promising drug delivery system. The mechanism for the formation of nanoliposomes is based on the hydrophilic-hydrophobic interaction between phospholipids and water molecules, such that they have an aqueous solution core surrounded by a hydrophobic membrane. Nanoliposomes enhance the effect of drugs by improving their solubility, availability to target, stability, and by decreasing their non-specific interactions [12]. Their hydrophilic drugs interact with the core; while hydrophobic ones interact with the outer phospholipid bilayer.

NLPs can effectively work against intracellular pathogens like Mtb, due to its uptake by macrophages. NLPs are susceptible to digestion by intestinal lipase, they are administered through respiratory or intravenous pathways. There are chances of nonspecific uptake by the spleen and liver mononuclear phagocyte system (MPS) which can be reduced by the addition of Polyethylene glycol (PEG) in the liposomal system (Nasiruddin et al., 2017).

In an experiment performed on mice, it was observed that anti-TB drugs like rifampicin or isoniazidalon encapsulated within liposomes were found to be more effective in clearing the infection compared to the pure drugs. The liposomal drugs were administrated to Mtb infected mice for six weeks twice a week. Apart from this, no significant cytotoxicity was observed when examined histopathologically (Pandey et al., 2004). However, NLPs have few drawbacks mainly regarding their physical stability and loading capacity, compared with other nanocarriers such as nanoparticles (Abed and Couvreur, 2014).

14.6.5.2 Nanoparticles

The advantage of using nanomaterials for drug delivery is that the increased surface area of the nanomaterial helps to bind and adsorb in therapeutic compounds or drugs. In some cases the drug particles are engineered to form nano-size particles (Singh and Lillard Jr, 2009).

Nanoparticles are taken up by cells more efficiently than larger molecules, so they could be used as compelling transport and delivery systems for drug delivery. Nanoparticles used for drug delivery are normally <100 nm in at least one dimension and made up of various biodegradable materials such as natural or synthetic polymers, lipids, or certain metals (Lin et al., 2015). Nanoparticles involved in drug delivery are generally composed of biocompatible and biodegradable substances or polymers, e.g. gelatin, albumin, polylactides, and polyalkylcyanoacrylates. The following are some nanoparticle-based drug delivery systems.

14.6.5.3 Ligand-Conjugated Anti-TB Drug

Nanopolymers have a bio-adhesive tendency in the gastrointestinal tract. Therefore, they are further modified by addition of bio-adhesive ligands. It has been studied that wheat germ agglutinin receptors are distributed on intestinal and alveolar epithelium, when covalently attached to poly D,L-glycolide (PLG), improves the efficacy of anti-TB medicines. When wheat germ agglutinin coated PLG nanoparticles are administrated orally or in the form of aerosol in mice, it resulted in longer retention of drugs like RIF, INH, and PYZ compared to uncoated PLG nanoparticles. This nanoparticle also found to have lower immunogenicity (Clark et al., 2000).

14.6.5.4 Microemulsions

Microemulsion is 'a system of water, oil and anamphiphile (surfactant and co-surfactant) which is a single optically isotropic and thermodynamically stable liquid solution'. The concept of microemulsion was described by Hoar and Schulman in 1943.

In recent years microemulsions have found wide application in the field of development and design of new drug delivery systems because of their characteristics like thermodynamic stability, high absorption capacity and solubility, resistance to enzymatic hydrolysis, and reduced toxicity. In their study Mehta et al. used a tween-based microemulsion system as a carrier for the drug, rifampicin. They found it to be a successful delivery system by various structural studies, i.e. electron microscopy, NMR, optical microscopy and dissolution,and release kinetics (Agrawal et al., 2009).

14.6.5.5 Solid Lipid Nanoparticles (SLNs)

SLNs are a crystalline nanosuspension in water. They have major applications in oral administration because of characteristics like steadiness and better encapsulation properties than those of liposomes and polymeric nanoparticles. In SLNs, the solid lipid matrix entraps the drug and creates lipid nanoparticles. SLNs are more stable and possess greater encapsulation ability.

In an experiment on Mtb H37Rv infected mice, anti-TB loaded SLNs were given orally, it was observed that the drug was retained in plasma for eight days and in tissues for 9–10 days. For complete clearance of the pathogen, f oral doses of SLN-loaded drug on every 10th day was required in comparison to 46 daily oral doses in the case of a free drug (Pandey et al., 2005).

14.6.5.6 Alginate Nanoparticles (ANPs)

Alginate is a natural polymer which has a large number of applications in various fields as a binding, suspending, and solidifying agent. The characteristics which make it a drug delivery agent are: aqueous background inside its matrix, adherence with intestinal epithelium, drug encapsulation without the use of organic solvent, high porosity of gel facilitating greater diffusion of larger drug molecules, degradable in physiological conditions, and no toxicity.

Antitubercular drugs (ATD)-loaded alginate nanoparticles were generated by González-Rodríguez et al. with high encapsulation efficiency in the range of 80–90% for rifampicin, 70–90% for isoniazid and pyrazinamide and 88–95% for ethambutol. Oral administration of this formulation showed the presence of drugs in plasma for more than a week compared to free drugs which were able to stay for 12–24 hours only (Segale et al., 2016).

When chitosan is added to alginate, the combination provides better pharmacokinetics, increased therapeutic efficacy, and reduced drug dose. ANPs have been combined and stabilized with poly(L-lysine)-chitosan, encapsulating anti-tubercular drugs like RIF, INH, pyrazinamide (PZA), and ethambutol (EMB). The ratio for drug to polymer was kept at 7.5:1, as compared to the ratio of 1:1 in the case of PLGNPs (Nasiruddin et al., 2017). Table 14.1.

14.7 CONCLUSION

In spite of advancements in science and technology in the current century, TB still remains the second most deadly infectious disease after HIV/AIDS. There are few efficient medications available which are effective against non-resistant tuberculosis and drug-resistant tuberculosis, viz. MDR and XDR-TB, is still a major concern requiring novel effective drugs and regimens. In this scenario, nanotechnology has paved a novel pathway through improved drug formulations and drug delivery systems to create effective therapy for TB, and at the same time the applications of nanotechnology are also associated with intellectual property law. To develop successful treatments involving nanotechnology will require collective effort and collaboration for the discovery of efficient medication as well as the approval of laws covering multiple patents and intellectual property.

TABLE 14.1

Recent Patents Involving Nanotechnology-Based Treatment of TB

Sl. No.	Patent Number	Title	About Nano-Product	References
1	EP2859345A1	A liposomal composition comprising a sterol-modified lipid and a purified mycobacterial lipid cell wall component and its use in the diagnosis of tuberculosis	It is a liposomal structure consisting of a sterol-modified lipid and a purified mycobacterial lipid cell wall material or its derivatives or analog.	Baumeister et al. (2015)
2	US20160038422A1	Tablet Composition for Anti-tuberculosis Antibiotics	The tablet consists of three components which are a saccharide/saccharide, a transition metal ion or a combination of metal ions which can bind to nitrogen and/or oxygen atom, and a water-soluble polymer capable of aggregating and enclosing the other constituents.	Manning et al. (2019)
3	WO2014068554A1	Sensor technology for diagnosing tuberculosis	A sensor-based technology consisting of a single nanomaterial, viz. gold nanoparticles and/or carbon nanotube-based sensor/sensors in conjunction with a pattern recognition algorithm for non-invasive and accurate diagnosis of tuberculosis caused by *M. tuberculosis*.	Haick and Gang (2013)
4	CN102914520A	Surface plasmon resonance biosensor for detecting tuberculosis, preparation method and application of surface plasmon resonance biosensor	It is a double-channel surface plasmon resonance sensor system; a CFP-10 antibody is fixed on the surface of a chip by protein coupling agent to prepare a needed immune chip.	Rich and Myszka (2008)
5	CN1368148A	Nano medicine for treating tuberculosis and its preparing process	A nanometer (0.1–200 nm) medicine for treating tuberculosis is prepared from the nanometer powders of 16 Chinese medicinal materials including tortoise plastron, rehmannia root, donkey-hide gelatin, Dragon's bone, etc.	Pandey and Khuller (2006)
6	WO2015116568A2	Muscle cell-targeting nanoparticles for vaccination and nucleic acid delivery, and methods of production and use thereof	The nanoparticles are composed of charged polymers or charged particles (e.g. polyamidoamide dendrimers) surface-modified with a muscle cell ligand (a targeting moiety) and an anionic enhancer (e.g. a phospholipid) resulting in a multinucleotide-carrying conjugate for introducing vaccines or genes into skeletal muscle cells of a subject.	Hirosue et al. (2001)

REFERENCES

Abed, N. and Couvreur, P. 2014. Nanocarriers for antibiotics: A promising solution to treat intracellular bacterial infections. *International Journal of Antimicrobial Agents*, 43(6), 485–496.

Agrawal, D., Udwadia, Z., Rodriguez, C. and Mehta, A. 2009. Increasing incidence of fluoroquinolone-resistant Mycobacterium tuberculosis in Mumbai, India. *The International Journal of Tuberculosis and Lung Disease*, 13(1), 79–83.

Baumeister, C., Shaw, W. A. and Verschoor, J. A. 2015. Liposomal composition comprising a sterol-modified lipid and a purified mycobacterial lipid cell wall component and its use in the diagnosis of tuberculosis. Google Patents.

Borysowski, J., Weber-Dąbrowska, B. and Górski, A. 2006. Bacteriophage endolysins as a novel class of antibacterial agents. *Experimental Biology and Medicine*, 231(4), 366–377.

Clark, M. A., Hirst, B. H. and Jepson, M. A. 2000. Lectin-mediated mucosal delivery of drugs and microparticles. *Advanced Drug Delivery Reviews*, 43(2–3), 207–223.

Comas, I., Coscolla, M., Luo, T., Borrell, S., Holt, K. E., Kato-Maeda, M., Parkhill, J., Malla, B., Berg, S. and Thwaites, G. 2013. Out-of-Africa migration and Neolithic coexpansion of Mycobacterium tuberculosis with modern humans. *Nature Genetics*, 45(10), 1176.

Cooper, A. M. 2009. Cell-mediated immune responses in tuberculosis. *Annual Review of Immunology*, 27, 393–422.

Daffé, M. 2015. The cell envelope of tubercle bacilli. *Tuberculosis*, 95, S155–S158.

Eldholm, V. and Balloux, F. 2016. Antimicrobial resistance in Mycobacterium tuberculosis: The odd one out. *Trends in Microbiology*, 24(8), 637–648.

Floyd, K., Glaziou, P., Zumla, A. and Raviglione, M. 2018. The global tuberculosis epidemic and progress in care, prevention, and research: An overview in year 3 of the End TB era. *The Lancet Respiratory Medicine*, 6(4), 299–314.

Greenblatt, D. J. 1985. Elimination half-life of drugs: Value and limitations. *Annual Review of Medicine*, 36, 421–427.

Haick, H. and Gang, P. 2013. Carbon nanotube structures in sensor apparatuses for analyzing biomarkers in breath samples. Google Patents.

Hari, B. V., Chitra, K. P., Bhimavarapu, R., Karunakaran, P., Muthukrishnan, N. and Rani, B. S. 2010. Novel technologies: A weapon against tuberculosis. *Indian Journal of Pharmacology*, 42(6), 338.

Hirosue, S., Mueller, B. G., Langer, R. S. and Mulligan, R. C. 2001. Sub-100nm biodegradable polymer spheres capable of transporting and releasing nucleic acids. Google Patents.

Lienhardt, C., Glaziou, P., Uplekar, M., Lönnroth, K., Getahun, H. and Raviglione, M. 2012. Global tuberculosis control: Lessons learnt and future prospects. *Nature Reviews. Microbiology*, 10(6), 407.

Lin, Y.-S., Lee, M.-Y., Yang, C.-H. and Huang, K.-S. 2015. Active targeted drug delivery for microbes using nano-carriers. *Current Topics in Medicinal Chemistry*, 15(15), 1525–1531.

Manning, T. J., Plummer, S. E. and Baker, T. A. 2019. Tablet composition for anti-tuberculosis antibiotics. Google Patents.

Mehta, D., Bassi, R., Singh, M. and Mehta, C. 2012. To study the knowledge about tuberculosis management and national tuberculosis program among medical students and aspiring doctors in a high tubercular endemic country. *Annals of Tropical Medicine and Public Health*, 5(3), 206.

Nasiruddin, M., Neyaz, M. and Das, S. 2017. Nanotechnology-based approach in tuberculosis treatment. *Tuberculosis Research and Treatment* 1 (2017), 1–12.

Pandey, R. and Khuller, G. 2006. Nanotechnology based drug delivery system (s) for the management of tuberculosis. *Indian Journal of Experimental Biology*, 44(5), 357–366.

Pandey, R., Sharma, S. and Khuller, G. 2004. Liposome-based antitubercular drug therapy in a guinea pig model of tuberculosis. *International Journal of Antimicrobial Agents*, 23(4), 414–415.

Pandey, R., Sharma, S. and Khuller, G. 2005. Oral solid lipid nanoparticle-based antitubercular chemotherapy. *Tuberculosis*, 85(5–6), 415–420.

Pontali, E., Sotgiu, G., D'ambrosio, L., Centis, R. and Migliori, G. B. 2016. Bedaquiline and multidrug-resistant tuberculosis: A systematic and critical analysis of the evidence. Eur Respiratory. *Società*, 47, 394–402.

Rich, R. L. and Myszka, D. G. 2008. Survey of the year 2007 commercial optical biosensor literature. *Journal of Molecular Recognition: An Interdisciplinary Journal*, 21(6), 355–400.

Sairam, A., Guthi, V. R. and Sarma, M. V. P. 2017. Influence of socio demographic factors on noncompliance among patients treated under RNTCP in a rural tuberculosis unit in Karimnagar district. *International Journal of Community Medicine and Public Health*, 4(12), 4538–4543.

Segale, L., Giovannelli, L., Mannina, P. and Pattarino, F. 2016. Calcium alginate and calcium alginate-chitosan beads containing celecoxib solubilized in a self-emulsifying phase. *Scientifica*, 2016, 1–8.

Sherman, S., Rohwedder, J. J., Ravikrishnan, K. and Weg, J. G. 1980. Tuberculous enteritis and peritonitis: Report of 36 general hospital cases. *Archives of Internal Medicine*, 140(4), 506–508.

Singh, R., Lillard, Jr. J. W. 2009. Nanoparticle-based targeted drug delivery. *Experimental and Molecular Pathology*, 86(3), 215–223.

Singh, R. and Nalwa, H. S. 2011. Medical applications of nanoparticles in biological imaging, cell labeling, antimicrobial agents, and anticancer nanodrugs. *Journal of Biomedical Nanotechnology*, 7(4), 489–503.

Skrahina, A., Hurevich, H., Zalutskaya, A., Sahalchyk, E., Astrauko, A., Van Gemert, W., Hoffner, S., Rusovich, V. and Zignol, M. 2012. Alarming levels of drug-resistant tuberculosis in Belarus: Results of a survey in Minsk. *European Respiratory Journal*, 39(6), 1425–1431.

Sparks, R. S. J., Young, S. R., Barclay, J., Calder, E. S., Cole, P., Darroux, B., Davies, M., Druitt, T., Harford, C. and Herd, R. 1998. Magma production and growth of the lava dome of the Soufriere Hills Volcano, Montserrat, West Indies: November 1995 to December 1997. *Geophysical Research Letters*, 25, 3421–3424.

Suri, S. S., Fenniri, H. and Singh, B. 2007. Nanotechnology-based drug delivery systems. *Journal of Occupational Medicine and Toxicology*, 2, 16.

15 A Patent Landscape of the Emerging Nanomolecules for the Treatment of Leishmaniasis

Priyanka Gautam, Abhishek Pathak,
and Sandeep Kumar Singh

CONTENTS

15.1 INTRODUCTION

A protozoan parasite of the genus *Leishmania* causes a vector-borne disease, leishmaniasis, which is prevalent in more than 98 countries at present. Out of 53 described species of *Leishmania* parasites, 20 are known to cause human pathogenesis that are spread by approximately 30 species of sand flies (Torres-Guerrero et al., 2017). The infection of *Leishmania* in the human may produce three clinical manifestations, i.e. cutaneous (CL), mucocutaneous (MCL), and visceral (VL) leishmaniasis, which differ in their immunopathology, degree of morbidity, and mortality. Between 12 and 15 million people in the world are infected, and 350 million are at risk of acquiring the disease. An estimated 1.5 to 2 million new cases occur each year, and it causes 70,000 deaths per year.

Visceral leishmaniasis (VL) (kala-azar), which is caused by *L. donovani*, is a latent threat to more than 147 million people living in the disease endemic South East Asia region and the Indian subcontinent. The estimates indicate about 100,000 cases per year that includes 15,000 reported cases (Karunaweera and Ferreira, 2018). Out of five VL affected countries, i.e. India, Bangladesh, Nepal, Thailand, and Bhutan in this region, India contributes more than 80% of reported cases whereas in Bhutan and Thailand reports are sporadic. In India, Bihar is the most VL-endemic state, with 90% of the Indian VL cases reported there. *L. infantum* causes VL in North Africa and Southern Europe while in Latin America the VL causing species is *L. chagasi,* and in East Africa *L. donovani* is responsible for VL infections (Thakur et al., 2018, Le Rutte et al., 2017).

15.2 IMMUNE BIOLOGY OF LEISHMANIASIS

Evasion of the parasite inside the host by adopting different strategies such as: (1) inactivation of effector molecules, one key strategy of immune evasion. The inhibition of complement activation is vital for many parasitic diseases. Antibodies can be rendered ineffective by highly specific

287

proteases secreted by parasites, and cytotoxic effector molecules of immune cells, such as reactive oxygen and nitrogen products, are counteracted by increased production of detoxifying parasite enzymes (glutathione-S-transferase, glutathione peroxidase, catalase, etc. (Hewitson et al., 2009; Evering and Weiss, 2006) and, (2) by secreting specifically-active products, parasites can also interfere with the cytokine network, enabling them to modulate local systemic host immune responses. Macrophages are proposed primary host cells for *Leishmania* but the role of these cells has not been well characterized either in disease prevention or in progression independent of T cells. The effector functions of macrophages for *Leishmania* have always been described in a T-dependent manner (Gupta et al., 2013; Soulat and Bogdan, 2017). The fate of infected macrophages in T cell activation is not well known. This is because of the T cells come later during infection.

There are two subtypes of Th cells which play roles in the balance of immune response and immune tolerance, i.e. Th1 and Th2. Th1 plays a protective role against pathogens and induces pro-inflammatory cytokines. While Th2 response comes to compensate excessive immune response generation, this is why they induce the release of anti-inflammatory cytokines. Sometimes, however, the parasite may also modulate the function of Th cells, ensuring their own survival inside the host. They will modulate the host in terms of signaling or antigen presentation for their own benefit, inducing factors that provide a disease progressive environment and prime T cells for Th2 differentiation (Kaiko et al., 2008; Spellberg and Edwards Jr, 2001). It is also possible that parasites start modulating macrophages at the time of entry and later on modulated parasitized macrophages interact with T-cells and may induce IL-4 and IL-10 (anti-inflammatory cytokines) in place of IL-12, IFN-γ (pro-inflammatory cytokines) and disease-inducing factors from T cells that help in disease progression and parasite survival in a susceptible host (Liu and Uzonna, 2012).

It is also known that *Leishmania*-parasitized macrophages produce IL-10 but not IL-4, suggesting the role of IL-10 before IL-4 in disease progression or in the susceptibility of the host. This suggests the crucial role of IL-10 in disease initiation independent of T cells and in disease progression later on in combination with IL-4 (Maspi et al., 2016). Many other molecules which are immune-inhibitory such as CD200, CD200R, and CD300a, etc., which are present on the surface of T cells, are also activated by parasites, which is why the function of effective T cells has changed (Vaine and Soberman, 2014; Cortez et al., 2011).

15.3 TREATMENT OF LEISHMANIASIS AND ITS LIMITATIONS

Soon after identification of the *Leishmania* parasite in 1901, pentavalent antimony compounds were used to treat VL cases, which were also considered as true anti-leishmanial. VL was controlled by such treatment until antimony resistant parasites appeared in the early 1980s in the Indian subcontinent and treatment failure has now reached up to 65% in a few endemic areas (Singh et al., 2016). In the early 1980s, an antimicrobial drug, pentamidine (1983), was used in antimony refractory cases and cured 99% of patients initially, however, within two decades its efficacy declined to approximately 70% of patients leading to its abandonment in VL treatment (Nagle et al., 2014).

During 1990–1998, an antifungal amphotericin B (AmpB) and an anticancer (miltefosin) became the first-choice drugs to treat VL cases. At present, single dose AmBisome, a liposomal formulation of AmpB, is a drug of choice for the VL elimination program and approved by the World Health Organization (WHO) as a preventive measure. At present, these drugs are effective with satisfactory results. However, recent reports indicate the parasites' potential for the development of resistance to AmpB and miltefosin (Berman, 2015).

In spite of significant knowledge of host-parasite relationships and leishmanial immunobiology, a vaccine candidate for either preventive or prophylactic is far from reality (Tiwari et al., 2018). Various antigenic formulations such as dead *Leishmania* parasites, native and recombinant vaccine antigens such as gp63, gp46, m2, PSA2, TSA, LACK, LmsT1, and Leish111f, etc., have been evaluated but all have failed to achieve long-lasting protective immunity (Singh et al., 2018). Among all vaccination approaches the live attenuated parasites have shown to produce the required magnitude of protective immunity, which provides the possibility of disease protection through vaccination (Barry and Arnott, 2014).

15.3.1 FAILURE OF TREATMENT OF LEISHMANIASIS

The magnitude of immune response generation depends on a pathogen's pathogenicity, because good antigens can only induce a good immune response, and good memory cell formation occurs for quick removal of pathogens during a future attack (Belkaid and Hand, 2014).

Most of the parasites have evolved several strategies to survive within the host macrophage by suppressing macrophages' microbicidal activities like decreased production of free radicals and pro-inflammatory cytokine (Arango Duque and Descoteaux, 2014). Parasites disrupt the homeostasis between immune response and immune tolerance, factors responsible for immune tolerance will definitely help to understand the role in parasite-induced macrophage dysfunction, and it will also be helpful to identify a potential vaccine (Ocaña-Guzman et al., 2018).

Although during recent years few factors have been discovered, the exact mechanisms of macrophage dysfunction, poor antigen presentation, and suppressed immune response are not clearly understood. Further due to suppressed immune response, it may be possible that parasites may modulate host inhibitory mechanisms for their survival, altering macrophage functions during poor antigen presentation and compromise the functions of T and B cells (Maizels and McSorley, 2016). Hence, in view of the capability of parasites to alter host macrophages, the linkage of parasite and host immune inhibitory mechanisms cannot be ignored. Table 15.1.

15.3.2 ALTERNATIVES FOR THE TREATMENT OF LEISHMANIASIS BY THE USE OF NANOPARTICLES

Since antileishmanial vaccines are still under development, control of the disease depends mostly on chemotherapy. Treatments available for VL have limitations such as parenteral administration, a long course of treatment, toxic side effects, and a high cost (Ghorbani and Farhoudi, 2018).

The first-line therapy for leishmaniasis includes sodium stibogluconate; however, its efficacy is now threatened by the emergence of drug-resistant parasites (Chappuis et al., 2007). As time progresses the drugs used to treat leishmaniasis changes, such as sodium stibogluconate being replaced with liposomized AmpB, but its use is limited due to very high cost and high toxicity. Its intravenous administration frequently causes rigor, chills, and fever, associated with myocarditis and nephrotoxicity (Laniado-Laborín and Cabrales-Vargas, 2009). AmpB formulations (the lipid complex, colloidal form, and the liposomal form) were developed to reduce adverse effects and improve pharmacokinetics and bioavailability (Singh et al., 2016, de Menezes et al., 2015).

TABLE 15.1

Distribution and Death Rate of Different Forms of Leishmania and Drugs Used in the Treatment of Leishmaniasis on Worldwide Level

S.No.	Disease	Disease burden	Death, people/ annually	Treatment	Availability of vaccine
1	Visceral leishmaniasis	0.2–0.4 million	20,000–40,000	Stibogluconate, Amphotericin B, Miltefosin, Paromomycin	No
2	Cutaneous leishmaniasis	0.7–1.2 million	20,000–30,000	Same	No
3	Mucocutaneous leishmaniasis	0.7–1.2 million	20,000–30,000	Same	No

A sudden major drastic change in chemotherapy of VL was the discovery of miltefosine, which was an analog of phosphatidylcholine, initially developed as an anticancer agent. It is an effective oral drug, but its use in women of child-bearing age is restricted due to teratogenicity (Tiwari et al., 2017). In addition, it has a long half-life, which might encourage the emergence of resistance. Interestingly, a few relapsed cases have already been reported. Consequently, there is an immediate requirement to search for new antileishmanial compounds (Sane et al., 2010). New drug delivery devices transport antileishmanial drugs to the target cell specifically by minimizing the toxic effects on normal cells (Akbari et al., 2017).

The antileishmanial drugs currently used have several limitations such as low efficacy, toxicity, adverse side effects, drug-resistance, length of treatment and cost lines. Consequently, there is an immediate requirement to search for new antileishmanial compounds (Sangshetti et al., 2015). New drug delivery devices transport antileishmanial drugs to the target cell specifically to reduce the toxic effect of chemotherapy, a new nano-based therapy used to treat leishmaniasis (Akbari et al., 2017; ud Din et al., 2017).

Nanotechnology must be applied for curing infectious diseases like leishmaniasis. Many drugs have been made and are still being discovered, but the protozoan is showing drug resistance against currently used medicine. There is an immediate and urgent need to develop effective nanotechnology devices in fighting this infectious disease. Different mechanisms were proposed for the antimicrobial property of nanoparticles (ud Din et al., 2017; Shahcheraghi et al., 2016; Mulani et al., 2019).

Some time ago, it was proved that the formulation of artemisinin-loaded nanoparticles improved its activity against *L. donovani* amastigote ex vivo. Nanoparticles of AmpB are more effective compared to conventional AmpB (Want et al., 2015; Want et al., 2017). This formulation may have a good safety profile, and if production costs are low, it may prove to be a feasible alternative to conventional AmpB in the treatment of VL (Mishra et al., 2013).

If nanoparticles are systemically administered, they will be clumped together after exposure to plasma, which is why their efficacy will be also decreased. Antileishmanial nanoparticles must be conjugated with biological compounds such as antibodies or lectins, which bind to specific targets, in order to exert more toxicity for parasites and less toxicity for normal cells (Bruni et al., 2017).

Some nanoparticles have high cytotoxicity on macrophages, which must be considered. The use of nanoparticles for the treatment of VL may have both positive and negative consequences. Some reports indicate that gold nanoparticles (AuNPs), titanium dioxide nanoparticles (TiO_2-NPs), zinc oxide nanoparticles (ZnO-NPs), magnesium oxide nanoparticles (MgO-NPs), etc. have antibacterial properties (Wang et al., 2017).

On the other hand, some nanoparticles have a photothermal effect after exposure to near infrared (NIR) light. These nanoparticles absorb NIR energy and convert it to heat, then the temperature is increased and cells will be damaged (de Boer et al., 2018). It is stated that *Leishmania* parasites are sensitive to heat, and heat therapy has been used in a new way with nanoparticles. Table 15.2.

15.4 CONCLUSION

In spite of good knowledge of the epidemiology and etiology of leishmaniasis, currently there is not any effective vaccine available against leishmaniasis and the percentage of affected population progressively increases at world level. Current antileishmanial drugs which are used to treat leishmaniasis have several limitations such as toxicity, high cost, and drug resistance.

Development of a new drug requires lots of money and takes a long time, about 10–12 years. Therefore, the search should continue for formulations that can be effective in treatment of leishmaniasis. In recent years, nanotechnology-based drug delivery systems have improved and advanced to integrate different aspects. For effective treatment and control of leishmaniasis different delivery technologies have been developed such as bioactive targeting to the parasitophorous vacuole of the macrophage where the amastigotes of *Leishmania* are localized. New drug delivery devices such as nanoparticles, microparticles, liposomes, and noisomes transport drugs to the target cell

TABLE 15.2

Recent Patents on Nanoparticle-Based Treatment for Leishmaniasis

S.no.	Patent no.	Topic name	About nanoparticles	References
1	BR102015030291A2	Process of obtaining nano particulate active ingredient from *ocotea duckei* Vattimo extract, said ingredient, pharmaceutical composition comprising said ingredient and uses thereof for leishmaniasis treatment.	An active ingredient Nano structuring process comprising lignoid rich extract obtained from the *ocotea duckei* Vattimo plant. Said process comprises the following steps: a) preparing a hot microemulsion comprising an oil phase, an aqueous and surfactant phase, wherein the oil phase comprises a lignoid fraction obtained from *ocotea duckei* Vattimo; b) pouring the hot microemulsion into distilled water while stirring; c) adding to the sucrose dry protectant suspension; d) lyophilizing the suspension obtained in step c). The resulting input from the process is a nanostructured lignoid fraction and can be used in drugs to treat leishmaniasis disease. This is used for the treatment of visceral and integumentary leishmaniasis.	(Oryan, 2015)
2	BR102013017881A2	Use of Chitosan and chondroitin nanoparticles for the treatment of leishmaniasis.	Chitosan and chondroitin nanoparticles are prepared by the pec technique (polyelectrolyte complexes) using two oppositely charged polymers both give synergistic effect against leishmaniasis. In addition nanoparticles were able to carry the amphotericin drug with good efficacy in the treatment of leishmaniasis as well as low toxicity.	(Ribeiro et al., 2014)
3	BR102017004291A2	Process for electrochemical detection of leishmaniasis using magnetic nanoparticles and magnetic/chitosan/polyaniline electroplated film.	The invention is based on the application of a short sequence of oligonucleotide (DNA probe) electrostatically bonded to the surface of magnetite (npsfe304), chitosan and polyaniline nanoparticles for pathogen detection in low concentration samples. In particular in the present invention a genosensor has been developed for the identification of specific DNA sequences from *leishmania chagasi* in sample from infected dog patients.	(MACIEL, 2017)
4	BR102018000232A2	Chitosan nanoparticles containing S-nitro-mercaptosuccinic acid (S-nitroso–MSA) for treatment of cutaneous leishmaniasis.	Chitosan nanoparticles containing S-nitroso–mercaptosuccinic acid to be used in the treatment of cutaneous leishmaniasis, due to the leishmanicidal action of its active principle, S-nitroso MSA. This alternative therapy for cutaneous leishmaniasis, in which the chitosan nanoparticles encapsulated S-nitroso–MSA possesses high stability in aqueous media bioavailability of the active principle and reduced side effects.	(Cabral et al., 2019)

(Continued)

TABLE 15.2 (CONTINUED)
Recent Patents on Nanoparticle-Based Treatment for Leishmaniasis

S.no.	Patent no.	Topic name	About nanoparticles	References
5	ES2388963A1	Development and use of lipid nanoparticles containing edelphosine and other phospholipid esters in anti-parasitary and anti-parasitary therapy.	Use of lipid nanoparticles containing edelphosine and other phospholipid ethers in anti-tumor and anti-parasitic therapy. The present invention provides compositions comprising nanoparticles of lipid origin containing phospholipid ether. This is formed as for oral treatment of cancer and leishmaniasis. The pharmaceutical composition of the present invention entails a marked increase in oral bioavailability and an improvement in the treatment against the use of free phospholipid ether.	(Puri et al., 2009)
6	WO2015177820A1	Nanoparticulate systems for vehiculating drugs for the treatment of leishmania.	Polymeric nanoparticle systems as effective vehicles for compounds containing antimony, also antimonials, such as N-methylglucamine antimoniate (Glucantime). In particular, the active principle, used for the treatment of pathologies related to infection with Leishmania, was entrapped in the aqueous core of nano capsules having an external layer, or shell of polylactic acid (PLA), that allows reducing cell toxicity and modulation of its biodistribution and of its biopharmaceutical properties, without altering its pharmacological effectiveness.	(Cosco et al., 2015)
7	WO2016113763A1	A novel antileishmanial formulation.	A novel antileishmanial formulation containing compounds belonging to the series of (2Z,2'Z)-3,3'-(alkane-1,2-diylbis(azanediyl))bis(6-hydroxy-1- aryl/ fluorenyl hex-2-en-1-ones) or their salts thereof showing antileishmanial activity under in vivo conditions against *L. donovani* in hamster model. These compounds or their formulation elicit antileishmanial activity both via oral and IV routes.	(Jain and Jain, 2013)
8	BR102013033866A2	Antileishmania compositions containing fullerol and use.	Compositions containing fullerenes, preferably fullerol, which are spherical carbon nanostructures, as novel tools for the treatment of visceral and cutaneous leishmaniasis. fullerol-containing antileishmanial compositions may comprise yet another leishmanicidal or immunomodulatory drug, being in free form or associated with a Nano carrier such as liposomes, which confers pharmacokinetic properties more favorable to the active ingredients, fullerol has surprisingly demonstrated antileishmanial activity, as well as an ability to reduce meglumine antimoniate hepatotoxicity when administered in combination in an animal model of visceral leishmaniasis.	(Mucke et al., 2015)

(*Continued*)

TABLE 15.2 (CONTINUED)
Recent Patents on Nanoparticle-Based Treatment for Leishmaniasis

S.no.	Patent no.	Topic name	About nanoparticles	References
9	US20100055201A1	Anti-parasitic methods and compositions utilizing diindolylmethane-related indoles.	Use of diindolylmethane-related indoles for the treatment of parasitic disease. Additive and synergistic interaction of diindolylmethane-related indoles with other known anti-parasitic and pro-apoptotic agents is believed to permit more effective therapy and prevention of protozoal parasitic infections. The methods and compositions described provide new treatment of protozoal parasitic diseases of mammals and birds including malaria, leishmaniasis, trypanosomiasis, trichomoniasis, neosporosis, and coccidiosis.	(Zeligs, 2013)
10	WO2013010238A1	Microparticle pharmaceutical compositions containing antiparasitics for prolonged subcutaneous therapy, use of the pharmaceutical compositions for producing a medicament, and method for treating parasitosis.	A pharmaceutical composition containing amphotericin B, nitrochalcone (CH8) or glucantime encapsulated in biodegradable, slow-release polymer microparticles, to the method of encapsulating the pharmaceutical drug inside the particles, to the use of these pharmaceutical compositions, to a medicament and to the treatment of parasitosis.	(Lavik et al., 2013).
11	BR102015014710A2	Antimony (v) organometallic compounds containing quinolone and fluoroquinolone binders as leishmanicidal chemotherapeutic agents.	The process of obtaining potential antimony (v) -based antimony organometallic metallopharmaceuticals containing quinoline and fluoroquinolone binders, as well as the use of such compounds in the treatment of leishmaniasis. These compounds act as prototypes for the development of new leishmanicidal drugs.	(Fedorowicz and Sączewski, 2018)
12	BR102017005445A2	Thiosemicarbazones indole derivatives useful for the treatment of leishmaniasis	Indole-Thiosemicarbazones derivatives useful for the treatment of leishmaniasis. The present invention is characterized by the production of indole-Thiosemicarbazones, tested in vitro experimental models against *Leishmania infantum*, which showed that there was an inhibition after 24 h of 100% of promastigote forms at concentrations 12.5 and 25 µg/ml. These results obtained in in vitro models place the indole-Thiosemicarbazones derivatives and their salts as candidates for new leishmanicidal agents.	(Taha et al., 2019)

specifically, minimizing the toxic effects to normal cells. Nano-based delivery systems have many known clinical advantages, e.g. liposomal formulation of AmpB is unquestionably effective, but could also be expensive. As was described in this review, in recent years there have been several reported studies as far as the treatment against leishmaniasis through nanotechnology is concerned.

15.5 FUTURE PROSPECTS

Some of them have shown encouraging results that suggest nanotechnology might play an effective role in the future, due to its many advantages, especially in comparison to the current antileishmanial treatment. Liposomes, nanoclusters (NCs), polymers, and nanoparticles have been the object of much study, because nanotechnology may not only be a valuable tool in antileishmanial treatment but in prevention (vaccines) as well. Interestingly, few nanotechnology-based formulations are being introduced in clinics for the treatment of leishmaniasis. However, none of the recent studies mentioned further in vivo and clinical tests, which could bring new challenges due to the large difference between the two scenarios.

Until today, AmBisome is the only nanotechnology-based antileishmanial drug, which is used in clinical practice, but, despite its proven efficacy and low toxicity among other features that make it an excellent alternative, its use remains restricted in some areas due to its high cost. Further efforts should be made in order to have more drugs based on this promising technology in the future antileishmanial drug arsenal.

ACKNOWLEDGEMENT

This work was performed with resources from the Non Government/Non-profit Organization; Indian Scientific Education and Technology Foundation, Lucknow, India and Indian Council of Medical Research (ICMR), New Delhi, India.

REFERENCES

Akbari, M., Oryan, A. and Hatam, G. 2017. Application of nanotechnology in treatment of leishmaniasis: A review. *Acta Tropica*, 172, 86–90.

Arango Duque, G. and Descoteaux, A. 2014. Macrophage cytokines: Involvement in immunity and infectious diseases. *Frontiers in Immunology*, 5, 491.

Barry, A. E. and Arnott, A. 2014. Strategies for designing and monitoring malaria vaccines targeting diverse antigens. *Frontiers in Immunology*, 5, 359.

Belkaid, Y. and Hand, T. W. 2014. Role of the microbiota in immunity and inflammation. *Cell*, 157(1), 121–141.

Berman, J. 2015. Amphotericin B formulations and other drugs for visceral leishmaniasis. *The American Journal of Tropical Medicine and Hygiene*, 92(3), 471–473.

Bruni, N., Stella, B., Giraudo, L., Della Pepa, C., Gastaldi, D. and Dosio, F. 2017. Nanostructured delivery systems with improved leishmanicidal activity: A critical review. *International Journal of Nanomedicine*, 12, 5289.

Cabral, F. V., Pelegrino, M. T., Sauter, I. P., Seabra, A. B., Cortez, M. and Ribeiro, M. S. 2019. Nitric oxide-loaded chitosan nanoparticles as an innovative antileishmanial platform. *Nitric Oxide: Biology and Chemistry*, 93, 25–33.

Chappuis, F., Sundar, S., Hailu, A., Ghalib, H., Rijal, S., Peeling, R. W., Alvar, J. and Boelaert, M. 2007. Visceral leishmaniasis: What are the needs for diagnosis, treatment and control? *Nature Reviews Microbiology*, 5(11), S7.

Cortez, M., Huynh, C., Fernandes, M. C., Kennedy, K. A., Aderem, A. and Andrews, N. W. 2011. Leishmania promotes its own virulence by inducing expression of the host immune inhibitory ligand CD200. *Cell Host and Microbe*, 9(6), 463–471.

Cosco, D., Britti, D. Fresta, M., Paolino, D. and Trapasso, E. 2015. (WO/2015/177820) Nanoparticulate systems for vehiculating drugs for the treatment of leishmania infection-related pathologies. PCT/IT2015/000134.

de Boer, W. D., Hirtz, J. J., Capretti, A., Gregorkiewicz, T., Izquierdo-Serra, M., Han, S., Dupre, C., Shymkiv, Y. and Yuste, R. 2018. Neuronal photoactivation through second-harmonic near-infrared absorption by gold nanoparticles. *Light: Science and Applications*, 7, 100.

de Menezes, J. P. B., Guedes, C. E. S., Petersen, A. L. D. O. A., Fraga, D. B. M. and Veras, P. S. T. 2015. Advances in development of new treatment for leishmaniasis. *BioMed Research International*, 2015, e815023.

Evering, T. and Weiss, L. M. 2006. The immunology of parasite infections in immunocompromised hosts. *Parasite Immunology*, 28(11), 549–565.

Fedorowicz, J. and Sączewski, J. 2018. Modifications of quinolones and fluoroquinolones: Hybrid compounds and dual-action molecules. *Monatshefte für Chemie-Chemical Monthly*, 149(7), 1199–1245.

Ghorbani, M. and Farhoudi, R. 2018. Leishmaniasis in humans: Drug or vaccine therapy? *Drug Design, Development and Therapy*, 12, 25.

Gupta, G., Oghumu, S. and Satoskar, A. R. 2013. Mechanisms of immune evasion in leishmaniasis. *Advances in Applied Microbiology*, 82(2013), 155–184.

Hewitson, J. P., Grainger, J. R. and Maizels, R. M. 2009. Helminth immunoregulation: The role of parasite secreted proteins in modulating host immunity. *Molecular and Biochemical Parasitology*, 167(1), 1–11.

Jain, K. and Jain, N. K. 2013. Novel therapeutic strategies for treatment of visceral leishmaniasis. *Drug Discovery Today*, 18(23–24), 1272–1281.

Kaiko, G. E., Horvat, J. C., Beagley, K. W. and Hansbro, P. M. 2008. Immunological decision-making: How does the immune system decide to mount a helper T-cell response? *Immunology*, 123(3), 326–338.

Karunaweera, N. D. and Ferreira, M. U. 2018. Leishmaniasis: Current challenges and prospects for elimination with special focus on the South Asian region. *Parasitology*, 145(4), 425–429.

Laniado-Laborín, R. and Cabrales-Vargas, M. N. 2009. Amphotericin B: Side effects and toxicity. *Revista Iberoamericana de Micología*, 26(4), 223–227.

Lavik, E., Kwon, Y. H., Kuehn, M., Saluja, S., Bertram, J. and Huang, J. 2013. Sustained intraocular delivery of drugs from biodegradable polymeric microparticles. Google Patents.

Le Rutte, E. A., Chapman, L. A., Coffeng, L. E., Jervis, S., Hasker, E. C., Dwivedi, S., Karthick, M., Das, A., Mahapatra, T. and Chaudhuri, I. 2017. Elimination of visceral leishmaniasis in the Indian subcontinent: A comparison of predictions from three transmission models. *Epidemics*, 18, 67–80.

Liu, D. and Uzonna, J. E. 2012. The early interaction of Leishmania with macrophages and dendritic cells and its influence on the host immune response. *Frontiers in Cellular and Infection Microbiology*, 2, 83. doi: 10.3389/fcimb.2012.00083.

Maciel, B. G. 2017. *Nanocompósitos magnéticos magnetita/quitosana/polianilina e seu uso na extração e purificação de DNA*. Recife, Brazil: Universidade Federal de Pernambuco.

Maizels, R. M. and Mcsorley, H. J. 2016. Regulation of the host immune system by helminth parasites. *The Journal of Allergy and Clinical Immunology*, 138(3), 666–675.

Maspi, N., Abdoli, A. and Ghaffarifar, F. 2016. Pro-and anti-inflammatory cytokines in cutaneous leishmaniasis: A review. *Pathogens and Global Health*, 110(6), 247–260.

Mishra, J., Dey, A., Singh, N., Somvanshi, R. and Singh, S. 2013. Evaluation of toxicity and therapeutic efficacy of a new liposomal formulation of amphotericin B in a mouse model. *The Indian Journal of Medical Research*, 137(4), 767.

Mucke, H. A., Consultancy, H. P., Strømgaard, K., Bach, A. and Nissen, K. 2015. Patent highlights June-July 2015. Future Sci Ltd United House, 2 Albert PL, London, N3 1QB, England.

Mulani, M. S., Kamble, E. E., Kumkar, S. N., Tawre, M. S. and Pardesi, K. R. 2019. Emerging strategies to combat ESKAPE pathogens in the era of antimicrobial resistance: A review. *Frontiers in Microbiology*, 10, 539.

Nagle, A. S., Khare, S., Kumar, A. B., Supek, F., Buchynskyy, A., Mathison, C. J., Chennamaneni, N. K., Pendem, N., Buckner, F. S. and Gelb, M. H. 2014. Recent developments in drug discovery for leishmaniasis and human African trypanosomiasis. *Chemical Reviews*, 114(22), 11305–11347.

Ocaña-Guzman, R., Vázquez-Bolaños, L. and Sada-Ovalle, I. 2018. Receptors that inhibit macrophage activation: Mechanisms and signals of regulation and tolerance. *Journal of Immunology Research*, 2018, 8695157.

Oryan, A. 2015. Plant-derived compounds in treatment of leishmaniasis. *Iranian Journal of Veterinary Research*, 16(1), 1.

Puri, A., Loomis, K., Smith, B., Lee, J.-H., Yavlovich, A., Heldman, E. and Blumenthal, R. 2009. Lipid-based nanoparticles as pharmaceutical drug carriers: from concepts to clinic. *Critical Reviews™ in Therapeutic Drug Carrier Systems*, 26(6), 523–580.

Ribeiro, T. G., Franca, J. R., Fuscaldi, L. L., Santos, M. L., Duarte, M. C., Lage, P. S., Martins, V. T., Costa, L. E., Fernandes, S. O. and Cardoso, V. N. 2014. An optimized nanoparticle delivery system based on chitosan and chondroitin sulfate molecules reduces the toxicity of amphotericin B and is effective in treating tegumentary leishmaniasis. *International Journal of Nanomedicine*, 9, 5341.

Sane, S. A., Shakya, N., Haq, W. and Gupta, S. 2010. CpG oligodeoxynucleotide augments the antileishmanial activity of miltefosine against experimental visceral leishmaniasis. *The Journal of Antimicrobial Chemotherapy*, 65(7), 1448–1454.

Sangshetti, J. N., Khan, F. A. K., Kulkarni, A. A., Arote, R. and Patil, R. H. 2015. Antileishmanial drug discovery: Comprehensive review of the last 10 years. *RSC Advances*, 5(41), 32376–32415.

Shahcheraghi, S., Ayatollahi, J., Lotfi, M., Bafghi, A. and Khaleghinejad, S. 2016. Application of nano drugs in treatment of leishmaniasis. *Glob J Infect Dis Clin Res*, 2(1), 018–020. doi:10.17352/2455,5363,1.5-2.

Singh, O. P., Singh, B., Chakravarty, J. and Sundar, S. 2016. Current challenges in treatment options for visceral leishmaniasis in India: A public health perspective. *Infectious Diseases of Poverty*, 5, 19.

Singh, R. K., Gannavaram, S., Ismail, N., Kaul, A., Gedda, M. R. and Nakhasi, H. 2018. Centrin deleted Leishmania donovani parasites help CD4+ T cells to acquire Th1 phenotype and multi-functionality through down regulation of CD200-CD200R immune inhibitory axis. *Frontiers in Immunology*, 9, 1176.

Soulat, D. and Bogdan, C. 2017. Function of macrophage and parasite phosphatases in leishmaniasis. *Frontiers in Immunology*, 8, 1838.

Spellberg, B., Edwards, J.E. 2001. Type 1/Type 2 immunity in infectious diseases. *Clinical Infectious Diseases*, 32(1), 76–102.

Taha, M., Uddin, I., Gollapalli, M., Almandil, N. B., Rahim, F., Farooq, R. K., Nawaz, M., Ibrahim, M., Alqahtani, M. A. and Bamarouf, Y. A. 2019. Synthesis, anti-leishmanial and molecular docking study of bis-indole derivatives. *BMC Chemistry*, 13(1), 102.

Thakur, L., Singh, K. K., Shanker, V., Negi, A., Jain, A., Matlashewski, G. and Jain, M. 2018. Atypical leishmaniasis: A global perspective with emphasis on the Indian subcontinent. *PLoS Neglected Tropical Diseases*, 12(9), e0006659.

Tiwari, B., Pahuja, R., Kumar, P., Rath, S. K., Gupta, K. C. and Goyal, N. 2017. Nanotized curcumin and miltefosine, a potential combination for treatment of experimental visceral leishmaniasis. *Antimicrobial Agents and Chemotherapy*, 61(3), e01169-16.

Tiwari, N., Kishore, D., Bajpai, S. and Singh, R. K. 2018. Visceral leishmaniasis: An immunological viewpoint on asymptomatic infections and post Kala Azar dermal leishmaniasis. *Asian Pacific Journal of Tropical Medicine*, 11(2), 98.

Torres-Guerrero, E., Quintanilla-Cedillo, M. R., Ruiz-Esmenjaud, J. and Arenas, R. 2017. Leishmaniasis: A review. *F1000Research*, 6, 750.

Ud Din, F., Aman, W., Ullah, I., Qureshi, O. S., Mustapha, O., Shafique, S. and Zeb, A. 2017. Effective use of nanocarriers as drug delivery systems for the treatment of selected tumors. *International Journal of Nanomedicine*, 12, 7291.

Vaine, C. A. and Soberman, R. J. 2014. The CD200–CD200R1 inhibitory signaling pathway: Immune regulation and host–pathogen interactions. *Advances in Immunology*, 121, 191–211.

Wang, L., Hu, C. and Shao, L. 2017. The antimicrobial activity of nanoparticles: Present situation and prospects for the future. *International Journal of Nanomedicine*, 12, 1227.

Want, M. Y., Islammudin, M., Chouhan, G., Ozbak, H. A., Hemeg, H. A., Chattopadhyay, A. P. and Afrin, F. 2017. Nanoliposomal artemisinin for the treatment of murine visceral leishmaniasis. *International Journal of Nanomedicine*, 12, 2189.

Want, M. Y., Islamuddin, M., Chouhan, G., Ozbak, H. A., Hemeg, H. A., Dasgupta, A. K., Chattopadhyay, A. P. and Afrin, F. 2015. Therapeutic efficacy of artemisinin-loaded nanoparticles in experimental visceral leishmaniasis. *Colloids and Surfaces, Part B: Biointerfaces*, 130, 215–221.

Zeligs, M. A. 2013. Anti-parasitic methods and compositions utilizing diindolylmethane-related indoles. Google Patents.

Section 4

Industrial Nanotechnology

16 Intellectual Property Issues in Industrial Nanotechnology

Saeideh Ebrahimias, Arash Ketabforoush Badri, and Parsa Ketabforoush Badri

CONTENTS

16.1 INTRODUCTION

Nanotechnology is the study of particles at atomic scales and control of them. In other words, the physical, chemical, and biological properties of nanomaterials are different from that in the mass of the matter (Chaturvedi et al., 2012). In such unique scales and properties, nanoparticles lead to new achievements in medical and engineering sciences (Velmurugan, 2019). Eric Drexler introduced the term "nanotechnology" for the first time in 1986 in a book titled "Engines of Creation". Nanotechnology is a general term that applies to all advanced technologies in the nanoscale field (Singh, 2017). The National Nanotechnology Initiative (NNI) gives a definition for nanotechnology, which includes the following: development of technology and research at atomic, molecular, or micro molecular surfaces at a scale of 1 to 100 nm; creation and use of structures and tools that have new properties and functions due to their small or mid-size; ability to control or manipulate at atomic levels (NNI, 2000). As an emerging science, nanotechnology promises the nanoscale manufacture of materials and machines made to atomic specifications. Nanotechnology is a field which is the junction of chemistry, physics, biology, computer science, and engineering (Bastani and Fernandez, 2002).

Applying this definition of nanotechnology and how countries invest and use nanoscience will now be discussed. Considering the financing of nanotechnology research across different parts of the world, there is a difference (Jia, 2005). In Europe, private investors are placed second after government institutions. In Japan and the United States, there is almost a balance between the private and the governmental sector, and both invest equally in the field of nanotechnology (Shand and Wetter, 2007). European private sector investment in this area is small but European investment in nanoscience worldwide is significant, indicating that the governments of European countries are paying attention to the new opportunities in this technology and they have quickly entered the nano race (Hsiao and Fong, 2004).

Based on previous experience, if nanotechnology can be developed as a general-purpose technology, its effects on economic growth and productivity will be much greater than current predictions (Adam and Youssef, 2019). The emerging nature of nanotechnology means that a better understanding of the origin, importance, and socioeconomic effects of its development is very important (Khan et al., 2019). The policies of science, technology, and innovation in this field should be based on reliable and comparable statistics and indicators and other systematic approaches (Naghdi et al., 2013). In this regard, the United States Patent and Trademark Office (USPTO) has presented a comprehensive report on the development of nanotechnology through systematic analysis of existing indicators and statistics. In this report, the top five countries with the highest number of nanotechnology patents granted in the period 2014 to 2019 have been included; the United States, South Korea, Japan, China, and Taiwan.

Figure 16.1 shows the number of nanotechnology patents granted by the USPTO in the period 2014 to 2019. The term "nanotechnology patents" is all of the patents that include at least one claim related to nanotechnology. According to this chart, among the top five countries, the United States has had the highest number of patents granted between 2014 and 2019, South Korea and China have had an increasing number of granted patents until 2017 and it then decreased, Japan and Taiwan have had a decreasing number of nanotechnology patents granted between 2014 and 2019.

The analysts believe that nanotechnology, biotechnology, and information technology (IT) are the three scientific fields that are going to form the third industrial revolution. In the industrial field, nanotechnology enables us to create and invent things that were unimaginable and this is the beginning of another industrial revolution (Azarm, 2016). Due to the interdisciplinary nature of nanotechnology, there is a possibility of the overlapping of patents and the difficulty of distinguishing between nano-based and traditional patents (Chowdhury et al., 2016). The growing competition requires careful attention to intellectual property rights and strategies (Zhang et al., 2013). Nanotechnology research and products have reached new heights which require them to be linked up with policy regulations. Intellectual property rights include patent, copyright, trademark, design rights, and trade secrets. The application of these laws can protect researchers and developers and their rights. The emergence of a high level of research and development in interdisciplinary subjects such as nanotechnology has led to an increase in reliance on intellectual property rights protection (Chowdhury et al., 2016).

16.2 NANOTECHNOLOGY IN MAJOR ECONOMIES

The subject of nanotechnology is one of the most important issues that play a key role in the development of countries. This technology has been able to participate in many fields of science and affects its programs in various aspects (Zhang, Sulfab and Fernandez, 2013). Table 16.1 refers to advances in nanotechnology.

As shown in Table 16.1, much progress has been made on nanotechnology in recent years. For this reason, looking at the structure of nanotechnology in different countries can be helpful. To achieve this goal, in this section, the United States, the European Union, and the United Kingdom are being examined.

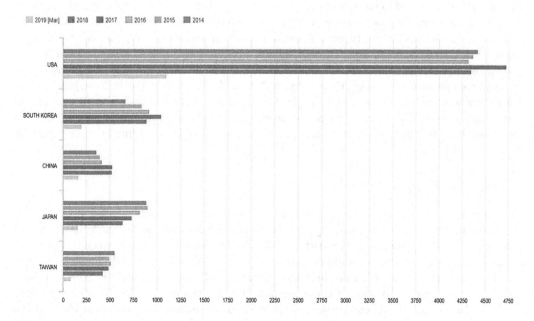

FIGURE 16.1   Top 5 countries in terms of nanotechnology patents. (Source: www.orbit.com.)

TABLE 16.1

Advances in Nanotechnology

Category	Field	Technology examples
Tools	Computation	Simulation of nanoparticle self-assembly
		Statistical analysis of nanostructures
	Nanocharacterization	High-spatial-resolution scanning probe-based microscopy measuring a wide range of phenomena
		Atomic-scale imaging of nanostructures by electron beam-based microscopy
	Nanomanufacturing	Scale-up synthesis and patterning of nanostructures
		Controlled integration of nanoelements into multiscale ensembles
Applications	Nanobiotechnology	Nanoparticle as carrier for drug delivery
		Nanostructured polymers for tissue repair
	Nanoelectronics	Transistors with features below 30 nm
		Carbon electronics featuring graphene
	Nanomagnetics	Magnetic random-access memory using quantum spin Hall effect
		Semiconductor spintronics
	Nanophotonics	Plasmonics-enabled ultrahigh-resolution imaging and targeted medical therapy
		Silicon-based light amplification and emission for transmitting information optically across a chip
	Sustainability	Nanocomposite membranes for water purification
		Nanostructured solar cells for solar energy conversion

Source: Adapted from Zhang, Sulfab and Fernandez, 2013.

16.2.1 United States

Every quarter the Food and Drug Administration (FDA) is responsible for evaluating and regulating different materials and products, rather than adopting a specific legal framework for different sciences (FDA, 2007). This group ensures coordination and communication. In September 2009, an FDA document was used to identify the sources of nanomaterials, how they behave in the environment, their problems for people, animals and plants, and how to prevent and mitigate these issues (SPIE, 2009). The Bush administration decided in 2007 that no specific regulation or markup of nanoparticles would be necessary. Critics swore this was a consumer treatment, treated as a "guinea pig", without proper knowledge of labeling. Berkeley, CA is currently the only city in the United States to regulate nanotechnology. Cambridge, MA, considered the same law in 2008, but the committee that was studying it for Cambridge in its final report recommends facilitating the collection of information. The potential effects of nanomaterials are presented in Isabella (2008). On December 10, 2008, the U.S. National Research Council issued a report calling for more regulation on nanotechnology.

16.2.2 European Union

The European Union has identified a group for reviewing the concepts of nanotechnology called the Scientific Committee on Emerging and Newly Identified Health Risks (SCENIHR), which published a list of dangers from nanoparticles (SCENIHR, 2013). As a result, manufacturers and importers of carbon products, including carbon nanotubes, should provide all health and safety information within a year to ensure biosafety. A number of European countries have requested national or European registration of nanomaterials. France, Belgium, Sweden, and Denmark have established a national enrollment register for nanomaterials. In addition, the European Commission has appointed the European Chemicals Agency (ECHA) to establish a European Union Observatory for Nanomaterials (EUON), which aims to gather general information on the safety and market of nanomaterials and nanotechnology.

16.2.2.1   United Kingdom

The 2004 report by the British Royal Society and the Royal Academy of Engineers states that regulations in the U.K. do not require further testing for the production of nanoparticles. The Royal Society suggested that such a regulation be revised to "use nanoparticles and nanotubes as new chemicals under this legal framework." They also recommended that existing regulations be prevented because it is expected that "chemical nanoparticles and nanotubes will not be poisoned by toxicity in a larger form ... In some cases, this poisoning from the same chemical chemistry is larger in shape" (The Royal Society, 2004). The Better Regulation Commission in 2003 recommended that the British government: "Through a deliberate discussion, it enables people to understand their dangers and help them deliver accurate information; Be open about how to make a decision and acknowledge when there is uncertainty; Communicate with the people and participate as much as possible in the decision making process; Make sure the two-way communication channels are developed; Strong leadership to address risk issues, especially providing information and implementation policies." These recommendations were initially accepted by the British government. Given that "there is no clear problem for public discussion of the kind provided by the workforce", the British government's response was to accept the suggestions. The 2004 report of the Royal Society identified two specific government issues: The roles and behavior of institutions and their ability to "minimize unintended consequences" through appropriate settings; the extent to which the public can trust and play a role in determining the trajectories that nanotechnologies may follow as they develop.

16.3   INTELLECTUAL PROPERTY AND RELATED CONCEPTS

Intellectual property is a broadly categorized description for a number of intangible assets that legally protects the use or implementation of that property by others without the consent of the owner (Bawa et al., 2005). Intellectual property can include patents, copyrights, trade markets, and trade secrets (Husovec et al., 2019). Table 16.2 shows the different sectors of intellectual property.

16.3.1   PATENTS

Patenting is done with the purpose of promoting innovation by granting exclusive rights to inventors (Chen and Zhang, 2019). In return, the inventor is supposed to provide a detailed description of how to make and use the invention which is made available to the public (Singh, 2019). This promotes innovation and competition, which brings about advancements in technology and industry (Olafusi et al., 2019). In order for an invention to be patentable, it must be human-made, not found in nature or discovered, and must have an actual physical embodiment (Bawa, 2005). One essential property of a patent is its utility, i.e. the invention must actually do something useful (Owoeye et al., 2017). Another feature of patentability is novelty, which means the invention must have never been made, used, seen, or discussed in public (Yu, 2017). The third standard of patentability is obviousness. This can be a major hurdle for inventions (Banerjee, 2013). If an invention simply consists of putting together aspects of things which already exist with no inventive step, it is not patentable (Baker, 2016; Singh et al., 2016; Singh et al., 2019). A minor change that would make an invention "novel" is not enough to prove it was not obvious. Figure 16.2 shows the geographic origin of the patents granted by the EPO based on the country of residence of the first patentee listed on the published patent.

16.3.2   COPYRIGHT

Copyright protection is referred to as written articles, drawings, photos, software, and any tangible creation describing an abstract idea (Story, 2002). Whenever something is drawn, written, or

TABLE 16.2

The Different Sectors of Intellectual Property

	Subject matter	Enforceable period	Cost of acquiring in US (USD, approximate)
Patent	Process, machine, manufacture, composition of matter, or an improvement of an existing idea that is new, useful, and nonobvious	20 years from the earliest effective filing date (EFD)	> 10,000 for filing
Copyright	Original and creative work that is fixed in a tangible medium of expression	Effect instantly and for the life of the creator plus 70 years (individual creator) or for 95–120 years (business creator)	<100
Trademark	Distinctive word, phrase, logo, symbol, or other device to distinguish a product or service from competitors	10 years following the date of registration, subject to renewal	1,000–3,000
Trade secret	Any information that has been kept in confidence and has business value	Infinite until information is disseminated publicly	>20,000 for issue and maintenance No formal filing procedure

Source: Adapted from Zhang, Sulfab and Fernandez, 2013.

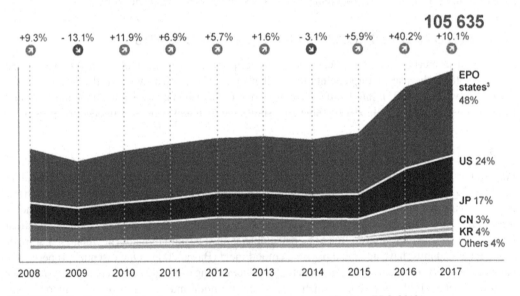

FIGURE 16.2   Top 5 countries in terms of nanotechnology patents. (Source: EPO 2018.)

saved on a computer device, it is automatically granted a specific level of copyright protection. It is also possible to register a copyright for a small fee (Bettig, 2018). Copyright registration makes it possible for the holder to collect additional damages in a lawsuit (Acharya and Tanna, 2018). Software might also use patent rights providing additional protection as well as additional costs (Hacker, 2018). Copyright can also be used for preventing copying a software program, artwork, or document, including reproducing different parts of it (Czerniewicz et al., 2018). This means that copyright does not provide protection against reverse engineering or works inspired by the original work. If software is written in a new language or a story is told from a different perspective, it is considered a new separate work and is protected by certain copyright rights

(Caso and Guarda, 2019). There have been many changes in copyright laws, but it generally lasts the lifetime of the author plus 70 to 95 years from the first publication, or 120 years from the creation, whichever is the shortest. Protection for the life of the copyright does not require any payment and is free.

16.3.3 TRADEMARK

Trademarks are indications that are used in trading to identify products manufactured by different companies (Correa, 2007). The trademark of any company is the symbol chosen by its customers (Bouchoux, 2012). In fact, the trademark of any company is a symbol that customers choose for their products (Dratler and McJohn, 2006). So, in this way, each company is recognized by its rivals. However, in order to protect trademarks, they should be registered (Kang et al., 2019). In some countries, even if the trademark is not registered, it is protected as long as it is used. However, the best way to protect it is registration (Rozek, 2018). The only condition that applies to a registered trademark is that it should be clearly defined. If this is not done correctly, trademark protection cannot be seen (Uzun, 2018). A trademark is a recognizable design, sign, or expression which is associated with distinguishing services or products of a specific trader from similar products or services of other traders (de Rassenfosse, 2019). Trademarks are applicable at the level of a state or nation, which legally adds strength to the claims. The value of a trademark is in the recognition derived from it. Hence, a trademark is best registered by a company after it can demonstrate public use.

16.3.4 TRADE SECRETS

A trade secret is a formula, practice, process, design, instrument, pattern, or compilation of information which is not generally known or reasonably ascertained, by which a business can obtain an economic advantage over its competitors (Contigiani et al., 2018). There is no formal protection granted by the government; instead, each business must take measures to guard its own trade secrets, such as the formula of Coca-Cola soft drinks, which is stored and protected by the company itself. On a certain level, a trade secret is the opposite of a patent. A patent application must contain a thorough description of how the invention is created. A trade secret, on the other hand, is commercially important information protected by restricting who has access to it. A famous example of a trade secret is the recipe of internationally sold sodas. Stealing a trade secret is considered a crime. In order to obtain a trade secret, a company must take a few steps to demonstrate they have their secret information protected (Menell, 2018). For instance, the information must never be kept on a computer which could be accessed by any unauthorized persons. The company must also prove that they have a formal trade secret policy to protect its information. Trade secret protection lasts as long as the information is kept secret and properly guarded (Menell, 2007). There are no added fees or additional steps. Table 16.3 is a comparison between patents and trademarks.

16.4 THE IMPORTANCE OF INTELLECTUAL PROPERTY IN NANOTECHNOLOGY

Emerging nanotechnologies have the potential to have a wide range of commercially important bioscience and medical applications and products. (Genieser and Gollin, 2007). Recognizing that nanotechnology is likely to be the next great technological frontier, U. S. government officials have considered the potential of nanotechnology to transform society and the economy on a scale comparable to the effects of the industrial revolution. In order to organize and promote nanotechnology research and development activities, the 21st Century Nanotechnology Research and Development Act was signed, authorizing USD 3.7 billion in funding for federal nanotechnology research and development from 2005 until 2008 (Tullis, 2012). The nanotechnology industry is already generating such varied technologies as stronger and lighter building materials, more durable coatings,

TABLE 16.3

Characteristics of Trade Secret and Patent Protections Compared

Element	Trade secrets	Patents
Subject matter must be patentable, novel, non-obvious and useful	No	Yes
Prior registration and examination by government agency is required	No	Yes
Public disclosure is required	No	Yes
Process of acquiring the right may take years	No	Yes
Has only a defined term of protection	No	Yes
Only dishonest or wrongful conduct is prohibited	Yes	No
Internal controls are required to establish the right	Yes	No

Source: Linton (2016).

efficient batteries and fuel cells, improved television display technology, and microscopic computer chips (Zhang et al., 2013). Someday, the nanotechnology industry is expected to enable environmental cleaning mechanisms for air and water, as well as injectable biosensors to detect the presence of infectious agents. Nowadays, medical researchers are actively exploring the potential of nanotechnology in drugs, drug delivery, diagnostics, devices, gene therapy, and tissue engineering (Tullis, 2012). With the increasingly competitive market and rapid generation of intellectual property, as well as the recent reform in the U.S. patent law system, more effort and attention must be paid to intellectual property (IP) protection strategies (Zhang et al., 2013).

It was projected for nanotechnology industry products and services to reach USD 3 trillion by 2020. Since the launch of the National Nanotechnology Initiative (NNI) in 2000 through 2012, Congress has dedicated approximately USD 15.6 billion for nanotechnology.

Vast differences exist between different countries and technology areas in nanotechnology industry development. A study was conducted to characterize and analyze the importance of specific nanotechnology domains for the east and the west. When comparing regional strengths and weaknesses, the United States leads significantly in nanobiotechnology, which shows potential support in the relevant research domain by public and private funding or the potential interests of a critical mass of expertise to explore the biological application. Although European regions have had strong activity in researching the nanomaterial domain, Asian regions have shown a strong research performance in the nanoelectronics domain, but they have lagged greatly behind in nanobiotechnology (Zhang et al., 2013).

The nanotechnology industry is a young field that focuses on two categories: basic research and materials and science products. In 2003, the United States had approximately 104 nanotechnology research institutions and 430 nanotechnology startups producing commercial nanotechnology products. Basic nanotechnology research undertaken in U.S. research institutions, including universities, public laboratories, and private laboratories, primarily focuses on areas such as chemistry, physics, computer science, and biology. The first successful wave of commercial nanotechnology products has been in materials science. Materials science companies are producing innovative products in areas such as coatings, powders, and particulates, nanoengineered chemicals, carbon nanotubes, clays, and biomedical devices. The commercial viability of more complex technologies like ultra-efficient batteries or molecular computer chips historically has been limited by the materials used to make them. However, with the basics being assembled at smaller and more stable levels, near-term developments in the nanotechnology industry should bring about remarkable advances in many significant areas of the industry (Larrimore Ouellette, 2015).

Considering all these facts, there is certainly a competitive market in nanotechnology industry products worldwide and many companies seek to be the pioneers of their own field of activity. The hard competition results in conflicts and debates among the inventors and developers of companies and corporations. There is also a possibility of coincidental overlapping of ideas and inventions. Therefore, there must be a legal way to bring all this into an organization. The first step in securing the commercial potential of nanotechnology is establishing intellectual property rights to protect innovation (Tullis, 2012).

Intellectual property (IP) plays an essential role in the modern-day economy by virtue of providing protection for ideas. Entrepreneurs and small start-up organizations thrive due to the protection provided by the legal framework. With the increasing move towards a knowledge-oriented economy, the role of IP is poised to increase wealth creation, growth, and development through the application of new technology across the world. IP describes the ownership of innovations, inventions, ideas, and creativity. Ownership of IP enables an individual or an organization to benefit from its application and any financial rewards to be reaped from encouraging other innovators. IP has been broadly classified into four areas – patents, trademarks, copyrights, and designs. Patents are related to inventions of products and processes that can be used within the industry. Trademarks are used for brand identity to distinguish between goods and services. Copyrights apply to materials such as literature, sound, films, broadcasts, software, and multimedia. Designs encompass the form of products such as appearance, color, shape, texture, or material composition (Singh, 2012).

Considering the information given so far, the type of IP that is suitable for the nanotechnology industry is patents. By obtaining patents on nanotechnology products, a company can exclude others from making, using, selling, or importing the patented invention for a period, generally 20 years, from the filing of an application in the PTO. In addition to the opportunity to file patents for innovations and strategic technologies and processes, early leaders have the opportunity to participate in the formation of particular standards set by industry groups or government agencies. These standards can make it possible for a patent holder to exert additional leverage over the marketplace in a way that is consistent with antitrust laws. The flip side of this proposition is that companies late to the game may find themselves blocked from producing materials that comply with such standards by companies who have already patented them (Genieser and Gollin, 2007).

According to StatNano, a total of 13,046 published patent applications related to nanotechnology were filed at the United States Patent and Trademark Office (USPTO) and the European Patent Office (EPO) in 2018, with the United States and East Asian countries having the greatest shares of the total patent applications.

According to the ISO/TS 18110 standard, a patent is considered as a nanotechnology patent only if it includes at least one claim related to nanotechnology or is registered with an International Patent Classification (IPC) classification code related to nanotechnology. Based on this definition, a total of 13,046 nanotechnology published patent applications were filed at the USPTO and EPO in 2018; 11,280 of which were issued at the USPTO, and the rest at the EPO. In 2018, the share of nanotechnology patents of the total patents filed at these two patent offices was relatively small, in so far as it reached 2.5 to 3% despite its slight growth during recent years, while the share of nanotechnology articles was around 10%. These statistics suggest that a significant proportion of scientific achievements in the field of nanotechnology still remains at the level of the publication of scientific articles and does not make it to the stage of innovation and technology. Table 16.4 lists the top 25 countries filing nanotechnology patents in 2018, together with the share of nanotechnology patents of their total patents.

Accordingly, more than half of the nanotechnology published patent applications at the USPTO belong to the United States, while the country takes the second spot at the EPO, following Japan which holds first place. Following the United States, the next places in the list are taken by East Asian countries, indicating their particular attention to the thriving nanotechnology market in the United States and Europe. Saudi Arabia's nanotechnology patents, ranking 10th at the USPTO, interestingly account for around 18% of the country's total patent applications. Amongst the top five countries in terms of nanoscience production, are India with 66 patent applications in 17th place, and Iran with 42 nanotechnology patent applications in 21st place of the USPTO's top nanotechnology patent holders. In view of the fact that these two countries are at the forefront of knowledge

TABLE 16.4

Top 25 Countries in Filling Nanotechnology Patents in 2018

Rank	Country	Nano published applications (USPTO)	Share (%) of USPTO	Nano published applications (EPO)	Share of nano in total (%)
1	USA	5,646	50.1	311	3.1
2	South Korea	1,004	8.9	247	4.3
3	China	913	8.1	169	3.9
4	Japan	792	7.0	327	1.4
5	Taiwan	532	4.7	19	4.0
6	Germany	424	3.8	153	1.9
7	France	295	2.6	106	3.5
8	UK	224	2.0	26	3.1
9	Canada	194	1.7	13	3.1
10	Saudi Arabia	158	1.4	0	17.8
11	Switzerland	140	1.2	64	2.6
12	Netherlands	115	1.0	31	2.3
13	Singapore	85	0.8	7	4.7
14	Belgium	82	0.7	44	5.5
15	Finland	71	0.6	14	3.9
16	Italy	71	0.6	28	2.1
17	India	66	0.6	5	4.0
18	Australia	60	0.5	1	3.0
19	Sweden	52	0.5	7	1.2
20	Spain	46	0.4	61	5.3
21	Iran	42	0.4	0	26.8
22	Luxembourg	32	0.3	2	4.9
23	Denmark	30	0.3	10	2.0
24	Turkey	29	0.3	16	8.7
25	Austria	28	0.2	21	2.2
26	Total (World)	11280	100	1766	2.8

Source: USPTO (2019).

production in the field of nanoscience, particularly in case of Iran, whose nanotechnology patents comprise approximately 27% of the country's total patent applications at the USPTO in 2018, in spite of their outstanding interest in this field, these pioneering countries have not been as successful in nanotechnology innovations and inventions as they are in nanoscience creation (Statnano, 2018).

16.5 APPLICATIONS OF NANOTECHNOLOGY IN INDUSTRY

Nanotechnology will produce vast economic benefits in various industries, especially in areas that have created global challenges. Different countries in the world have invested in research and development in this field, and many companies are active in this field.

The socio-economic perspective of nanotechnology has led to a rapid growth in government investment in research and development of this field. In fact, the level of government investment in nanotechnology research and development is very extensive compared to other technologies and at the same time, the investment by the private sector is rapidly increasing. According to a forecast of the market volume of nanotechnology, the return on investment in this area has been estimated at USD 150 billion in 2010 to USD 3,100 billion in 2015. This process will lead to the creation of two million jobs globally (Palmberg et al., 2009).

Despite these widespread developments, government investment and the participation of companies in the development of nanotechnology is still not recorded properly. While it seems that due to the difficulties in measuring nanotechnology and its effects, these predictions are somewhat unrealistic, but analysts believe that nanotechnology is considered a so-called general-purpose technology, (the growth engine of the century). A general-purpose technology promotes the rapid and massive upgrading of existing technologies and applies in various areas. The development of these types of technologies requires complementarity technologies and the participation of different organizations and institutions (Hermann et al., 2005).

It should be noted that nanotechnology research in Europe focuses on the peaceful uses of nanotechnology, while in the United States a high share of the public nanotechnology budget is spent on military applications (Dang et al., 2010). Another positive aspect is the financing of nanotechnology through government and its social dimension; nanotechnology has a positive impact on economic development, and it is only from this perspective that people will support nanotechnology as community consumers, pressure groups, and regulatory institutions. If the environmental and social issues of nanotechnology are taken into account and society's expectations of this technology are met, nanotechnology will have a positive economic impact. So, in this area, market volume is considered an appropriate indicator for assessing its economic significance. Thus, in case of the success of nanotechnology, it will significantly improve the performance of many products, or produce new products (Naghdi et al., 2013). According to the National Science Foundation (NSF), the market for nanotechnology products will rise to USD 1,000 billion in 2015. In accordance with the definition of nanotechnology and its contribution to the value-added of final products and the rate of optimism, the predictions made vary between USD 150 billion in 2010 to USD 6.2 trillion in 2014 (Mitsubishi Institute, 2002). The predictions made by Lux Research Institute show that the market for nanotechnology products will be ten times more than the IT market. It also shows a significant increase in the volume of the nanotechnology product market since the early 2010s (Lux Research, 2007).

Nanotechnology industrial products can include pharmaceutical and medical products, sanitation and health, optics and electronics, water, sewage and environment, building, paint, resin and composite, household goods, textile and apparel, automotive and transportation, oil and related industries, nanomaterials, nanocoating, plasma surface processing equipment, and nanofiber production equipment. This section will deal with the products of the sanitation and health, water, sewage and environment, paint, resin and composite, and nanocoating sectors.

16.5.1 Sanitation and Health Sector Products

One of the biggest challenges that modern medicine is facing is the fact that the body does not entirely absorb the given dose of the drug. Thanks to nanotechnology, scientists can now make sure that drugs are better formulated and delivered to the required parts of the body. Rich countries are now investing heavily in nanotechnology health materials, some of which have already been approved by the FDA. However, there is still a long road ahead, and the fact that whether these developments will be available for everyone is still unknown. Furthermore, there are some debates about the toxicity of nanoparticles to human health and the environment (Misra et al., 2010). Many emerging economies have done some ambitious research and development in nanotechnology. Still, many poor countries have a responsibility to strengthen healthcare systems and provide easier and more effective treatments. Scientists argue that the use of nanotechnology and nanomedicine is vital for the developing world. World health goals can be met by the use of nanotechnology, and researchers claim that nanotechnology can greatly contribute to meeting the Millennium Development Goals for health, specifically those to reduce child mortality, improve maternal mortality and fight HIV and malaria and other such diseases (Azamal Moshed et al., 2017).

Several studies have been conducted in this field, of which the most important are Kumari et al. (2019); Tsou et al. (2019; Ajaz et al. (2019); Rasouli et al. (2018); Kojabad and Ebrahimiasl (2017); and Ebrahimiasl et al. (2014). Kumari et al. (2019) investigated the application of the combinatorial approaches of medicinal and aromatic plants with nanotechnology and its impacts on healthcare. Tsou et al. (2019) in their study examined nanotechnology mediated drug delivery for the treatment of obesity and its related co-morbidities. Rasouli et al. (2018) studied nanofibers for biomedical and healthcare applications. Kojabad and Ebrahimiasl (2017) examined pencil graphite electrode modified nanosensors for the detection and determination of tramadol in blood serum. Ebrahimiasl et al. (2014) investigated novel conductive polypyrrole/zinc oxide/chitosan bionanocomposite: synthesis, characterization, antioxidant, and antibacterial activities.

Different products are produced using nanotechnology in the sanitation and health field. Including these products can be referred to as a filter respirator mask, SilvoSept products, silver nanocrystalline dressing, nanosunglasses, ecofriendly cleaners with plant nanoemulsion, antibacterial wound spray, antibacterial waxing pad, antibacterial filtering harmful tobacco compounds, antibacterial waste bag, antibacterial leather waxes, dust absorbing mask, hospital disinfection (nanobiocide), fluids, sunscreens, lotions, brightening and anti-wrinkle creams, antibacterial insoles containing silver nanoparticles, nanosilver colloid surface disinfectant, and rodenticide powder. In this part, dust absorbing masks, hospital disinfection (nanobiocide), sunscreens, brightening and anti-wrinkle creams, antibacterial insoles containing silver nanoparticles, and nanosilver colloid surface disinfectants are introduced.

16.5.1.1 Dust-Absorbing Mask

The dust-absorbing mask is a flexible pad for respiratory protection against dust caused by construction and cleaning, dust caused by soil, wood, fiberglass, silica from glass and ceramic products, or vacuum cleaners. It is also used for protection against all types of allergens. Today, nylon fibers are used to make such masks to improve their performance. Due to the surface-to-volume ratio and high porosity, these nanofibers increase the filtration efficiency and, according to the standards, the addition of nanofibers to the filtration efficiency filter improves the absorption rate of dust. This product is manufactured by Khavaran Nano Technologies Co., established in 2012.

16.5.1.2 Hospital Disinfection (Nanobiocide)

Nanosilver has been used as colloidal silver for over 100 years. Colloidal silver containing silver with various concentrations and particle sizes has been used for a long time to treat wounds and infections. Silver nanoparticles can prevent bacteria from growing on or sticking to the surface. This can be particularly useful in the surgery room, where all surfaces in the patient's body should be sterilized. The antibacterial effect of silver increases dramatically in nanometers so that they can kill more than 650 bacterial species. These nanoparticles also exhibit severe anticoagulant activities against some fungal infections. Silver nanoparticles react with sulfur and phosphorus compounds of membrane proteins and affect the morphology and cell structure of the bacteria and cause it to die. By reducing particle size, the release of silver ions has increased, which increases their antibacterial activity. Nanocoating Metal Company, established in 2008, has been able to supply this product in two disinfectant concentrations containing 100 ppm silver and disinfectant nanoparticles containing 2,000 ppm silver nanoparticles.

16.5.1.3 Sunscreens, Brightening, and Anti-Wrinkle Creams

The skin of the human body naturally lacks the ability to deal with the effects of sunlight. The most widely known types of ultraviolet rays are divided into three wavelengths: UVA, UVB, and UVC. UVA interferes with the natural process of skin regeneration through the formation of free radicals and causes symptoms such as premature aging of the skin and its appearance, and darkness. UVB has less permeability in skin layers and has direct effects such as redness, inflammation, sunburn, and in cases of prolonged exposure it increases the likelihood of a variety of skin cancers. Pars Hayan Company, established in 1982, has been able to produce three products of 35 SPF sunscreen, 90 SPF color sunscreen, and 46 SPF colorless fat-free sunscreens that do not contain chemical absorbents.

16.5.1.4   Antibacterial Insoles Containing Silver Nanoparticles

With the increasing importance of foot health, nanosilver-plated footwear provides consumers with benefits. When the foot sweats, the bacterium grows in raw fiber, and where the bacterium grows, it will also eliminate bad smell. Odorless insoles are healthier than typical ones and are much more convenient when used. These insoles do not cause any allergic reactions and their contents do not have any harm to health. When nanosilver contacts bacteria and fungi, it affects the cellular metabolism and inhibits cell growth, which ultimately results in the destruction of almost 100% of bacteria and fungi. By combining an antimicrobial substance into the polymer, its performance will remain the same until the expiration of the insole. Amin Sot Manufacturing Co., established in 1991, has been able to produce these insoles. The company's products include local antibacterial silicone insoles containing silver nanoparticles, antibacterial magnetic massage masks containing silver nanoparticles and antibacterial silicone heel pads containing silver nanoparticles.

16.5.1.5   Nanosilver Colloid Surface Disinfectant

Silver nanoparticles are severely antifungal against some infections. Silver nanoparticles destroy the structure of the fungus and cause severe damage to its cells. It has also been proven that nanoparticles of 5 to 20 nanometers of silver destroy the cells of the first type of AIDS virus infected. The electrochemical properties of modified silver nanoparticles result in faster response times and lower detection limits in nanoscale sensors. Due to the high levels of silver nanoparticles, nanosilver colloid surface disinfectant has the ability to eliminate the hardiest microorganisms that any antiseptic cannot eliminate. This product can be used in areas that are very vulnerable to contamination and are free from any microbial contamination. This product is produced at high and low doses. The Nano Part Khazar Co., established in 2009, produces nanosilver colloid disinfectant surfaces.

16.5.2   Water, Sewage, and Environment Sector Products

The sewage system has an important role in life as it protects both health and the environment (Anjum et al., 2016). The reduction and degradation of water resources associated with increasing interest in public health and hygiene draws special attention to the sewage system (Amin et al., 2014). People use water on a daily basis in most regular activities such as cooking, bathing, etc.; once used, drinking water becomes wastewater (Betsholtz et al., 2019). The sewage system is used to collect this water and clean it before returning it to the environment, keeping the environment safe from various types of pollution (Regkouzas and Diamadopoulos, 2019).

Several studies have been conducted in this field, of which the most important are Loosli et al. (2019); Chaturvedi and Dave (2019); Hassan and Mahmood (2019); Zhang et al. (2019); Ebrahimiasl et al. (2017); Ketabforoush Badri et al. (2016); Taheri et al. (2016); Khazaee et al. (2015); and Khatamian et al. (2012). Loosli et al. (2019) found sewage spills are a major source of titanium dioxide engineered nanoparticle release into the environment. Chaturvedi and Dave (2019) studied water purification using nanotechnology emerging opportunities. Hasan and Mahmood (2019) examined the use of iron oxide nanoparticles and application in the removal of heavy metals from sewage water. Zhang, Zhang, and Liang (2019) studied nanotechnology in the remediation of water contaminated by poly- and perfluoroalkyl substances. Ebrahimiasl et al. (2017) examined Ppy/nanographene modified pencil graphite electrode nanosensors for the detection and determination of herbicides in agricultural water. Taheri et al. (2016) investigated the results of a hospital waste survey in Tabriz hospitals. Ketabforoush Badri et al. (2016) investigated the quality of the environment in MENA countries using ICT. Khazaee et al. (2015) examined the challenge of medical waste management in Alborz province, Iran. Khatamian et al. (2012) studied heterogeneous photocatalytic degradation of 4-nitrophenol in aqueous suspension by Ln (La^{3+}, Nd^{3+} or Sm^{3+}) doped zinc oxide (ZnO) nanoparticles.

Different products are produced using nanotechnology in the water, sewage, and environment sector. Included in these products are: water and wastewater treatment system by plasma cavitation; electro-dialysis selective selection system for nitrogen desalination; nanocoating system for arsenic dehydration; arsenic dehydration machine; halophilic ultrafiltration nanocomposite membranes; Ultrafiltration (UF), Microfiltration (MF), and Nanofiltration (NF) ceramic membranes; domestic gray water purifier using ceramic nanostructured membranes; and silica-rubber nanocomposite for use in the manufacture of pipe and fittings gaskets. Next, halophilic ultrafiltration nanocomposite membranes, UF, MF, and NF ceramic membranes, domestic gray water purifier using ceramic nanostructured membranes are introduced.

16.5.2.1 Halophilic Ultrafiltration Nanocomposite Membranes

Ultrafiltration is the technology of separating water pollutants by membranes. In general, the uses of ultrafiltration membranes include drinking water, primary industrial purification, pre-purification of ultra-pure water, pre-treatment for desalination of seawater, and purification of water-cooling towers. Halophilic membranes (hollow thin fibers) are the membranes used in the ultrafiltration process, which are used extensively in various industries, especially drinking water and sewage water treatment, pharmaceutical industries, dairy industry, and oil and gas industry. Ifa Pajouhesh Co., established in 2003, has been able to produce this product. Halophilic ultrafiltration nanocomposite membranes range from 0.2 to 0.02 microns and are made of polyethylene (PE), polypropylene (PP), polyvinyl chloride (PVC) composites with nano- and nano-ZnO polymers.

16.5.2.2 UF, MF, and NF Ceramic Membranes

UF, MF, and NF ceramic membranes are used in a wide range of temperatures and pressures for various applications like bacterial separation in industries such as dairy, juice, suspending matter and polymer materials, separation of water and oil emulsions, and metal recovery. These filters are made of alpha-alumina engineering ceramics and their cavities are in the range of 0.1 to 1 micrometer. In other words, these filters do not pass through particles and macromolecules larger than 1 nm. Danesh Pajouhan San'at Nano Company, founded in 2010, produces this product.

16.5.2.3 Domestic Gray Water Purifier Using Ceramic Nanostructured Membranes

Ceramic nanostructured membranes are of particular importance in the process of water treatment of sewage and household waste because of their specific structure, since these membranes have a very high resistance against the phenomenon of clogging and chemical washing, which for these properties are evaluated in poor polymer membranes. In this project, ceramic nanostructured membranes are used in gray water purification. This type of membrane removes the majority of detergents from gray water, which makes it possible to reuse the treated water so that water consumption is reduced by 50%. Danesh Pajouhan San'at Nano Company has produced this product with a minimum capacity of 8 liters per hour. The range of performance of this product is in all pH ranges, does not require an operator, and the functionality Total dissolved solids/electrical conductivity (TDS/EC) range of the system does not bring about any limitations.

16.5.3 PAINT, RESIN, AND COMPOSITE PRODUCTS

One of the important areas in the nanotechnology industry is paint, resin, and composite. Paint is most commonly used to protect color or provide texture to objects. Paint is usually stored, applied and sold in liquid form, but most types of it dry into solid form after application. In chemistry, resin is a solid or highly viscous substance of plant or synthetic origin that is typically convertible into polymers in the process of curing. Many materials are produced via the conversion of synthetic resins to solids. A composite material (shortened to composite) is a material made from two or more constituent materials with significantly different physical or chemical properties, producing a material which has different characteristics from the individual components.

Several studies have been conducted in this field, of which the most important are (Du et al., 2019); (Majeed et al., 2019); (Zheng et al., 2019); (Prasomsin et al., 2019); (Trompeta et al., 2019); (Sharma et al., 2019); (Ebrahimiasl and Younesi, 2018); (Ebrahimiasl and Rajabpour, 2015); (Zarei et al., 2014). Du et al. (2019) examined the synthesis and characterization of nano-TiO$_2$/SiO$_2$-acrylic composite resin. Majeed, Saleh and Abdulmajeed (2019) investigated the effect of nanoparticles on the thermal conductivity of the epoxy resin system. Zheng, Han and Ou (2019) studied nano-composites for structural health monitoring. Prasomsin et al. (2019) examined multiwalled carbon nanotube reinforced bio-based benzoxazine/epoxy composites with Near infrared (NIR)-laser stimulated shape memory effects. Trompeta et al. (2019) assessed the critical multifunctionality threshold for the optimal electrical, thermal, and nanomechanical properties of carbon nanotubes/epoxy nanocomposites for aerospace applications. Sharma et al. (2019) examined the commercial application of cellulose nanocomposites. Ebrahimiasl and Younesi (2018) investigated the shelf life extension of packages using copper/biopolymer nanocomposite (BNC) produced by a one-step process. Ebrahimiasl and Rajabpour (2015) studied the synthesis and characterization of novel bactericidal Cu/HPMC BNCs using a chemical reduction method for food packaging. Zarei, Ebrahimiasl and Jafarirad (2014) investigated the development of bactericidal Ag/chitosan nanobiocomposites for active food packaging.

Different products are produced using nanotechnology for paint, resin, and composite. Included in these products are: anticorrosion powder paint containing silica nanoparticles (improved corrosion resistance); electrostatic powder paint with smoke reduction features; waterborne self-cleaning paint; inkjet waterborne ink formulated with pigment nanoparticles; antibacterial waterborne acrylic paint; thermoplastic cold traffic paint; thermoplastic acrylic nanocomposite resin; UV curing acrylate coating containing silica nanoparticles; nanosil; megastone; waterborne acrylic lacquer; antibacterial meat board; human-fitting nanocomposite gate with high mechanical properties; trash containing nanoclay for increased wear resistance; antibacterial geomembrane; polypropylene granules containing antibacterial zinc oxide nanoparticles; nanoclay containing traffic color for increased abrasion resistance; nanoinsulating paint; plasma-processed polyethylene to improve color absorption; Acrylonitrile Butadienestyrene (ABS) granules containing zinc oxide nanoparticles; PVC adhesive high viscosity oil filter containing nanoparticles; epoxy paint containing nanoscale materials to increase corrosion resistance; alkyd paint containing nanoscale materials to increase corrosion resistance; nanocomposite tubes and silent connections; nanoadditive PVC; polypropylene granules containing antibacterial silver nanoparticles; masterbatch containing nanoparticles polyethylene-based with high mechanical properties (ADFIL 230); polyamide nanocomposite carbon nanotubes to improve electrical conductivity; and antibacterial polyamide (PA) masterbatch containing nanoparticles.

In the following part are introduced: anti-corrosion powder paint containing silica nanoparticles (improved corrosion resistance); self-cleaning waterborne paint; antibacterial waterborne acrylic paint; UV curing acrylate coating containing silica nanoparticles; nano clay containing traffic color for increased abrasion resistance; epoxy paint containing nanoscale materials to increase corrosion resistance; masterbatch containing nanoparticles polyethylene-based with high mechanical properties (ADFIL 230); and antibacterial polyamide (PA) masterbatch containing nanoparticles.

16.5.3.1 Anticorrosion Powder Paint Containing Silica Nanoparticles (Improved Corrosion Resistance)

Corrosion is a natural process that gradually converts pure metal into a more stable form of oxidation or hydroxide. Depending on the type of metal and the surrounding environment, corrosion-resistant coatings reduce corrosion. Silica, due to its very low free electrons, has excellent resistance to corrosion, thermal stability, and low conductivity. Adding silica to paint creates a quality texture without affecting the paint. Corrosion-resistant paint protects buildings from building corrosion against foreign corrosive agents such as water or oxygen that slowly eliminate steel. They are also used on steel structures in the oil, gas sectors, in addition to power stations, bridges, cellulose production plants, canned fish mills, and so on. Anticorrosive powder paints containing silica nanoparticles are known as powdery paint containing silica nanoparticles with an amorphous crystalline structure. The average size of silica particles is 48 nm. This product has improved corrosion resistance due to the presence of silica nanoparticles. Adding silica nanoparticles to colors can improve its micro compatibility, abrasion resistance, scratch resistance, and weather conditions. Antiseptic powdery paint containing silica nanoparticles is produced by Gostaresh Kimia Corporation, established in 2012.

16.5.3.2 Self-Cleaning Waterborne Paint

Self-cleaning waterborne paint is a clear, self-cleaning coating that is designed and supplied with the use of titanic photocatalytic properties as well as nanotechnology. In the presence of light, the photocatalyst will be able to decompose environmental pollutants. In addition, due to the lack of adhesion of these materials on the surface, the remaining pollutants are easily removed from the surface by rain. This product has various applications such as in the building industry to protect building surfaces with self-cleaning properties, in the food and pharmaceutical industries as 100% sanitary coatings applied on different surfaces, even pre-painted surfaces, hospitals, and health centers. Therapies to prevent the growth of microbes and harmful bacteria on building surfaces and to protect the exterior of buildings against environmental pollution can maintain beauty and initial durability more than 95% for up to 10 years. Gostaresh Kimia Corporation, established in 2012, produces this product.

16.5.3.3 Antibacterial Waterborne Acrylic Paint

One of the most important factors in the transmission of diseases is the surface of walls, floors of buildings, recreation centers, and public utilities. Nowadays in advanced countries, the use of antibacterial coatings is considered essential, including in hospitals, nurseries, restaurants, etc., as well as in buildings under construction. This product is good quality that has added antibacterial properties using silver nanoparticles. The application of this product can be used to color the walls and all gypsum, cement, old painted surfaces in public places, such as hospitals, health centers and laboratories, childcare centers, offices and organizations, transport stations, and so on. Antibacterial waterborne acrylic paint has been made based on 10,900 standards for *Escherichia coli* bacteria (gram-negative) and reduced from 10 and to 10 for the bacteria. This product is produced by Gostaresh Kimia, established in 2012.

16.5.3.4 UV Curing Acrylate Coating Containing Silica Nanoparticles

UV curing acrylate coating containing silica nanoparticles paints are enhanced by silica nanoparticles. This product is usable in various industries, including car, transport, civil and building, paint, and resin. UV curing acrylic coatings containing silica with nanoparticles have the same properties such as high resistance to scratches, high transparency, resistance to pebbles, and high resistance to stains. The anti-smoke and anti-honeycomb are coatings among some of the advantages of this nano-friendly product in the transportation industry. UV coated acrylate coatings containing silica nanoparticles are produced by Radisic Poushesh Co., founded in 2010.

16.5.3.5 Nanoclay Containing Traffic Color for Increased Abrasion Resistance

Paints are a mixture of pigments and fillers that are placed in a resin or connector field. A liquid is also added to the mixture to provide flexibility in the product for use in different applications. Ordinary street traffic nanocolor is used for general line painting use. This paint, that has been improved by nanotechnology, is used in various lines, including flood lines in urban and interstellar streets, carving lines, pedestrian lines, and traffic signs on the road. The nanocolored traffic paint to increase the abrasion-resistant surface as a scratch-resistant finish has mineral additives in nano-sized dimensions that increase the abrasion resistance of this paint. The main add-on is to create this nanoclay property. Pishgaman Fannavari Asia Company, established in 2003, produces this product.

16.5.3.6 Epoxy Paint Containing Nanoscale Materials to Increase Corrosion Resistance

In advanced technologies, corrosion is still a major problem, as the main cause of industrial breakdowns and the loss of billions of dollars annually for preventive maintenance, repair, and restoration of metal materials. Corrosion-resistant metal paint protects against damage caused by moisture, fog, oxidation, or exposure to industrial chemicals. Anticorrosion coating protects excess metal surfaces and prevents contact with chemical compounds and corrosive material. By adding nanomaterials as fillers in polymer paint, its corrosion resistance increases while the mechanical properties of the material improve. Such benefits provide these materials for industrial applications. Corrosion resistance of nanocoating epoxy can be attributed to the nanoparticulate properties. Epoxy is a substrate for nanoparticles and improves the adhesion between epoxy and metal surfaces by adding nanoparticles. By reducing the cavities in the coating and increasing the penetration path for the corrosive solution, these particles provide better sealing properties in the polymer coatings. The Fateh Shimi Gostaran Company, established in 2018, produces this product.

16.5.3.7 Masterbatch Containing Nanoparticles Polyethylene-Based with High Mechanical Properties (ADFIL 230)

In this product, nanosized calcium carbonate is distributed in a nanopolythene substrate, resulting in a much better optical property than calcium carbonate in micron. Properties of these materials are created while the antiadhesion properties and better cooling of the product are also achieved by preserving the physical and mechanical properties. Using this masterbatch, the finished product module will increase, while the length to the tear point increases. The passing of light from the particle of nanoparticles emitted in calcium carbonate will be better, and less cloudy. In cases where the film needs to be welded, the presence of this rapid reaction in the formulation generates a boiling temperature in the material. Among the features of high density polyethylene nanoparticles, are factors such as the distribution of calcium nanocarbonate in a polyethylene substrate with the possibility of passing of light, creating anti-block properties and increasing the production cycle, increasing the physical properties and a mechanical coating with coating capability and surface glossiness, reduction of collapse and preventing deformation of the unit, increase of thermal stability, increase of modulus and tensile strength, decrease of sewing temperature, and welding. Pouya Polymer Tehran Company, established in 1998, produces this product.

16.5.3.8 Antibacterial Polyamide (PA) Masterbatch Containing Nanoparticles

Synthetic polyamides are commonly used in the textile, automotive, carpet, and apparel industry due to their durability and high strength. New plastic materials with intrinsic antibacterial properties can be made for this purpose by polymerizing and copolymerizing the new monomers, or by chemical modification or the combination of polymers. Masterbatch is a liquid or solid additive used to produce colored paintbrushes (color masterbatches) or transfer other properties to polymeric materials (additive masterbatches). The most commonly used method for the production of antibacterial polymers is the use of different organic and inorganic materials such as tea extract, chitosan, copper, silver, zinc, etc. in the polymeric matrix. The antibacterial activity of these nanomaterials has proved to function against pathogenic bacteria such as *E. coli* and *S. aureus*. Polyamides are commonly used in the textile industry, cars, carpets, sports shoes, and so on. Nanoparticles produce more active varieties of oxygen in comparison to the bulk sample, and so the antibacterial activity of nanoparticles is more visible. Parsa Polymer Sharif Company, established in 2007, produces antibacterial masterbatch of polyamide (PA) containing nanoparticles. Antibacterial polyethylene mixes containing zinc oxide nanoparticles, antibacterial baseball PP masterbatch, and ABS antibacterial masterbatch are among other products manufactured by this company.

16.5.4 Nanocoating Products

Coating is an engineering process in which coatings with suitable materials allow the durability of parts, machines, and equipment to be extended. With the emergence of nanotechnology, the process of coating nanoparticles developed using this technology, which boosts efficiency, persistency, and productivity. Nanocoatings are a type of coating that have dimensions of 1 to 100 nanometers in one of their components and also have unique properties compared with ordinary and traditional coatings. Nanotechnology coatings will create unique properties at the surface of the coating due to the nature of the nanocoatings (NDSH, 2018). Figure 16.3 shows coating methods.

As shown in Figure 16.3, coating methods can be placed into three categories of molten or semi-molten mode, dissolved mode, and gas mode. The molten or semi-molten mode itself is divided into three sections of welding, thermal spray, and plasma laser. The dissolved mode takes place in three batches of sol-gel spray/immersion, electrochemical, and chemical. The gas model also includes the physical and vapor sediment. Table 16.5 shows the classification of all types of coatings and nano-resistant structural imperatives. Nanocoating technology in the industry has many applications that increase the importance of this sector more than ever before. To become more familiar with the applications of nanocoating technology in the industry, Table 16.6 refers to the applications of nanocoating technology in various industries.

Several studies have been conducted in this field, of which the most important are Delfini et al. (2019); Bakhy et al. (2018); Sharif Hossain et al. (2018); and Bellisario et al. (2018). Delfini et al. (2019) examined the evaluation of atomic oxygen effects on nanocoated carbon-carbon structures for re-entry applications. Bakhy, Zidan and El Din Abdoul (2018) studied the effect of nanomaterials on edible coating and film improvement. Hossain et al. (2018) investigated a nanocellulose based nanocoating biomaterial dataset using corn leaf biomass. Bellisario et al. (2018) examined the manufacturing of antibacterial additives by nanocoating fragmentation.

Different products are produced using nanotechnology in the nanocoating field. Included in these are: physical coating steam device with hybrid technology; super hard coating devices with physical vapor deposition (PVD) method; industrial coating device with 2,300 cathode arc method; industrial coating device with 1,600 cathode arc method; industrial coating device with 1,600 sputtering

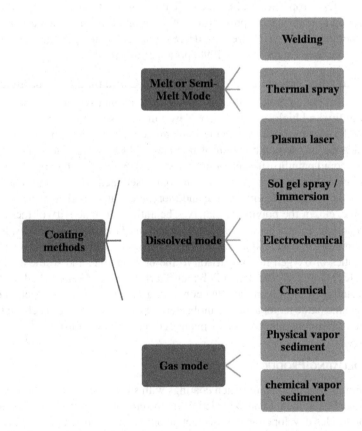

FIGURE 16.3   Coating methods.

TABLE 16.5

All Types of Coatings and Nano-Resistant Structural Imperatives

Nano component		Example	Properties
Common cover	No nanometer component	VPS-HVOF Plating	Improve the functional properties relative to the base piece
Nano-structuring	Nanometer constructive structure	PVD, PACVD, EB-PVD, ESD Nanoplating	Increase corrosion resistance Increased thermal stability Increase wear resistance
Nano compositing	Dispersion of nanoparticles	Nano-HVOF, PVD, PACVD Nanoplating	Increase surface hardness Increased wear resistance
Nano Layering	Coating with a nanometer thickness	PVD PAVCD	Increased fracture toughness Reduction of friction coefficient Improved wear resistance
Two-step process	One of the above + Nitrified layer	Plasma Nitriding + PAVCD or PVD	Increase mechanical performance Increase adhesion of the coating

Source: NDSH (Nanotechnology Development Specialist Headquarters) 2018.

method; nanocomposite nanolayer coatings, ultra-hard, wear-resistant and erosion-resistant on the surfaces of a variety of cutting and shearing tools; nanolayer and nanocomposite coatings extremely hard and erosion resistant, on the surfaces of all types of metal molding materials; nanolayer and nanocomposite coatings, extremely hard and resistant to wear and corrosion on the surfaces of cold-blown parts of turbine engines; hard nanoscale coating on aluminum extrusion molding; Golfa nanocoating; nanodecorative coatings resistant to corrosion and scratches; zirconium-based nanoceramic conversion coatings and nanozirconium conversion coating (SurBond).

In the next part, physical coating steam device with hybrid technology; industrial coating device with 1,600 sputtering method; nanocomposite nanolayer coatings, ultra-hard, wear-resistant and erosion-resistant on the surfaces of a variety of cutting and shearing tools; and hard nanoscale coating on aluminum extrusion molding, are introduced.

16.5.4.1 Physical Coating Steam Device with Hybrid Technology

Physical coating steam devices with hybrid technology can be used as a tool for making various coatings due to the use of sputtering and thermal evaporation technology with a cathodic arc. In the sputtering method, most of the focus is on tangent ion bombardment, which is basically unhealthy. While in the cathode arc line, an arc is responsible for the local heating function of the target. Physical vapor deposition layering with a cathodic arc is a powerful and effective way to cover a wide range of super-hard coatings, nanocomposite coatings and multi-layer thin films such as TiN, TiAIN, CrN, ZrN, AlCrTiN, and TiAlSiN. This device is manufactured by Mohandesi Sath Sevin Plasma Company founded in 2009.

16.5.4.2 Industrial Coating Device with 1,600 Sputtering Method

Sputtering is one of the methods of physical steam mode coating. In this method, high-acceleration ions, after colliding with the target material, transfer their energy by separating the target material and spreading it in space. With this method, it is possible to cover all kinds of optical components such as car mirrors, high-temperature sensitive plastic materials, and so on. This machine is technically cylindrical with a 160×180 cm cylinder with a final pressure of 10–5 torr, with two rectangular sputterers and a rotating speed controller, and a full-color touch screen control system and plasma cleaning system. The industrial coating machine is manufactured by the Yarnikan Saleh Co., established in 2006.

TABLE 16.6

The Applications of Nanocoating Technology in Various Industries

Industrial field	Introducing the uses and benefits of nanocoatings
Oil and energy	Corrosion and abrasion in the operating environment of the oil and energy industries destroy the surface of the components and equipment. Nanocoatings, in addition to extending the life of these components, reduce the duration of interruptions in various operations.
Building and home	Due to the variety of colors, proper properties, ease of use, a variety of materials, appearance, stainlessness, and high precious deposits, nanocoatings are widely used in these industries including valves, fittings, tiles, and ceramics.
Automobile manufacturing	Nanocoatings are used in automotive components and tools and molds. Increasing engine performance and efficiency, improving internal and external appearance, reducing fuel consumption, biocompatibility and cost-effective results from the use of this technology.
Agriculture and food/Mold and tools	In the maintenance and repair of equipment such as motors, gears, and bearings of machines, as well as cutting blades for combines, nanoscale coating technology is applied to abrasion, corrosion, and impact resistance.
Manufacturing	Maintaining and repairing tools and molds in various industries is critical to wear, corrosion and impact. Nanocoating technology enhances the life and quality of the molds, tools, and finished products.
Polymer and plastic	Various plastic and polymer industries, including cylindrical and rotary extruders, plastic injection molds, blow molds and various blades, have improved properties, and quality using nanoparticles.
Aerospace	With this technology, the reliability and life expectancy of equipment and components will increase. Application of these types of coatings results in high temperature resistance, corrosion and wear of engine components, and other equipment of this industry.
Textile	Parts of knitting and spinning machines, such as straps, wheels, rapiers, types of needles, rings and nails, parts of printing, and finishing dyeing machines, are all degraded for various reasons and need to use nanocoatings.
Medical and pharmaceutical	Medical tools, implants, pharmaceutical industry parts, processing and packaging, and many electronic and mechanical components are strictly coated to enhance hardness, chemical neutrality, biocompatibility, and antibacteriality.

Source: NDSH (Nanotechnology Development Specialist Headquarters) 2018.

16.5.4.3 Nanocomposite Nanolayer Coatings, Ultra-Hard, Wear-Resistant, and Erosion-Resistant on the Surfaces of a Variety of Cutting and Shearing Tools

Nanocomposite nanolayer coatings, ultra-hard, wear-resistant, and erosion-resistant on the surfaces of a variety of cutting and sharing tools are for methods such as pine and matrix, punch, form and deep drawing; cold and hot forgings; extrusion; and die casting. Plastic injection will increase the surface quality and longevity of the molds by up to 10 times and improve the performance of the molds. Technically, these coatings have extremely high hardness (5,000 – 2,500 WC) with ideal tofu; ideal surface fineness (raffles less than 0.2 microns); suitable tribological properties to prevent adhesion of materials to mold; temperature tolerance is up to 1,100°C. This product is manufactured by Mohandesi Sath Sevin Plasma Company founded in 2009.

16.5.4.4 Hard Nanoscale Coating on Aluminum Extrusion Molding

These TiN/CrN coatings with a thickness of three microns and extremely high hardness increase the wear resistance of the molds against aluminum ceilings from the molded groove and increase the molding production by three times. One of the characteristics of nanoscale coatings can be the lack of a tendency of aluminum sticking to the mold, reduction of machine pressure due to reduced friction of mold and melt, and the possibility of a three-fold increase in production without increasing the weight of the profile. This product is manufactured by Fannavari Sakht Ara Company, established in 2014.

16.6 INFRINGEMENT OF INTELLECTUAL PROPERTY IN NANOTECHNOLOGY INDUSTRIES

Generally, there are two types of violation of intellectual property rights. Violating intellectual property rights, "infringements" in relation to patents, copyright, and trademarks and "misappropriations" related to trade secrets, are violations of civil or criminal law, depending on the intellectual property, the jurisdiction, and the nature of the act.

16.6.1 PATENT INFRINGEMENT

A patent infringement is typically due to the use or sale of the patented inventions without the permission of the inventor. The scope of the patented invention or the level of protection is defined in patent claims. In many jurisdictions, there is a safe way to use a patent for research. This safe way does not exist in the United States unless research is conducted for purely philosophical purposes or for collecting information to request approval for a drug. In general, cases involving civil cases (for example, in the United States) are being investigated, but several jurisdictions also include criminal offenses such as Argentina, China, France, Japan, Russia, and South Korea (Cutler, 2008).

16.6.2 COPYRIGHT INFRINGEMENT

Copyright infringement is the production, distribution, display or performance of a work or derivative works without the permission of the copyright holder, which is usually the publisher or other company that represents or the creator of the work. This is often referred to as "piracy". While copyrights are being created, copyright holders can only be harmed if the copyright owner registers it (Panethiere, 2005). Copyright enforcement is generally the responsibility of the copyright owner. The Anti-Counterfeiting Trade Agreement (ACTA), signed in May 2011 by the United States, Japan, Switzerland, and the European Union, is authorized by its parties to impose criminal penalties, including imprisonment and punishment for copyright and trademark infringement. There are copyright restrictions and exceptions that allow limited use of copyrighted works as long as they do not violate the holders' rights (Bitton, 2012).

16.6.3 TRADEMARK INFRINGEMENT

Trademark infringement happens when one party uses a trademark that is identical or confusingly similar to a trademark of the other party, in relation to the same products or services, which are similar to the other party's products or services. In many countries, a trademark is protected without registration, but trademark registration provides legal advantages for enforcement. Violations can be solved by civil law and in several jurisdictions under criminal law (Manta, 2011).

16.6.4 TRADE SECRET MISAPPROPRIATION

Misappropriation of trade secrets is different from other intellectual property laws, since, by definition, trade secrets are secret, while patents and trademarks and registered copyrights are publicly available. In the United States trade secrets are protected under state laws, and states have almost universally approved a privacy law. The United States also provides federal law in the form of the Economic Information Act of 1996 (18 U.S.C.), which categorizes theft or misappropriation of a trade secret as federal crimes. The law contains two legal provisions of two types of activities. First, 18 U.S.C. defines theft of trade secrets in favor of foreign powers. Second, 18 U.S.C. legalizes this robbery for commercial or economic purposes. Under Commonwealth common law jurisdictions, confidentiality and trade secrets are considered as equal rights, not property rights, but theft is almost the same as in the United States.

REFERENCES

Acharya R, Tanna G (2018) Intellectual Property Rights and Licensing in India. *Journal of the Licensing Executives Society*. 6(2): 167–172.

Adam H, Youssef A (2019) Economic Impacts of Nanotechnology Industry: Case Study on Egypt. *Cairo, International Conference on Digital Strategies for Organizational Success*.

Ajaz A et al. (2019) Nanomedicines: Challenges and Perspectives for Future. *Scientific Research and Essays*. 14(5): 32–38.

Amin M T, Alazba A, Manzoor U (2014) A Review of Removal of Pollutants from Water/Wastewater Using Different Types of Nanomaterials. *Advances in Materials Science and Engineering*. 24(4): 1–24.

Anjum M et al. (2016) Remediation of Wastewater Using Various Nano-Materials. *Arabian Journal of Chemistry*. 12(8): 4897–4919.

Azamal Moshed A M, Islam Sarkar M K, Khaleque A (2017) The Application of Nanotechnology in Medical Sciences: New Horizon of Treatment. *American Journal of Biomedical Sciences*. 4(2): 1–14.

Azarm B (2016) Nanotechnology in Engineering Sciences. *New Ideas in Science and Technology*. 1(3): 53–64.

Baker B K (2016) Rans-Pacific Partnership Provisions in Intellectual Property. *PLoS Medicine*. 1(3): 1–7.

Bakhy E, Zidan N, El Din Aboul H (2018) The Effect of Nano Materials on Edible Coating and Films' Improvement. *International Journal of Pharmaceutical Research and Allied Sciences*. 7(3): 20–41.

Banerjee R (2013) The Success of, and Response to, India's Law against Patent Layering *Harvard. Journal of International Law*. 54(4): 204–232.

Bastani B, Fernandez D (2002) Intellectual Property Rights in Nanotechnology. *Thin Solid Films*. 420(1): 472–477.

Bawa R (2005) Will Nanomedicine Patent Land Grab Thwart Commercialization. *Nanomedicine Biology and Medicine*. 1(1): 346–350.

Bawa R et al. (2005) Protecting New Ideas and Inventions. *Biology and Medicine*. 1(1): 150–158.

Bellisario D, Quadrini F, Santo L, Tedde G (2018) Manufacturing of Antibacterial Additives by Nano-Coating Fragmentation. *Texas, Manufacturing Engineering Division*. 1(2): 4–11.

Betsholtz A, Jacobsson S, Haghighatafshar S, Jönsson K (2019) *Sewage Sludge-Based Activated Carbon – Production and Potential in Wastewater and Storm Water Treatment*. Lund: VA-Teknik Södra.

Bettig R V (2018) *Copyrighting Culture the Political Economy of Intellectual Property*. New York: Routledge.

Bitton M (2012) Rethinking the Anti-Counterfeiting Trade Agreement's Criminal Copyright Enforcement Measures. *The Journal of Criminal Law and Criminology*. 102(1): 67–117.

Bouchoux D E (2012) *Intellectual Property: The Law of Trademarks, Copyrights, Patents, and Trade*. New York: Cengage learning.

Caso R, Guarda P (2019) *Copyright Overprotection Versus Open Science: The Role of Free Trade Agreements*. Singapore: Springer.

Chaturvedi S, Dave P N (2019) Water Purification Using Nanotechnology and Emerging Opportunities. *Chemical Methodology*. 3(1): 115–144.

Chaturvedi S, Dave P, Shah N (2012) Applications of Nano-Catalyst in New Era. *Journal of Saudi Chemical Society*. 16(3): 307–325.

Chen Z, Zhang J (2019) Types of Patents and Driving Forces behind the Patent Growth in China. *Economic Modelling*. 80(5): 294–302.

Chowdhury P, Gogoi M, Das S, Zaman A, Hazarika P, Borchetia S, Bandyopadhyay T (2016) Intellectual Property Rights for Nanotechnology Application in Agriculture. In: Ranjan S, Dasgupta N, Lichtfouse E (eds) *Nanoscience in Food and Agriculture, Sustainable Agriculture Reviews*, vol. 21. Springer, Berlin, pp. 1–25. doi:10.1007/978-3-319-39306-3_1.

Contigiani A, Hsu D H, Barankay I (2018) Trade Secrets and Innovation: Evidence from the Inevitable Disclosure Doctrine. *Strategic Management Journal*. 39(11): 2921–2942.

Correa C (2007) *Trade Related Aspects of Intellectual Property Rights: A Commentary on the TRIPS Agreement*. Oxford: Oxford University Press.

Cutler M L (2008) International Patent Litigation Survey: A Survey of the Characteristics of Patent Litigation in 17 International Jurisdictions Archived 2013-09-22 at the Wayback Machine.

Czerniewicz L, Deacon A, Walji S (2018) Educators and open education resources in Massive Open Online Courses. Proceedings of the 11th International Conference on Networked Learning 2018: 1–9.

Dang Y et al. (2010) Trends in Worldwide Nanotechnology Patent Applications: 1991 to 2008. *Journal of Nanoparticle Research: An Interdisciplinary Forum for Nanoscale Science and Technology.* 12(3): 687–706.

de Rassenfosse G (2019) On the Price Elasticity of Demand for Trademarks. *Journal of Industry and Innovation.* doi:10.1080/13662716.2019.1591939.

Delfini A et al. (2019) Evaluation of Atomic Oxygen Effects on Nano-Coated Carbon-Carbon Structures for Re-Entry Applications. *Acta Astronautica.* 161: 276–282.

Dratler J, McJohn S M (2006) *Intellectual Property Law: Commercial, Creative and Industrial Property.* New York: Law Journal Press.

Du B et al. (2019) Synthesis and Characterization of Nano-TiO2/SiO2-Acrylic Composite Resin. *Advances in Materials Science and Engineering*: 1–7. doi:10.1155/2019/6318623.

Ebrahimiasl S, Nahli R, Zakaria A (2017) Ppy/nanographene Modified Pencil Graphite Electrode Nanosensor for Detection and Determination of Herbicides in Agricultural Water. *Science of Advanced Materials.* 9(12): 2045–2053.

Ebrahimiasl S, Rajabpour A (2015) Synthesis and Characterization of Novel Bactericidal Cu/HPMC BNCs Using Chemical Reduction Method for Food Packaging. *Journal of Food Science and Technology.* 52(9): 5982–5989.

Ebrahimiasl S, Seifi R, Eftekhar Nahli R, Zakaria A (2017) Ppy Nanographene Modified Pencil Graphite Electrode Nanosensor for Detection and Determination of Herbicides in Agricultural Water. *Science of Advanced Materials.* 9(12): 2045–2053.

Ebrahimiasl S, Younesi A (2018) Shelf Life Extension of Package's Using Cupper/ (Biopolymer Nanocomposite) Produced by One-Step Process. *Journal of Food Biosciences and Technology.* 8(1): 47–58.

Ebrahimiasl S, Zakaria A, Kassim A, Basri S (2014) Novel Conductive Polypyrrole/Zinc Oxide/Chitosan Bionanocomposite: Synthesis, Characterization, Antioxidant, and Antibacterial Activities. *International Journal of Nanomedicine.* 10: 217–227.doi:10.2147/IJN.S69740.

FDA (2007) *Task Force Report.* US FDA: Nanotechnology: A report of the U.S. Food and Drug Administration Nanotechnology Task Force. 25 July 2007. Available at http://www.fda.gov/nanotechnology/taskforce/report2007.pdf (accessed 30 January 2020).

Genieser L, Gollin M (2007) Intellectual Property Issues in nanotechnology. *Journal of Commercial Biotechnology.* 13(3): 195–198.

Hacker P (2018) Lessons from IP Markets for Data Markets. On Moral Rights, Property Rules, and Resale Royalties. *Intellectual Property Quarterly.* 7: 45–67. [Accessed https://ssrn.com/abstract=3125702].

Hassan D F, Mahmood M B (2019) Using of Iron Oxide Nanoparticles and Application in the Removing of Heavy Metals from Sewage Water. *Iraqi Journal of Science.* 60(4): 732–738.

Hermann K et al. (2005) Protecting Nanotechnology Intellectual Property (Nano-IP) in China. *Nanotechnology Law and Business.* 2(1): 96–109.

Hsiao J C, Fong K (2004) Making Big Money from Small Technology. *Nature.* 428(6979): 218–220.

Husovec M et al. (2019) *Making the Rules: The Governance of Standard Development Organizations and Their Policies on Intellectual Property Rights.* Tilburg: Joint Research Center.

Isabella J (2008) Big Worries About Micro Particles. The Tyee. thetyee.ca [Online] [Accessed https://thetyee.ca/News/2008/04/07/NanoParticles/].

Jia L (2005) Global Governmental Investment in Nanotechnologies. *Current Nanoscience.* 1(3): 263–266.

Kamei S (2002) *Promoting Japanese Style Nanotechnology Enterprises.* Mitsubishi Research Institute, Tokyo, Japan.

Kang R, Jung T, Lee K (2019) Intellectual Property Rights and Korean Economic Development: The Roles of Patents. Utility Models and Trademarks. *Journal of Area Development and Policy.* doi:10.1080/23792949.2019.1585889.

Ketabforoush Badri A, Ketabforoush Badri A, Yahyavi R (2016) Examination of the Quality of Environment in MENA Countries with Emphasizing Using of ICT. *Noble International Journal of Social Sciences Research.* 1(1): 11–15.

Khan Z et al. (2019) Nanotechnology: An Elixir to Life Sciences. *Preprints*: 202–235. 2019020235. doi:10.20944/preprints201902. 0235.v1.

Khatamian M et al. (2012) Heterogeneous Photocatalytic Degradation of 4-Nitrophenol in Aqueous Suspension by Ln (La3+, Nd3+ or Sm3+) Doped ZnO Nanoparticles. *Journal of Molecular Catalysis A: Chemical.* 365(4): 120–127.

Khazaee M et al. (2015) The Challenge of Medical Waste Management: A Case Study in Alborz. *Acta Velit.* 1(3): 70–79.

Kojabad R, Ebrahimiasl S (2017) Pencil Graphite Electrode Modified Nano Sensor for Detection and Determination of Tramadol in Blood Serum. *Medellín-Colombia.* 1: 597–604.

Kumari P, Luqman S, Meena A (2019) Application of the Combinatorial Approaches of Medicinal and Aromatic Plants with Nanotechnology and Its Impacts on Healthcare. *DARU: Journal of Pharmaceutical Sciences.* 27(2): 1–15.

Larrimore Ouellette L (2015) Nanotechnology and Innovation Policy. *Harvard Journal of Law and Technology.* 29(1): 34–75.

Linton K (2016) The Importance of Trade Secrets: New Directions in International Trade Policy Making and Empirical Research. *Journal of International Commerce and Economics*: 1–17. Published electronically September. 2016. http://www.usitc.gov/journals.

Loosli F et al. (2019) Sewage Spills Are a Major Source of Titanium Dioxide Engineered (Nano)-Particle Release into the Environment. *Environmental Science: Nano.* 6(3): 763–777.

Lux Research (2007) *The Nanotech Report.* New York: Lux Research.

Majeed N S, Salih S M, Abdulmajeed B A (2019) Effect of Nanoparticles on Thermal Conductivity of Epoxy Resin System. *IOP Conference Series: Materials Science and Engineering.* 518: 1–10.

Manta I D (2011) The Puzzle of Criminal Sanctions for Intellectual Property Infringement. *Harvard Journal of Law and Technology.* 24(2): 469–518.

Menell P S (2007) The Property Rights Movement's Embrace of Intellectual Property. *Ecology Law Quarterly.* 34: 713–754.

Menell P S (2018) Economic Analysis of Intellectual Property Notice and Disclosure. *UC Berkeley Public Law Research Paper.* 1: 1–53.

Misra R, Acharya S, Sahoo S K (2010) Cancer Nanotechnology: Application of Nanotechnology in Cancer Therapy. *Drug Discovery Today.* 15(19): 842–850.

Naghdi Y, Kaghaziyan S, Mohseni N, Parhizi H (2013) The Effects of Nanotechnology Expansion on Economic Growth in Selected Countries. *Economic Modeling Journal.* 1: 85–99.

NDSH (2018) *Nanotechnology Products.* Tehran: Mehrvision Technology Development.

NSTC/NSET (Nanoscale Science, Engineering, and Technology Subcommittee of the National Science and Technology Council) (2000) *National Nanotechnology Initiative: The Initiative and Its Implementation Plan.* Washington, DC.

Olafusi O S, Sadiku E R, Snyman J, Ndambuki J M (2019) Application of Nanotechnology in Concrete and Supplementary Cementitious Materials: A Review for Sustainable Construction. *SN Applied Sciences.* 1(580).

Owoeye O, Olatunji O, Faturoti B (2017) Patents and the Trans-Pacific Partnership: How TPP-Style Intellectual Property Standards May Exacerbate the Access to Medicines Problem in the East African Community. *The International Trade Journal.* 33(2): 197–218.

Palmberg C, Dernis H, Miguet C (2009) *Nanotechnology: An Overview Based on Indicators and Statistics.* OECD Publishing, Paris, France.

Panethiere D (2005) *The Persistence of Piracy: The Consequences for Creativity, for Culture, and for Sustainable Development.* UNESCO, Paris, France.

Prasomsin W et al. (2019) Multiwalled Carbon Nanotube Reinforced Bio-Based. *Nanomaterials.* 9(881): 1–17.

Rasouli R, Barhoum A, Bechelany M, Dufresne A (2018) Nanofibers for Biomedical and Healthcare Applications. *Macromolecular Bioscience.* 1: 1–18.

Regkouzas P, Diamadopoulos E (2019) Adsorption of Selected Organic Micro-Pollutants on Sewage Sludge Biochar. *Chemosphere.* 224(4): 840–851.

Rozek R P (2018) Valuing the Residual Intellectual Property in Mature Pharmaceutical Products. *European Journal of Risk Regulation.* 9(2): 337–350.

SCENIHR (2013) The Scientific Committee on Emerging and Newly Identified Health Risks [Online] [Accessed https://ec.europa.eu/health/home_en#7].

Shand H, Wetter K J (2007) Trends in Intellectual Property and Nanotechnology: Implications for the Global South. *Journal of Intellectual Property Rights.* 12: 111–117.

Sharif Hossain A, Uddin M, Veettil V, Fawzi M (2018) Nano-Cellulose Based Nano-Coating Biomaterial Dataset Using Corn Leaf Biomass: An Innovative Biodegradable Plant Biomaterial. *Data in Brief.* 17: 162–168.

Sharma A et al. (2019) Commercial Application of Cellulose Nano-Composites–A Review. *Biotechnology Reports.* 1: 1–20.

Singh H B et al. (2016) *Intellectual Property Issues in Biotechnology.* Wallingford, UK: CABI, 304 pages, ISBN-13: 78–1780646534.

Singh H B et al. (2019) *Intellectual Property Issues in Microbiology.* Singapore: Springer, 425 pages, ISBN- 978-981-13-7465-4.

Singh K A (2012) Intellectual Property in the Nanotechnology Economy. NANOFO-RUM, pp. [Accessed http: //www.nanoforum.org/dateien/temp/Article%20on%20Intellectual %20Property%20-%2012%20JAN .pdf?05032011233715].

Singh K K (2019) Legal Issues in Nanotechnology. In: A. Kumar, A. K. Singh and K. K. Choudhary, eds. *Role of Plant Growth Promoting Microorganisms in Sustainable Agriculture and Nanotechnology.* Woodhead Publishing, Sawston, UK, 107–120.

Singh N A (2017) Nanotechnology Innovations, Industrial Applications and Patents. *Environmental Chemistry Letters.* 15(2): 189–191.

Society T R (2004) *Nano Manufacturing and the Industrial Application of Nanotechnologies.* London: The Royal Society and The Royal Academy of Engineering.

SPIE (2009) Small Concerns: Nanotech Regulations and Risk Management. *SPIE Newsroom.*

StatNano (2018) https://statnano.com/news/65159/Nanotechnology-Patents-of-2018-at-the-USPTO-and-EPO -through-the-Lens-of-Statistics (accessed 05 January 2020).

Story A (2002) Don't Ignore Copyright, the 'Sleeping Giant' on the TRIPS and International Educational Agenda. In: P. Drahos and R. Mayne, eds. *Global Intellectual Property Rights*, vol. 1, Palgrave Macmillan, London, UK, 125–143.

Taheri M, Hamidian A H (2016) Results of a Hospital Waste Survey in Tabriz Hospitals. *The Journal of Middle East and North Africa Sciences.* 2(3): 70–77.

The Royal Society (2004) *Nanoscience and Nanotechnologies: Opportunities and Uncertainties.* London: The Royal Society and The Royal Academy of Engineering.

Trompeta A F et al. (2019) Assessing the Critical Multifunctionality Threshold for Optimal Electrical, Thermal, and Nanomechanical Properties of Carbon Nanotubes/Epoxy Nanocomposites for Aerospace Applications. *Aerospace.* 6(7): 1–18.

Tsou Y et al. (2019) Nanotechnology-Mediated Drug Delivery for the Treatment of Obesity and Its Related Comorbidities. *Advance Healthcare Materials.* doi:10.1002/adhm.201801184.

Tullis T K (2012) Current Intellectual Property Issues in Nanotechnology. *Nanotechnology Reviews.* 1(2): 189–205.

USPTO Performance and Accountability Report (2019) https://www.uspto.gov/sites/default/files/documents/ USPTOFY19PAR.pdf (accessed 05 January 2020).

Uzun A S (2018) Exhaustion Principle in the Intellectual Property Law. Electronic Copy. [Accessed: https:// ssrn.com/abstract=3279357].

Velmurugan C (2019) Mapping of Research Productivity on Nanotechnology in Canada: A Scientometric Profile. *Library Philosophy and Practice.* 4(15): 1–24.

Yu P (2017) The Investment-Related Aspects of Intellectual Property Rights. *The American.* 66: 829–910.

Zarei A, Ebrahimiasl S, Jafarirad S (2014) Development of Bactericidal Ag/Chitosan Nanobiocomposites for Active Food Packaging. *International Multidisciplinary Microscopy Congress.* Antalya, 255–260.

Zhang W, Zhang D, Liang Y (2019) Nanotechnology in Remediation of Water Contaminated by Poly- and Perfluoroalkyl Substances: A Review. *Environmental Pollution.* 247: 266–276.

Zhang Y, Sulfab M, Fernandez D (2013) Intellectual Property Protection Strategies for Nanotechnology. *Nanotechnology Reviews.* 2(6): 725–742.

Zheng Q, Han B, Ou J (2019) Nanocomposites for Structural Health Monitoring. In: P.-T. Fernando, M.V. Diamanti, A. Nazari, C. Goran-Granqvist, A. Pruna, and S. Amirkhanian, eds. *Nanotechnology in Eco-Efficient Construction.* Woodhead Publishing, Sawston, Cambridge, 227–259.

17 Nanotechnology in the Philippines
Status and Intellectual Property Rights Issues

*Marilyn B. Brown, Cristine Marie B. Brown,
and Robert A. Nepomuceno*

CONTENTS

17.1 INTRODUCTION

Nanotechnology is an emerging multidisciplinary field dealing with the manipulation of materials in the nanoscale to elicit new and unique properties. Despite the common notion that the field is in its infancy and is currently classified as an emerging field of study, it is not new as there has already been several kinds of research encompassing nanoscience phenomena decades ago, and nanoparticles have existed in nature since time immemorial. Specifically, nanotechnology's roots can be traced back to 3rd-century artisans who were adept in what currently is known as nanocomposites, or the application of nanoparticles to enhance the general properties of a composite material. Perhaps one of the most famous examples of the application of nanotechnology in the era is the Lycurgus cup, a Roman-era decorative iridescent artifact that changes color depending on the light source. It reveals a brilliant red color when light passes through it, while it appears as emerald green under dim lighting. The effect is the product of the application of gold-silver alloy nanoparticles in the composite material. The Middle East between the 3rd and 17th centuries saw the fabrication and production of an exceptionally sharp and shatter-resistant sword – the Damascus steel. The blade is widely regarded as the inspiration for the Valyrian steel in George R.R. Martin's magnum opus, *A Song of Ice and Fire*. Originally from India and Sri Lanka from where the Middle Eastern blacksmiths imported their ingots, they brought the Damascus steel to Damascus where it became popular, and its production thrived. The blade is characterized by distinctive patterns reminiscent of raindrops or flowing water. Recently, it was discovered that the steel blades contain nanowire and carbon nanotubules conferring superplasticity and hardness, making it extraordinary for its time. In a more recent example, the 1925 Nobel Laureate in physics Richard Adolf Zsigmondy, first observed and measured nanoparticles through studying gold colloids in the microscope. Through that, he devised the 'nanometer' concept for characterizing particle size of 10^{-9} meters in length. At the 1959 American Physical Society meeting at Caltech, a physicist named Richard Feynman presented a lecture entitled "There's Plenty of Room at the Bottom" in which he hypothesized that matter could be manipulated at the atomic level (Santamaria 2012). This process of engineering enables scientists to exploit and modify the physicochemical properties of nanomaterials resulting in unique attributes and overall enhanced performance compared to their bulk counterparts (Jeevanandam et al. 2018).

In 1974, a publication written by Norio Taniguchi first used the term 'nanotechnology' which he described as, "… mainly consists of the processing of separation, consolidation, and deformation of materials by one atom or one molecule." Advancements in the field of nanotechnology were pronounced during the 1980s wherein Kim Eric Drexler published his futuristic book "Engines of Creation: The Coming Era of Nanotechnology." (Hulla et al. 2015). Feynman's vision of molecular manufacturing inspired Drexler in conceptualizing a radical and advanced form of nanotechnology involving the construction of nonliving matter through naturally occurring biological machinery upon instruction. Molecular-sized machinery, called assemblers, was said to be capable of building any product so long as it follows the law of nature. Over the years, growing understanding and numerous perceived potential uses of nanoscience prompted the significant increase of interest from several industrial sectors particularly medicine, food, cosmetics, agriculture, and electronics for the field of nanotechnology.

17.2   APPLICATIONS

Multifaceted applications of nanotechnology are set to revolutionize specific industrial processes and medical practices and to craft entirely new tools. Nanomedicine, an emerging area of nanotechnology, was shown useful in developing solutions for archaic difficulties in medical research, and its applications are primarily concentrated in the areas of drug delivery, detection of pathogens and tumors, tissue engineering, diagnostics technology, and in vivo imaging by fluorescent dyes (Lee Ventola 2012). In the food industry, growing food demand and heightened concerns of public for food quality and safety urged researchers to develop innovative techniques on extending shelf life, enhancing chemosensory properties, and maintaining the nutritional value of food products (Singh et al. 2017). For example, biosensors composed of fluorescent and biospecific materials can operate as a highly sensitive and rapid detection tool for foodborne pathogens (Khezri et al. 2016) like *Escherichia coli* O157:H7 and *Staphylococcus aureus*. Furthermore, biosensors are not limited to the detection of microbial contamination; they can also work in quantifying macromolecules, including carbohydrates, amino acids, alcohols, and acids present in foods.

17.3   PHILIPPINE ROADMAP ON NANOTECHNOLOGY (2017–2022)

Nanotechnology is identified as integral in the advancement of the technology revolution of the 21st century. In 2011, the recorded global investment for nanotechnology was USD 17.8 billion, according to Lux Research, wherein funding is increasing at an estimated rate of 20% annually (McManus 2014). The most substantial government-supported nanotechnology funding was documented for countries like Japan, the United States, China, Taiwan, South Korea, and Germany. In the Philippines, the Department of Science and Technology-Philippine Council for Advanced Science and Technology Research and Development Council (DOST-PCASTRD) fronted extensive R&D initiatives for nanotechnology in the country by launching a 10-year strategic roadmap, crafted by a group of interdisciplinary local scientists in 2009 (Figure 17.1). The first batch of research projects was granted a preliminary budget of USD 60 million and have received government and private sector support since then. In an effort to fortify economic development and technological advancement

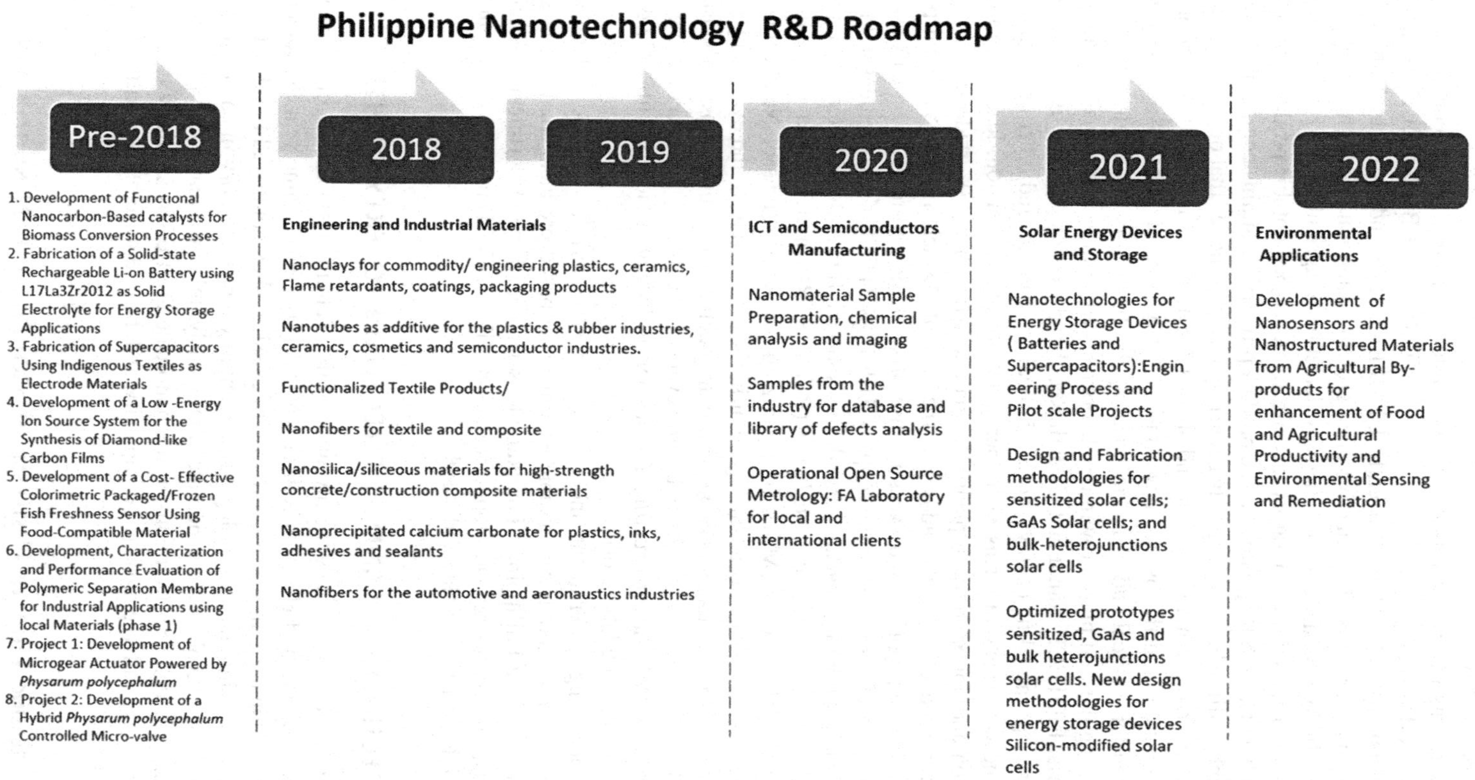

Philippine Nanotechnology R&D Roadmap

Pre-2018

1. Development of Functional Nanocarbon-Based catalysts for Biomass Conversion Processes
2. Fabrication of a Solid-state Rechargeable Li-on Battery using L17La3Zr2012 as Solid Electrolyte for Energy Storage Applications
3. Fabrication of Supercapacitors Using Indigenous Textiles as Electrode Materials
4. Development of a Low -Energy Ion Source System for the Synthesis of Diamond-like Carbon Films
5. Development of a Cost- Effective Colorimetric Packaged/Frozen Fish Freshness Sensor Using Food-Compatible Material
6. Development, Characterization and Performance Evaluation of Polymeric Separation Membrane for Industrial Applications using local Materials (phase 1)
7. Project 1: Development of Microgear Actuator Powered by *Physarum polycephalum*
8. Project 2: Development of a Hybrid *Physarum polycephalum* Controlled Micro-valve

2018

2019

Engineering and Industrial Materials

Nanoclays for commodity/ engineering plastics, ceramics, Flame retardants, coatings, packaging products

Nanotubes as additive for the plastics & rubber industries, ceramics, cosmetics and semiconductor industries.

Functionalized Textile Products/

Nanofibers for textile and composite

Nanosilica/siliceous materials for high-strength concrete/construction composite materials

Nanoprecipitated calcium carbonate for plastics, inks, adhesives and sealants

Nanofibers for the automotive and aeronautics industries

2020

ICT and Semiconductors Manufacturing

Nanomaterial Sample Preparation, chemical analysis and imaging

Samples from the industry for database and library of defects analysis

Operational Open Source Metrology: FA Laboratory for local and international clients

2021

Solar Energy Devices and Storage

Nanotechnologies for Energy Storage Devices (Batteries and Supercapacitors):Engineering Process and Pilot scale Projects

Design and Fabrication methodologies for sensitized solar cells; GaAs Solar cells; and bulk-heterojunctions solar cells

Optimized prototypes sensitized, GaAs and bulk heterojunctions solar cells. New design methodologies for energy storage devices Silicon-modified solar cells

2022

Environmental Applications

Development of Nanosensors and Nanostructured Materials from Agricultural By-products for enhancement of Food and Agricultural Productivity and Environmental Sensing and Remediation

FIGURE 17.1 Roadmap of Philippine Nanotechnology Research and Development Program of the Department of Science and Technology (DOST) for the years 2017 to 2022.

of the country, DOST-PCASTRD prioritized the areas of semiconductor, information and communication technology, energy, agriculture, health, and environment (Asian and Pacific Centre for Transfer of Technology 2010).

Among the recipients of project funding, is the National Institute of Molecular Biology and Biotechnology of the University of the Philippines Los Baños (UPLB BIOTECH) which houses its own nanobiotechnology laboratory. Currently, UPLB BIOTECH endeavors are leaning towards the formulation of nanofertilizers for effective plant growth promotion and higher crop yield, and nanobiosensors for biological and chemical activity detection in agriculture and food safety areas. The institute's nanobiotechnology laboratory has already developed notable products including the BIOTECH NanoPlant Growth Regulator™. The product primarily contains compounds like auxin, cytokinin, and gibberellin which are naturally occurring plant hormones that operate harmoniously as chemical messengers in stimulating root and shoot formation, stem elongation, seed germination, flowering initiation, and the overall developmental processes of plants. Nanoencapsulation, commonly adapted for medical and pharmaceutical applications, was the technological principle utilized in packaging the said plant growth regulators derived from locally isolated plant growth-promoting bacteria. BIOTECH NanoPlant Growth Regulator™ tested effective in several agriculturally important crops including cassava, coffee, jambalaya, banana, and other ornamental plants through its enhanced delivery in terms of higher solubility and dispersion, leading to improved nutrient absorption and uptake ratio in plants (Palma and Almeda 2018). Aside from its promising impact on crop production, it also addresses the current obstacle in environmental pollution as a consequence of inappropriate and excessive use of synthetic and toxic chemicals as fertilizers. Additionally, in an effort to widen the scope of UPLB BIOTECH's involvement in Philippine nanotechnology, the institute has partnered with the National Tuberculosis Reference Laboratory of the Research Institute for Tropical Medicine (RITM-NTRL) and Alocilja Nano-Biosensors Lab of Michigan State University (MSU) in organizing the Applications of Nanotechnology in Tuberculosis Detection Workshop as a starting point in raising awareness and evaluating potential benefits of nanotechnology in rapid detection of, and drug delivery for, tuberculosis, (Enciso 2017). In 2018, UPLB-BIOTECH also collaborated with several UP colleges, in particular, the College of Arts and Sciences (CAS), in constructing a nanoscience-dedicated facility with fully equipped instrumentation and research laboratories.

Locally, the increasing government and private sector support for nanotechnology inclined studies have resulted in numerous remarkable nanomaterial-based products attending to some of the critical problems of the Philippines. For example, scientists of the Department of Science and Technology-Industrial Technology Development Institute (DOST-ITDI) have developed an organic packaging material consisting of clay and starch.

17.4 NANOTECHNOLOGY AND INTELLECTUAL PROPERTY RIGHTS

Access to the nanoscale was paved by the invention of the scanning tunneling electron microscope which earned Gerd Binnig and Heinrich Rohrer a Nobel Prize in the same year. Nanotechnology differs from other technology in that it is difficult to classify when compared to other technology. Moreover, patents involving nanotechnology are potentially a new territory in patent applications. For one, the technology involves the nanoscale, a previously unknown realm when the patent laws were formulated. Although the inventions in the nanoscale are treated similarly to conventional inventions, the issue lies with the intricacy, complexity, and the multidisciplinary nature of nanotech patents.

17.4.1 THE INCREASING TREND OF UPSTREAM RESEARCH PATENTS IN NANOTECHNOLOGY

Research in nanotechnology requires close collaboration with multitudes of seemingly unrelated fields such as engineering, biology, and computer science. Thus, the patent application for a nanotechnology invention most commonly involves a team of researchers coming from different fields of expertise. The researchers claiming a patent for an invention usually involves multiple components which may necessitate individual patents for each component. Given the substantial market of nanotechnology innovations in a wide field of applications, patents involving nanotechnology are highly coveted. However, since nanotechnology patents mostly involve a team of researchers from a particular institution, institutional or university patents are currently becoming a trend, especially in the United States. Moreover, since most universities are doing basic research in nanotechnology, there is also an increasing trend of filing for patents involving upstream research in nanotechnology. An increase in the patent application for upstream nanotech research poses a problem in the succeeding downstream nanotechnology research and future innovation endeavors. Access to upstream research, after all, is critical in research and development efforts. As a result of this trend, the university will, therefore, have a prerogative in giving licenses to their technology which could, for instance, potentially result in the monopoly of a particular technology, driving the cost upward for the consumers.

17.4.2 The Multidisciplinary Nature and Multi-Industry Application of Nanotechnology Patents

Patents in nanotechnology are increasing rapidly in the United States, from 125 patents involving nanotechnology in 1985, it rose to around 6,000 in 2003. As nanotechnology is a relatively new field, patent evaluators have an understandable lack of expertise in the field. Most commonly, patent applications are centralized based on the field of discipline or its industrial application with a team of patent evaluator experts assigned in that particular field, but due to the multidisciplinary nature and multi-industrial application of nanotechnology innovations, classification of a particular patent claim becomes a complicated task. Thus, an instance of duplicate patent claims using the same technology in two completely different fields of industrial application is a possibility. Also, due to the wide range of applications of a seemingly basic nanotechnology patent, a small number of patent holders will own rights to other industries as well. For instance, the nanomaterial carbon nanotubes can be used in the electronic, energy, material, and biological industries. The multidisciplinary and wide range of potential applications in multiple industries of nanotechnology innovation is also a hindrance in patent claims in courts and patent offices since prior art will not be exclusive to one particular industry, it will be scattered in a sea of filed patents and prior arts in several industries. A patentee will, therefore, have the burden of additional cost due to the necessity of multiple patent evaluators specializing in their respective fields. Moreover, since nanotechnology is a relatively new field, standardized terms for nanotechnology-related terminology also poses a problem. Prior art search, therefore, is more complicated, and could potentially invalidate a particular nanotechnology patent claim on grounds of prior art if caution is not observed. This absence of standardized terms also contributes to overlaps and broad coverage of a particular patent.

17.4.3 Complexity in Classifying Nanotechnology Inventions

Under Philippine law (Section 22.2 of the Intellectual Property Code), inventions that can be patented are "any technical solution of a problem in any field of human activity which is new, involves an inventive step and industrially applicable". Moreover, the right to a patent belongs to the inventor, or if more than one person made the invention, the right would jointly belong to them. To satisfy the above requirements of the Philippine Intellectual Property Code, three criteria should be met: 1) The invention should be new and to be considered as new, the invention should not form part of prior art; 2) The invention must involve an 'inventive step' such that it should not be obvious to a person skilled in the art at the time of patent application and; 3) The invention must be industrially applicable. Most intellectual property (IP) laws are predicated on the preceding criteria, and most patents on nanotechnology application fail to satisfy the 'novelty' and 'not obvious to a person skilled in the art'. For instance, most patents in nanotechnology particularly in pharmaceutical and agricultural application would increase the availability of non-soluble active compounds by linking it to nanoparticules; essentially, they made the active compound smaller for better absorption. However, the U.S. courts, the most liberal in their interpretation of IP laws, ruled against the patent of a nanotechnology with such a feature, citing that it failed to satisfy the 'novelty' criterion since the only difference was in size. Moreover, changing the size is a modification 'obvious' to a person skilled in the art.

17.5 NANOTECHNOLOGY PATENTS IN THE PHILIPPINES

With more countries investing in nanotechnology research, the number of applications for nanotech patents has been increasing exponentially worldwide. The website, StatNano (https://statnano.com/), provides the current global status of nanotechnology research such as statistics from the Orbit database on the number of nanotech patent applications published and granted in the two major patent offices in the world: United States Patent and Trademarks Office (USPTO) and European Patent Office (EPO). According to the site, 139,444 nanotech patent applications have been filed in the USPTO while 19,740 in the EPO. Of these, 85,993 and 19,685 nanotech patents have already been granted by USPTO and EPO, respectively, as of March 2019. Among the countries, the United States has consistently led the nanotech patent race. At present, the USPTO has 74,432 nanotech patent applications and 47,836 nanotech granted patents. More than half of the filed applications for nanotech patents in USPTO are from the United States. This is six times higher than South Korea, the country with the second highest number of nanotech patent applications. In the EPO, the United States has 3,912 published nanotech applications and 5,817 nanotech granted patents. This is not surprising since the leading universities and companies in the field of nanotechnology are in the United States.

The Philippines has a total of 17 nanotech patent applications in the USPTO. From 2001 to 2014, the increase in the number of applications has been really slow with only seven nanotech patent applications over a decade. Realizing the importance and potential of nanotechnology, the Philippine government has started funding nanotechnology projects over the last five years. The recent extensive promotion of nanotechnology in the country resulted in an additional 10 nanotech patent applications filed from 2015 to 2018. To date, however, there are still no nanotech granted patents given to the Philippines both from USPTO and EPO. Thus, IP issues regarding nanotechnology in the Philippines are not as pronounced as those of the United States, but due to the steadfast support of the Philippine government in advancing nanotechnology research in the Philippines, patent applications and awards are expected to increase in the coming years.

REFERENCES

Asian and Pacific Centre for Transfer of Technology (APCTT). 2010. Innovation in nanotechnology: An Asia-pacific perspective. Retrieved from: http://apctt.org/nis/sites/all/themes/nis/pdf/Nanotech_Report_Fin al.pdf.

Clunan, A. L., and Hsueh, R. (2014). Nanotechnology in a globalized world: Strategic assessments of an emerging technology. Project report of U.S. Naval Postgraduate School (NPS).

Enciso, A. 2017. NTRL pilots nanotechnology application in detecting tuberculosis. Retrieved from: http://rit m.gov.ph/ntrl-pilots-nanotechnology-application-detecting-tuberculosis/.

Hulla, J., Sahu, S., and Hayes, A. 2015. Nanotechnology: History and future. *Human & Experimental Toxicology*, 34(12), 1318–1321. doi:10.1177/0960327115603588

Jeevanandam, J., Barhoum, A., Chan, Y. S., Dufresne, A., and Danquah, M. K. 2018. Review on nanoparticles and nanostructured materials: History, sources, toxicity and regulations. *Beilstein Journal of Nanotechnology*, 9, 1050–1074. doi:10.3762/bjnano.9.98

Jia, L. 2005. Global governmental investment in nanotechnologies. *Current Nanoscience*, 1(3), 263–266.

Khezri, S., Moghaddaskia, E., Mehdi Seyedsaleh, M., Abedinzadeh, S., and Dastras, M. 2016. Application of nanotechnology in food industry and related health concern challenges. *International Journal of Advanced Biotechnology & Research (IJBR)*, 7, 1370–1382.

Palma, A., and Almeda, R. 2018. Scaling things up: 8 "Small" UPLB technologies that can impact big on people's lives. Retrieved from: https://ovcre.uplb.edu.ph/press/features/item/502-scaling-things-up-8-small-uplb-technologies-that-can-impact-big-on-people-s-lives.

Santamaria, A. 2012. Historical overview of nanotechnology and nanotoxicology. *Nanotoxicity*, 1–12. doi:10.1007/978-1-62703-002-1_1

Singh, T., Shukla, S., Kumar, P., Wahla, V., and Bajpai, V. K. 2017. Application of nanotechnology in food science: Perception and overview. *Frontiers in Microbiology*, 8, 1501. doi:10.3389/fmicb.2017.01501

STATNANO: Nano science, technology and industry information. 2019. Retrieved from: www.statnano.com. Accessed. August 2019.

Taniguchi, N. 1974. On the basic concept of nano-technology. In *Proceedings of the International Conference of Production Engineering*. Japan Society of Precision Engineering, Tokyo.

Ventola, C. L. 2012. The nanomedicine revolution: Part 1: Emerging concepts. *P & T: A Peer-Reviewed Journal for Formulary Management*, 37(9), 512–525.

18 Measurement Techniques for Thermal Conductivity of Nanofluids

Kelvii Wei GUO

CONTENTS

18.1 INTRODUCTION

For fluid heating and cooling, one of the extremely crucial aspects of various applications is heat transfer, for instance in microelectronics, transportation, power generation, manufacturing, chemical processes, etc. It is well known that a tiny enhancement in the practical applications will increase equipment life, save energy, and reduce processing time.

To date, for the requirements of current industry, applications such as microprocessors need to be more powerful with a smaller size, and the urgent need to reduce the vehicle weight requires heat transfer with high efficiency; it demands heat flow with good properties to adequately meet the practical requirements.

It is well known that the present traditional fluids (water, oil, kerosene, ethylene glycol (EG), etc.) are widely adopted in various fields. However, because of the low thermal conductivity of these kinds of fluids, the efficiency of the heat transfer is usually too low to apply in devices with high heat flux, such as electronics cooling, solar thermal collectors, and material processing. In order to improve the thermal conductivity of traditional fluids, nanomaterial technology, which is one of the state-of-art technologies, is used. It should be noted that due to the attractive characteristics of a large surface area and quantum effects of the nanomaterials, they possess unique mechanical, electrical, physical, chemical, magnetic, and optical properties.

Nanofluids are the mixtures of solid and liquid along with suspended metallic or nonmetallic nanoparticles to exert a beneficial influence on the heat transfer performance of the fluid. Exploration of the deep understanding of the flow of nanofluids and the characteristics of heat transfer, together with the underlying mechanisms for the practical applications of nanofluids, is urgently required. To date, investigation of nanofluids has been exploited in both experimental and theoretical aspects.

Owing to nanoparticles being small in size with a large surface area, nanofluids definitely possess higher energy savings, longer stability, better thermal properties, lower pumping power, less clogging in flow passages and greater reduction in friction, and are more homogeneous than base fluids without nanoparticles. Nowadays various kinds of nanofluids have been exploited in a wide range of practical applications [1–4]. For example, in order to enhance the thermal conductivity of traditional fluids, the nano solid particles (metallic and non-metallic particles of nano size (>100 nm)) with superior thermal properties are often dispersed into the fluid.

Currently, many kinds of nanofluids have been researched and can be categorized approximately in two classes: (1) basic metals e.g. silver (Ag), aluminum (Al), copper (Cu), nickel (Ni), etc., (2) metal oxides, aluminum oxide (Al_2O_3), cerium oxide (CeO_2), copper oxide (CuO), iron oxide (Fe_3O_4), silicon dioxide (SiO_2), silicon carbide (SiC), titanium oxide (TiO_2), zinc oxide (ZnO), zirconium dioxide (ZrO_2), etc. [5–32].

However, all results obtained relating to nanofluids are still deficient and contradictory. Therefore, accurate measurement techniques are urgently needed to implement or support the theory of nanofluids while fully understanding the effect and mechanism of various parameters.

18.2 PREPARATION OF NANOFLUIDS

It is especially significant to prepare a stable suspension of nanofluids before they are successfully applied in the practical applications. It is known that good preparation is essential to achieve better performance and results in terms of higher efficiency, smaller size, lower cost, and better safety of manufactured devices.

At present, nanofluids are prepared by two methods. One is the single step method; another is the two-step method. Nanoparticles are synthesized and dispersed at the same time into the base fluid in the single step method. Due to their characteristics, nanofluids made by the single step method are suitable for the high thermal conductivity of metal nanoparticles to prevent oxidation. The drawbacks of this method include the requirement of a vacuum, small scale production, and low rate. As a result, due to the above-mentioned factors, the technique related to the single step method is complicated and expensive. For the two-step method, nanofluids are prepared in two stages. Nanoparticles are primarily fabricated in a powder form. After that, nanoparticles are dispersed into the base liquid to form a stable solution. In this method, the key processing parameters influencing the properties are the size, shape, concentration of the particles, temperature, pH, surfactants, the material properties, various base fluids, nanoparticles types, sonication, solid volume fraction, and various mechanisms of thermal conductivity enhancement. Investigation shows that nanofluid thermal conductivity increases with the increment of concentration and temperature. Moreover, different surfactants employed for the dispersion of the particles and pH also influence the thermal conductivity and thermal stability of the fabricated nanofluid [33–55].

18.3 MEASUREMENT TECHNIQUES FOR THERMAL CONDUCTIVITY

As the nanoparticles' outstanding thermos-physical properties resulted in no or low penalty in pressure drop, along with significantly enhancing the fabricated nanofluids' anomalous thermal conductivity, nanofluids have attracted more and more research focus. Therefore, nanofluid preparation with longer stability is imperative to meet the requirements of more efficient application in the field of heat transfer with better sustainability. Currently, many contradictory results in the literature show the influence of effective parameters on thermophysical properties due to the above-mentioned parameters. Consequently, accurate measurement techniques are essential to implement or support the theory of nanofluids while fully understanding the effect and mechanism of various parameters [56–62].

As is generally known, the thermal conductivity enhancement of nanofluid is extremely important, due to its efficiently influencing the convective heat transfer of nanofluids, even with nanoparticles on a small scale. Hence, detailed exploration of the measurement techniques of thermal conductivity of nanofluids is imminently required [63–66].

Until now, various techniques have been used for measuring the thermal conductivity of nanofluids. The kinds of methods explored for determining the thermal properties of nanofluids in the laboratory are the hotwire method, thermal constants analyzer, steady-state parallel-plate, and 3ω method.

18.3.1 Transient Hotwire Method

In 1962 Horrocks and McLaughlin [67] proposed a transient hotwire method to determine thermal conductivity with absolute accuracy. The schematic of the transient hotwire apparatus is shown in Figure 18.1, where a thin, long platinum wire is taken as a dual line heat source and the temperature sensor. The heating wire is adopted as a resistance thermometer for measuring temperature instantaneously. By the electrically heated wire immersed axially into the liquid through a cylindrical cell keeping the thermal equilibrium at the first stage, the temperature can be observed successfully. The thermal conductivity of nanofluids is heated by resistive heating with the electric power increased by the preset step and measured by the immersed sensor, where heat from the sensor will be transferred to the surrounding fluid through conduction. The transient hotwire method shows a good solution to nanofluid thermal transport characterization [68–80].

Recently, Pryazhnikov et al. [81] used the transient hotwire method, systematically measuring the thermal conductivity coefficient of nanofluids at room temperature. In the research, more than 50 different nanofluids were adopted based on water, engine oil and ethylene glycol (EG) containing particles of Al_2O_3, CuO, SiO_2, TiO_2, ZrO_2 and diamond, where the volume concentration of the nanoparticles ranged from 0.25 to 8% and the size of the particles ranged from 10 to 150 nm. Results show that the thermal conductivity coefficient of the nanofluids not only relied on the concentration of the particles, but also on the size of the particles, the property of the materials, and the kind of base fluid. It indicates that the nanofluid's thermal conductivity coefficient is improved with the increment of the size of nanoparticles, and the direct correlation between the thermal conductivity of the nanoparticle material and the thermal conductivity of nanofluid containing these particles does not exist convincingly. Results also show that the base liquid plays a crucial role in the thermal conductivity of the nanofluids. The interesting result is that when the thermal conductivity of the base fluid is lower the relative thermal conductivity coefficient of the nanofluids is higher.

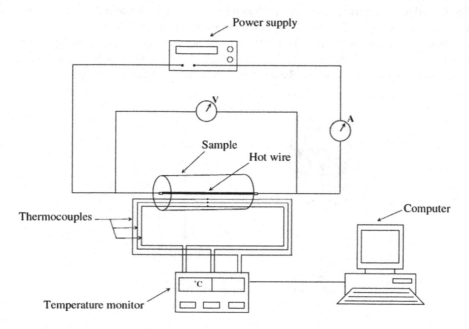

FIGURE 18.1   Schematic of the transient hotwire apparatus.

18.3.2 Thermal Constants Analyzer Technique

The thermal constants analyzer technique adopts the theory of the transient plane source (TPS) to obtain nanofluid thermal conductivity [82]. The transient plane source in this method is taken as both the heat source and temperature. From the schematic of the experimental setup of measuring nanofluid thermal conductivity, as shown in Figure 18.2, one can see that the thermal constants analyzer technique is easily controllable with a quick testing procedure and wide testing ranges, and the range of the nanofluid thermal conductivity is from 0.02 to 200 W/m-K. Additionally, it does not need sample preparation, and the size of the sample is very flexible. Generally, the thermal constants analyzer technique consists of a thermometer, a thermal constant analyzer, a constant temperature bath, and a vessel, where the testing probe is vertically immersed in the nanofluid vessel with constant temperature by keeping it in the heating bath. Meanwhile, the temperature sensor is immersed in the nanofluid vessel to measure the temperature. After that, the nanofluid's thermal conductivity can be gained by measuring the resistance of the probe [83–91].

18.3.3 Steady-State Parallel-Plate Method

Challoner and Powell [92] explored the steady-state parallel-plate method in 1956. The schematic of the experimental setup for the steady-state parallel-plate method is shown in Figure 18.3, where the testing sample is set in between two round parallel Cu plates, and the sample temperature is measured by each thermocouple. The thermal conductivity of nanofluids passing through two Cu plates, considering the effect of the spacer of the glass, is calculated by the 1D heat conduction equation of the main heater power because of the total heat supplied by the main heater across the liquid between the upper plate and the lower Cu plate. Furthermore, for preventing heat loss from the fluid to the surroundings, the guard heaters are used to keep the nanofluids at a constant temperature [93–104].

18.3.4 3ω Method

The 3ω method is also exploited as an ideal alternative to the thermal transport characterization for nanofluids, and its schematic of the setup is shown in Figure 18.4. The definite attractive advantage of the 3ω method is that both the thermal conductivity and thermal diffusivity can be simultaneously achieved from a single measurement. Because the hotwire and hotstrip techniques are tightly connected, the 3ω method utilizes a radial flow of heat from a single element taken both as the heater and thermometer [105]. In the 3ω method, a heat wave of frequency 2ω will be generated

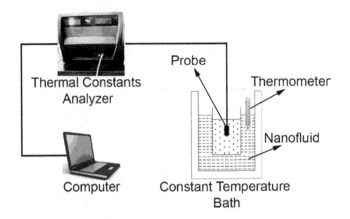

FIGURE 18.2 Schematic of the experimental setup for transient plate source method.

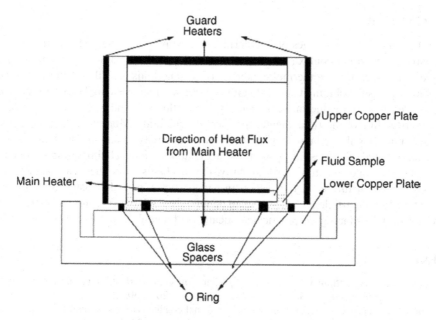

FIGURE 18.3   Schematic of experimental setup for steady-state parallel-plate method.

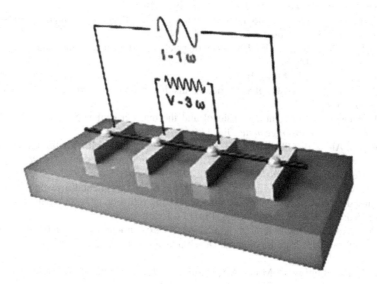

FIGURE 18.4   Schematic of the setup for 3ω method.

by passing a sinusoidal current of frequency ω through the metal wire and will be reduced to the frequency 3ω by the voltage component. The main captivating characterization of this method is the usage of the frequency dependent on the temperature oscillation substituted for the time-domain response. In the experimental setup for the 3ω method, thermocouples, test fluid, Peltier element, etc. are needed.

Compared with the traditional measurement methods of thermal conductivity, accurate measurements for homogeneous mixing can be achieved for the stationary nanofluid with only a small sample volume by the 3ω method. Moreover, the insight into the thermophysics of the aggregation process in the nanofluids can be also achieved by the spatial variation of the thermal conductivity measurement [106–114].

18.4 CONCLUSION

Owing to the unique characteristics of the thermal conductivity of nanofluids, it attracts more and more research focus in the contexts of various applications related to heat transfer. The methods presently taken to measure the thermal conductivity of nanofluids are all still in the experimental stage because the results obtained are still controversial and inconsistent. Therefore, the measurement techniques used for the thermal conductivity of nanofluids should be improved on the basis of the full understanding of the fundamental theories of nanofluids. Moreover, the new methods for significantly enhancing the reliability of the thermal conductivity of nanofluids should be explored with the experimental advances, along with the accurate atomic-level simulation development on nanoscale materials and structures. Additionally, the mechanisms of thermal conductivity (such as Brownian motion, nature of heat transport in nanoparticles, particle clustering, liquid layer in/nanolayer at the liquid/particle interface, etc.) in nanofluids need to be urgently investigated further to remove the obstacles facing the practical applications of nanofluids.

REFERENCES

1. Devendiran DK, Amirtham VA. A review on preparation, characterization, properties and applications of nanofluids. *Renewable and Sustainable Energy Reviews* 2016; 60: 21–40.
2. Tawfik MM. Experimental studies of nanofluid thermal conductivity enhancement and applications: A review. *Renewable and Sustainable Energy Reviews* 2017; 75: 1239–1253.
3. Bellos E, Said Z, Tzivanidis C. The use of nanofluids in solar concentrating technologies: A comprehensive review. *Journal of Cleaner Production* 2018; 196: 84–99.
4. Liang GT, Mudawar I. Review of pool boiling enhancement with additives and nanofluids. *International Journal of Heat and Mass Transfer* 2018; 124: 423–453.
5. Sundar LS, Singh MK, Sousa ACM. Enhanced thermal properties of nanodiamond nanofluids. *Chemical Physics Letters* 2016; 644: 99–110.
6. Sharifpur M, Tshimanga N, Meyer JP, Manca O. Experimental investigation and model development for thermal conductivity of α-Al_2O_3-glycerol nanofluids. *International Communications in Heat and Mass Transfer* 2017; 85: 12–22.
7. Wang RT, Wang JC. Intelligent dimensional and thermal performance analysis of Al_2O_3 nanofluid. *Energy Conversion and Management* 2017; 138: 686–697.
8. Alkasmoul FS, Al-Asadi MT, Myers TG, Thompson HM, Wilson MCT. A practical evaluation of the performance of Al_2O_3-water, TiO_2-water and CuO-water nanofluids for convective cooling. *International Journal of Heat and Mass Transfer Part A* 2018; 126: 639–651.
9. Hamid KA, Azmi WH, Nabil MF, Mamat R, Sharma KV. Experimental investigation of thermal conductivity and dynamic viscosity on nanoparticle mixture ratios of TiO_2-SiO_2 nanofluids. *International Journal of Heat and Mass Transfer* 2018; 116: 1143–1152.
10. Keyvani M, Afrand M, Toghraie D, Reiszadeh M. An experimental study on the thermal conductivity of cerium oxide/ethylene glycol nanofluid: Developing a new correlation. *Journal of Molecular Liquids* 2018; 266: 211–217.
11. Moldoveanu GM, Huminic G, Minea AA, Huminic A. Experimental study on thermal conductivity of stabilized Al_2O_3 and SiO_2 nanofluids and their hybrid. *International Journal of Heat and Mass Transfer Part A* 2018; 127: 450–457.
12. Sati P, Shende RC, Ramaprabhu S. An experimental study on thermal conductivity enhancement of DI water-EG based ZnO(CuO)/graphene wrapped carbon nanotubes nanofluids. *Thermochimica Acta* 2018; 666: 75–81.
13. Iacobazzi F, Milanese M, Colangelo G, Lomascolo M, de Risi A. An explanation of the Al_2O_3 nanofluid thermal conductivity based on the phonon theory of liquid. *Energy* 2016; 116(1): 786–794.
14. Charab AA, Movahedirad S, Norouzbeigi R. Thermal conductivity of Al_2O_3+TiO_2/water nanofluid: Model development and experimental validation. *Applied Thermal Engineering* 2017; 119: 42–51.
15. Agarwal R, Verma K, Agrawal NK, Singh R. Sensitivity of thermal conductivity for Al_2O_3 nanofluids. *Experimental Thermal and Fluid Science* 2017; 80: 19–26.
16. Satti JR, Das DK, Ray D. Investigation of the thermal conductivity of propylene glycol nanofluids and comparison with correlations. *International Journal of Heat and Mass Transfer* 2017; 107: 871–881.

17. Serebryakova MA, Zaikovskii AV, Sakhapov SZ, Smovzh DV, Sukhinin GI, Novopashin SA. Thermal conductivity of nanofluids based on hollow γ-Al_2O_3 nanoparticles, and the influence of interfacial thermal resistance. *International Journal of Heat and Mass Transfer* 2017; 108, Part B: 1314–1319.

18. Heyhat MM, Irannezhad A. Experimental investigation on the competition between enhancement of electrical and thermal conductivities in water-based nanofluids. *Journal of Molecular Liquids* 2018; 268: 169–175.

19. Jain S, Borah R, Saikia N, Gogoi S, Trinavee K. The effect of stability on the thermal conductivity of copper nanofluid at low particle loadings. *Materials Today: Proceedings* 2018; 5(9), Part 3: 20652–20659.

20. Akhgar A, Toghraie D. An experimental study on the stability and thermal conductivity of water-ethylene glycol/TiO_2-MWCNTs hybrid nanofluid: Developing a new correlation. *Powder Technology* 2018; 338: 806–818.

21. Dalkılıç Ahmet Selim, Yalçın Gökberk, Küçükyıldırım Bedri Onur, Öztuna Semiha, Eker Ayşegül Akdoğan, Jumpholkul Chaiwat, Nakkaew Santiphap, Wongwises Somchai. Experimental study on the thermal conductivity of water-based CNT-SiO_2 hybrid nanofluids. *International Communications in Heat and Mass Transfer* 2018; 99: 18–25.

22. Dehkordi BAF, Abdollahi A. Experimental investigation toward obtaining the effect of interfacial solid-liquid interaction and basefluid type on the thermal conductivity of CuO-loaded nanofluids. *International Communications in Heat and Mass Transfer* 2018; 97: 151–162.

23. Aparna Z, Michael M, Pabi SK, Ghosh S. Thermal conductivity of aqueous Al_2O_3/Ag hybrid nanofluid at different temperatures and volume concentrations: An experimental investigation and development of new correlation function. *Powder Technology* 2019; 343: 714–722.

24. Ranjbarzadeh R, Moradikazerouni A, Bakhtiari R, Asadi A, Afrand M. An experimental study on stability and thermal conductivity of water/silica nanofluid: Eco-friendly production of nanoparticles. *Journal of Cleaner Production* 2019; 206: 1089–1100.

25. Akilu S, Baheta AT, Kadirgama K, Padmanabhan E, Sharma KV. Viscosity, electrical and thermal conductivities of ethylene and propylene glycol-based β-SiC nanofluids. *Journal of Molecular Liquids* 2019; 284: 780–792.

26. de Oliveira LR, Ribeiro SRFL, Reis MHM, Cardoso VL, Filho EPB. Experimental study on the thermal conductivity and viscosity of ethylene glycol-based nanofluid containing diamond-silver hybrid material. *Diamond and Related Materials* 2019; 96: 216–230.

27. Kalantari A, Abbasi M, Hashim AM. Enhancement of thermal conductivity of size controlled silver nanofluid. *Materials Today: Proceedings* 2019; 7: 612–618.

28. Sedeh RN, Abdollahi A, Karimipour A. Experimental investigation toward obtaining nanoparticles' surficial interaction with basefluid components based on measuring thermal conductivity of nanofluids. *International Communications in Heat and Mass Transfer* 2019; 103: 72–82.

29. Safaei MR, Ranjbarzadeh R, Hajizadeh A, Bahiraei M, Afrand M, Karimipour A. Effects of cobalt ferrite coated with silica nanocomposite on the thermal conductivity of an antifreeze: New nanofluid for refrigeration condensers. *International Journal of Refrigeration* 2019; 102: 86–95.

30. Omrani AN, Esmaeilzadeh E, Jafari M, Behzadmehr A. Effects of multi walled carbon nanotubes shape and size on thermal conductivity and viscosity of nanofluids. *Diamond and Related Materials* 2019; 93: 96–104.

31. Esfe MH, Esfandeh S, Amiri MK, Afrand M. A novel applicable experimental study on the thermal behavior of SWCNTs (60%)-MgO (40%)/EG hybrid nanofluid by focusing on the thermal conductivity. *Powder Technology* 2019; 342: 998–1007.

32. Nikam AV, Dadwal AH. Scalable microwave-assisted continuous flow synthesis of CuO nanoparticles and their thermal conductivity applications as nanofluids. *Advanced Powder Technology* 2019; 30(1): 13–17.

33. Goudarzi K, Nejati F, Shojaeizadeh E, Yousef-abad SKA. Experimental study on the effect of pH variation of nanofluids on the thermal efficiency of a solar collector with helical tube. *Experimental Thermal and Fluid Science* 2015; 60: 20–27.

34. Zarringhalam M, Karimipour A, Toghraie D. Experimental study of the effect of solid volume fraction and Reynolds number on heat transfer coefficient and pressure drop of CuO-Water nanofluid. *Experimental Thermal and Fluid Science* 2016; 76: 342–351.

35. Al-Balushi LM, Rahman MM, Pop I. Three-dimensional axisymmetric stagnation-point flow and heat transfer in a nanofluid with anisotropic slip over a striated surface in the presence of various thermal conditions and nanoparticle volume fractions. *Thermal Science and Engineering Progress* 2017; 2: 26–42.

36. Alipour H, Karimipour A, Safaei MR, Semiromi DT, Akbari OA. Influence of T-semi attached rib on turbulent flow and heat transfer parameters of a silver-water nanofluid with different volume fractions in a three-dimensional trapezoidal microchannel. *Physica E: Low-Dimensional Systems and Nanostructures* 2017; 88: 60–76.

37. Gupta M, Singh V, Kumar R, Said Z. A review on thermophysical properties of nanofluids and heat transfer applications. *Renewable and Sustainable Energy Reviews* 2017; 74: 638–670.

38. Maheshwary PB, Handa CC, Nemade KR. A comprehensive study of effect of concentration, particle size and particle shape on thermal conductivity of titania/water based nanofluid. *Applied Thermal Engineering* 2017; 119: 79–88.

39. Sasmal C. Effects of axis ratio, nanoparticle volume fraction and its size on the momentum and heat transfer phenomena from an elliptic cylinder in water-based CuO nanofluids. *Powder Technology* 2017; 313: 272–286.

40. Mikkola V, Puupponen S, Granbohm H, Saari K, Ala-Nissila T, Seppälä A. Influence of particle properties on convective heat transfer of nanofluids. *International Journal of Thermal Sciences* 2018; 124: 187–195.

41. Ambreen T, Kim MH. Heat transfer and pressure drop correlations of nanofluids: A state of art review. *Renewable and Sustainable Energy Reviews* 2018; 91: 564–583.

42. Buschmann MH, Azizian R, Kempe T, Juliá JE, Martínez-Cuenca R, Sundén B, Wu Z, Seppälä A, Ala-Nissila T. Correct interpretation of nanofluid convective heat transfer. *International Journal of Thermal Sciences* 2018; 129: 504–531.

43. Kakavandi A, Akbari M. Experimental investigation of thermal conductivity of nanofluids containing of hybrid nanoparticles suspended in binary base fluids and propose a new correlation. *International Journal of Heat and Mass Transfer* 2018; 124: 742–751.

44. Gangadevi R, Vinayagam BK, Senthilraja S. Effects of sonication time and temperature on thermal conductivity of CuO/water and Al$_2$O$_3$/water nanofluids with and without surfactant. *Materials Today: Proceedings Part 3* 2018; 5(2): 9004–9011.

45. Alirezaie A, Hajmohammad MH, Alipour A, Salari M. Do nanofluids affect the future of heat transfer? "A benchmark study on the efficiency of nanofluids". *Energy* 2018; 157: 979–989.

46. Agarwal R, Verma K, Agrawal NK, Duchaniya RK, Singh R. Synthesis, characterization, thermal conductivity and sensitivity of CuO nanofluids. *Applied Thermal Engineering* 2016; 102: 1024–1036.

47. Babita Sharma SK, Gupta SM. Preparation and evaluation of stable nanofluids for heat transfer application: A review. *Experimental Thermal and Fluid Science* 2016; 79: 202–212.

48. Yazid MNAWM, Sidik NAC, Mamat R, Najafi G. A review of the impact of preparation on stability of carbon nanotube nanofluids. *International Communications in Heat and Mass Transfer* 2016; 78: 253–263.

49. Fuskele V, Sarviya RM. Recent developments in nanoparticles synthesis, preparation and stability of nanofluids. *Materials Today: Proceedings* 2017; 4(2), Part A: 4049–4060.

50. Kumar DD, Arasu AV. A comprehensive review of preparation, characterization, properties and stability of hybrid nanofluids. *Renewable and Sustainable Energy Reviews* 2018; 81(2): 1669–1689.

51. Mirbagheri MH, Akbari M, Mehmandoust B. Proposing a new experimental correlation for thermal conductivity of nanofluids containing of functionalized multiwalled carbon nanotubes suspended in a binary base fluid. *International Communications in Heat and Mass Transfer* 2018; 98: 216–222.

52. Asadi Amin, Aberoumand Sadegh, Moradikazerouni Alireza, Pourfattah Farzad, Żyłae Gaweł, Estellé Patrice, Mahian Omid, Wongwises Somchai, Nguyen Hoang M, Arabkoohsar Ahmad. Recent advances in preparation methods and thermophysical properties of oil-based nanofluids: A state-of-the-art review. *Powder Technology* 2019; 352: 209–226.

53. Babar H, Ali HM. Towards hybrid nanofluids: Preparation, thermophysical properties, applications, and challenges. *Journal of Molecular Liquids* 2019; 281: 598–633.

54. Arshad A, Jabbal M, Yan YY, Reay D. A review on graphene based nanofluids: Preparation, characterization and applications. *Journal of Molecular Liquids* 2019; 279: 444–484.

55. Graves JE, Latvytė E, Greenwood A, Emekwuru NG. Ultrasonic preparation, stability and thermal conductivity of a capped copper-methanol nanofluid. *Ultrasonics Sonochemistry* 2019; 55: 25–31.

56. Das PK. A review based on the effect and mechanism of thermal conductivity of normal nanofluids and hybrid nanofluids. *Journal of Molecular Liquids* 2017; 240: 420–446.

57. Mohamad AA. Thermal contact theory for estimating the thermal conductivity of nanofluids and composite materials. *Applied Thermal Engineering* 2017; 120: 179–186.

58. Sarviya RM, Fuskele V. Review on thermal conductivity of nanofluids. *Materials Today: Proceedings* 2017; 4(2), Part A: 4022–4031.

59. Alirezaie A, Hajmohammad MH, Ahangar MRH, Esfe MH. Price-performance evaluation of thermal conductivity enhancement of nanofluids with different particle sizes. *Applied Thermal Engineering* 2018; 128: 373–380.

60. Ebrahimi R, de Faoite D, Finn DP, Stanton KT. Accurate measurement of nanofluid thermal conductivity by use of a polysaccharide stabilising agent. *International Journal of Heat and Mass Transfer* 2019; 136: 486–500.

61. Zendehboudi A, Saidur R, Mahbubul IM, Hosseini SH. Data-driven methods for estimating the effective thermal conductivity of nanofluids: A comprehensive review. *International Journal of Heat and Mass Transfer* 2019; 131: 1211–1231.

62. Sedaghat F, Yousefi F. Synthesizes, characterization, measurements and modeling thermal conductivity and viscosity of graphene quantum dots nanofluids. *Journal of Molecular Liquids* 2019; 278: 299–308.

63. Jabbari F, Rajabpour A, Saedodin S. Thermal conductivity and viscosity of nanofluids: A review of recent molecular dynamics studies. *Chemical Engineering Science* 2017; 174: 67–81.

64. Alawi OA, Sidik NAC, Xian HW, Kean TH, Kazi SN. Thermal conductivity and viscosity models of metallic oxides nanofluids. *International Journal of Heat and Mass Transfer* 2018; 116: 1314–1325.

65. Naddaf A, Heris SZ. Experimental study on thermal conductivity and electrical conductivity of diesel oil-based nanofluids of graphene nanoplatelets and carbon nanotubes. *International Communications in Heat and Mass Transfer* 2018; 95: 116–122.

66. Gupta Munish, Singh Vinay, Kumar Satish, Kumar Sandeep, Dilbaghi Neeraj, Said Zafar. Up to date review on the synthesis and thermophysical properties of hybrid nanofluids. *Journal of Cleaner Production* 2018; 190: 169–192.

67. Horrocks JK, McLaughlin E. Non-steady state measurements of thermal conductivities of liquids polyphenyls. *Proceedings of the Royal Society of London. Series A – Mathematical and Physical Sciences* 1963; 273: 259–274.

68. Hachey MA, Nguyen CT, Galanis N, Popa CV. Experimental investigation of Al_2O_3 nanofluids thermal properties and rheology – Effects of transient and steady-state heat exposure. *International Journal of Thermal Sciences* 2014; 76: 155–167.

69. Lee JY, Lee HS, Baik YJ, Koo JM. Quantitative analyses of factors affecting thermal conductivity of nanofluids using an improved transient hot-wire method apparatus. *International Journal of Heat and Mass Transfer* 2015; 89: 116–123.

70. Azarfar Sh, Movahedirad S, Sarbanha AA, Norouzbeigi R, Beigzadeh B. Low cost and new design of transient hot-wire technique for the thermal conductivity measurement of fluids. *Applied Thermal Engineering* 2016; 105: 142–150.

71. Wilk J, Smusz R, Grosicki S. Thermophysical properties of water-based Cu nanofluid used in special type of coil heat exchanger. *Applied Thermal Engineering* 2017; 127: 933–943.

72. Ilyas SU, Pendyala R, Narahari M, Lim SS. Stability, rheology and thermal analysis of functionalized alumina- thermal oil-based nanofluids for advanced cooling systems. *Energy Conversion and Management* 2017; 142: 215–229.

73. Aparna Z, Michael MM, Pabi SK, Ghosh S. Diversity in thermal conductivity of aqueous Al_2O_3– and Ag-nanofluids measured by transient hot-wire and laser flash methods. *Experimental Thermal and Fluid Science* 2018; 94: 231–245.

74. Guo WW, Li GN, Zheng YQ, Dong C. Measurement of the thermal conductivity of SiO2 nanofluids with an optimized transient hot wire method. *Thermochimica Acta* 2018; 661: 84–97.

75. Akyürek EF, Geliş K, Şahin B, Manay E. Experimental analysis for heat transfer of nanofluid with wire coil turbulators in a concentric tube heat exchanger. *Results in Physics* 2018; 9: 376–389.

76. Yavari M, Mansourpour Z, Shariaty-Niassar M. Controlled assembly and alignment of CNTs in ferrofluid: Application in tunable heat transfer. *Journal of Magnetism and Magnetic Materials* 2019; 479: 170–178.

77. Prajapati H, Deshmukh NN. Design and development of thin wire sensor for transient temperature measurement. *Measurement* 2019; 140: 582–589.

78. Hapenciuc CL, Negut I, Borca-Tasciuc T, Mihailescu IN. A steady-state hot-wire method for thermal conductivity measurements of fluids. *International Journal of Heat and Mass Transfer* 2019; 134: 993–1002.

79. Palacios A, Cong L, Navarro ME, Ding YL, Barreneche C. Thermal conductivity measurement techniques for characterizing thermal energy storage materials – A review. *Renewable and Sustainable Energy Reviews* 2019; 108: 32–52.

80. Daouas N, Fguiri A, Radhouani MS. Solution of a coupled inverse heat conduction–radiation problem for the study of radiation effects on the transient hot wire measurements. *Experimental Thermal and Fluid Science* 2008; 32(8): 1766–1778.

81. Pryazhnikov MI, Minakov AV, Rudyak VYa, Guzei DV. Thermal conductivity measurements of nanofluids. *International Journal of Heat and Mass Transfer* 2017; 104: 1275–1282.

82. Paul G, Chopkar M, Manna I, Das PK. Techniques for measuring the thermal conductivity of nanofluids: A review. *Renewable and Sustainable Energy Reviews* 2010; 14(7): 1913–1924.

83. El-Brolossy TA, Saber O. Non-intrusive method for thermal properties measurement of nanofluids. *Experimental Thermal and Fluid Science* 2013; 44: 498–503.

84. Alberola JA, Mondragón R, Juliá JE, Hernández L, Cabedo L. Characterization of halloysite-water nanofluid for heat transfer applications. *Applied Clay Science* 2014; 99: 54–61.

85. Warzoha RJ, Fleischer AS. Determining the thermal conductivity of liquids using the transient hot disk method. Part II: Establishing an accurate and repeatable experimental methodology. *International Journal of Heat and Mass Transfer* 2014; 71: 790–807.

86. Angayarkanni SA, Philip J. Review on thermal properties of nanofluids: Recent developments. *Advances in Colloid and Interface Science* 2015; 225: 146–176.

87. Haghighi EB, Utomo AT, Ghanbarpour M, Zavareh AIT, Nowak E, Khodabandeh R, Pacek AW, Palm B. Combined effect of physical properties and convective heat transfer coefficient of nanofluids on their cooling efficiency. *International Communications in Heat and Mass Transfer* 2015; 68: 32–42.

88. Agarwal DK, Vaidyanathan A, Kumar SS. Experimental investigation on thermal performance of kerosene–graphene nanofluid. *Experimental Thermal and Fluid Science* 2016; 71: 126–137.

89. Bahiraei M. Particle migration in nanofluids: A critical review. *International Journal of Thermal Sciences* 2016; 109: 90–113.

90. Verma SK, Tiwari AK, Chauhan DS. Experimental evaluation of flat plate solar collector using nanofluids. *Energy Conversion and Management* 2017; 134: 103–115.

91. Mashali F, Languri EM, Davidson J, Kerns D, Johnson W, Nawaz K, Cunningham G. Thermo-physical properties of diamond nanofluids: A review. *International Journal of Heat and Mass Transfer* 2019; 129: 1123–1135.

92. Challoner AR, Powell RW. Thermal conductivity of liquids: New determinations for seven liquids and appraisal of existing values. *Proceedings of the Royal Society of London. Series A – Mathematical and Physical Sciences* 1956; 238: 90–106.

93. Singh SK, Mishra M, Jha PK. Nonuniformities in compact heat exchangers—Scope for better energy utilization: A review. *Renewable and Sustainable Energy Reviews* 2014; 40: 583–596.

94. Agostino FJ, Krylov SN. Advances in steady-state continuous-flow purification by small-scale free-flow electrophoresis. *TrAC Trends in Analytical Chemistry* 2015; 72: 68–79.

95. Cerón JF, Pérez-García J, Solano JP, García A, Herrero-Martín R. A coupled numerical model for tube-on-sheet flat-plate solar liquid collectors. Analysis and validation of the heat transfer mechanisms. *Applied Energy* 2015; 140: 275–287.

96. Tahseen TA, Ishak M, Rahman MM. An overview on thermal and fluid flow characteristics in a plain plate finned and un-finned tube banks heat exchanger. *Renewable and Sustainable Energy Reviews* 2015; 43: 363–380.

97. Ahmed S, Zueco J, López-González LM. Numerical and analytical solutions for magneto-hydrodynamic 3D flow through two parallel porous plates. *International Journal of Heat and Mass Transfer Part A* 2017; 108: 322–331.

98. Benenti G, Casati G, Saito K, Whitney RS. Fundamental aspects of steady-state conversion of heat to work at the nanoscale. *Physics Reports* 2017; 694: 1–124.

99. Izurieta EM, Borio DO, Pedernera MN, López E. Parallel plates reactor simulation: Ethanol steam reforming thermally coupled with ethanol combustion. *International Journal of Hydrogen Energy* 2017; 42(30): 18794–18804.

100. Furukawa M. Mathematical model of parallel-plate-channeled electromagnetic-driven pulsating dream pipe for rapid heat removal. *International Journal of Heat and Mass Transfer* 2017; 104: 1048–1059.

101. Quintero AE, Vera M. Laminar counterflow parallel-plate heat exchangers: An exact solution including axial and transverse wall conduction effects. *International Journal of Heat and Mass Transfer* 2017; 104: 1229–1245.

102. Ahmed HE, Salman BH, Kherbeet ASh, Ahmed MI. Optimization of thermal design of heat sinks: A review. *International Journal of Heat and Mass Transfer* 2018; 118: 129–153.

103. Joshi SS, Dhoble AS. Photovoltaic -Thermal systems (PVT): Technology review and future trends. *Renewable and Sustainable Energy Reviews* 2018; 92: 848–882.

104. Gómez-Castro FM, Schneider D, Päßler T, Eicker U. Review of indirect and direct solar thermal regeneration for liquid desiccant systems. *Renewable and Sustainable Energy Reviews Part 1* 2018; 82: 545–575.

105. Oh DW, Jain A, Eaton JK, Goodson KE, Lee JS. Thermal conductivity measurement and sedimentation detection of aluminum oxide nanofluids by using the 3ω method. *International Journal of Heat and Fluid Flow* 2008; 29(5): 1456–1461.

106. Yiamsawasd T, Dalkilic AS, Wongwises S. Measurement of the thermal conductivity of titania and alumina nanofluids. *Thermochimica Acta* 2012; 545: 48–56.

107. Ding TT, Jannot Y, Degiovanni A. Theoretical study of the limits of the 3ω method using a new complete quadrupole model. *International Journal of Thermal Sciences* 2014; 86: 314–324.

108. Akilu S, Sharma KV, Baheta AT, Mamat R. A review of thermophysical properties of water based composite nanofluids. *Renewable and Sustainable Energy Reviews* 2016; 66: 654–678.

109. Eggers JR, Kabelac S. Nanofluids revisited. *Applied Thermal Engineering* 2016; 106: 1114–1126.

110. Dasgupta D, Mondal K. Thermal circuits based model for predicting the thermal conductivity of nanofluids. *International Journal of Heat and Mass Transfer* 2017; 113: 806–818.

111. Poplaski LM, Benn SP, Faghri A. Thermal performance of heat pipes using nanofluids. *International Journal of Heat and Mass Transfer* 2017; 107: 358–371.

112. Boussatour G, Cresson PY, Genestie B, Joly N, Brun JF, Lasri T. Measurement of the thermal conductivity of flexible biosourced polymers using the 3-omega method. *Polymer Testing* 2018; 70: 503–510.

113. Ahmadi MH, Mirlohi A, Nazari MA, Ghasempour R. A review of thermal conductivity of various nanofluids. *Journal of Molecular Liquids* 2018; 265: 181–188.

114. Sajid MU, Ali HM. Thermal conductivity of hybrid nanofluids: A critical review. *International Journal of Heat and Mass Transfer Part A* 2018; 126: 211–234.

115. Sobhani M, Tighchi HA, Esfahani JA. Attenuation of radiative heat transfer in natural convection from a heated plate by scattering properties of Al_2O_3 nanofluid: LBM simulation. *International Journal of Mechanical Sciences* 2019; 156: 250–260.

116. Doganay S, Turgut A, Cetin L. Magnetic field dependent thermal conductivity measurements of magnetic nanofluids by 3ω method. *Journal of Magnetism and Magnetic Materials* 2019; 474: 199–206.

19 Nanotechnology Industry
Scenario of Intellectual Property Rights

Rakesh A. Afre

CONTENTS

19.1 INTRODUCTION

Nanotechnology is the engineering of matter at the atomic and molecular scale where size is measured in billionths of a meter (one nanometer = one-billionth of a meter). Nanotechnology is a very diversified field with a range of technologies at the nanoscale, such as materials, pharmaceuticals, biotechnology, genomics, neuroscience, robotics, and information technology, etc. In the last three decades, developed countries like the United States, Japan, Germany, France, and Korea have invested several billion dollars in research and development in this atomic-scale technology. All this enthusiasm played an important role resulting in several thousand patents of revolutionary new products and applications. Furthermore, there will be new developments which will make efficient energy sources at nanoscale with potential storage capacity like batteries and solar cells, and drug delivery systems, chemotherapy, stents, etc. While nanotechnology as a revolutionary domain has a number of unique attributes that are particularly susceptible to the development of patent bottlenecks, the problem with it is not new. A patent does not guarantee the right to make something or do anything with it. Instead, a patent gives the owner the right to exclude others from making, using, or selling anything that embodies the technology covered by the patent. When an organization has all of the necessary patents to develop a given technology, it can proceed without intellectual property entanglements.

The important role of intellectual property (IP) is in generating assets and job opportunities to the scientific community. In general, the well developed countries have a stronger, more serious and secure approach to IP and strictly maintain intellectual property rights (IPR) issues. In particular, it can be said that IP has become one of the most important assets of knowledge-based economies.

Creativity is essential to economic growth. It is feared that the latest technologies and thereby the progress of societies could be halted if there is no IPR. IPRs encourage the development of new technologies. IPRs aim to create a harmonious relationship among investors, inventors, and consumers. The ultimate goal of patent rights is to promote invention and encourage further development of that invention for the benefit of society.

19.1.1   CRITERIA FOR NANOTECHNOLOGY PATENT INVENTION

The word patent derived from Latin 'patere' which means 'to lay open'. The patenting of engineering science inventions is not identical to that of different technologies. To publish a patent in this interdisciplinary field of nanotechnology the following conditions must be fulfilled:

a) The novelty of the inventions/developments in the interdisciplinary field of nanotechnology is very difficult to determine.

b) Due to the interdisciplinary nature of this field, the innovative products of nanotechnology inventions have several applications. An emerging technology, the inventions produced are very difficult to identify and there is also a perceived risk of overlapping patents.

c) In order for a patent to be granted, under the universal patent laws, it has to refer to an invention that further fulfills the three criteria of uniqueness, innovation, and applicability to industry.

d) As regards the inventive step, the invention may not follow logically from what is already known to a person skilled in that particular art.

e) In general, getting a patent for nanotechnology is more likely to pose obstacles for proving non-obviousness.

f) Nanobiotechnology inventions must consider environmental jurisprudence across various jurisdictions.

Before going into the detail of nanotechnology patents, it will be very useful to discuss the rights of the patent. The first factor of patent rights is to protect its technological inventions. Patents are known because of the outcome indicators of applied analysis and technological advancement. Patenting any analysis or technology is meant to protect the invention, technology, process, devices, and biotechnological/nanotechnological compositions rather than conventional concepts. In general, a patent is defined as the exclusive right granted by statute to a party who protects functional concept, processes, methods, or apparatus that are non-obvious and novel, to use and develop that invention, to prevent others from manufacturing, selling, or using the invention for a limited time, which depends on the inventions and jurisdictions. The term for patent rights is typically up to 20 years. The invention must be unique at the global level. The purpose of the patent is to advance the innovation through disclosure and sharing knowledge of the invention to the public, and in exchange, the inventor or owner is rewarded with legal rights of ownership. In particular, the legal rights refer to not replicating the invention, selling, using, or importing to any entity, but it gives the owner exclusive rights to capitalize on the invention. Patents are obtained through a critical, lengthy, and costly process. In the process of patent applications, it should be filed within one year of the invention otherwise it is void.

It is very obvious that nanotechnology will be one of the vital technologies of the 21st century and has the potential to create new opportunities and prosperity. The inherently interdisciplinary nature of nanotechnology presents significant challenges in granting patents. Nanotechnology patents should not be treated differently but it is clear that such a complex technology creates more complex problems within the system of patents. This rapidly growing technology will challenge the conventional regulatory system in patent lawsuits. However, nanotechnology will generate both evolutionary and revolutionary products for the future, thereby improving every aspect of life. The current trends in nanotechnology and the genuine requirement of patents can help in understanding the impact of this technology on the future.

19.2 UNDERSTANDING PATENT POTENTIALITY AND LIMITATIONS IN THE NANOTECHNOLOGY DOMAIN

The patent system design is to stimulate innovation. Three important characteristics need to be considered: (1) nanotechnology is science-oriented and it is obvious; (2) it is interdisciplinary and it has revolutionized the industrial sector; and (3) patent filing is important to avoid the duplication of research work.

19.2.1 THE IMPORTANCE OF IPR

In general, IPR is defined as a scientific or technological invention that is protected from being replicated and having the properties of technological development. IPR is protected in different forms such as patents, copyrights, and trademarks, which enable people to obtain recognition as well as motivation and obviously financially benefit from what they have achieved or invented. IPR is only protected under the IPR laws in the country where the patent is filed. In order to obtain protection of IPR in more than one country, separate applications must be made to the patent office of that particular country. There are several ways to file the patent in other countries which is quite expensive due to the government rules and regulations, charges, and costs of translation (Singh et al. 2016, 2019).

A definition of copyright is a legal terminology to describe one's right over its artistic work. The works which are covered under copyright are books, music, films, databases, advertisements, drawings, and paintings. In some cases, trade secrets can be maintained in an inexpensive way to protect IPR that is not filed or published but should not be disclosed for certain reasons. Very important for any nanotechnology firm or industry, and especially so to new upcoming industries, is planning to prioritize the marketplace to encompass their patent strategy. In certain situations, copyrights and trademarks can be very important to plan for the nanotechnology industry. However, a company may want to register trademarks for its brand name and logo as well as its various products before they launch. If it is required to apply for trademark protection in other countries it is easy and possible under the Madrid Protocol and a single application is sufficient for a group of countries. Other possible methods for protection can be considered for each country in which the company is interested because trademark application costs are moderate in comparison to patents.

Copyright always protects the work expressed, not the idea that created it which is why it is used for protection against piracy. In other words, any idea can only be protected under patents for long term use. In nanotechnology applications, like computational tools, software is developed for research and development, and technical information copyright plays an important role.

19.2.2 THE INTERDISCIPLINARY NATURE OF THE NANOTECHNOLOGY DOMAIN

Nanotechnology researchers have also used forward and backward citations to determine the relationship between academic and patent literature, particularly to show knowledge-sharing trends. Nanotechnology deals with very smallest particles like atoms and molecules of any materials and changing the properties at nanoscale which cannot be seen with the naked eye. After its inception in the 20th century, many new properties of metals, semiconductors, and insulators have been discovered. These properties are unique which is why these materials have found many new applications in physics, chemistry, biology, electronics, and materials science. This versatile nature of the nanotechnology domain creates many problems in understanding the patents. Nanotechnology enables spectacular advances and newly developed products in a wide spectrum of applications. All electronic devices and their components such as PC chips, transistors, diodes, and capacitors are decreasing in size and increasing in performance, increasing the speed, which follows Moore's law that is why all the devices are getting faster, thinner, and better. Nanotechnology is playing a greater role in other areas of health, biotechnology, environment, water purification systems, automobiles, and medicines, etc.; the list is endless. It is very useful in diagnosis, and drug delivery to target cancerous cells where harmful cells are killed locally without harming the healthy cells. Carbon nanomaterials and their composites are those which are incredibly light in weight with strong tensile strength and good mechanical properties, and electrical properties which make them useful in sports equipment, cars, airplanes, fighter jets, self-cleaning glass windows, paints, and coatings, and more effective in batteries, solar cells, etc.

19.2.3 COMPLICATED IP ISSUES IN NANOTECHNOLOGY

There are some consequences of patenting nanotechnology inventions which pose more critical issues than other technologies that create complex issues. As nanotechnology innovations are multidisciplinary in nature, they involve a range of disciplines like physics, chemistry, biology, health, medicines, materials science, and various other engineering domains. Also, there are many broad spectrums of industries; that's why nanotechnology is defined by its size, hence the industries developing technology at nanoscale. Carbon and its nanoforms such as carbon nanotubes have applications in electronics, biotechnology, materials, and energy, owing to their unique optical, electrical, thermal, and mechanical properties. Other types of nanomaterials such as fullerenes and dendrimers contribute an important tool to the electrical, chemical and medical industries. Disputes in nanotechnology patents do not stand out as stemming from nano-specific issues but include those associated with patent thickets. To give a very popular example of a patent having multi-industry application, U.S. Patent No. 5874029 (University of Kansas), 1999, is a patent granted by the United States Patent and Trademark Office (USPTO) for methods for nanonization by recrystallization from organic solutions sprayed on anti-solvents. Another example is Nanosys, a young U.S. start-up in 2001, which within a short period of time had gathered a significant patent portfolio and claimed infringement of their quantum dot patents by Nanoco Technologies (U.K.). This issue of infringement was settled in 2009 and Nanoco had terminated its business in the United States. Some more examples: U.S. Patent No. 6322901 (to MIT), again in 2001, on quantum dots used for optoelectronics devices; quantum dot patent (U.S. Patent No. 6653080 to Quantum Dot Corp., USA, 2003) for a novel and rapid method used for mapping genetic variations in DNA and RNA; and one used in computing to the Wisconsin Alumni Research Foundation, 2003. These are the cases as an example of a patent thicket or of a set of 'strong foundational patents of broad scope that have been shown to stifle innovation'. There are many such situations that were made worse by the lack of understanding at the patent offices, which led to rejections as well as the acceptance of overly broad claims. Worldwide analysis of nano-patents has revealed that many of the claims are very broad in nature. Carbon nanotube patents have very broad claims because of the properties that have been studied are multidisciplinary in nature. Many such strategies have been used to address this problem, such as patent pooling using a reference system to provide standard mechanisms for clearing the patents and nanoproducts to the market. There are some other suggestions, in continuously growing nano patent thickets, and arguments considering the information on physical, chemical, and biological science streams, as an open discussion model where the formation of patent pools is impractical.

19.2.4 DIVERSIFICATION OF THE NANOTECHNOLOGY FIELD AND UNDERSTANDING PATENT LAW

Patent law describes four general areas of invention: methods or processes, machines or equipment, methodology or manufacturing, and composition of matter. The advancement and revisions of patent law in the science and technology domains have moved along different trajectories. Science and technology have moved ahead a long way to keep pace with development in the basic understanding of industries, and therefore need the patent law to be conserved in its original form. The frequent revisions of patent law have not identified any strong and serious concerns in line with the emerging technological developments for the 21st century such as nanoscience and nanotechnology, digitization, biotechnology, applied biology, and materials, etc. Now, in many judicial systems the patent law is infinite and anti-competitive, creates interests and ignores up to date knowledge and the primary aspects of IPR. The protection and enforcement of IPR must contribute to promoting technological inventions and dissemination of technology, however, the mutual advantage of users of technical knowledge with social and economic benefit is protected in Article 7 of Trade-Related Aspects of Intellectual Property Rights (TRIPS). There are nine potential categories of nanotechnology, among them three are major: materials sciences, computer technology, and medical applications.

19.2.5 Advantages of Patents

The method of obtaining a patent protection is a more robust system of protecting an innovation than protection by trade secrets. After obtaining a patent on nanotechnology-based research and development, an industry/institution/university cannot allow others from duplicating, using, or selling the patented work for a period of at least 20 years from the date of filing an application at the patent office. A patent is granted on the basis of uniqueness, novelty, an inventive step, and applicability in industry. An inventor has to disclose the invention when applying for a patent; the inventor has to present all the information to investors, customers, and industry without claiming protection in order to bring this new technology/process/new material to the market. By selling the patent rights to others an inventor can earn a profit, though royalty earned from patents is only 5–10%. The idea of patenting can help in restricting competition in the market. Patents also help all sizes of company to expand their market share, and they can file patents in other countries where they are planning to capture the market and increase company share value.

Most people think about utility/usage/productivity of patents which are concerned with a product, methodology, equipment etc., and under the right conditions, a designed patent can be a very valuable item. The design of the patent does not protect the structural or utility features of an article of manufacture, it has a very broad and distinct meaning. A designed patent has one or many claims to fulfil the judgement criteria. In general, the specifications are short and most of them are in redefined form. The drawings/figures/charts in a patent application describe the scope of the claimed subject matter; the drawings/figures/charts, however, show the ornamental features of the invention rather than its utility aspects. The description of rights covers the aesthetical aspects. Both forms of patent are wide-ranging and qualify for further consideration. A filed patent application provides the inventor the right to exclude others from practicing an invention, the early innovators can block patents on important technology during an 'IP gold rush' and this is applicable to emerging technological inventions. The blocking of patents may prevent other patentees from practicing their area of invention and allow other researchers/scientists whose inventions depend on new technologies in diversified markets to apply cross-licensing and obtain royalties. Similarly, companies which delay protecting patent rights have reduced profits through license fees to block patent holders or have been advised to move entirely from the market.

19.2.6 Patent Practice for Nanotechnology

The actual fact of a nanotechnology patent is it cross-cuts between various technologies that create numerous issues for the IP laws. When analyzing an application for a nanotechnology patent, the patent attorney has to find out about novel information and obviousness. A patent examiner may come up with an assertion that a nanomaterial product lacks novelty because the relevant nanomaterial is already in a current product, even though the nanomaterial was not identified. A nanomaterial may have been identified with desirable technical information, even though the mechanism behind it was not understood. In these situations, the patents and trade office (PTO) will not grant a patent for later identifying the mechanism underlying the characteristic of a known material. The patent examiner is not responsible for identifying specific characteristics of a material to get it patented if they actually arise from structural features already present in a known material.

19.3 SCENARIO OF NANOTECHNOLOGY PATENTS IN VARIOUS INDUSTRIES AND THEIR IMPACT

In present scenarios of technological developments such as nanotechnologies, an emerging and interdisciplinary field of research, there is a risk that broad patents are being granted. A breakthrough in scientific research fields such as materials science, computer technology, pharmaceuticals, agriculture, environmental, electronics, nano- and biotechnologies, and medicine are listed as priority. The role of nanotechnologies has widened its scope at global level, as with its precision manipulation at atomic level new avenues can be explored in information science, molecular biology, genetic engineering, and medicine. Recently the investment in nanotechnology research and development (R&D) pertaining to new emerging technological inventions has been consistently increasing in developing countries.

19.3.1 Patents in Materials Science Industry

The development of materials and their unique properties is applicable to every technology and industrial application, wherein the automobile, medical, electronics, materials, energy, environment, and agriculture, etc. fields are some of the examples in which materials science plays an important role. The nanotechnology sector has grown extensively over the past few decades and incorporates changes in modern technologies. The fundamental properties of materials at atomic level have led to the creation of materials with unique and useful properties: coatings on surfaces to make them harder and glossier, mobile telephones and laptops are faster, medicines are more effective, UV-sunscreen lotions can be made transparent, and vehicles are getting smarter, and many more to come. Industries with 'innovation' as a tagline are currently developing faster and tinier batteries, self-cleaning windows, magnetically-driven microbots, targeted drug delivery, energy-saving street lights, and nanofilament-impregnated contact lenses capable of enabling high definition (HD) virtual reality (VR) or augmented reality (AR) displays. These advances in nanotechnology are the motivation to innovate in a diverse way to create technology from smart windows to consumer electronics, and from medical to micromachining. As this advancement in nanotechnology has made inventors protect their inventions through IPR, i.e. with patent filing, it is important seek protection against it. Recent studies have found that the patenting of nanotechnology-related inventions is increasing and has more than tripled since the 2000s. Further, the increase in the rate of patents issued in this area is nearly 25%. This comes in understanding that patent applications related to nanotechnology may be finding relatively less resistance from the patent offices. There are many factors which are contributing to this phenomenon. For example, the fundamental properties of nanotechnology inventions can mean that the patent offices have less prior knowledge to rely upon in making rejections of claims mentioned in patent applications. In addition, patents may be available for various well-known materials where, for example, the nanoscale manipulation to these materials leads to unique or exemplified properties. While metal oxide-based sunscreens have been available for years (very few new patents available in this area), this manipulation of metal oxide particle sizes at nanoscale for sunscreens is optimized to average particle sizes of less than 100 nm. There may be tremendous opportunities in obtaining a patent in this arena of nanotechnology, and the inventors of the patent must remain vigilant in getting patents that will matter to others and will not lead to challenges of litigation. This approach merely focuses on getting patents without considering how it remains in the market or will also end up blocking future competition with others.

19.3.2 Patents in Nanomedicine and the Health Industry

The domain of medicine and health is booming and on the verge of a revolution by implementing nanotechnology-based treatments and medicines. Globally more than 300 companies are engaged in R&D on nanomedicine and a majority of these are new start-ups or small to medium-sized industries rather than large ones. There are more than five different nanotechnology-based therapies with numerous applications ranging from early detection of medical imaging to regenerative bone and tissue repair. With the outstanding properties of the submicron level materials, researchers/scientists are able to develop more effective, precise, and easily administered treatments which make it so revolutionary, but it also makes it problematic and potentially dangerous. This emerging technology creates several issues relating to ethical, privacy, intellectual property (IP), and legal matters. For example, can a trial of nanomedicine with a participant's informed consent be valid even then researchers who conduct the study do not really know the extent or magnitude of the risks? Is there an e-healthcare system foolproof enough to make the personalized medical data from this system integrated, and medical devices based on nanotechnology worth the risk of unauthorized access? Do these IP law systems sufficiently protect nanomedicine without altering the environment, which ultimately fulfils the definition of innovation? The existing industries/institutions are capable of regulating nanomedicine, but there is often conflict of interest, under the jurisdictions of several regulatory authorities. Finally, all the issues cannot be dealt with in a comprehensive way without cooperation and understanding among the people, scientists, pharmaceutical and biotechnology industries, and regulatory authorities. The ethical issues related to nanomedicine are too important to be dismissed without consequential discourse between scientists and the public. Private issues will only be adequately considered and addressed from the root level by the biotechnology and healthcare industry itself, instead of a major legalized solution from the regulatory bodies. Understanding between industry and legislators needs to be established only when nanomedicine innovations can be protected without hampering future progress. The nontechnical issues related to nanomedicine can be divided into roughly four different categories. These four categories both simplify and conclude the major issues related to ethical, privacy, IP, and other legal issues. This emerging technology list is not completely agreed by many readers, and analysis will produce additional issues raised by people. Ethical issues arising in nanotechnology are broken down and are based on a treatment's lifecycle from R&D and distribution, to its final application and use. For the analysis and clarification of these issues, available and developed medical ethics paradigms are used.

19.3.3 Patents in the Biotechnology Industry

A emerging field of biotechnology connected with nanotechnology is nanobiotechnology and is actively spreading around the world, especially when it concerns patents. This new technology creates lots of challenges and opportunities in adapting the patent regime to its particular context. There are some consequences in obtaining patents of nanobiotechnology innovations that pose problems when compared to other technologies, owing to their multi-dimensional nature, cross-sectorial applications, and broad criteria, fulfilling those of novelty, non-obviousness, and applicability to industry. The patents in nanobiotechnology can cover the process of preparing the metal nanoparticles, the process of drug delivery through nanoparticles, or the biomedical devices used. The distinctive classification of various methods such as diagnostic therapeutic, medicinal, surgical, and curative, and the subject matter of each of them is covered here. On the other side of patentability, it is argued that exempting medical methods, in conjunction with public policy, allowing patents in this field, would draw ethical, moral and practical problems, as nanoparticles constitute diagnostic, surgical or therapeutic methods. Concerning medical methods, the amendments to the Indian Patents Act 1970, sec 3(i), can be imported from the European jurisdiction regarding understanding of medical methods.

For more than two decades the role of IPR in science and technology has been exaggerated at global level – primarily due to the prescribed rules of the TRIPS and by bilateral/regional trade agreements. In the TRIPS agreement all World Trade Organization (WTO) member countries are obligated to adopt and enforce minimum standards of IPR. As mentioned in the TRIPS agreement, it requires member countries to make patents available for inventions, whether related to products or processes, in the fields of technology without any discrimination, subject matter to the standard patent criteria i.e. novelty, inventive step, and application to industry. In the negotiations on the TRIPS agreement, consensus was not reached on biotechnology inventions, which are a controversial area of research. It was decided by the United States and some developed countries to push for no exclusions to patentability, while some developing countries preferred to exclude all biological inventions from IPR laws. On the basis of some documents, it is mentioned that biological things like plants, animals, and essential biological methods may be excluded from patentability. However, WTO members must protect plant species either by patents or by an effective sui generis system. The controversial statement surrounding the patenting of biotechnological inventions at the WTO is valid for the discussion on nanobiotechnology patents because nano- and bio-scale materials and methods – inventions that claim living things – raise many of the basic queries. The vast number of nanobiotechnology patents are already being granted to many industries and include claims on basic building blocks of the entire bio-world. The TRIPS agreement obligates all developing countries to enforce patents on nanobiotechnology inventions, even those that incorporate plants and animals. From the last few decades, governments, the U.N. Human Rights Commission, civil bodies and social awareness drives, have warned of the inequities of IP for the world.

19.3.4 PATENTS IN THE ENVIRONMENT INDUSTRY

Globally many research groups are working on environmental health and safety concerning the use of nanomaterials. There are certain nanomaterials which are studied primarily according to their toxic nature, and by understanding the potential for exposure to the materials, i.e. they will be released into the environment. Nanomaterials which show high toxicity and exposure are certainly the most environmentally relevant for future study. This diversified nature of the environment would be considered, for both patent and consumer products when prioritizing nanomaterials for analysis and study. For example, the nanotechnology domain indicates silver (Ag) is the most pervasive nanomaterial in nanotechnology products. In nanotechnology inventions, silicon and carbon with their nanoforms are the most cited materials. Some other nano-functionalized metals and their oxides, ceramics, semiconductors, and biomaterials are also patented, other than silver nanomaterials. Concerning these nanomaterials, many of them are photoactive, magnetic, superconducting, contain low earth abundance transition metals, and/or contain elements with a toxic nature to study for environmental implications. Common classification of nanomaterials and their life cycles is essential to understand exposure potential, and the emerging nanomaterials and IP to identify a wide, diversified range of nanotechnology applications. The consumer products which contain highly hygienic, anti-corrosion and antifouling coating applications were in the fields of electronics and computer hardware, however, most of these nanotechnology inventions are between 2000 and 2015. According to the USPTO, patents granted on semiconductor and solid-state device patents account for more than 40% but in the last few years innovation has boomed for nanomaterial-based biotechnology, manufacturing processes, layering, and coating technologies. This diversified nature of nanotechnology applications across consumer and patent databases shows all aspects of the industry, and identifies environmental health and safety (EHS) research by environmental scientists. On the other hand, near to 50% of patents studied by the World Intellectual Property Organization (WIPO) and the USPTO were nanotechnology IP dealing with nanoscale layers, coatings, or other surface modifications, but used no common nanoparticles (with 2 or 3 nm dimensions). Most of the patents in layering and coating nanotech applications have not been with corresponding research on the lifecycle of these nano-based products. Thin polymeric, ceramic, or semiconductor coated layers may be subject to flaking during manufacturing of products, their use, and end-of-life phases such as factors affecting nanomaterial, release potential harm to humans or continuous exposure to the environment are not yet well understood. In other cases such as nanomaterial-based electronics and information storage devices (which also contain nanoscale layers), there is little opportunity for physical contact between the nanomaterials and humans or the environment during their use, however, more attention is to be addressed in such a scenario.

19.3.5 PATENTS IN AGRICULTURE AND FOOD INDUSTRY

Nowadays, population growth and food demand is increased, so food safety is a major concern. This concern has led to developments in innovative techniques for better production of crops and preservation of food. Therefore, publications and patents show a rise in nanoscience and technology research for food and agriculture. Food nanotechnology helps in improving shelf life and prevents food from spoiling and losing nutrients. This study has emphasized various trends in nanotechnology patent applications in food and agriculture in the last decade as revealed by a Google search for patents in Europe, Germany, the United States, Canada, Japan, and China. The United States is well known to be active in the international market among all patent offices, followed by China and other countries. The United States Patent and Trademark Office (USPTO published more patents on nanofood for the period 2011–2015, while China is more inclined towards nanoagriculture patents. Germany shows fewer patents in both nanofood and agriculture, however, there is a recent rise in the trend. Furthermore, the increase has not been very significant in nanofood patents with a small increase from 6 to 17 but it is a larger gain than for nanoagriculture, i.e. from 3 to 24 published in the years 2011 and 2017 respectively, although the nanofood patents issued are greater in number. In comparison to the number of patent applications for this period, a tremendous increase in the applications in recent years has been embraced and advancements in nanotechnology derived food and agriculture is anticipated. China showed profound growth, with a large number of nanotechnology patents published internationally. It is also shown growth in the number of publications, followed by the United States, Germany, Canada, and Europe has shown a small contribution to this sector. China has always inclined towards benefit to society and understood the ability to control the research at nanoscale level that leads to revolution in this emerging technology. In 2015, the highest number of patents (18) under the 'Nano in Agriculture' stream was published by China. Only one patent showed both the United States and Canada with a slow rate in research and development in nanotechnology in 2011, whereas other countries have not contributed to this field. In 2012–13, the United States demonstrated its contribution to the nanotechnology field of research. The contribution of China and Germany is very low in this field whereas Canada and Europe were not active for this period. China's contribution is emerging with exponential growth in recent years and climbing steeply year by year. The United States showed consistent variation in contribution to this area in 2011–15. Europe and Canada stand out in third and fourth places because they have shown some contributions in this field, however Germany has remained relatively inactive with fewer publications. The United States has shown a growth of about 28% in the number of publications for 2011–15 in the agriculture sector. A number of patents are issued, notably, to fertilizers and pesticides at nanoscale which are specific for certain crops such as, eucalyptus, sugarcane, spinach, etc., which are environmentally friendly and require less fertilizer than the conventional crops. The highest number of nanoagriculture patents was issued in 2015 which certainly indicates an increase in nanoagriculture research for the coming years.

Strong emphasis has been placed on development of biodegradable and compostable food packaging material and nanotechnology-based sensing materials for determination of food safety. On the other hand, food preservation is also an important factor and for that purpose nanosensors are designed to show fluorescence in various colors when in close contact with food pathogens. These nanosensors function as an 'electronic tongue or nose' that are placed directly into the packaging material to detect chemicals released during food spoilage. Such inventions are very useful in this field and creating new ideas to enable future generations to live healthily and happily.

19.3.6 PATENTS IN THE PHARMACEUTICAL INDUSTRY

The interesting, unique, and far-ranging properties of nanomaterials and nanotechnologies have paved the way to facilitate tremendous breakthroughs in the field of pharmaceuticals and medical devices. The integration of biotechnology, genetics, and nanotechnology, have opened new avenues for nanobiotechnology, introduced targeted drug delivery, minimized side effects, and enabled cutting-edge cancer treatments. Because of the high value of the pharmaceutical industry to national and multinational industries, higher R&D costs, hard clinical trials and requirements of data, a potential drug market, and strong competition, patents are generally critical to innovation and the market. While claiming patents for pharmaceutical products which relate to active contents such as salts, drugs, and isomers, etc., or cover any one of them separately, it may also individually consider a process of manufacturing or include both a process and a final product. Some other considerations are included for the evaluation criteria of various types of claims that are very typical for this area. It is important to keep in mind that while developing a new molecule for pharmaceutical use may require various inventive steps, techniques for the preparation of medicines in various forms and dosages are very well known. There is a narrow range of developments that could be genuinely considered inventive in this area in view of the state-of-the-art medicines. The same active ingredient in the forms of different dosages, as tablets, capsules, ointments or liquids for parenteral requirement, which can be formulated using different pharmaceutically acceptable excipients such as fillers, binders, disintegrants, and lubricants. Since last decade, drug companies have been dependent on patented drugs – these patented drugs can generate annual sales of more than USD 1 billion. Nanotechnology techniques are employed and patented for drug delivery, using nanoparticles, dendrimers, nanoemulsions, and to improve drug solubility/bioavailability. Many of the exceptionally great patents related to treatments employ nucleic acids, like gene therapy, interfering RNA (RNAi) techniques, and antisense molecules. Despite the nanometer size range of nucleic acid molecules no significant technological processing has been carried out on the molecules themselves. It is however found that all different types of applicants, starting from universities to small/medium-sized companies, to large pharmaceutical industries, are filing patents relating to pharmaceutical nanotechnology. It is not surprising that the universities and small/medium-sized enterprises obtain broader platform technology patents, whereas the bigger enterprise pharmaceutical companies often file narrow range specific formulation-type patents. As mentioned earlier, a great many patent applications based on nanotechnology, relate to drug delivery using nanoparticles and refer to patent WO0132139, which is concerned with coating of drugs with less water solubility and lipophilicity. This process combines a lipid system with a drug, i.e. cisplatin, having low concentration and mixing to one or more freeze/thaw cycles to form lipid encapsulated nanoparticulate compounds (WO03017987) from McGill University, contains biodegradable polymeric nanocapsule membrane compounds of hemoglobin and polyhemoglobin, which enhances in vivo circulation. The compounds are formed by water-insoluble copolymers of polyethylene glycol and polylactic acid to make selectively permeable membranes. For the treatment of various skin conditions, patent WO03093243 from Aventis Research and Technologies is for the delivery of anti-androgenic compounds using nanoparticle formulations. It is conveyed here that the formulation of nanoparticles further ensures enrichment of the compounds at the follicles and to aqueous or alcoholic formulations, it will not precipitate the active substance on hair and therefore there is less risk of exposure to pregnant women. Another patent (WO0002590) for the enhancement of drug delivery in solid tumors describes the use of electromagnetic pulses or ultrasonic radiation and nanoparticles. Electromagnetic radiation, ultrasonic waves, or local heating of the particles results in damage to the tumor cells and allows delivery of therapeutic agents from blood into cancer cells. The many patents granted are related to formulation of calcium phosphate in a nanocrystalline form, for example from Biosante Pharmaceuticals,

Inc., relating to therapeutic calcium phosphate nanoparticles. The above examples of patents merely relate to the use of nanoparticles to give an example of the wide variety of applications which nanoparticles are being used for in the pharmaceutical research area. Nevertheless, innovative drug formulation patents are attractive to the pharmaceutical industry, because they offer the strong possibility of extending the market life of a drug.

19.3.7 PATENT LANDSCAPING

A patent is a legal document that certifies authorship of an invention or prototype and the exclusive right to use it. It is a very important and informative source of science and technology, and is not limited to a description of the invention but gives a useful indication of the level of current research and innovation before the product launches on the market. To evaluate present trends and selected areas of support to nanotechnology, statistical data analysis of patents is used to compare the number of applications by region, area of research, and citations. Another way to study the characteristic information of patents surrounding any particular technology in a specific area or at global level is to analyze the patent landscape, referred as patent landscaping [WIPO, 2015]. This method compiles statistical information of bibliographical data and analyzes a large set of patent information to visualize the results. Patent landscaping describes technological solutions of the patent documentation and is mapped in the form of isolated 'islands'. The map gives clear understanding of how close the islands of certain patent holders are to one another and how are they distributed on the technological solutions spectrum.

A very good example of effective patent landscaping is the open-access publications from the United Kingdom Intellectual Property Office (IPO) which analyze innovations in nanotechnology and detail the patents of industries and universities in nanotoxicity, among other things [IPO, 2009]. Another interesting example of the IPO's studies on the activity of patents of a versatile material, graphene [IPO, 2013], where patent landscaping clearly shows the dynamics of technologies such as the synthesis of graphene, LEDs, semiconductor devices, and applicants' geographical location by year. Here the focus is on the classes and sections of technologies where microstructure and nanotechnology innovations are used. As mentioned in the table of technology, the registered technologies under International Patent Classification (IPC) subgroups are classified in 'nanotechnologies':

- B82B — nanomaterials prepared by manipulating individual atoms, molecules, or limited collections of atoms or molecules and manufacturing methods,
- B82Y — applications or specified uses of nanomaterials; analysis of nanomaterials; manufacture or treatment of nanomaterials.

19.3.8 SCENARIO OF NANOTECHNOLOGY PATENTS AT THE GLOBAL LEVEL

The number of patent applications filed in the top 50 depositories were about 13,000 in 2008 compared to about 1,200 in 2000 (Huang et al. 2004, 2005). During 1990–2010 a study of Indian publications and patents in nanotechnology (in foreign patent offices like USPTO, EPO, Japan Patent Office) are summarized with the following conclusions (Gupta, 2008):

- Total of 167 patents, of those, 64 patents (39% of the total patents) belong to government organizations, 45 patents (27% of the total) to industry and 10 patents (6% of the total) to academia and research institutes. There are 37 patents (22% of the total) that belong to individual inventors. The remaining patents are collaborative patents between government organizations or industry.
- The lead contributors from the government organizations such as research laboratories of the Council of Scientific and Industrial Research (CSIR), Defence Research and Development Organisation (DRDO).

- Industry or companies like Panacea Biotech Limited, Ranbaxy Laboratories, Arrow Coated Products Limited, and Stempeutics Research Private Limited are leading in industry patents.
- Academic institutions are more focused on publications than patents, and it is vice-versa for industry.

The annual rate of increase for all of the patent publications is profound for the years 2000 to 2008 (34.5%). As per the Science Citation Index, this rate of increment is higher for article publication, at 20–25% for the same time period when the keyword 'title–abstract' approach for patent applications is searched (Dang et al. 2010). Nanotechnology is an emerging field and advanced in the 20th century by leaps and bounds. Research funding and infrastructure rank the United States in first place among the other top twenty developed countries; however, India ranks in nineteenth place in the USPTO nanotechnology research. For scientific publications in nanotechnology research, the United States again is in first position among the top twenty developed countries, while India is placed in eighth position (Chen et al. 2013).

19.4 CONCLUSION

Nanotechnology merges many disciplines of science and technology to make its multidisciplinary nature critical to issues of intellectual property rights. A patent application related to a nanotechnology invention requires careful understanding of the potential end uses so that all claims are adequately covered by the patent, an exercise in drawing up expertise in a multidisciplinary field of research. The emergence of nanotechnology in various services brought human capability to a new level in approaches to invention and innovation. It has also enhanced the intellectual property scenario in terms of authentication in outcomes of nanotechnology patents. The scope of nanotechnology has also expanded from single to complex nanosystems. Unique legal issues that require proper understanding of case law from distinct areas of science and technology will arise. Nanotechnology will pose some tough challenges to current concepts of law and lawsuits.

Strong patent protection may spur research and invention, which may lead to a patent thicket and expensive litigation. Patent thickets, a peculiar phenomenon in patents especially for nanotechnology, refer to a situation where there are tough decisions in granting a patent for a nanotechnology-based invention when it is obviously not workable due to a previously granted patent for a similar invention. Subsequent patent applicants, in order to make their patent workable, have to first obtain licenses from the previous patent holders.

Concerning the issues related to patent applications, many authorized bodies from various countries need to be established to manage the nanotechnology patents in products and services. Issues pertaining to nanotechnology-based patents should become a prime feature among the various bodies involved to make sure that any problems arising can be solved in a proper and timely manner. To achieve competitive advantage of a patent it should act as a strategic weapon and be considered as a priority within companies. The responsible authorized members and management board members of the company and the community should be aware of the hidden potential embedded in a patent. One of the major concerns of nanotechnology patent outcomes is dealing with duplication of products and services. It is important to avoid conflicts especially in determining major criteria of patents: novelty, inventive step, and industrial application. Nanotechnology offers high potential in many research fields such as materials, manufacturing processes, electronics, health and well-being, environmental sustainability, energy, and so on.

REFERENCES

Chen H, Roco MC, Son J, Jiang S, Larson CA, Gao Q (2013) Globalnanotechnology development from 1991 to 2012: Patents,scientific publications, and effect of NSF funding. *The Journal of Nanoparticle Research* 15(9), 1951. doi:10.1007/s11051-013-1951-4.

Gupta VK (2008) Indian publications and patents in Nanotechnologyduring 1990–2007. *Presented at DSDS Special Event on Capabilities and Governance Issues in Emerging Technologies, Held at TERI*, New Delhi.

Huang Z, Chen H, Che ZK, Roco MC (2004) Longitudinal patent analysis for nanoscale science and engineering: Country, institution and technology field analysis based on USPTO patent database. *Journal of Nanoparticle Research* 6(4), 325–354.

Huang Z, Chen HC, Yan L, Roco MC (2005) Longitudinalnanotechnology development (1991–2002): National science foundation funding and its impact on patents. *Journal of Nanoparticle Research* 7(4–5), 343–376.

IPO (2009) Patent Informatics Project Report: UK innovation nanotechnology patent landscape analysis. Available at https://assets.publishing.service.gov.uk/government/uploads/system/uploads/attachment_data/file/312326/informatic-nanotech.pdf (accessed 30 January 2020).

IPO (2013) *Graphene. The Worldwide Patent Landscape in 2013*. Newport: Intellectual Property Office. Retrieved from http://www.ncl.ac.uk/curds/documents/informatics-graphene-2013.pdf.

Singh HB, Jha A, Keswani C (2016) *Intellectual Property Issues in Biotechnology*. Wallingford, UK: CABI. 304 pages, ISBN-13: 978-1780646534.

Singh HB, Keswani C, Singh SP (2019) *Intellectual Property Issues in Microbiology*. Singapore: Springer-Nature. 425 pages, ISBN- 978-981-13-7465-4.

Wang G, Guan J (2012) Value chain of nanotechnology: A comparative study of some major players. *Journal of Nanoparticle Research* 14(2), 702–716.

WIPO (2015) Patent landscape reports. Retrieved from http://www.wipo.int/patentscope/en/programs/patent_landscapes/.

20 A Mass Approach towards Carbon Nanotubes in the Field of Nanotechnology

Yashfeen Khan and Anees Ahmad

CONTENTS

ABBREVIATIONS

CMNC	Ceramic matrix nanocomposites
CNTs	Carbon nanotubes
CVD	Chemical vapor deposition
DWCNT	Double-walled carbon nanotubes
EPO	European Patent Office
IPR	Intellectual Property Right
MMNC	Metal or metal oxide matrix nanocomposites
MWCNT	Multi-walled carbon nanotubes
PMNC	Polymer matrix nanocomposites
SWCNT	Single-walled carbon nanotubes
USPTO	United States Patent and Trademark Office

20.1 CARBON NANOTUBES

Awareness of various carbon allotropes is not new chemistry. The widely known crystalline allotropes of carbon are diamond and graphite. These have sp^3 and sp^2 hybridized carbons, respectively. Diamond forms an extended 3D network, whereas graphite forms planar sheets. Fullerenes, also called buckyballs (zero dimension closed cage carbon molecules), are the third allotropic members of carbon. These buckyballs were discovered in 1985 by Kroton, Smalley, Curl, and co-workers at Rice University while they were investigating the nature of carbon in interstellar space [3]. Hybridization of fullerenes is neither planar nor completely pyramidal; instead, it exhibits slightly pyramidal characteristics (i.e. sp^3 characters present in sp^2 carbons [4, 5]. The already prepared but undiscovered carbon nanotubes, also called buckytubes, were discovered by Iijimaetal of NEC Laboratory in Tsukuba, Japan, using high-resolution transmission electron microscopy (HRTEM) in 1991. These materials are lighter than aluminum and stronger than steel [6]. These are the fourth allotropic members of carbon clan [7]. During the direct current arcing of graphite, these carbon nanotubes were found deposited on the negative electrode. This arcing was being done for the preparation of fullerenes, and luckily, multi-walled carbon nanotubes (MCNTs) were first discovered. It was less than two years before Bethune and co-workers found single-walled carbon nanotubes (SWCNTs) at the IBM Almaden Laboratory in 1993. Since then carbon nanotubes (CNTs) have undoubtedly become one of the headlines of nanotechnology due to their fascinating properties, contributing it to various applications like in electronic devices as field-emission sources, in supercapacitors and actuators, gas and hydrogen storage, sensors and probes, in batteries, for gaining knowledge about water in animal cells, tissue engineering, in drug delivery applications, and many more [8–10].

20.2 STRUCTURE, SPECTRA, AND CHARACTERIZATION

CNTs are cylindrical nanostructures manufactured with a length-diameter ratio of up to 132,000,000:1 [7]. Earlier, when discovered, CNTs were called elongated and concentric layered microtubules. The structure of CNTs was imagined to be made out of graphene, by either rolling or distorting the two-dimensional sheets to form one dimensional carbon nanotubes and zero-dimensional fullerenes. Applied quantum chemistry explains the sp^2 hybridization in CNTs similar to graphene. These provide a unique strength to CNTs [11, 12]. Each strand of a nanotube is bonded together by van der Waals forces, i.e. Π-stacking. CNTs are screened reliably using Raman and Fourier transform infrared spectroscopy. Scanning electron microscopy is responsible for a surface overview, while transmission electron microscopy helps in identifying the internal morphology of the CNTs [13].

20.3 TYPES OF CNTS

CNTs are divided into three categories depending upon the rolling of the graphene sheet. Single-walled CNTs (SWCNTs), double-walled CNTs (DWCNTs), and multi-walled CNTs (MWCNTs). An SWCNT possesses a single graphene sheet folded upon itself, has a diameter of 1–2 nm and a tube length millions of times longer (Figure 20.1). The length varies depending on the preparation methods. The structural description of SWCNTs is conceptualized by their chiral vectors (n, m), which build three different types of CNTs: armchair, zigzag, and chiral, as shown (Figure 20.2). If $m = 0$, the nanotubes are called zigzag nanotubes, and if $n = m$, the nanotubes are called armchair nanotubes. Otherwise, they are called chiral [4].

DWCNTs consist of two concentric carbon nanotubes; the outer tube is inscribing the inner tube (Figure 20.3). MWCNTs have multiple layers of graphene whirling around. Diameters vary from 2 to 50 nm depending on the number of graphene tubes. The numerous layers of MWCNT are approximately at an inter-layer distance of 0.34 nm, each one forming a single tube, with all the tubes having a larger outer diameter of 2–100 nm(Figure 20.4) [4, 10].

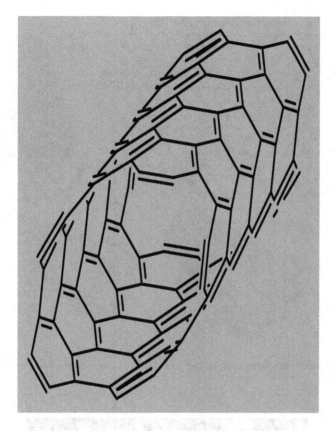

FIGURE 20.1 Schematic diagram of SWCNT.

The nanotube structures or their diameters are represented with the parameters shown below in Eqs. 1.3.1–1.3.8 [5,7,14].

$$\text{Translational vector} = \mathbf{T} = \mathbf{t}_1 a_1 + \mathbf{t}_2 a_2 = \left(\mathbf{t}^1, \mathbf{t}_2\right)$$

$$\text{Chiral vector} = C_h = na_1 = na_2 = (n,m) \quad \text{Chiral vector} = C_h = na_1 = na_2 = (n,m) \quad (1.3.1)$$

$$\text{Chiral angle} = \cos\theta = (2n+m)/\left(2\times\left(n_2+m_2+n\times m\right)^{\frac{1}{2}}\right) \quad (1.3.2)$$

$$\text{Length of chiral vector} = \mathbf{L} = \mathbf{a}\times\left(n_2+m_2+n\times m\right)^{\frac{1}{2}} \quad (1.3.3)$$

$$\text{Diameter} = \mathbf{d}_t = \mathbf{L}/\pi \quad (1.3.4)$$

$$\text{Number of hexagons in the unit cell} = \mathbf{N} = (2\times\left(n_2+m_2+n\times m\right)/\mathbf{Dr} \quad (1.3.5)$$

$$\text{Symmetry vector} = \mathbf{R} = pa_1 + qa_2 = \left(p,q\right) \quad (1.3.6)$$

$$\text{Pitch of the symmetry vector} = \zeta = (m\times p - n\times q)\times\mathbf{T})/\mathbf{N} \quad (1.3.7)$$

$$\text{Rotation angle of the symmetry vector} = \Psi = 2\pi/\mathbf{N}\left(\text{in radian}\right) \quad (1.3.8)$$

Where $\mathbf{t}_1 = (2m + \mathrm{N})/\mathbf{d}_r$; $\mathbf{t}_2 = -(2n + m)/d\mathrm{R}$; $\mathbf{d}_r = \gcd(2n = 2m + n)$ in m are lengths of chiral vectors.

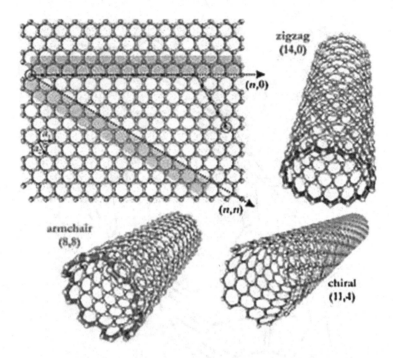

FIGURE 20.2   Different types (armchair, zigzag and chiral) of carbon nanotubes on the basis of chirality. (Reproduced from [1].)

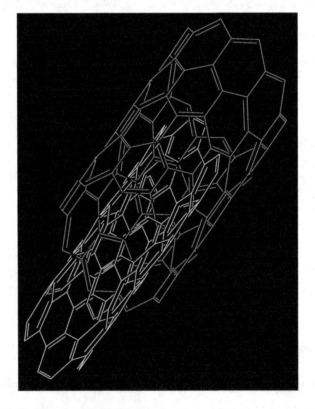

FIGURE 20.3   Schematic diagram of DWCNT.

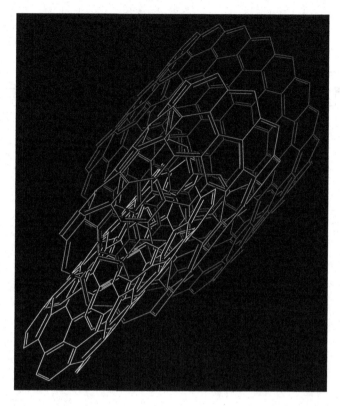

FIGURE 20.4 Schematic diagram of MWCNT.

20.4 PRODUCTION STRATEGIES FOR CNTS

In the past thirty years, thousands of methods have been reported in several reviews for synthesizing CNTs, out of which, arch-discharge, chemical vapor deposition (CVD), and laser ablation are the main three techniques (Figure 20.5) [5, 7].

Among these, the CVD method has the advantages of being simple and low cost, requiring relatively mild synthesis conditions (such as normal pressure and moderate temperature), and providing high yields. It also offers reasonable control over the number of walls, nanotube diameter, length, and alignment of CNTs. In the CVD method, a carbon feedstock (in vapor phase) is converted into CNTs and other byproducts upon coming in contact with a heated catalyst. CNTs synthesized by CVD retain small quantities of impurities from the catalysts (iron, cobalt, molybdenum), which can affect the electronic properties of CNTs, and this poses a substantial barrier to many applications. Hence, the effective removal of impurities is essential [15]. The amount of residual iron in CNTs has a massive impact on their electrochemical properties, however, with the help of controlled microwave treatment >90% of the iron catalyst present in the CNTs can to be removed [5], but the presence of iron residue strongly tarnishes the electrochemical properties. In addition to these classical methods, recent trends have also been observed in the synthesis of CNTs.

20.5 DISPERSION OF CNTS: THE BRIDGE FROM RAW CNTS TO RIPENED CNTS

It is clear that CNTs hold different chemical, mechanical, optical, and electronic properties that make them the center of attraction for all scientists worldwide. However, the major problem that is linked with CNTs is that the outer walls of pristine CNTs are found to be chemically inert. The weak

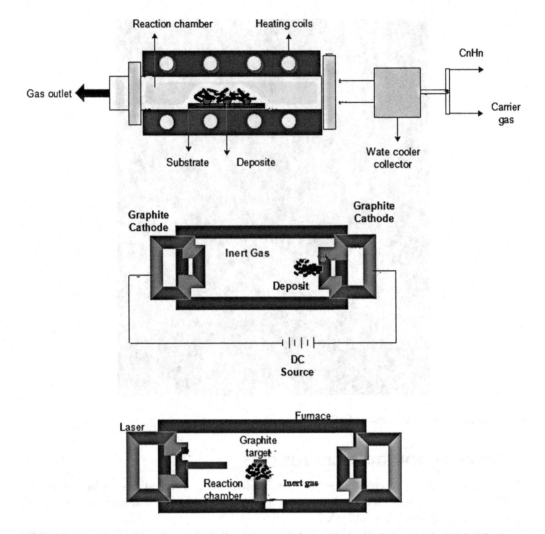

FIGURE 20.5   Schematic representation of methods used for carbon nanotube synthesis: (a) chemical vapor deposition method, (b) arc discharge method (c) laser ablation method [2].

intermolecular forces between carbon atoms of CNTs cause tube agglomeration and hence, result in poor dispersion of CNTs in various solvents and polymeric matrices. One of the most suitable routes to overcome this problem is to modify the walls of pristine CNTs, i.e. to functionalize CNTs (Figure 20.6). Functionalization makes the non-reactive sites of pristine CNTs chemically reactive, enhances their properties and hence, broadens their application potential [4,14].

20.6   TYPES OF FUNCTIONALIZATION OF CNTS

Researchers reported different methods for the functionalization of CNTs (Table 20.1), which are as follows:

20.6.1   NON-COVALENT FUNCTIONALIZATION

Here, the desired functional groups are attached to the walls of CNTs but through weak van der Waals forces. Besides these forces, hydrophobic or Π-interactions are also involved. This kind

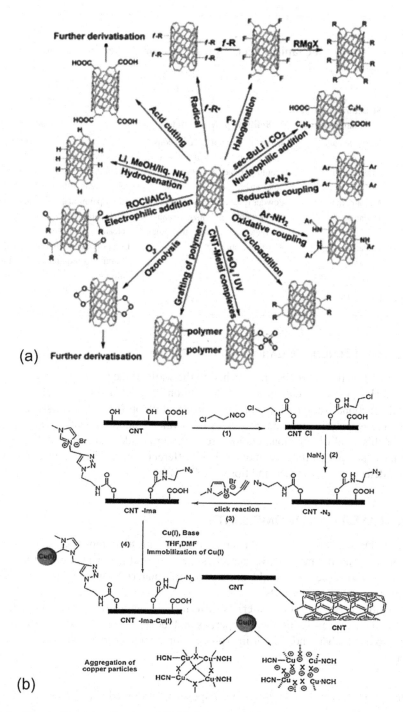

FIGURE 20.6 Types of functionalization of CNT. (Reproduced from [2].)

of functionalization is relevant and exciting because it does not induce any structural damage or transformations in the CNTs, i.e. it does not destroy the conjugated system of the CNT sidewalls. Some common examples of performing the non-covalent type of functionalization are the use of polymers, polynuclear aromatic compounds, surfactants, CNT wrapping, and non-covalent protein interaction, employing π-π stacking or hydrophobic interactions [14, 17].

TABLE 20.1

Comparison of Various Mechanical Techniques Used for Dispersing Carbon Nanotubes in a Polymer Material

	Factor		
Technique	Structural damage to nanotubes	Befitting polymer matrix	Use
Ultra sonication	Major damage	Water-soluble polymer, viscous polymer or oligomer, monomer.	low laboratories, easy operations, and after use cleaning.
Calendaring	Undamaged	Liquid polymer or oligomer, monomer.	Operation training is required, after use cleaning is difficult.
Ball Milling	Major damage	Powder (polymer or monomer).	Easy operations, needs cleaning after use.
Shear mixing	Undamaged	Water-soluble polymer, viscous polymer or oligomer, monomer.	low in laboratories, easy operation, and cleaning after use.
Extrusion	Undamaged	Thermoplastics.	Large-scale preparation, operation training is required, difficult to clean after use.

Source: Reprinted from [16].

20.6.2 COVALENT FUNCTIONALIZATION

Here, the desired functional groups are attached to the walls of the CNTs permanently in an inevitable manner. Halogenation, end group, cycloaddition, radical addition, nucleophilic addition, hydrogenated functionalization, electrophilic addition of chloroform, ozonolysis, i.e. the addition of ozonoids, and osmolysis are some examples of covalent functionalization [14, 18]. The major advantages of this kind of functionalization are a strong covalent attachment of CNTs with the polymeric matrix and proper dispersion of CNTs in different solvents. The main disadvantage is the production of defects on the surface and tips of CNTs [18].

20.7 CNT-BASED NANOCOMPOSITES

Just as the way Darwinism explains, 'the survival of the fittest', composite theory also favors the strongest, lightest, and the most active materials, i.e. a hybrid material. Combining two or more materials on a macroscopic level results in a new hybrid material having better properties in comparison to their parental counterparts.

A combination of two or more materials differing in composition on a macro scale forms a composite. The constituents retain their properties and identities, i.e. they do not dissolve or merge completely into each other, and exhibit interphase. Components can be physically identified from one another.

Example: mortar, concrete, metal-matrix composites, reinforced plastics, etc.

Among different composite materials, a nanocomposite is a kind of composite material, having nanosized particles embedded into a polymer, metal, or ceramic matrix material. Inclusion of these nanoparticles in the matrices results in the formation of a nanomaterial with entirely different chemical, mechanical, and morphological domain structure.

Examples of nanocomposites:

Class Example
Polymer Thermoplastic/thermoset polymer, layered silicates, CNT-polymer, etc.
Ceramic SiO_2/Ni, Al_2O_3/ TiO_2,Al_2O_3/CNT
Metal FeMgO, CNT/Al, CNT/$AgNO_3$

20.7.1 TYPES OF CNT COMPOSITES

20.7.1.1 Polymer Matrix Nanocomposites (PMNC)

The matrix of PMNC is a polymer like polyaniline, polypyrrole, etc. Enhancing the advantages of CNTs as fillers, these CNTs must be adequately dispersed and must have good interfacial interaction with the polymer r matrices. Liu et al. registered the successful filling of CNTs with polymers by using supercritical carbon dioxide (an excellent solvent). They incorporated polystyrene into hollow MWCNT having a length of approximately 2–3 nm and an outer diameter of 40–50 nm [19]. PMNCs have a wide range of applications, from the aircraft industry to household appliances [20].

20.7.1.2 Ceramic Matrix Nanocomposites (CMNC)

The matrix of ceramic matrix nanocomposites (CMNCs) is a ceramic-like SiC, Al_2O_3, etc. Niihara proposed the concept of CMNCs in 1991 [21]. CMNCs have potential applications in the field where high temperature and corrosion resistance are required. This class of materials offers high strength to density ratios. The fabrication of such composites using nanotechnology can make the composites able to be used in long-lasting applications [19, 22].

20.7.1.3 Metal and Metal Oxide Matrix Nanocomposites (MMNC)

The matrix of metal and metal oxide nanocomposites (MMNCs) is a metal or a metal oxide. Incorporation of CNTs into the metals improves the mechanical properties due to the high stiffness and strength of CNTs. The analysis shows that when metals like Cu, Ni, etc. [23,24] and metal oxides like Fe_3O_4 [25], TiO_2 [26], and MnO_2 [27] have been doped into CNT/polymer composites, the properties of these composites are improved, yielding a hybrid ternary composite (Figure 20.7).

20.8 CNTS: FINALLY MOVING OUT OF THE LAB; PATENTING AND INDUSTRIALIZATION

CNTs are already almost 30 years old. In these years, CNTs have passed through a complete Gartner curve representing its rise in the academic field to its absorption in the commercial sector. However, as per the May 2019 report of IDTechEx, the CNT market has again entered an era of rapid growth. Thus, keeping track of the present and the future range of this emerging technology appears to be a necessity [28]. Although the public may not see it, CNTs are becoming a core contributor to our everyday lives. Therefore academicians, researchers, innovators, and investors are obliged to become aware of the patent regime, the value of protecting their patents, issues, challenges, and risk management involved in patenting. The current patent landscape study shows that there are four major technical routes for CNT patenting:

1. Fabrication methods of CNTs
2. Modification of CNTs
3. Properties of CNTs
4. Application potential of CNTs [29].

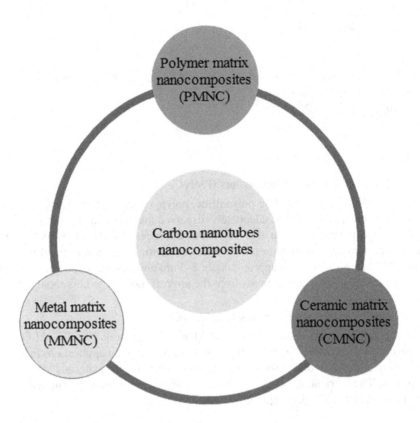

FIGURE 20.7   Types of CNT based nanocomposites.

20.8.1   PATENT CLAIM VALIDITY AND SCOPE

There are several key arguments related to patent validity and scope that will likely be raised by managers and lawyers assessing the patent landscape. Four patent law doctrines likely to be used to challenge the validity of nanotube patents are:

- Whether carbon nanotubes are 'patentable subject matter' under 35 U.S.C. § 101;
- Whether prior art 'anticipates' (or inherently anticipates) the patent under 35 U.S.C. § 102;
- Whether the patent is 'obvious' in light of the prior art under 35 U.S.C. § 103; and
- Whether the patent adequately discloses and enables the invention under 35 U.S.C. § 112 [6]

20.8.2   GLOBAL SKETCH OF PATENTING CNTs

In this section, the outline of the overall development of the IP concerning the growth of CNTs is discussed. The 'patents' search was conducted using a series of keywords including carbon nanotube, SWCNT, MWCNT, film, buckypaper, bucky, United States Patent and Trademark Office (USPTO), European Patent Office (EPO), etc., based on publication date up to June 2019. This search was conducted for patents in the United States, Europe (U.K.), Japan, and South Korea.

20.8.2.1   Production Analysis

After the discovery of SWCNTs in 1993, the NEC Corporation in 2004 declared itself the authority for manufacturing and selling CNTs. Sumitomo Corporation, Japan, came up first to negotiate for permission. IBM patented an incredible SWCNT with U.S. Patent Number 5424054. It is regarded as one of the most critical patents that can shape the future developments of CNTs. Later, Carbon Nanotechnologies, Inc. took over IBM's patent and became the leading producer of CNTs [30]. The table below gives the details of companies and universities involved in patenting CNTs. Figure 20.8, Table 20.2, and Table 20.3 give the bar graph details of top IPR-contributing companies for CNTs, where Hon Hai Precision Industry Co. Ltd. holds the maximum number of patents (204) in the overall portfolio.

20.9 CONCLUSION

In light of the continuous progress of carbon nanotubes over the last 30 years, CNT based materials have opened new pathways for developing, patenting, and industrializing hybrid functional materials which are found to have much higher application potential compared to their counterparts. The chapter discussed all the global aspects of carbon nanotubes from their past, to present, patents, intellectual property rights, issues, and strategies. In a nutshell, CNTs were always regarded as a new and exciting type of material, especially for energy and electronic applications. All types of CNTs were being investigated and will be further utilized in countless applications worldwide with equal importance.

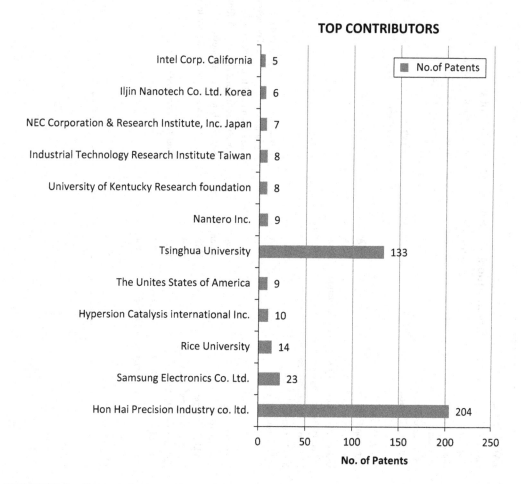

FIGURE 20.8 Key contributing assignees for IPR of CNTs.

TABLE 20.2

List of Companies and Organizations Contributing to the IPR of CNTs

Company	Origin		Patent Portfolio Focus	Link
Hon Hai Precision Industry Co. Ltd. (Foxconn)	Taiwan	204	Touch Screens Displays (iPad, iPhone), capacitors, transistors, diodes	https://www.foxconn.com/
Tsinghua University	Taiwan	133		http://www.nthu.edu.tw/
Samsung Electronics Co. Ltd.	South Korea	23	Touch screen displays (used in LCDs, smartphones, etc.), transistors, capacitors, diodes	
Rice University	USA	14		https://www.rice.edu/
Hyperion Catalysis	USA	10		http://hyperioncatalysis.com
Nantero Inc.	USA	9		
University of Kentucky Research Foundation	USA	8		http://www.itri.org.tw
Industrial Technology Research Institute	Taiwan	8		http://www.nec.com
NEC Corporation and research Institute, Inc	Japan	7		http://www.iljinnanotech.co.kr/
Iljin Nanotech CO. Ltd.	Korea	6		http://intel.com/
Intel Corporation	USA	5	Motherboard chips, integrated circuits, flash memory, graphics cards, embedded processors, and other devices related to communications and computing.	
Toray Industries	Japan	6	Touch screen panels using CNT films	http://www.toray.com/
Fuji Xerox	Japan	5	Solar cells to electronic devices	http://www/fujixerox.com.jp/eng/
Lockheed Martin	USA	5	Livid fabric, capacitors, and two terminal switches	https://www/lockheedmartin.com/
Battelle Memorial Institute	USA	5		http://www.battelle.org/
Eastman Kodak Company	USA	4	Touch screen panels	http://www.kodak.com/corp/default.htm
The Reagents of the University of California	USA	4		http://regents.universityofcalifornia.edu/
Agency of Industrial Science and Technology	Japan	4		
Hitachi, Ltd.	Japan	4		http://www.hitachi.com/
LG Electronics, Inc	South Korea	4		http://lg.com/

TABLE 20.3
List of Current Patents Obtained from the Official Site of the European Patent Office (2011–2019)

S.No.	Year	Description	IPC	CPC
1	2018	Apparatus And Method For Synthesizing Vertically Aligned Carbon Nanotubes	C01B32/162 C23C16/02 C23C16/26 (+2)	B82Y30/00 B82Y40/00 C01B2202/08 (+3)
2	2018	Method For Dispersing Carbon Nanotubes In Monomer And Reaction Injection Moulding Method And Apparatus Using The Same	B29C45/00 C01B32/174 C08K3/04	
3	2018	Method Of Producing Strong And Current-Conducting Fiber By Drawing Films From Carbon Nanotubes	B82B3/00 B82Y40/00 D01F9/12	B82B3/0038 B82Y40/00 D01F9/12
4	2018	Thermal Method Of Carbon Nanotubes Cleaning	B82B3/00 B82Y40/00 C01B32/17	B82B3/0009 B82B3/0095 B82Y40/00 (+1)
5	2018	Self-Healing Asphalt Mixture Using Carbon Nanotubes	C04B14/02 C04B18/02 C04B20/10 (+2)	
6	2017	Methods For Producing Advanced Carbon Materials From Coal	C01B32/154 C01B32/16 C01B32/184 (+1)	C01B3/02 C01B32/05 C01B32/154 (+22)
7	2017	Carbon Nanotube Composition And Manufacturing Method Therefor	B01J23/847 C01B32/158	
8	2017	Entangled Type Carbon Nanotubes And Manufacturing Method Therefore	C01B32/158 C01B32/162	
9	2017	Carbon Nanotubes Purification Method, Thin Film Transistor Thin Film Transistor Preparation Method	C01B32/17 H01L51/05 H01L51/30	
10	2017	Bonding Method Using A Carbon Nanotube Structure	B32B2250/05 B32B2264/10 B32B2405/00 (+9)	B32B37/12 B32B5/16 B32B7/00 (+3)
11	2017	Annealed Metal Nanoparticle Decorated Nanotubes	C01B32/168 G01N33/00	C01B32/168 C01B32/176 G01N27/12 (+1)
12	2017	In Situ Bonding Of Carbon Fibers And Nanotubes To Polymer Matrices	C08J3/20 C08J3/22 C08J5/06 (+4)	

(Continued)

TABLE 20.3 (CONTINUED)

List of Current Patents Obtained from the Official Site of the European Patent Office (2011–2019

S.No.	Year	Description	IPC	CPC
13	2017	Carbon Nanotube And Method For Fabrication Thereof	C01B32/166	
14	2017	Control Of Trion Density In Carbon Nanotubes For Electro-Optical And Opto-Electric Devices	G16C10/00 G16C20/40 G16C60/00	C01B2202/02 C01B2202/22 C01B32/159 (+4)
15	2017	Fin Field Effect Transistor Using Carbon Nanotubes As Conductive Trenches And Preparation Method Thereof	H01L21/336 H01L29/78	
16	2017	Mixtures Of Discrete Carbon Nanotubes	C01B32/168 C01B32/174 C08J5/00 (+3)	C01B2202/06 C01B32/168 C01B32/174 (+11)
17	2017	Single-Walled Carbon Nanotube-Based Slurry For Improved Nuclear Fuel Cladding Coatings And Method Of Fabrication Of Same	C01B32/15 C01B32/158 C01B32/159 (+2)	
18	2017	System That Utilizes Carbon Nanomaterial In Polymer Matrix With Specific Features Of Surface Tube And Surrounding Polymeric Interactions For Improved Aggregate Stability	C08J5/04 C08K3/04 C08K5/19 (+1)	C01B2202/04 C08J2361/10 C08J5/042 (+4)
19	2017	Conjugated Molecule For Selective Separation Of Carbon Nanotubes	C01B32/159 C01B32/17 C08G65/00	
20	2016	Processes For Controlling Structure And Properties Of Carbon And Boron Nanomaterials	B82Y40/00 C01B32/156 C01B32/176	B82Y40/00 C01B21/064 C01B21/068 (+9) B01J23/745
21	2016	Carbon Nanotube Aggregate	C01B32/162	C01B2202/08 C01B2202/26 (+5)
22	2016	Method For Preparing Single-Wall Carbon Nanotube Fiber Assembly	C01B32/162C01B2202/0 C01B32/164C01B32/16 D01D7/00	C01B32/162 (+3)
23	2015	Practical Method And Apparatus Of Plating Substrates With Carbon Nanotubes (CNTs)	B05D3/00	B05D3/207 B82Y30/00 C01B32/158 B82Y30/00
24	2011	Method For Making Carbon Nanotube Film	C01B32/168	C01B32/168 Y10T428/249924

Source: https://www.epo.org/index.html

REFERENCES

1. Gao C, Gua Z, Liu J-H, Huang X-J. Nanoscale The new age of carbon nanotubes : An updated review of functionalized carbon nanotubes in electrochemical sensors. *Nanoscale*, 2012;4:1948–63. RSC.
2. Sireesha M, Babu VJ, Ramakrishna S. Functionalized carbon nanotubes in bio-world : Applications, limitations and future directions. *Mater Sci Eng B* [Internet], 2017;223:43–63. Elsevier B.V. doi:10.1016/j.mseb.2017.06.002.
3. Kroto HW, Heath JR, O'Brien SC, Curl RF, Smalley RE. C60: Buckminsterfullerene. *Nature*, 1985;318(6042):162–163.
4. Ibrahim, Saeed K. Carbon nanotubes-properties and applications: A review. *Carbon Lett* [Internet], 2013;14:131–44. Available from: http://koreascience.or.kr/journal/view.jsp?kj=HGTSB6&py=2013&vnc=v14n3&sp=131.
5. Rao CNR, Govindaraj A. *Carbon Nanotubes. RSC Nanosci Nanotechnol No 18 Nanotub Nanowires*, 2nd ed. Cambridge, UK: The Royal Society of Chemistry, 2011. p. 1–189.
6. Harris DL. Carbon nanotube patent thickets. In: Fritz Allhoff, Patrick Lin, editor. *NanotechnolSocCurrEmerg Ethical Issues*. Kalamazoo, San Luis Obispo: Springer; 2009. p. 163.
7. Oueiny C, Berlioz S. Carbon nanotube – polyaniline composites. *Prog Polym Sci*, 2014;39:707. Elsevier.
8. Hirlekar R, Yamagar M, Garse H, Vij M. Review article carbon nanotubes and its applications: A review. *Asian J Pharm Clin Res*, 2009;2:17–27.
9. Khare R, Bose S. Carbon nanotube based composites- a review. *J Miner Mater Charact Eng*, 2005;4:31–46.
10. Hassan J, Diamantopoulos G, Homouz D, Papavassiliou G. Water inside carbon nanotubes: Structure and dynamics. *Nanotechnol Rev*, 2016;5:341–54.
11. Iijima S. Helical microtubules of graphitic carbon. *Lett to Nat*, 1991;354:56–8.
12. Iijima, Sumio, Ichihashi T. Single-shell carbon nanotubes of 1-nm diameter. *Lett to Nat*, 1993;363:603–5.
13. Buang NA, Fadil F, Majid ZA, Shahir S. Characteristic of mild acid functionalized multiwalled carbon nanotubes towards high dispersion with low structural defects. *Dig J Nanomater Biostructures*, 2012;7:33–9.
14. Chen L, Xie H, Yu W. Functionalization methods of carbon nanotubes and its applications. In: L Chen, editor. *Carbon NanotubAppl Electron Devices*. Rijeka, Croatia: InTech, 2008. p. 213–32.
15. Su M, Zheng B, Liu J. A scalable CVD method for the synthesis of single-walled carbon nanotubes with high catalyst productivity.*ChemPhys Lett*, 2000;322:321–6. Elsevier.
16. Chen J. A review of the interfacial characteristics of polymer nanocomposites containing carbon nanotubes. *R Soc Chem*, 2018;8(49):28048–85.
17. Hirsch A. Minireview functionalization of single-walled carbon nanotubes. *Angew Chem IntEd*, 2002;41:1853–9. Wiley.
18. Jeon I-Y, Chang DW, Kumar NA, Baek J-B. Functionalization of carbon nanotubes. In: S Yellampalli, editor. *Carbon NanotubPolym Nanocomposites*. Rijeka, Croatia: InTech, 2011.
19. Zhang XS, Yang LW, Liu HT, Zu. M. A novel high-content CNT-reinforced SiC matrix composite-fiber by precursor infiltration and a novel high-content CNT-reinforced SiC matrix composite-fiber by precursor infiltration and pyrolysis process composite-fiber by precursor infiltration and pyrol. *RSC Adv*, 2017;7:23334–41.
20. Zhang Q, Huang J, Qian WZ, Wei F. The road for nanomaterials industry : A review of carbon nanotube production, post-treatment, and bulk. *Carbon Nanotub Small-J*, 2013;9:1237–65. Wiley-VCH.
21. Samal SS, Bal S. Carbon nanotube reinforced ceramic matrix composites- a review. *J Miner Mater Charact Eng*, 2008;7:355–70.
22. Sharma S, Shashi SK, Tomar V. Ceramic matrix composites with nano technology–an overview ceramic matrix composites with nano technology–an overview. *Int Rev ApplEng Res*, 2014;4:99–102.
23. Janas D, Liszka B. Copper matrix nanocomposites based on carbon nanotubes or graphene. *RSC, Mater Chem Front*, 2018;2:22–35. Royal Society of Chemistry.
24. Wei Y, Wang X, Zhang J, Liu H, Lv X, Zhang M, et al. Facile approach of Ni / C composites from Ni / cellulose composites as broadband microwave absorbing materials †. *RSC Advan*, 2017;7:31129–32. Royal Society of Chemistry.
25. Pham GV, Trinh AT, Xuan T, To H, Nguyen TD, Nguyen TT, et al. Incorporation of Fe 3 O 4 / CNTs nanocomposite in an epoxy coating for corrosion protection of carbon steel. *Adv Nat Sci Nanosci Nanotechnol*, 2014;5:035016–24. IOP Publishing.
26. Shaban M, Ashraf AM, Abukhadra MR. TiO 2 nanoribbons / carbon nanotubes composite with enhanced photocatalytic activity ; fabrication, characterization, and application. *Sci Rep* [Internet], 2018;8:1–17. Springer US. doi:10.1038/s41598-018-19172-w.

27. Razak S Izwan Abd, Ahmad A Latif, Zein Sharif Hussein S. Polymerisation of protonic polyaniline/ multi-walled carbon nanotubes-manganese dioxide nanocomposites. *J Phys Sci*, 2009;20:27–34.

28. Genieser L, Gollin M. Intellectual property issues in nanotechnology. *J Commercial Biotechnol*, 2007;13:195–8.

29. Solutions RL. Patent landscape analysis. 2012.

30. Covered T, Nanotubes PC. Carbon nanotubes – who owns the patents and IP rights for carbon nanotubes? Carbon nanotechnologies hold the majority of the patenting strategy employed by carbon nanotechnologies in the carbon nanotube sector. 2005.

Index

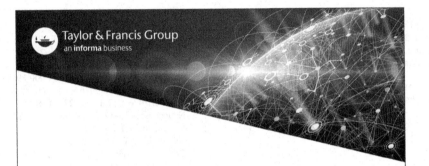

Printed in the United States
By Bookmasters